Animal Growth Regulation

Animal Growth Regulation

Edited by
Dennis R. Campion
United States Department of Agriculture
Agricultural Research Service
Midwest Area
Peoria, Illinois

Gary J. Hausman
United States Department of Agriculture
Agricultural Research Service
Richard B. Russell Agricultural Research Center
Athens, Georgia

and
Roy J. Martin
University of Georgia
Athens, Georgia

Plenum Press • New York and London

Library of Congress Cataloging in Publication Data

Animal growth regulation / edited by Dennis R. Campion, Gary J. Hausman, and Roy J. Martin.
    p.   cm.
    Bibliography: p.
    Includes index.
    ISBN 0-306-42978-0
    1. Growth—Regulation. 2. Domestic animals—Growth—Regulation. I. Campion, Dennis R. II. Hausman, Gary J. III. Martin, Roy J.
QP84.A55   1989   88-31736
636.089′26—dc19   CIP

© 1989 Plenum Press, New York
A Division of Plenum Publishing Corporation
233 Spring Street, New York, N.Y. 10013

Printed in the United States of America

Contributors

DALE E. BAUMAN • Department of Animal Science, Cornell University, Ithaca, New York 14853-4801

PETER J. BECHTEL • Muscle Biology Laboratory, University of Illinois, Urbana, Illinois 61801

D. H. BEERMANN • Department of Animal Science, Cornell University, Ithaca, New York 14853-4801

J. LEE BEVERLY • Department of Foods and Nutrition, University of Georgia, Athens, Georgia 30602

DOUGLAS J. BOLT • United States Department of Agriculture, Agricultural Research Service, Beltsville Agricultural Research Center, Reproduction Laboratory, Beltsville, Maryland 20705

R. DEAN BOYD • Department of Animal Science, Cornell University, Ithaca, New York 14853-4801

DENNIS R. CAMPION • United States Department of Agriculture, Agricultural Research Service, Midwest Area, Peoria, Illinois 61604

WILLIAM R. DAYTON • Department of Animal Science, University of Minnesota, St. Paul, Minnesota 55108

THEODORE H. ELSASSER • United States Department of Agriculture, Agricultural Research Service, Beltsville Agricultural Research Center, Livestock and Poultry Science Institute, Beltsville, Maryland 20705

C. L. FERRELL • United States Department of Agriculture, Agricultural Research Service, Roman L. Hruska U.S. Meat Animal Research Center, Clay Center, Nebraska 68933

J. JOE FORD • United States Department of Agriculture, Agricultural Research Service, Roman L. Hruska U.S. Meat Animal Research Center, Clay Center, Nebraska 68933

DARREL E. GOLL • Muscle Biology Group, University of Arizona, Tucson, Arizona 85721

MARCIA R. HATHAWAY • Department of Animal Science, University of Minnesota, St. Paul, Minnesota 55108

G. J. HAUSMAN • United States Department of Agriculture, Agricultural Research Service, Richard B. Russell Agricultural Research Center, Athens, Georgia 30613

E. J. HENTGES • National Livestock and Meat Board, Chicago, Illinois 60070

D. E. JEWELL • Purina Mills, St. Louis, Missouri 63131

WILLIAM KELLY JONES, JR. • Nutrition Division, Kraft, Inc., Glenview, Illinois 60025

WILLIAM C. KLEESE • Muscle Biology Group, University of Arizona, Tucson, Arizona 85721

JOHN KLINDT • United States Department of Agriculture, Agricultural Research Service, Roman L. Hruska U.S. Meat Animal Research Center, Clay Center, Nebraska 68933

ROY J. MARTIN • Department of Foods and Nutrition, University of Georgia, Athens, Georgia 30602

H. J. MERSMANN • United States Department of Agriculture, Agricultural Research Service, Roman L. Hruska U.S. Meat Animal Research Center, Clay Center, Nebraska 68933

WILLIAM J. MILLARD • Department of Pharmacodynamics, College of Pharmacy, University of Florida, Gainesville, Florida 32610

DEBRA M. MORIARITY • Department of Biological Sciences, University of Alabama, Huntsville, Alabama 35899

JAN NOVAKOFSKI • Muscle Biology Laboratory, University of Illinois, Urbana, Illinois 61801

VERNON G. PURSEL • United States Department of Agriculture, Agricultural Research Service, Beltsville Agricultural Research Center, Reproduction Laboratory, Beltsville, Maryland 20705

PETER J. REEDS • Children's Nutrition Research Center, Department of Pediatrics, Baylor College of Medicine, Houston, Texas 77030

CAIRD E. REXROAD, JR. • United States Department of Agriculture, Agricultural Research Service, Beltsville Agricultural Research Center, Reproduction Laboratory, Beltsville, Maryland 20705

HOLLY E. RICHTER • Department of Biological Sciences, University of Alabama, Huntsville, Alabama 35899

E. MARTIN SPENCER • Laboratory of Growth and Development, Children's Hospital of San Francisco, San Francisco, California 94118

NORMAN C. STEELE • United States Department of Agriculture, Agricultural Research Service, Beltsville Agricultural Research Center, Livestock and Poultry Science Institute, Beltsville, Maryland 20705

ADAM SZPACENKO • Muscle Biology Group, University of Arizona, Tucson, Arizona 85721

GARY E. TRUETT • Department of Foods and Nutrition, University of Georgia, Athens, Georgia 30602

JEFFREY D. TURNER • Department of Animal Science, MacDonald College of McGill University, Quebec, Canada H9X-ICO

RONALD B. YOUNG • Department of Biological Sciences, University of Alabama, Huntsville, Alabama 35899

Preface

The biotechnological advances of recent years have put us on the brink of unprecedented gains in animal productivity. Manipulation of animal growth rate and composition of gain is now possible by a variety of techniques. Examples include ingestion of beta-adrenergic agonists, injection of somatotropin, castration, immunization, and gene insertion. *Animal Growth Regulation* addresses modern concepts of growth regulation with an emphasis on agriculturally important animals. This emphasis is not exclusive, as many situations exist in which the only information available was generated in other species, and this information has been included for the sake of clarity and completeness. However, because of the overall orientation of this volume, particular attention has been given to the regulation of skeletal muscle, adipose tissue, and bone growth. Certain hormones and growth factors have a profound influence on growth regulation and this basic physiological knowledge is being harnessed to manipulate growth. Thus, considerable emphasis has been given to growth hormone–somatomedin/insulinlike growth factor regulation of cell and tissue growth. The involvement of peptides coded by protooncogenes and of negative growth regulators, such as transforming growth factor-β, represents an emerging area of molecular biology wherein basic knowledge offers potential exploitation for growth manipulation. Opportunities also exist for regulation of protein turnover, especially from the standpoint of protein degradation. Therefore, a place was reserved for these topics in order to provide relevant basic knowledge. Discussions of molecular genetics, especially gene expression, were incorporated in cases where regulation of gene expression could be linked to specific growth factors and inhibitors, or to differentiation of the three major tissues mentioned above.

    The authors and their contributions represent a wide range of backgrounds and disciplines, including biochemistry, anatomy, histology, physiology, endocrinology, and nutrition. In their respective chapters, the authors have captured the relevance and emphasized the recency of studies appropriate to animal growth regulation. The last two chapters relate specifically to the status of new and novel methods of growth regulation in the intact animal. The end result is a monograph that should be extremely useful to current and future graduate students, to postdoctoral fellows, and to seasoned investigators.

Dennis R. Campion
Gary J. Hausman
Roy J. Martin

Contents

CHAPTER 3

Endocrine Regulation of Adipogenesis

G. J. HAUSMAN, D. E. JEWELL, AND E. J. HENTGES

CHAPTER 4

Autocrine, Paracrine, and Endocrine Regulation of Myogenesis

WILLIAM R. DAYTON AND MARCIA R. HATHAWAY

CHAPTER 8
Skeletal Muscle Proteases and Protein Turnover
DARREL E. GOLL, WILLIAM C. KLEESE, AND ADAM SZPACENKO

CHAPTER 9
Regulation of Protein Turnover
PETER J. REEDS

CHAPTER 12
Mechanisms of Action for Somatotropin in Growth
R. DEAN BOYD AND DALE E. BAUMAN

CHAPTER 13
Regulation of Somatomedin Production, Release, and Mechanism
of Action
NORMAN C. STEELE AND THEODORE H. ELSASSER

CHAPTER 14
Sexual Differentiation and the Growth Process
J. JOE FORD AND JOHN KLINDT

CHAPTER 15
Potential Mechanisms for Repartitioning of Growth by
β-Adrenergic Agonists
H. J. MERSMANN

CHAPTER 16
Gene Transfer for Enhanced Growth of Livestock
VERNON G. PURSEL, CAIRD E. REXROAD, JR., AND DOUGLAS J. BOLT

CHAPTER 17

Status of Current Strategies for Growth Regulation
D. H. BEERMANN

Placental Regulation of Fetal Growth

C. L. FERRELL

1. Introduction

Growth of the fetus should be of paramount concern to those involved in animal production. The influence of fetal growth on animal production is manifest in several ways. Birth weight is typically related to neonatal mortality by a "U"-shaped curve. Birth weights lower than optimum are associated with reduced energy reserves, lowered thermoregulatory capability, and increased perinatal mortality. In addition, low birth weight is correlated with low rates of growth postnatally and decreased mature size. Conversely, birth weights that are greater than optimum result in increased dystocia and perinatal mortality as well as decreased rebreeding performance of the dam.

Fetal growth is influenced to varying degrees by numerous factors including fecundity, sex, parity, breed or breed cross, heat or cold stress, and maternal nutrition. The importance of these and other effectors of fetal growth differs among species. In general, however, birth weight of each fetus decreases with increased number of fetuses, is greater for males than females, and increases with increased parity of the dam. Birth weights are decreased by heat stress or poor maternal nutrition and increased by cold stress. Some of the possible explanations for these observations will be discussed in subsequent sections.

Typical patterns of fetal weight versus stage of gestation for the bovine (Ferrell *et al.*, 1976) and ovine (Koong *et al.*, 1975) are shown in Fig. 1. Similar patterns are seen in swine and other species. Fetal weight increases slowly during early gestation and quite rapidly during the later stages. About 90% of birth weight is achieved during the last 40% of gestation in both the

C. L. FERRELL ● United States Department of Agriculture, Agricultural Research Service, Roman L. Hruska U.S. Meat Animal Research Center, Clay Center, Nebraska 68933.

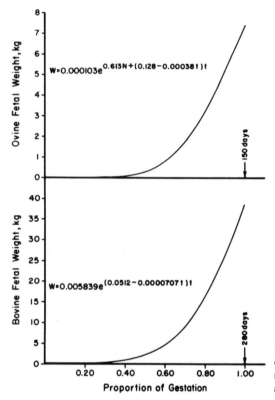

Figure 1. Pattern of fetal weight change of the ovine and bovine. Data recalculated from Koong *et al.* (1975) and Ferrell *et al.* (1976).

bovine and ovine. In addition to the rapid changes in weight, percentage of dry matter, protein, and fat increase by twofold or more during this interval. The rapid changes in fetal weight and composition suggest the need for substantial changes in maternal circulation and metabolism as well as major changes in uteroplacental (uterus plus placenta) function to meet fetal needs for growth and development.

The classical studies of Walton and Hammond (1938) demonstrated the importance of fetal genotype on birth weight and by inference on fetal growth. In addition, they demonstrated the importance of "the maternal environment" to fetal growth. Although the regulation of fetal growth is not yet well understood, three major components are involved: the fetal genome, the placenta, and the maternal system. Obviously, these components are closely interrelated. However, the placenta plays a central role in the regulation of fetal growth; some have suggested that the fetus rarely expresses its full genetic potential for growth because of placental and maternal constraints (Gluckman and Liggins, 1984). Placental involvement in the regulation of fetal growth through its regulation of substrate supply and its production and transport of hormones and other substances to the fetal and maternal systems will be emphasized in this chapter.

2. Placental Anatomy and Pattern of Blood Flow

By definition, the placenta is the region of contact between chorion and maternal tissues. Classical classification of placental types follows the scheme proposed by Grosser in the early 1900s (Amoroso, 1952), which grouped placentas according to the number of membranes or cell layers that intervene between the maternal and fetal circulations. In the simplest form, according to Grosser, an intact uterine mucosa is apposed to the chorion and the maternal and fetal bloods are separated by the maternal endothelium, connective tissue, uterine epithelium, fetal (chorionic) epithelium, connective tissue, and endothelium. In more complex types, these layers, especially the maternal constituents, are successively broken down until the chorionic epithelium comes into direct contact with maternal blood. Consequently, Grosser, using the name of the maternal tissue contiguous with the chorion as the denominator, distinguished four types of placentas: epitheliochorial, syndesmochorial, endotheliochorial, and hemochorial. Details of the structure and development of the placenta of various species have been presented by Amoroso (1952) and Ramsey (1982).

Pig and horse placentas have been considered representative of the diffuse epitheliochorial type, which many consider a primitive structure. Closer evaluation, however, has revealed a high degree of organization, and in the horse, thousands of microcotyledons, with specialized areas of respiratory gas exchange, active transfer, and biochemical activity (Silver *et al.*, 1973). The placentas of sheep, goats, cattle, and deer, thought by Grosser to represent a syndesmochorial type, are now classified as cotyledonary epitheliochorial (Steven, 1975). In animals having this type of placenta, highly vascularized, oval structures known as caruncles are distributed over the mucosal surface of the uterus. During pregnancy, caruncles form connections with localized villous projections of the chorion, termed cotyledons. The caruncle and cotyledon combined are referred to as placentomes. Placentomes number about 80 to 100 in sheep, 160 to 180 in goats, and 70 to 150 in cows (Amoroso, 1952).

Various theoretical models of the patterns of maternal and umbilical blood flow within the placenta have been reviewed by Meschia (1983). The simplest models consist of a homogeneous membrane separating streams of maternal and fetal blood flowing either in the same direction (concurrent) or in opposite directions (countercurrent). A third model, in which maternal and fetal blood flow perpendicular to each other, has been termed crosscurrent. An important property of each of the elementary models is that their clearance (ratio of transfer rate to maternal artery – fetal artery concentration difference), thus their effectiveness, tends to a flow-limited maximum as the ratio of permeability to flow approaches infinity. At relatively high permeabilities, the countercurrent arrangement is the most favorable to placental diffusion, but the direction of flows become less relevant if the permeability of the placenta to the substance under consideration is low or if one flow is much smaller than the other.

The hypothesis that placental clearance of inert molecules is flow-limited has been supported by evidence in several species. Equivalent clearances were obtained when antipyrine was compared to ethanol or tritiated water in sheep, when antipyrine was compared to deuterium oxide in cattle, and when nitrous oxide was compared to tritiated water in guinea pigs (Meschia, 1983; Reynolds *et al.*, 1985a).

The arrangement of vessels within the ovine placentome was described by Barcroft and Barron (1946) and from those observations, they hypothesized a countercurrent arrangement of maternal and fetal flows. Subsequent studies, however, shed doubt on the countercurrent hypothesis and led to a reexamination of the vascular structure (Makowski, 1968; Silver *et al.*, 1973). Those investigators hypothesized a crosscurrent pattern of blood flows in placentomes of the sheep. In addition, clearances of antipyrine, ethanol, and tritiated water in sheep are considerably lower than predicted by a countercurrent model and, in fact, are somewhat lower than those of a concurrent exchange (Meschia, 1976). These and other observations are most readily explained by the crosscurrent hypothesis and support the idea that the ovine placenta is a relatively inefficient organ of diffusional exchange.

Tsutsumi and Hafez (1964) described arteries and veins within the walls of the maternal crypts in the cow and suggested blood flowed in opposite directions, i.e., countercurrent flow. However, data reported by Reynolds *et al.* (1985a) and Reynolds and Ferrell (1987) indicated that neither the concurrent nor countercurrent model of the arrangement of maternal and fetal microvasculatures explained the low clearance rates of D_2O or antipyrine across the bovine placenta. In addition, concentrations of oxygen or oxygen partial pressures differ between uterine and umbilical veins of cows (Silver *et al.*, 1973; Reynolds *et al.*, 1986) to a similar or greater extent than observed in sheep. The most appropriate model of the arrangement of the maternal and fetal vasculature in the placentome of the cow has not been determined. Available data, however, suggest uneven distribution of maternal and fetal flows to the placenta and/or significant shunting of placental flows away from the areas of exchange.

Anatomical information reviewed by Silver *et al.* (1973) suggested a countercurrent arrangement of flow in the microcotyledons of the mare. In addition, Silver and Comline (1975) demonstrated a much smaller uterine–umbilical vein PO_2 difference under hypoxic or normoxic conditions than has been observed in ruminants. They also showed that in hyperoxic conditions, umbilical vein PO_2 was greater than uterine vein PO_2. These findings are indicative of a rather high effectiveness of diffusion across the equine placenta, which is in keeping with a countercurrent arrangement. Data obtained to date suggest a countercurrent arrangement of the rabbit placenta. The human placenta (hemochorial) and the sheep placenta (epitheliochorial), although very different structurally, are both somewhat less effective than predicted by a concurrent model (Meschia, 1983), suggesting a poor relationship between gross structure and function.

In summary, it should be noted that even though placentas may have the same or differing numbers of tissue layers between maternal and fetal blood, tissue layer thickness, permeability, and metabolic rate of each layer may vary from one species to another, from one breed to another within species, and from place to place or time to time within the same placental unit (Faber and Thornburg, 1986). Furthermore, although permeability of a placental tissue may affect the transfer of many substances such as oxygen, carbon dioxide, urea, or water, several other factors are of importance for all types of placental transfer. In addition to vascular architecture, diffusion distances and total surface area for diffusion, the rate and distribution of blood flow on either side, concentration gradients across the exchange area, usage by the placenta, and any special feature such as carrier-mediated transport may facilitate the transfer of a given substance.

3. Placental Growth

In all of the species for which data are available, the weight of the placenta increases much more rapidly than that of the fetus in early gestation. The growth of the placenta subsequently slows, while that of the fetus increases. Thus, about midgestation, depending on the species, the weight of the fetus exceeds that of the placenta. Placental weight reaches a maximum just over halfway through gestation in the sheep and goat, then declines (Bell, 1984). In the mouse, rabbit, and guinea pig, placental weight reaches a maximum somewhat later in gestation and appears to decline near term, whereas it continues to increase until near term in man, rhesus monkey, and cattle (Dawes, 1968; Prior and Laster, 1979).

Moderate to severe undernutrition during pregnancy results in reduced placental weight in sheep (Alexander and Williams, 1971; Mellor, 1983). Neither the specific nutrients nor the mechanisms involved in these nutritional effects have been examined in sheep. Protein deficiency appears to be more potent than energy deficiency in reducing placental weight in guinea pigs. Conversely, supplemental protein in early pregnancy increased the weight and cellularity of rat placentas at term. Birth weight and possibly placental growth, in cattle, swine, or horses, are less sensitive to nutritional deficiencies than in sheep, rats, or guinea pigs.

Prolonged elevation of environmental temperature can cause remarkable reductions in placental weight in ewes (Alexander and Williams, 1971). The combined weight of twin placentas is usually greater than that of a single placenta in sheep because greater proportions of available caruncles are occupied and individual placentomes are heavier (Alexander, 1964a). Weights of placentas of individual twins, however, are usually less than those of single fetuses, indicating that increased placentome size does not completely compensate for the decrease in placentome number available for each twin. These general relationships extend to triplets and quadruplets in prolific breeds of sheep and apparently apply to other species. Placental weight also tends to increase with parity (or age) of the ewe.

The relationship between placental size and function is not well documented. Limited data indicate that patterns of change in placental DNA content, total number of placental nuclei, volume of connective tissue, and total villous surface area tend to follow patterns similar to that of placental weight. Conversely, the number of endothelial nuclei in the fetal components of the placenta apparently continue to increase in sheep during gestation as does the degree of vascularization in both the caruncle and cotyledon. It appears that although functional capacity of the placenta is related to weight, it is more closely related to placental perfusion.

4. Growth of Uteroplacental Blood Flow

Increased blood flow through the uterine arteries of the cow (Ford *et al.*, 1979), sow (Ford and Christenson, 1979), and ewe (Greiss and Anderson, 1970) has been measured in early pregnancy. Those studies demonstrated an early and local effect of the blastocyst on uterine blood flow in the cow and pig. Uterine blood flow continues to increase rapidly during early pregnancy in the cow and sow.

Total blood flow to the uterus increases exponentially during the second half of gestation in the cow (Reynolds *et al.*, 1986). The relationship of uterine blood flow *(F)* to day of gestation *(t)* was:

$$F \text{ (liters/min)} = 0.4792e^{0.0129t} \quad (N = 31, R^2 = 0.91)$$

which suggested that the instantaneous rate of increase in uterine blood flow (1.29% per day) was somewhat less than the instantaneous rate of increase in fetal weight (about 3 to 5% per day; Ferrell *et al.*, 1976). Data for sheep (Rosenfeld *et al.*, 1974) and goats (Cotter *et al.*, 1969) suggest similar patterns in those species as gestation advances. These data are consistent with the concept of fetal growth constraint, especially during late gestation. Blood flow does not increase uniformly to all uterine tissues, however. Data reported by Rosenfeld *et al.* (1974) indicated that blood flow to caruncular tissues was about 27% of total uterine blood flow in early pregnancy but was about 82% of the total in late pregnancy.

During late pregnancy, uterine blood flows attain levels of about 10–14 liters/min in cows and 1.5–2.0 liters/min in ewes. Blood flows to the uterus per kilogram of gravid uterine tissue (about 300 ml/kg per min) of the mare (Silver and Comline, 1975), goat (Cotter *et al.*, 1969) and sow (Ford *et al.*, 1984) are similar to those of the cow and ewe. These values represent approximately 10 to 20% of cardiac output. In animals having more than one fetus, total uterine blood flow increases with the number of fetuses, but not proportionally.

Fetal as well as maternal perfusion of the placenta is an important determinant of nutrient availability to the fetus. In cows, 85% of the variation in placental D_2O clearance was attributed to umbilical blood flow during the second half of pregnancy (Reynolds and Ferrell, 1987). When umbilical flow is much smaller than uterine blood flow, the lesser flow is the primary regulator of diffusion in a flow-limited system. In the cow, the ratio of umbilical to uterine flow increased from 0.10 at about 135 days of gestation to 0.44 at 250 days. In the sheep, this ratio was 0.14 at 75 days of gestation (Bell *et al.*, 1986) and 0.51 at 130 to 145 days of gestation, suggesting that the pattern of increase with gestational age is similar to that of the cow.

Umbilical blood flow, like uterine blood flow, increased exponentially during pregnancy (Reynolds *et al.*, 1986). The regression of umbilical flow (*f*) on day of gestation (*t*) was:

$$f \text{ (liters/min)} = 0.0114e^{0.0245t} \quad (N = 24, \ R^2 = 0.94)$$

which indicated that the instantaneous rate of increase in umbilical blood flow was nearly twice as great as that in uterine blood flow (2.45% versus 1.29% per day) but was slightly less than that in fetal weight.

Available data in cows suggest that umbilical blood flow per kilogram fetus is relatively constant during the second half of gestation in the cow (Reynolds *et al.*, 1985b, 1986). Values varied from about 160 to 230 ml/kg per min. Similar values have been reported during late gestation in the mare (Silver and Comline, 1975), sow (Macdonald *et al.*, 1985), and ewe. In contrast, Bell *et al.* (1986) reported values from 263 to 844 (mean 468) ml/kg per min at midgestation in the sheep. The high flow estimates might be related to the high fetal heart rate in midgestation as compared to late gestation (209 versus 140 beats/min). In cows, guinea pigs, and humans, fetal heart rate does not appear to decrease during this interval; however, a comparable decrease has been reported in swine.

Several lines of evidence suggest an important involvement of uterine and umbilical blood flows, probably through their involvement in placental metabolite transfer, in fetal growth. Both uterine and umbilical blood flows were greater in Charolais cows, which typically have larger calves at birth, than in Hereford cows and were greater per fetus in cows bearing single than in those bearing twin fetuses (Ferrell and Reynolds, 1987). Data for single or twin fetus-bearing ewes support these observations. Similarly, uterine and umbilical blood flows per fetus decrease with increased number of fetuses in swine. In these types of studies, however, placental perfusion, placental weight, and fetal weight are, in general, highly correlated. Cause-and-effect relationships are thus difficult to establish.

Data reported by Bell *et al.* (1987) and Reynolds *et al.* (1985b) demonstrated that fetal and placental weight as well as uterine and umbilical blood flows were reduced in chronically heat-stressed sheep and cows. In sheep, placental clearance paralleled placental weight, but umbilical blood flow per kilogram fetal weight was reduced by 30%. Umbilical and uterine blood flows and fetal weights were reduced 24, 34, and 18%, respectively, in heat-stressed cows. When expressed relative to fetal weight, umbilical blood flow was 82% that of control cows. It was hypothesized that during heat stress, greater proportions of cardiac output were routed to the periphery and lungs to dissipate body heat, thus reducing the volume of blood and nutrients supplied to the gravid uterus.

Mechanical restriction of the increase in uterine blood flow associated with increasing gestational age has been shown to decrease fetal growth in swine and guinea pigs. Mechanical restriction of uterine blood flow in sheep resulted in no acute change in umbilical blood flow or placental oxygen consumption, but resulted in a substantial (30%) reduction in fetal oxygen uptake in sheep. Similarly, mechanical reduction of umbilical blood flow (Anderson *et al.*, 1986) during late pregnancy in sheep resulted in a linear decrease in fetal oxygen uptake, a result consistent with reduction in fetal growth by a chronic decrease in umbilical flow (Anderson and Faber, 1984).

An alternative experimental method to study the effects of reduced placental size was first reported by Alexander (1964b) who surgically removed caruncles from the uterus of sheep. Placental size was reduced and fetal growth, especially after 100 days, was decreased. Similar results were reported by Owens *et al.* (1986) who also showed that uterine and umbilical blood flows and placental clearance of antipyrene, as well as placental and fetal weights, were reduced.

The above results, taken together, support the hypothesis that placental size and placental perfusion from both the maternal and fetal systems are related and that fetal growth may, in fact, be limited by placental perfusion and function.

5. Regulation of Blood Flow

Although numerous researchers have acknowledged the importance of uterine and umbilical blood flows to fetal growth, the factors involved in regulating the described sequence of changes are not known, but several possibilities have been suggested. The pattern of changes in uterine blood flow during the estrous cycle and early pregnancy is likely in response to the pattern of circulating estrogens. Several researchers have demonstrated increased uterine blood flow in response to estrogen in early pregnancy (Rosenfeld et al., 1976; Resnik, 1981). The pattern of circulating estrogen concentrations during these intervals is consistent with the observed changes in uterine blood flow (Ferrell and Ford, 1980); however, blood flow and estrogen concentrations are not parallel during subsequent stages of pregnancy. Rosenfeld et al. (1976) found that injection of estradiol-17β in pregnant ewes increased blood flow to the myometrium, endometrium, and caruncles but the response depended on stage of pregnancy and varied among regions of the uterus. Myometrial and endometrial blood flows increased in response to estradiol-17β injection throughout pregnancy whereas flow to caruncular tissue did not respond to the injection after the second month. Evidence is also available to suggest a modulation of the response to estrogens by progesterone. Although estrogens may be the primary factor causing the pattern of blood flow change during estrus and early pregnancy, they do not appear to be of primary importance in the exponential increase in uterine blood flow during the second half of gestation.

Infusion of the catecholamines, epinephrine and norepinephrine, into the pregnant animal causes a decrease in uterine blood flow; however, the effect is more pronounced in the myometrium and endometrium than in the caruncle (Meschia, 1983). Neurogenic control of blood flow to these tissues appears to be mediated by α_1- and α_2-adrenergic receptors. The α_1 receptors are found on vascular smooth muscle membranes and stimulate vasoconstriction when activated by norepinephrine. Conversely, norepinephrine synthesis and release by adrenergic neurons are inhibited by presynaptically located α_2-adrenergic receptors. Sauer et al. (1987) have demonstrated in cows that α_1-receptor numbers were similar in caruncular, intercaruncular, and uterine arterial tissues, but that caruncular tissues contained much greater numbers of α_2 receptors than did the other tissues. They also showed a progressive decline in norepinephrine concentrations indicating a progressive decline in innervation toward the caruncular vascular bed. Their results suggested a preferential decrease in the sensitivity of the caruncular bed to neuronally mediated vasoconstrictions while the ability of all uterine tissues to respond to circulating catecholamines of adrenal origin is retained.

The role of prostaglandins in the regulation of uteroplacental blood flow is still obscure despite numerous attempts to define it. Injection of PGE_2 and $PFG_{2\alpha}$ into the maternal circulation caused uterine contractions and a decrease in placental blood flow, but when PGE_2 was injected into the fetus, a modest increase in maternal–placental blood flow was observed. Meschia (1983) reasoned that PGE_2 might cause contractions as well as vasodilation but that the predominant effect depends on local concentrations. Injection of PGI_2 and PGD_2 into the uterine artery of pregnant sheep caused an increase in uterine blood flow, but a decrease when injected into the left ventricle. The latter result is possibly a secondary effect of hypertension resulting from the ventricular injection.

When the placental cotyledons begin to grow, a rapid and substantial increase in blood flow occurs to the site of implantation, indicating that maternal angiogenesis is stimulated by local factors produced by the growing villi. A heat-labile angiogenic factor of greater than 100,000 kDa and present in caruncular tissue in cows and cotyledonary tissue of sheep has been reported (Reynolds et al., 1988).

As noted by Meschia (1983), a distinction should be made between chronic regulation of blood flow, which occurs as a result of growth and development of placental circulation, and acute regulation, which occurs primarily as a result of changing the diameter of vessels within the uterine and/or placental vasculature. Some of the factors known to influence uterine blood flow, such as estrogens and catecholamines, appear to be involved in both acute and chronic regulatory functions whereas others seem to act directly on the uterine vasculature as short-term regulators. Because of the different histological events underlying the increase in placental perfusion at different stages of gestation, it seems likely that different stimuli are involved at different stages of gestation. Also, the extent and mechanisms of fetal regulation of placental blood flow are an area of speculation. Each conceptus promotes growth of maternal blood flow through its own placenta, thus the chronic regulation of blood flow must result from a complex interaction between fetal and maternal control mechanisms.

6. Placental Transport

Mechanisms by which metabolites enter and cross the placenta have been classified as passive diffusion, facilitated diffusion, active transport, and solvent drag. In addition to these, some metabolites may gain access to the fetus through contact of the fetal trophoblast with the decidua of the uterus at other than placental sites.

It is important to recognize that most of the metabolites transferred across the placenta are metabolized and/or produced by placental tissues. At about 180 days of gestation in the cow, for example (Fig. 2), 26% of the oxygen,

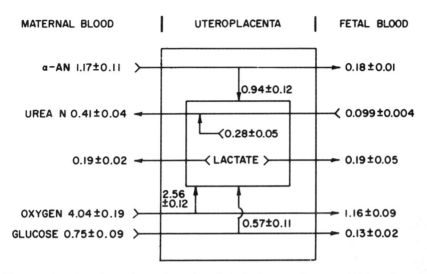

Figure 2. Summary of net substrate fluxes through the bovine uteroplacenta at 165 days of gestation. Urea nitrogen and α-amino nitrogen fluxes are in meq/min. All other fluxes are in mmole/min. From Ferrell *et al.* (1983)

17% of the glucose, and 15% of the amino acids taken up by the gravid uterus from the bovine maternal circulation reached the fetus (Ferrell *et al.*, 1983). Lactate was produced by placental tissues and transferred to both fetal and maternal circulations.

Water is readily transported across the placenta and is accumulated to ensure adequate expansion of fluid spaces (Munro *et al.*, 1983). Several hypotheses have been proposed to explain water transport including a hydrostasis theory, a bicarbonate–osmotic control theory, and a localized osmotic control theory; however, each requires more data for validation.

Studies with inhibitors of respiratory enzymes in the human placenta (Guiet-Bara, 1980) indicated that placental utilization of the oxygen depends on respiration of that organ, whereas transfer to the fetus is by simple diffusion. The predominance of evidence to date supports the concept of simple diffusion of oxygen, carbon monoxide, and carbon dioxide. However, concentration gradients from the mother to fetus and the greater affinity of fetal hemoglobin for oxygen compared to maternal hemoglobin facilitate that diffusion. Further, increased perfusion of the placenta can improve oxygen exchange.

Oxygen utilization or consumption rate is a useful index of oxidative metabolism. Thus, numerous reports documenting rates of oxygen uptake by the gravid uterus, uteroplacenta, and fetus of the ewe and cow as well as several other species are available. Available data suggest similar patterns of oxygen consumption in sheep and cattle and demonstrate three important concepts. First, uteroplacental and fetal oxygen consumption increase rapidly during the second half of gestation and, in the cow, tend to parallel increases in wet weight of those tissues. Second, uteroplacental tissues utilize a large proportion of the total oxygen taken up from the maternal circulation but the proportion varies from about 0.70 at midgestation to about 0.40 during late gestation. Third, the oxygen consumption rate of uteroplacental tissues is high relative to that of the fetus (0.5 versus 0.25 mmole/kg per min; Reynolds *et al.*, 1986) and is approximately equal to that of the central nervous tissue. A high proportion of blood flow to the uteroplacenta perfuses the cotyledonary and caruncular tissues during this interval, and thus the rate of oxidative metabolism of these tissues is very high indeed.

Glucose is utilized extensively by the uteroplacenta and fetus of all species of mammals studied and serves as both an energy substrate and an important source of carbon for growth of those tissues. During gestation, uptake of glucose by gravid uterine tissues has a major impact on the glucose economy of the maternal system. This is especially true in ruminants, which depend primarily upon *in vivo* synthesis to meet glucose needs.

Glucose transport by the placenta is carrier-mediated (Munro *et al.*, 1983; Battaglia and Hay, 1984). Fetal glucose concentrations are lower than maternal concentrations and increase linearly with increasing concentrations in the maternal artery of the cow, horse, sheep, and primates, within normal ranges of maternal arterial glucose concentrations.

Only 15 to 20% of the glucose taken up from the maternal circulation in the cow and about 30% in the ewe is transferred to the fetus as glucose (Battaglia and Meschia, 1981; Ferrell *et al.*, 1983; Reynolds *et al.*, 1986). Net glucose uptake is sufficient to account for 50 to 75% of the oxidative metabolism of the fetus in cattle and sheep. Gluconeogenic enzymes are present in the liver of fetal sheep and cattle (Prior, 1982), indicating that additional glucose may be supplied by gluconeogenesis. Several reports suggest that gluconeogenesis is an important source of glucose, but only when the exogenous supply is limited.

Placental tissues of the cow and sheep produce lactate from glucose and there is a net transfer of lactate from placental tissue to both the fetus and mother. Lactate production in placental tissues accounts for 25 to 30% of gravid uterine glucose uptake. Fetal lactate uptake from placental tissues is sufficient to account for up to 20 to 50% of fetal oxygen consumption, assuming that all of the lactate is completely oxidized. Bassett (1986) suggested that lactate does not readily cross the ovine placenta and that cotyledonary and caruncular tissue produce lactate from glucose originating from fetal and maternal vascular pools, respectively. He interpreted those data to reflect the operation of a substrate cycle within the fetus–fetal placenta rather than an exogenous source of lactate as often suggested.

Fetal–placental nitrogen metabolism has been reviewed by Lemons (1979) and others. The concentration of most free amino acids is greater in fetal blood than in maternal blood in the guinea pig, human, sheep, and cow, supporting the concept of active transport of amino acids across the placenta. The role of the placenta and fetus in utilizing these substrates has been reported for the sheep (Battaglia and Meschia, 1981) and cow (Ferrell *et al.*, 1983). In the sheep, almost all of the amino acids taken up by the gravid uterus are transferred to the fetus where a portion (about 25%) are apparently degraded and the nitrogen excreted as urea. Urea is transferred to the maternal circulation via passive diffusion. Nitrogen, released from the small amount of amino acids degraded by placental tissues, diffuses to the maternal and fetal systems as ammonia. Conversely, in the cow, a large proportion of the amino acids taken up by the gravid uterus appear to be metabolized by placental tissues. In the cow, both the fetus and placenta appear to produce urea from the metabolism of amino acids. This observation requires further verification but is supported by observations indicating the presence of the urea cycle enzymes in cow placentomes (Ferrell *et al.*, 1985). In both the sheep and cow, amino acid degradation by the fetus is sufficient to account for about 25% of fetal energy expenditures.

In ruminants, the volatile fatty acids (acetate, propionate, and butyrate) represent about 65% of the energy supply postnatally, and there have been several studies to quantitate the contribution of the VFAs to fetal metabolism. The available data suggest that acetate transfer may account for 10 to 20% of fetal oxygen consumption (Char and Creasy, 1976). Longer chain fatty acids, however, do not appear to cross the sheep placenta in energetically significant amounts. Transport of other important nutrients and metabolites has been discussed by Munro *et al.* (1983).

7. Hormone Secretion

In addition to supplying nutrients to and transporting waste products from the fetus, the placenta is a complex endocrine organ that synthesizes and metabolizes a number of peptide and steroid hormones as well as numerous other compounds. The precise role of placental and fetal hormones and the manner in which they coordinate fetal and placental growth and coordinate activities of those tissues with changes in the maternal system are not known. Therefore, the following discussion will primarily concern some of the hormones produced by the placenta that are thought to have important roles. For more detailed discussions, interested readers are referred to the reviews of Buster (1984), Gluckman and Liggins (1984), and Longo (1985).

Chorionic somatomammotropin or placental lactogen is a polypeptide hormone structurally similar to pituitary growth hormone and prolactin. Its secretion increases with increased placental size and it is secreted into both the fetal and maternal circulations, although the pattern of secretion appears to differ somewhat among species (Gluckman and Liggins, 1984). In humans, placental lactogen may act as a lipolytic, glucose-sparing agent that decreases peripheral sensitivity to insulin, thus sparing glucose for gravid uterine use. It is also thought to restrict maternal utilization of protein, thereby increasing amino acid availability to the conceptus. These effects are not evident in the ewe (Waters *et al.*, 1985). However, the short duration of those studies precludes conclusions regarding homeorhetic effects. Freemark *et al.* (1986) have observed specific receptor binding of ovine placental lactogen to fetal sheep hepatic membranes. The binding increased progressively during the second half of gestation, reaching a maximum at 3 to 7 days prepartum. Their results as well as observations that ovine placental lactogen has potent stimulatory effects on amino acid transport, glycogenesis, and somatomedin production have led to the hypothesis that placental lactogen may serve as a fetal "growth hormone" in the sheep. This hypothesis requires further verification. In addition, Freemark *et al.* (1986) reported that no growth hormone receptors were present in ovine fetal hepatic tissue, which supports other observations regarding the limited prenatal involvement of growth hormone in the regulation of growth.

Prolactin, produced by placental tissue of several species, has been proposed to have several roles in fetal development. These roles include serving as a fetal "growth hormone," stabilization of uterine prostaglandin production, and perhaps in the regulation of amniotic fluid volume. Other peptide hormones have been identified or claimed to be products of placental synthesis. Most, like placental lactogen, are analogues of corresponding pituitary or hypothalamic hormones and include ACTH, FSH, TSH, and α-MSH.

Estrone, estradiol, estriol, and progesterone are produced by placental tissues in many species, although the sites of production and patterns of secretion vary tremendously among species. In the human, chorionic gonadotropin in conjunction with placental lactogen appears to promote placental production of progesterone from maternal cholesterol. Part of the progesterone passes to the maternal circulation where it supplements progesterone produced by the corpus luteum. Progesterone and pregnenolone pass to the fetus, which apparently uses these steroids to produce dehydroepiandrosterone and androstanedione, products that are subsequently converted to estradiol and estriol by the placenta. In the ewe, cow, and mare, maintenance of pregnancy is dependent on ovarian progesterone secretion until about 50–60, 210–230, and 150–200 days of gestation, respectively. The goat and sow are dependent on corpus luteum production of progesterone throughout pregnancy. These observations suggest that progesterone production by conceptus tissues is low or does not occur to a substantial degree until late in gestation, and the ontogeny differs among species. Estrone and estradiol, the estrogens thought to be of primary importance in cows, sheep, and swine, are produced by placental tissues in the cow and ewe and secreted into both the maternal and fetal circulations, primarily as the sulfated forms. In ewes, androstenedione is produced by the placenta from progesterone and is produced by the fetal adrenal gland. Androstanedione is aromatized in the placenta to estrogens (Mitchell *et al.*, 1986) and negative feedback has been proposed as a mechanism of regulation.

The precise roles of steroids in the regulation of fetal growth are unclear. There is controversial evidence suggesting (1) steroid involvement in growth of placental tissue and perhaps in the increased placental vascularity during later stages of gestation, (2) steroid involvement, especially during early gestation, in the regulation of uterine blood flow and perhaps in the increased cardiac output observed in most species during gestation, (3) steroid involvement in altered maternal metabolism via, for example, their somatotropiclike effects, and (4) progesterone involvement in protection of the conceptus against homograft rejection. The mechanisms of the variety of actions of the steroids are not known but the concept that specific metabolic products of estrogen metabolism, such as catechol estrogens, exert their actions at specific target sites has received support.

As suggested previously in this chapter, several prostaglandins are produced from arachidonic acid by uteroplacental tissues. Their roles are varied, but many involve local effects on vascular resistance and uterine contractility (Rankin, 1976).

It must be stressed that the fetus, in part, controls its own destiny through its interactions with the placental and maternal systems. For example, rates of fetal insulin and IGF-II secretion vary depending upon circulating levels of certain nutrients including glucose and amino acids. Placental receptors for these fetal hormones have been demonstrated and their promotion of placental growth and nutrient transfer has been indicated. In addition, placental estrogen synthesis appears to be enhanced by fetal adrenal synthesis of precursors, which in turn appears to be influenced by both fetal and placental ACTH.

8. Conclusions

A simplified schematic representation of the regulation of fetal growth is shown in Fig. 3. In concept, the fetal genome can be viewed as the primary determinant of fetal growth. Fetal growth is suggested to have a direct effect on placental growth, in part due to its production of hormones such as insulin and IGF-II, hormone precursors such as androstenedione, and through its regulation of umbilical blood flow. Conversely, the placenta may directly impact fetal growth via its production of various hormones such as placental lactogen.

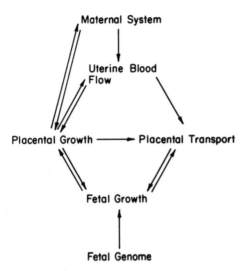

Figure 3. Schematic representation of the primary factors thought to be involved in the regulation of fetal growth. See text for details.

The placenta, likewise, is directly involved in maternal cardiovascular and metabolic changes such as increased cardiac output and lipolysis, and decreased peripheral glucose and amino acid utilization. These effects are apparently hormonally mediated although the hormones involved and the mechanisms of actions are not well understood. Conversely, although not well documented, it is evident that the maternal system is directly involved in the regulation of placental growth, although part of this effect may be mediated through its influence on placental perfusion. Placental growth can be viewed as having direct effects on uterine blood flow. These effects seem to be the result of several factors including placental production of angiogenic factors that cause growth of the vascular bed and production of vasodilatory substances such as prostaglandins. Placentally produced estrogens may be involved in both vascular growth and vasodilation. Decreased catecholamine receptors may be involved in a decreased response to acute stresses. Uterine blood flow, in turn, impacts placental growth and placental transport, probably through alteration of substrate availability. Certainly, placental transport is directly impacted by placental size, maternal and fetal perfusion of the placenta, and placental structure. Placental transport, via its control of substrate availability, has a direct impact on fetal growth. The fetus, through its production of metabolic hormones and its regulation of circulating metabolite concentrations, directly influences placental transport. Part or all of these components may be influenced by various external factors such as nutritional or thermal environment, disease, and so forth. Unfortunately, our understanding of this system is at best limited. Many more questions remain to be answered than have been.

References

Alexander, G. 1964a. Studies on the placenta of the sheep (Ovis aries L.); placental size. J. Reprod. Fertil. 7:289.

Alexander, G. 1964b. Studies on the placenta of the sheep (Ovis aries L.); effect of surgical reduction in the number of caruncles. J. Reprod. Fertil. 7:307.

Alexander, G. and D. Williams. 1971. Heat stress and development of the conceptus in domestic sheep. J. Agr. Sci. (Camb.) 76:53.

Amoroso, E. C. 1952. Placentation. In: A. S. Parkes (Ed.) Marshall's Physiology of Reproduction. Vol. II, pp 127–311. Longmans, Green and Co., New York.

Anderson, D. F. and J. J. Faber. 1984. Regulation of fetal placental blood flow in the lamb. Amer. J. Physiol. 247:R567.

Anderson, D. F., C. M. Parks and J. J. Faber. 1986. Fetal O_2 consumption in sheep during controlled long-term reductions in umbilical blood flow. Amer. J. Physiol. 250:H1037.

Barcroft, J. and D. J. Barron. 1946. Observations upon the form and relations of the maternal and fetal vessels in the placenta of the sheep. Anat. Rec. 94:569.

Bassett, J. M. 1986. Nutrition of the conceptus: aspects of its regulation. Proc. Nutr. Soc. 45:1.

Battaglia, F. C. and W. W. Hay, Jr. 1984. Energy and substrate requirements for fetal and placental growth and metabolism. In: R. W. Beard and P. W. Nathanielsz (Ed.) Fetal Physiology and Medicine. pp 601–628. Marcel Dekker, Inc., New York.

Battaglia, F. C. and G. Meschia. 1981. Fetal and placental metabolisms: Their interrelationships and impact upon maternal metabolism. Proc. Nutr. Soc. 40:99.

Bell, A. W. 1984. Factors controlling placental and fetal growth and their effects on future production. In: D. R. Lindsay and D. T. Pearce (Ed.) Reproduction in Sheep. pp 144–152. Cambridge Univ. Press, London.

Bell, A. W., J. M. Kennaugh, F. C. Battaglia, E. L. Makowski and G. Meschia. 1986. Metabolic and circulatory studies of fetal lambs at mid-gestation. Amer. J. Physiol. 250:E538.

Bell, A. W., R. B. Wilkening and G. Meschia. 1987. Some aspects of placental function in chronically heat stressed ewes. J. Dev. Physiol. 9:17.

Buster, J. E. 1984. Fetal, placental and maternal hormones. In: R. W. Beard and P. W. Nathanielsz (Ed.) Fetal Physiology and Medicine. pp 559–599. Marcel Dekker, Inc., New York.

Char, V. C. and R. K. Creasy. 1976. Acetate as a metabolic substrate in the fetal lamb. Amer. J. Physiol. 230:357.

Cotter, J. R., J. N. Blechner and H. Prystowsky. 1969. Blood flow and oxygen consumption of pregnant goats. Amer. J. Obstet. Gynecol. 103:1098.

Dawes, G. S. 1968. Foetal and Neonatal Physiology. Year Book Medical Publ., Inc., Chicago.

Faber, J. J. and K. L. Thornburg. 1986. Fetal nutrition: supply, combustion interconversion and deposition. Fed. Proc. 45:2502.

Ferrell, C. L. and S. P. Ford. 1980. Blood flow, steroid secretion and nutrient uptake of the gravid bovine uterus. J. Anim. Sci. 50:1113.

Ferrell, C. L., S. P. Ford, R. L. Prior and R. K. Christenson. 1983. Blood flow, steroid secretion and nutrient uptake of the gravid bovine uterus and fetus. J. Anim. Sci. 56:656.

Ferrell, C. L., W. N. Garrett and N. Hinman. 1976. Growth, development and composition of the udder and gravid uterus of beef heifers during pregnancy. J. Anim. Sci. 42:1477.

Ferrell, C. L. and L. P. Reynolds. 1987. Blood flow and nutrient uptake of the gravid uterus, fetus and uteroplacenta of Hereford or Charolais cows bearing single or twin fetuses. J. Anim. Sci. 65 (Suppl. 1):148.

Ferrell, C. L., D. A. Robertson and L. P. Reynolds. 1985. Urea production and urea cycle enzyme activities of maternal liver, fetal liver and placental tissues of cows. J. Anim. Sci. 61 (Suppl. 1):261.

Ford, S. P., J. R. Chenault and S. E. Echternkamp. 1979. Uterine blood flow of cows during the estrus cycle and early pregnancy: effect of the conceptus on the uterine blood supply. J. Reprod. Fertil. 56:53.

Ford, S. P. and R. K. Christenson. 1979. Blood flow to uteri of sows during estrous cycle and early pregnancy: local effect of the conceptus on the uterine blood supply. Biol. Reprod. 21:617.

Ford, S. P., L. P. Reynolds and C. L. Ferrell. 1984. Blood flow, steroid secretion and nutrient uptake of the gravid uterus during the periparturient period in sows. J. Anim. Sci. 59:1085.

Freemark, M., M. Comer and S. Handwerger. 1986. Placental lactogen and GH receptors in sheep liver: striking differences in ontogeny and function. Amer. J. Physiol. 251:E328.

Gluckman, P. D. and G. C. Liggins. 1984. Regulation of fetal growth. In: R. W. Beard and P. W. Nathanielsz (Ed.) Fetal Physiology and Medicine. pp 511–557. Butterworths, London.

Greiss, F. C., Jr. and S. G. Anderson. 1970. Uterine blood flow during early ovine pregnancy. Amer. J. Obstet. Gynecol. 106:30.

Guiet-Bara, A. 1980. Human placental oxygen transfer and consumption. Dissociation by cooling or use of respiratory enzyme inhibitors. Eur. J. Obstet. Gynecol. Reprod. Biol. 10:83.

Koong, L. J., W. N. Garrett and P. V. Rattray. 1975. A description of the dynamics of fetal growth in sheep. J. Anim. Sci. 41:1069.

Lemons, J. A. 1979. Fetal–placental nitrogen metabolism. Seminars in Perinatology. Vol. 3, No. 2. pp 177–190.

Longo, L. D. 1985. The role of the placenta in the development in the embryo and fetus. In: C. T. Jones and P. W. Nathanielsz (Ed.) Physiological Development of the Fetus and Newborn. pp 1–9. Academic Press, New York.

Macdonald, A. A., M. A. Heymann, A. J. Llanos, E. Pesonen and A. M. Rudolph. 1985. Distribution of cardiac output in the fetal pig during late gestation. In: C. T. Jones and P. W. Nathanielsz (Ed.) Physiological Development of the Fetus and Newborn. pp 401–404. Academic Press, New York.

Makowski, E. L. 1968. Maternal and fetal vascular nets in placentas of sheep and goats. Amer. J. Obstet. Gynecol. 100:283.

Mellor, D. J. 1983. Nutritional and placental determinants of foetal growth rate in sheep and consequences for the newborn lamb. Brit. Vet. J. 139:307.

Meschia, G. 1976. Physiology of transplacental diffusion. Obstet. Gynecol. Annu. 5:21.

Meschia, G. 1983. Circulation to female reproductive organs. In: S. R. Geiger (Ed.) Handbook of Physiology. Vol. 3, Sect. 2. Circulation. pp 241–269. American Physiological Society, Bethesda, MD.

Mitchell, B. F., S. J. Lye, L. Lukash and J. R. G. Challis. 1986. Androstenedione metabolism in the late gestation sheep fetus. Endocrinology 118:63.

Munro, H. N., S. J. Pilistine and M. E. Fant. 1983. The placenta in nutrition. Annu. Rev. Nutr. 3:97.

Owens, J. A., J. Falconer and J. S. Robinson. 1986. Effect of restriction of placental growth on umbilical and uterine blood flows. Amer. J. Physiol. 250:R427.

Prior, R. L. 1982. Gluconeogenesis in the ruminant fetus: evaluation of conflicting evidence from radiotracer and other experimental techniques. Fed. Proc. 41:117.

Prior, R. L. and D. B. Laster. 1979. Development of the bovine fetus. J. Anim. Sci. 48:1546.

Ramsey, E. M. 1982. The Placenta. Human and Animal. Praeger Pubs., New York.

Rankin, J. H. G. 1976. A role for prostaglandins in the regulation of the placental blood flows. Prostaglandins 11:343.

Resnik, R. 1981. The endocrine regulation of uterine blood flow in the nonpregnant uterus: A review. Amer. J. Obstet. Gynecol. 140:151.

Reynolds, L. P. and C. L. Ferrell. 1987. Transplacental clearance and blood flows of bovine gravid uterus at several stages of gestation. Amer. J. Physiol. 253:R735.

Reynolds, L. P., C. L. Ferrell and S. P. Ford. 1985a. Transplacental diffusion and blood flow of the gravid bovine uterus. Amer. J. Physiol. 249:R539.

Reynolds, L. P., C. L. Ferrell, J. A. Nienaber and S. P. Ford. 1985b. Effects of chronic environmental heat stress on blood flow and nutrient uptake of the gravid bovine uterus and foetus. J. Agr. Sci. (Camb.) 104:289.

Reynolds, L. P., C. L. Ferrell, D. A. Robertson and S. P. Ford. 1986. Metabolism of the gravid uterus, foetus and utero-placenta at several stages of gestation in cows. J. Agr. Sci. (Camb.) 106:437.

Reynolds, L. P., D. S. Millaway, J. D. Kirsh, J. E. Infeld and D. A. Redmer. 1988. Secretion of angiogenic activity by placental tissues of cows at several stages of gestation. J. Reprod. Fertil. 88:497.

Rosenfeld, C. R., F. H. Morriss, Jr., F. C. Battaglia, E. L. Makowski and G. Meschia. 1976. Effect of estradiol-17β on blood flow to reproductive tissues of the ewe. Amer. J. Obstet. Gynecol. 124:618.

Rosenfeld, C. R., F. H. Morriss, Jr., E. L. Makowski, G. Meschia and F. C. Battaglia. 1974. Circulatory changes in the reproductive tissues of the ewe during pregnancy. Gynecol. Invest. 5:252.

Sauer, J. J., A. J. Conley, D. B. Farley, D. E. Van Orden and S. P. Ford. 1987. Sympathetic control of blood flow to different vascular components of the gravid bovine uterus. J. Anim. Sci. 65 (Suppl. 1):147.

Silver, M. and R. S. Comline. 1975. Transfer of gasses and metabolites in the equine placenta: a comparison with other species. J. Reprod. Fertil. Suppl. 23:589.

Silver, M., D. H. Steven and R. S. Comline. 1973. Placental exchange and morphology in ruminants and the mare. In: K. S. Comline, K. W. Cross, G. S. Dawes and P. W. Nathanielsz (Ed.) Foetal and Neonatal Physiology. pp 245–271. Cambridge Univ. Press, London.

Steven, D. H. 1975. Anatomy of the placental barrier. In: D. H. Steven (Ed.) Comparative Placentation. pp 25–56. Academic Press, New York.

Tsutsumi, Y. and E. S. E. Hafez. 1964. Comparative study on maternal–fetal vascular relationships. Fifth International Congress for Animal Reproduction and Artificial Insemination, Trento. Vol. 2, pp 195–200.

Walton, A. and J. Hammond. 1938. The maternal effects on growth and conformation in Shires horse–Shetland pony crosses. Proc. Roy. Soc. London Ser. B125:311.

Waters, M. J., V. H. Oddy, C. E. McCloghy, P. D. Gluckman, R. Duplock, P. C. Owens and M. W. Brinsmead. 1985. An examination of the proposed roles of placental lactogen in the ewe by means of antibody neutralization. J. Endocrinol. 106:377.

CHAPTER 2

Endocrinology of Bone Formation

E. MARTIN SPENCER

1. Introduction

Bone is a metabolically active, multifunctional tissue. The study of its endocrine regulation is especially complicated because many calciotropic hormones and growth factors affect bone. These hormones and growth factors may have direct and indirect actions and may also elicit counterregulatory effects. For several, the different modes of action have not been clearly separated.

The growth of bone is probably affected by more hormones and growth factors than any other tissue. These can act by endocrine, paracrine, and autocrine mechanisms. Although not discussed in this chapter, bone formation is influenced by genetic constitution and ionic environment.

1.1. Unique Aspects of Bone Physiology

1.1.1. Turnover and Coupling

Bone is a vital tissue which is constantly being formed, removed (resorbed), and synthesized throughout life by a process called turnover. Not all bones turn over at the same rate; trabecular bone has a higher rate than cortical bone. Normally, bone formation and resorption are tightly coupled events. The mechanism of how this is accomplished has never been elucidated. It must be flexible enough to allow for remodeling, repair, and growth, especially during puberty. After the menopause in humans, coupling fails and the resorption rate is greater, resulting in postmenopausal osteoporosis. More than likely, because

E. MARTIN SPENCER • Laboratory of Growth and Development, Children's Hospital of San Francisco, San Francisco, California 94118.

of the critical importance of the coupling phenomena, more than one effector system has evolved to maintain this delicate balance between the rates of bone formation and resorption. Prostaglandins, interleukins, and transforming growth factor-β (TGF-β) have been suggested as couplers.

Constant bone turnover serves many functions. It enables bone to be remodeled according to the requirements of different stages of development. It allows for the periodic replacement of bony tissue. Major factors in this process of remodeling are the stresses and strains of weightbearing and muscle pulls which dictate the most functional structure for bone. How repeated stress causes bone to fatigue is unexplained. Were it not for continual reconstruction, microfractures would accumulate, impairing the functional integrity of bone. While microfractures might be minor for meat animals with a short growing cycle, the accumulation of microfractures would be quite detrimental for humans and other animals. Additional functions that bone turnover serves are to permit bone to be a ready reservoir for maintenance of plasma calcium, phosphorus, and magnesium levels and to protect the body from acidosis.

1.1.2. Differentiation

A corollary of the constant turnover of bone is that cells that make up bone, such as osteoblasts, osteoclasts, endothelial cells, and chondrocytes (until linear growth ceases), must be replaced continually throughout life by a process of differentiation from precursors. Trabecular osteoblasts probably are derived from undifferentiated mesenchymal cells in the bone marrow. Cortical osteoblasts originate from the external fibrous membrane of the cortex, the periosteum. The outer layer of the periosteum is comprised of fibroblasts and undifferentiated mesenchymal cells. The latter have the ability to differentiate into the preosteoblasts that form the inner layer of the periosteum. Preosteoblasts are induced to differentiate into osteoblasts which proliferate until they are end-differentiated and synthesize the organic matrix. The matrix becomes calcified, engulfing the osteoblasts and transforming them into osteocytes which still maintain contact with surface osteoblasts through thin processes in canaliculi. New bone formation also requires angiogenesis because blood vessels traverse bone in channels, nourishing the osteocytes and perhaps transmitting signals. Chondrocytes at the growth plate differentiate from prechondrocytes and subsequently undergo an orderly process of proliferation and maturation to accomplish linear growth. The current theory is that osteoclasts originate from blood monocytes/macrophages or earlier precursors that subsequently mature and fuse into osteoclasts. A failure of osteoclast differentiation results in a serious disease, osteopetrosis, which is characterized by a small marrow cavity because bone is not resorbed.

1.1.3. Hematopoiesis

The significance of the location of hematopoietic tissue in the bone marrow has not been adequately investigated. After fetal life, essentially the only normal place where elements of blood can be formed is in the bone marrow. Growth factors such as somatomedin C, colony-stimulating factors, and TGF-β synthesized by marrow osteoblasts could stimulate hematopoiesis. Such conditions as oxygen tension, temperature, ionic environment, presence of bony trabeculae, the rate of blood flow, as well as other subtle features, may very well be factors. The bone marrow also places blood precursor cells in the best position to differentiate into osteoclasts and initiate endosteal (the inner cortical surface) and trabecular bone resorption. One attribute of this distribution of the blood-forming elements is that the body is protected from the effects of the loss of a single blood-forming organ.

1.2. Bone Structure

In order to understand the endocrinology of bone formation, a short summary of the significant features of bone structure will be presented. For a detailed discussion, the readers are referred to standard textbooks of histology and embryology.

Bone can be classified in several ways: (1) by location—cortical, trabecular, or medullary bone; (2) by origin—endochondral or membranous; and (3) by specialized function—the epiphysis or chondrocyte growth plate which leads to increasing length of bone.

The cortex is the dense osseous tissue on the outer layer of bone and is especially prominent on long bones. Cortical bone is constantly being formed on the outside, or periosteal, surface and removed from the inner, or endosteal, surface. Preosteoblasts in the periosteum differentiate into osteoblasts which proliferate and synthesize the matrix when maturation is complete. The matrix consists of proteins and proteoglycans. The major protein is type I collagen, but osteoblasts also synthesize osteonectin, osteocalcin, proteoglycans, and growth factors which include TGF-β macrophage colony-stimulating factor, insulinlike growth factor, and somatomedin C. Several matrix proteins, such as albumin and (α) HS glycoprotein, are derived from plasma. Matrix is mineralized by the deposition of calcium and phosphorus to form a hydroxyapatite crystal structure. The calcium and phosphorus crystallize in matrix vessels and become associated with collagen molecules which are the basic structural framework of bone. The deposition of calcium may be facilitated by osteonectin which has binding sites for both calcium and collagen. The role of osteocalcin is uncertain. It contains the calcium-binding amino acid γ-carboxyglutamic acid, which may limit the mineralization process by binding to the ends of calcium crystals. Magnesium, sodium, and other minerals are also present in bone. As the bony matrix is secreted and calcified, osteoblasts become trapped inside and are thereby transformed into osteocytes. The latter are live cells that maintain contact with each other as well as with osteoblasts on the surface of bone by thin processes in channels and thus could transmit signals to bony surfaces. The function of osteocytes, however, is undefined since typically it is only the osteoblasts or preosteoblasts that can be studied *in vitro*. The products of osteocyte synthesis could be circulated, because they are nourished by blood vessels that are entrapped in canals within the bony matrix during formation of cortical bone.

Trabecular bone consists of bony spicules in a three-dimensional array surrounded by cortical bone. The bone marrow is located in the spaces between the trabeculae. The predominant orientation of the trabeculae is dictated by the lines of stress. The trabeculae are lined by resting (flat) osteoblasts which assume a plump morphology after activation. Bone is resorbed by osteoclasts, which are large multinucleated cells thought to be derived from precursors in the bone marrow by fusion. The signal for resorption is unclear. Before the osteoclasts can attach to bone, osteoblasts must be induced to leave the surface and osteoid must be eroded down to the mineral surface. The digestion of osteoid is probably facilitated by osteoblast production of collagenase which is stimulated by some growth factors that increase resorption: platelet-derived growth factor (PDGF), epidermal growth factor (EGF), and TGF-β. The area of osteoclast membrane associated with the mineral surface assumes a characteristic ruffled appearance bounded by a clear zone. Pits are excavated in the bone by a mechanism that is only partially understood, but involves the secretion of hydrogen ions and acid hydrolases. The pits subsequently are filled with new bone by osteoblasts after the osteoclasts detach.

Although the epiphysis is technically cartilage, it is an integral part of endochondral bone growth. Longitudinal bone growth is the result of growth at the epiphysis. The latter consists of zones of resting, proliferating, maturing, and calcifying cartilage which are regulated by a variety of hormones. The *resting zone* contains chondrocyte precursors. Their differentiation and the subsequent clonal proliferation form even columns of cells separated by newly secreted matrix; this is the *zone of proliferation*. In the *zone of maturation*, the chondrocytes secrete alkaline phosphatase, carbohydrates, and lipids. In the *zone of calcification*, the chondrocytes degenerate and ultimately die while the cartilaginous matrix becomes calcified and is reorganized into trabecular bone. As long as the epiphysis is present, successive differentiation, proliferation, maturation, and calcification of this tissue lead to an increase in bone length.

1.3. Methods of Investigating Bone Formation

A variety of methods are available for studying the endocrinology of bone. A combination usually must be applied to obtain a satisfactory interpretation. These involve direct and indirect measurements as well as *in vivo* and *in vitro* studies. Plasma is analyzed for calcium, phosphorus, magnesium, pH, calciotropic hormones, and growth factors. Urine is tested for its pH and content of hydroxyproline, creatinine, and minerals. The kinetics of bone formation have been studied using radioisotopes, e.g., calcium or strontium. Radioactive calcium is now used principally to quantitate intestinal calcium absorption. Histological examination of bone biopsies allows direct visualization of several processes. In recent years the utility of bone biopsies has greatly increased by a refinement called quantitative histomorphometry. The latter, which is coupled with double tetracycline labeling, is able to give information on a variety of dynamic parameters such as rates of bone formation, resorption, and osteoid maturation. The number of osteoblasts and osteoclasts, and the presence of woven bone (disorganized architecture) can be determined. Bones can also be analyzed for mineral, collagen, and protein contents. Bone and mineral density can now be reliably estimated in humans by noninvasive techniques such as dual photon absorptiometry (DPA) and quantitative computed tomography (QCT) with a precision approaching 1.2%. Ultrasound is also being utilized experimentally to determine bone mineral content. Recently, our laboratory participated in the development of an *in vivo* method to investigate the direct effects of continuous administration of calciotropic agents for extended periods of time on bone histomorphometry (Schlechter *et al.,* 1986). In conscious, unrestrained small laboratory animals, the right femoral artery is selectively perfused by an Osmotic Minipump (ALZA) with a concentration of a drug that elevates the local but not the systemic level. Since this does not initiate the counterregulatory effects that confuse systemic administration, the bones of the left leg can serve as a control for the right leg.

Many different cell culture systems have been developed to study osteoblasts and osteoclasts. For osteoblasts, virtually all the data in culture have come from embryonic and neonatal cells in primary culture or transformed cells in continuous culture. Cells for primary culture are typically removed from embryonic rodent or chick calvaria by enzymatic (collagenase) digestion. This produces a heterogeneous population which can be fractionated into periosteal cell-rich fractions containing an abundance of preosteoblasts and mesenchymal cells, and osteoblast fractions which have differentiated properties. Recent advances in culture methodology have allowed a few laboratories to successfully grow primary osteoblasts from adult tissue. This obviates the tenuous extrapolation of results from embryonic and malignant cells to normal adult or aging bone cells and it promises to produce valuable new data. Neonatal rat tibia in organ culture is also used when an organized, mixed population of cells is desirable. An advantage of this system is that it evidences coupling, i.e., a relationship between bone resorption and bone formation, that serves to maintain bone mass. Osteoclast activity is measured by the release of ^{45}Ca from bones prelabeled with this isotope. New systems are being developed to study the regulation of osteoclast formation from bone marrow and splenic cells.

2. Hormones

2.1. Somatomedins

Somatomedins are probably the most important regulators of bone formation. They are known primarily for mediating the growth-promoting effect of growth hormone (GH) on skeletal growth (Salmon and Daughaday, 1957). However, several calciotropic hormones interact with somatomedins heightening their pivotal role in bone formation. These substances include parathyroid hormone, calcitriol, thyroxine, insulin, corticosteroids, estrogens, PDGF, fibroblast growth factor (FGF), and EGF (Stiles *et al.*, 1979; Clemmons and Shaw, 1983; Copeland *et al.*, 1984). Although somatomedins are discussed in this section, they may act predominantly as local growth factors—see below.

Somatomedins, which are also called insulinlike growth factors, are polypeptides of 7500 Da. There are two classes of somatomedins: (1) a basic one called somatomedin C (insulinlike growth factor I, IGF-I) which is the mediator of the growth-promoting effect of GH postnatally (Rinderknecht and Humbel, 1978a; Spencer *et al.*, 1983); and (2) IGF-II which is less GH dependent and may have a role in fetal growth (Rinderknecht and Humbel, 1978b). The regulation of the plasma concentration of somatomedin C is primarily under the influence of GH but calciotropic hormones such as insulin, corticosteroids, thyroxine, and estrogen also influence the plasma level.

Somatomedins are potent mitogens and stimulate a variety of cell types of epithelial, endothelial, and mesenchymal origin. The basis of the insulinlike activity of somatomedins derives from their structural homology to insulin (Blundell *et al.*, 1978). They are transported in plasma bound to a carrier protein of 150 kDa which prolongs their activity and maintains them in an inactive state during their transport through plasma (Hintz and Liu, 1977). This binding protein, like the somatomedins, is GH dependent. Upon exposure to acid it liberates an acid-stable unit (50 kDa) which retains binding activity and an acid-labile unit. An additional unrelated binding protein of approximately 32 kDa also circulates but its function is unknown. It is not GH dependent, and in some cell systems potentiates the mitogenic action of somatomedin C (Elgin *et al.*, 1987).

Isaksson *et al.* (1982) were the first to demonstrate that direct injections of GH into the tibial epiphysis stimulated longitudinal bone growth. Russell and Spencer (1985) repeated this work and also found in subsequent studies that somatomedin injected directly into the epiphysis could stimulate growth. This effect of GH could be blocked by the simultaneous administration of a blocking antibody to somatomedin C (Schlechter *et al.*, 1986). It was concluded that GH acted on epiphyseal chondrocytes by stimulating somatomedin C production which stimulated growth in an autocrine manner. These and other studies have questioned the significance of hepatic production of somatomedin C and its endocrine action. It is possible that local tissue production is the major source of somatomedin C for growth promotion and it acts in autocrine and/or paracrine fashion. Green *et al.* (1985) have suggested a dual effector role for GH. They hypothesized that GH initiates the differentiation of mesenchymal cells and that only the differentiated cells are sensitive to the action of somatomedins which results in clonal growth. This is supported by immunohistochemical data from our laboratory showing that GH stimulates somatomedin C production primarily in the zone of proliferation and maturation but not in the zone of resting chondrocytes.

The somatomedin C level from cord blood has been shown to correlate with birth weight. The serum somatomedin C concentration is low at birth and progressively rises throughout childhood. During the pubertal growth spurt, in spite of little change in the level of GH, that of somatomedin C rises transiently into the range commonly found in patients with GH-producing tumors (acromegaly). However, after puberty, the somatomedin C level falls to the normal adult value. Thereafter it remains relatively constant until it begins to decline at the menopause in women but much later in men (Bennett *et al.*, 1984). The fall in the somatomedin C level at the time of menopause suggests a possible relationship to postmenopausal osteoporosis.

In patients with clinical abnormalities of growth, the plasma somatomedin C level correlates much better to actual growth than does that of GH hormone. For example, patients with impaired growth due to chronic liver disease, malnutrition, chronic renal disease, and Laron dwarfism have significantly lower somatomedin C values in spite of elevated GH levels. Somatomedin C plasma concentration also correlates with the rate of weight gain of pigs and cattle.

Somatomedins have been shown to stimulate skeletal growth *in vivo* in GH-deficient rodents (Holder *et al.*, 1981; Schoenle *et al.*, 1983). *In vitro*, they stimulate the proliferation of neonatal rat calvarial and embryonic chick osteoblasts and the synthesis of collagen by osteoblasts (Canalis, 1983; Schmid *et al.*, 1983; Howard and Spencer, 1984). Physiological concentrations of parathyroid hormone (PTH), which are inactive, potentiate the mitogenic activity of somatomedin C both *in vitro* and *in vivo* (Si *et al.,* 1988). This offers a potential explanation for the anabolic effects of PTH on bone. Somatomedins were not observed to affect osteoclast activity (Howard and Spencer, unpublished observations).

2.2. PTH

PTH is generally one of the first hormones considered when discussing bone formation. Although PTH stimulates bone turnover, it actually plays a relatively minor role in causing net bone formation. PTH's principal role is the maintenance of plasma calcium. Low levels of plasma calcium induce the parathyroid gland to release PTH which raises plasma calcium by stimulation of: (1) osteoclasts to resorb bone, (2) renal tubular reabsorption of calcium, and (3) 1-25-$(OH)_2D_3$ (calcitriol) production which increases intestinal calcium absorption and bone resorption. PTH lowers the plasma phosphorus level by fostering its renal excretion; this action prevents the deleterious precipitation of calcium phosphate in tissues since PTH releases both calcium and phosphorus from bone.

The actions of PTH suggest that it should be primarily catabolic on bone. This is borne out by the fact that most patients with advanced hyperparathyroidism do have osteoporosis, and those with hypoparathyroidism may have increased bone formation (Seeman *et al.,* 1982). However, PTH must also have some *in vivo* anabolic effects. Occasional patients with hyperparathyroidism have osteosclerosis (Gennant *et al.,* 1975) and there have been several clinical studies using appropriate doses of PTH either alone or with calcitriol that have shown stimulation of bone formation (Parfitt, 1976; Reeve *et al.,* 1980, 1981). The probable reason for the contrasting *in vivo* responses to PTH is that it has direct opposing actions on bone formation and resorption and also indirect actions on both. Although the catabolic effects tend to predominate, the multiplicity of actions of PTH on calciotropic hormones, growth factors, mineral balance, and metabolism may shift the outcome. The result that prevails in any given biological setting depends on many factors and is difficult to predict. Nevertheless, PTH is now being investigated as a possible therapy for postmenopausal osteoporosis.

PTH stimulation of osteoblasts results in a striking morphological change within a few hours: polygonal cells are transformed to spindle-shaped cells which no longer form a continuous lining. PTH stimulates osteoblasts to produce a substance required for osteoclastic bone resorption. This factor may be macrophage colony-stimulating factor (M-CSF), prostaglandins, and/or coupling factor (Elford *et al.,* 1987). PTH specifically inhibits osteoblast synthesis of type I collagen and alkaline phosphatase without affecting general protein synthesis (Dietrich *et al.,* 1976; Bringhurst and Potts, 1981).

The intracellular mediators of PTH are thought to be cAMP and calcium. PTH interacts with high-affinity plasma membrane receptors to stimulate adenylate cyclase, increase the intracellular cAMP content, and activate cAMP-dependent protein kinase. Concomitantly, PTH also increases intracellular calcium. Both calcium ionophores and analogues of cAMP can mimic the effect of PTH on osteoblasts.

Because PTH could not reliably be shown to directly increase osteoblast proliferation, its action on bone formation was thought to occur secondary (or be coupled) to the action of PTH on bone resorption. A PTH-responsive factor, which coupled bone formation to resorption, has been partially purified, but unfortunately, has never been verified (Howard et al., 1981). Recently, Howard and Spencer (1984) postulated an alternate mechanism. They found that physiologic concentrations of PTH, which have no effect on DNA synthesis, synergistically increased the somatomedin C stimulation of DNA synthesis by chick calvarial osteoblasts. This work could explain the synergistic stimulatory effect that PTH and calcitriol have on bone formation (Slovik et al., 1987). Since calcitriol has been shown to increase somatomedin C production by adult human osteoblasts in culture (Valentin-Opran et al., 1986), PTH could potentiate the action of the somatomedin C released locally, possibly by up-regulating the somatomedin C receptor.

2.3. Vitamin D

Although Vitamin D, specifically calcitriol, is necessary for bone growth, considerable controversy still exists concerning its mechanism of action. Calcitriol stimulates the resorption of bone by osteoclasts, the intestinal absorption of calcium and phosphorus, and the renal retention of phosphate (Raisz et al., 1972a). It functions primarily to maintain a normal plasma calcium, for in states of calcium deficiency, calcitriol acts to keep the plasma calcium normal rather than to stimulate bone formation. It may affect bone formation by providing an adequate supply of calcium and phosphorus for bone mineralization.

Calcitriol is formed by two sequential hydroxylation steps of vitamin D_3. The first occurs at the 25 position in the liver and the second at the 1-α position in the mitochondria of the kidney. Turner et al. (1980) showed that embryonic bone cells are also capable of the 1-α hydroxylation reaction; macrophages and keratokinocytes also possess this activity. Calcitriol acts like a classical steroid hormone by interacting with a 3.5 S cytoplasmic receptor in sensitive tissues.

Calcitriol stimulates bone resorption by promoting the differentiation of precursor monocytes to osteoclasts. Osteoblasts are required for this activity. Calcitriol has been suggested to stimulate the production by osteoblasts of M-CSF which mediates the differentiating action of calcitriol (Elford et al., 1987).

Calcitriol has several direct effects on osteoblasts. It increases osteocalcin (bone Gla protein) and decreases collagen synthesis. The latter results from an action on transcription because of the concomitant decrease in procollagen mRNA (Rowe and Kream, 1982). Calcitriol increases somatomedin C production by osteoblasts (Valentin-Opran *et al.*, 1986) and rapidly elevates the low plasma somatomedin C level in vitamin D deficiency (Spencer, unpublished observations). However, a direct stimulatory role of calcitriol on osteoblast proliferation has not been proven. In vitamin D deficiency, osteoid (uncalcified matrix) is in excess and bone growth soon stops. Whether the cessation of growth is due to a deficient mineral supply, calcitriol, or both has never been adequately answered. There have been studies suggesting that the maintenance of normal calcium in vitamin D deficiency allows normal bone formation. Calcitriol could have indirect effect on bone formation which may be overshadowed by its resorptive effect. These could be mediated by factors released from resorbed bone matrix such as somatomedins, coupling factor, or TGF-β.

Bone cells can synthesize $24,25-(OH)_2D_3$ and are thought to have high-affinity binding sites for this vitamin D metabolite (Henry and Norman, 1978; Turner *et al.*, 1980; Somjen *et al.*, 1982). However, it is still controversial whether $24,25-(OH)_2D_3$ is necessary for bone formation. To add to the confusion, occasional patients with renal osteodystrophy will only respond to therapy when $24,25-(OH)_2D_3$ is added to their therapeutic regimen. Corvol *et al.* (1978) presented evidence that $24,25-(OH)_2D_3$ increases the synthesis of glycosaminoglycans by chondrocytes. However, others have questioned the role of $24,25-(OH)_2D_3$ in bone formation by substituting a fluoro group at the 24 position which prevents 24 hydroxylation; yet this compound is fully active metabolically (Halloran *et al.*, 1981; Miller *et al.*, 1981; Ameenuddin *et al.*, 1982).

2.4. Calcitonin

Calcitonin is a polypeptide hormone of 3700 Da produced by the C cells of the thyroid gland. Its function is to maintain a normal plasma calcium level. An increase in this normal level triggers a release of calcitonin, which leads to renal excretion of calcium and decreased resorption of bone. The latter is accomplished by a direct inhibition of osteoclasts; calcitonin is the only substance known to do this, and osteoblasts are not required. Calcitonin does not appear to have any effect on osteoblasts. An excess or deficiency of calcitonin has little effect on mineral metabolism in humans.

2.5. Insulin

Insulin was originally implicated in normal skeletal growth primarily because uncontrolled juvenile (type I) diabetes leads to decreased growth, but an understanding of insulin's biochemical actions still eludes us.

Heath *et al.* (1979) questioned the importance of insulin in bone formation. They concluded that adult humans with diabetes mellitus have no major problem with bone formation because their fracture incidence was normal. Previous reports of decreased bone mass in diabetics need to be reevaluated with a larger number of patients employing the more sophisticated methods (e.g., DPA or QCT) of determining bone mineral content (Levin *et al.*, 1976; McNair *et al.*, 1979; Wiske *et al.*, 1982). Even if a small amount of bone loss in diabetics can be demonstrated, insulin deficiency *per se* may not be responsible. It could be attributed to a variety of factors such as: the effect of concomitant malnutrition, the metabolic derangements associated with insulin deficiency, decreased somatomedin C production, or a genetic predisposition linked to diabetes.

There is a considerable amount of data on the effects of insulin on chemically induced diabetes in rats. The extent to which this can be extrapolated to other animals is uncertain. Humans with diabetes mellitus do not develop abnormalities of calcium and vitamin D (Heath *et al.*, 1979), whereas rats with diabetes induced by alloxan or streptozotocin have an acute illness with major abnormalities in calcium metabolism (Schneider *et al.*, 1977; Spencer *et al.*, 1980). In diabetic rats the plasma levels of calcium, calcitriol, somatomedin C, and calcitonin are decreased and PTH levels are increased. They have decreased intestinal calcium absorption, ketosis, acute weight loss, and possibly toxic effects on tissues other than islet cells. All of these changes are reversed by insulin. The significance of the decreased calcitriol is moot because bone biopsies reveal no osteomalacia but rather a decreased turnover. Diabetic rats have also been reported to have impaired ossification and inhibition of cartilage proliferation (Weiss and Reddi, 1980).

The question of insulin's action has been approached *in vitro* using primary cultures of embryonic osteoblasts or chondrosarcoma cells (Stuart *et al.*, 1979; Foley *et al.*, 1982); however, the relevance to adult cells and diabetes must still be established. These studies have shown that insulin stimulates amino acid transport and the synthesis of protein, type I collagen, and RNA but has no effect on DNA synthesis (Peck and Messinger, 1970; Hahn *et al.*, 1971; Canalis *et al.*, 1977). The problem with these observations is that the high level of insulin used, 10^{-8} to 10^{-9} M, is in the range where insulin could actually be acting on the type 1 somatomedin receptor. This mechanism is also consistent with the observation of Daughaday *et al.* (1975) that insulin can replace GH in hypophysectomized animals. Since insulin is a known stimulator of somatomedin C production, a good part of its action could be mediated by somatomedin C.

2.6. Thyroid Hormones

Thyroid hormones are important in bone formation, for in virtually every species, hypothyroidism results in decreased bone growth. In spite of these observations, there is a paucity of data on thyroid hormone action. Its role in bone formation is generally thought to be permissive or osteogenic.

Bone turnover correlates directly with thyroid status (Kragstrup *et al.*, 1981). Hyperthyroidism has been shown to increase bone resorption but not to have any effect on net bone growth. Thyroid hormone can directly stimulate bone resorption (Mundy *et al.*, 1976), but its other effects on bone may be indirect, since, for example, it can stimulate the production of somatomedin C. In *in vitro* studies, several groups have shown that thyroid hormone does not affect either collagen or noncollagen protein synthesis in cultured calvarial cells and does not act synergistically with somatomedins (Halme *et al.*, 1972; Askenasi and Demeester-Mirkinen, 1975; Mundy *et al.*, 1976; Canalis, 1980).

The effect of thyroid hormone on cartilage has been studied more extensively. Simpson *et al.* (1950) found that thyroid administration to hypothyroid animals increased the growth plate. In normal animals, long-term administration of thyroxine increased differentiation, although there was no effect on linear growth. In a study on Snell dwarf mice, Smeets and van Buul-Offers (1983) demonstrated that concurrent GH and thyroxine administration had an additive effect on epiphyseal growth. They postulated that thyroxine was osteogenic, stimulating the proliferation of the resting zone of the epiphysis, whereas GH and somatomedin were chondrogenic, stimulating predominately the proliferative and degenerative zones. This is consistent with the generally accepted concept that thyroid hormone is important in differentiation of chondrocytes and interacts cooperatively with GH.

Chondrocytes in the growth plate have been shown to have receptors for triiodothyronine (T_3) (Kan *et al.*, 1984). Burch and Lebovitz (1982) examined thyroxine's effects on porcine chondrocytes and determined that protein and proteoglycan synthesis were increased as was the hypertrophic zone of the epiphysis. Froesch *et al.* (1976) also observed an effect on proteoglycan synthesis and demonstrated that thyroxine acted synergistically with somatomedins. An *in vitro* effect of thyroid hormone on cartilage cells has not always been observed (Hill, 1979). This could be due to the experimental conditions or species differences.

2.7. Sex Steroids

Sex steroids have a dramatic effect on skeletal growth as evidenced by the pubertal growth spurt, the larger bone structure in males, and the association of postmenopausal osteoporosis with estrogen lack. There is considerable information in the literature, but little definitive data to clarify their mechanism of action. The reasons for this are several: Both male and female sex hormones can have differential effects depending on whether one is studying low or high physiologic levels (e.g., at puberty) or pharmacologic levels. There may be species specificity and different genetic susceptibilities to sex hormones. Some of their anabolic action may depend on GH, since the administration of either sex hormone to castrates restores the periodic pulses of GH to normal or higher levels.

2.7.1. Testosterone

Testosterone, which must be converted to dihydrotestosterone before it is metabolically active, has been shown to increase linear growth by stimulating cell proliferation at the growth plate. Its action must be more complicated because deficiency of testosterone results in an increase in young chondrocytes and delayed maturation, and castration prior to puberty leads to greater than normal growth. The increased growth in prepubertal castrated males has been linked to the disappearance of both the stimulatory effect of testosterone, as well as the greater suppressive effect of estrogen (Silberberg and Silberberg, 1971). High physiological doses of testosterone, such as those which occur during puberty, are thought to enhance maturation of the growth plate and to inhibit linear growth (Silbermann, 1983).

Testosterone interacts with other hormones. Jansson et al. (1983) showed that GH is required for testosterone's action. It is generally held that testosterone has no effect on somatomedin C synthesis, but studies that might reveal additive or synergistic effects on interaction have not been done. It is also possible that some of the effect of testosterone could be secondary to its effect on muscle mass (Canalis and Raisz, 1978).

2.7.2. Estrogen

Estrogens have important effects on bone formation, but few direct actions are described. In both birds and mice, estrogens stimulate the formation of medullary bone (Simmons, 1966; Miller, 1978). Estrogens have also been reported to increase bone weight in fetal lambs (Gee and Preston, 1957). Ovariectomy decreases bone mass (Raisz and Kream, 1981). Estrogens have not been observed to stimulate osteoblasts directly, but prior studies were done under conditions where cellular receptors for estrogens were blocked by an indicator dye in the culture media. Alteration of experimental conditions has led to the demonstration of estrogen receptors on human bone (Eriksen et al., 1987). The high levels of estrogens that occur at puberty and pharmacologic doses inhibit linear growth. This occurs because of the accelerated aging of the chondrocyte growth plate which results from inhibition of cartilage growth and enhanced maturation (Van Sickle, 1985). Part of the actions of estrogens on linear growth and bone mass may also be due to interaction with GH and somatomedin C.

Estrogens can also stimulate osteoblasts indirectly. High doses have been reported to lower serum somatomedin C levels (Wiedemann *et al.*, 1976; Clemmons *et al.*, 1980); the inhibitory effects of estrogens on the growth plate and bone formation could be due to this. On the contrary, physiologic doses of estrogens stimulate somatomedin C production which is anabolic to chondrocytes and osteoblasts. Thus far, there is no information on whether estrogens have any action on the synthesis of somatomedin binding proteins which are major factors in determining the concentration of the free somatomedin and its biological activity. Since estrogens modulate the levels of other binding proteins, this should be investigated. Estrogens have other actions that could also affect bone formation indirectly: augmenting calcitriol production, intestinal calcium absorption, and the plasma levels of PTH and calcitonin (Gallagher *et al.*, 1980).

Estrogens inhibit bone resorption by osteoclasts. The biochemical mechanism has not been elucidated but direct actions have not been observed *in vitro*. The lack of estrogens after menopause is generally believed to be responsible for the increased bone resorption that leads to postmenopausal osteoporosis in many women.

2.8. Glucocorticoids

Glucocorticoids are excellent examples of hormones that have markedly different effects on bone formation when studied *in vivo* as opposed to *in vitro*. This discrepancy results because glucocorticoids have multiple *in vivo* effects on other calciotropic hormones which cannot be observed *in vitro*. For example, hyperadrenalcorticolism or chronic glucocorticoid therapy can lead to osteoporosis with decreased bone formation, increased bone resorption, and very thin osteoid seams. However, glucocorticoids directly inhibit bone resorption *in vitro* (Stern, 1969; Raisz *et al.*, 1972b), possibly by stimulating osteoclasts.

Although glucocorticoids in osteoblast culture systems have been shown to stimulate synthesis of collagen and alkaline phosphatase (Canalis, 1982) longer-term studies reveal that glucocorticoids decrease synthesis of both collagen and noncollagen protein. Thus, the early effects are direct effects on osteoblasts, but the longer-term effects result from decreased formation of new osteoblasts from precursors (Peck *et al.*, 1967; Canalis, 1980).

Glucocorticoids have other actions on calciotropic hormones. They have been reported to regulate the calcitriol receptor concentration, but this has been noted to vary with the species and other factors (Manologas *et al.*, 1979; Chen *et al.*, 1982). Glucocorticoids *in vivo* decrease intestinal calcium absorption which triggers an increase in PTH (Au, 1976; Hahn *et al.*, 1979). They have been shown to enhance the effect of somatomedin C on bone formation (Canalis *et al.*, 1980; Canalis, 1980), probably by up-regulation of the type 1 somatomedin receptor (Bennett *et al.*, 1982). Although both elevated and depressed levels of somatomedin have been reported secondary to glucocorticoid administration or Cushing's syndrome, these may reflect peculiarities of the assays for somatomedin . Gourmelen *et al.* (1982) showed that glucocorticoids have no real effect on somatomedin plasma levels when total somatomedin is measured after removal of the carrier protein. They have been reported to modulate the liver synthesis of somatomedins; however, the physiologic importance of this is uncertain. Glucocorticoids have been found to have a direct effect on cartilage which blunts the effects of somatomedins (Tessler and Salmon, 1975; Gourmelen *et al.*, 1982).

3. Local Regulators of Bone Formation

3.1. Prostaglandins

There is considerable evidence to suggest that prostaglandins (PGs) may play a role in the local regulation of bone formation. Both PGE_2 and prostacyclin (PGI_2) are produced by osteoblasts (Raisz *et al.*, 1979; Voelkel *et al.*, 1980), and cloned osteosarcoma (osteoblastlike) cells have been shown to synthesize prostaglandins (Rodan *et al.*, 1981). EGF, glucocorticoids, bradykinin, interleukin-1, TGFs, PDGF, and mechanical stress have been reported to stimulate the release of prostaglandins (Tashjian and Levine, 1978; Tashjian *et al.*, 1982). Prostaglandins augment both the formation and resorption of bone, and stimulate the synthesis of collagen by fetal chick and rat bone cells although at high doses inhibitory effects are noted. There is considerable evidence that prostaglandins are coupling factors. They appear to be involved as mediators of local bone metabolism in a wide variety of physiologic and pathologic conditions such as ectopic bone formation, induction of new bone formation, immobilization, periodontal disease, inflammatory arthritis, and some types of humoral hypercalcemia of malignancy. In ovariectomized and estrogen-treated rats, the synthesis of prostaglandins is decreased and increased, respectively.

3.2. Noncollagen Proteins

Several proteins and peptides produced by osteoblasts potentially regulate bone formation. Osteonectin is a 32-kDa protein that may have a role in initiating mineral precipitation by binding to both collagen and hydroxyapatite crystals. Osteocalcin (bone Gla protein, BGP) is a vitamin K-dependent 50-residue peptide containing γ-carboxyglutamic acid that may inhibit further mineral deposition by binding to hydroxyapatite crystals (Poser and Price, 1979). A larger 79-residue peptide, matrix Gla protein, has recently been shown to be related to BGP but is a product of a different gene. Proteoglycan II is one of several proteoglycans synthesized by osteoblasts.

3.3. Growth Factors

In the last few years, there has been an extraordinary amount of interest in growth factors as we have become aware of their importance in biology. Several have been shown to be produced by osseous tissue and more are possibly still to be discovered. Research has been aided by the advent of pure recombinant materials. Although considerable information on growth factors is available, their mechanisms of action are in general less well understood than those of hormones. This section will concentrate on only the few that at present seem to have the most importance (Table I). The somatomedins have been discussed in Section 2.1, and probably are the most important local growth factors regulating bone formation.

Growth factors are principally peptides or proteins which usually operate as paracrines or autocrines but may also act as endocrines. They interact with specific cell surface receptors and initiate intracellular events that lead to mitosis or differentiation. However, other specialized cell functions such as collagen synthesis can be stimulated or inhibited. Thus far, no effects on mineralization have been reported. Our knowledge is increasing about the mode of

Table I
The Effects of Growth Factors on Bone

	Osteoblast synthesis		Osteoclastic resorption
	DNA	Collagen	
Somatomedin C (IGF-I)[a]	+	+	−
IGF-II	+	+	
Prostaglandins[a]	+	+[b]	+
EGF	+	−	+
TGF-α			+
TGF-β_1	+	−	+
PDGF	+	+[c]	+
FGF (α, β)	+	−	
IL-1 (α, β)[a]	+	+	+
TNF			−
Interferon-γ	−	−	−
M-CSF[a]			+
Bone-derived growth factor (β_2-microglobulin)	+?		
Bone morphogenic protein	+	+	

[a] Produced by osteoblasts.
[b] High doses (−).
[c] Nonspecific effect.

action of some growth factors. A close relationship to oncogenes has been documented. The receptors of several growth factors—EGF, somatomedin C, PDGF—are protein kinases. Ligand binding activates the receptor kinase to autophosphorylate the receptor and/or intracellular substrates. This has been an attractive mechanism of action because: phosphorylation reactions have been shown to control many intracellular regulatory processes, several oncogene products are kinases, and these reactions can relatively easily be studied. Exactly how the kinase activity is linked to the mitogenic or other actions of growth factors has not been established.

Growth factor binding to cell surface membranes initiates very rapid changes in intracellular calcium and pH. Intracellular calcium, which can be measured by Quin 2 fluorescence, may be increased by liberation from intracellular stores, or by plasma membrane transport in an exchange reaction for H^+ by a calcium/ magnesium-dependent ATPase. The role of intracellular calcium as a second messenger can be investigated using calcium channel blockers, such as verapamil, to inhibit or using an ionophore, such as A23187, to reproduce calcium-mediated actions. The action of calcium may be linked to phosphatidylinositol turnover, another important intracellular regulating pathway. Phosphatidylinositol 4,5-bisphosphate hydrolysis by phospholipase C (which may be activated by ligand binding) produces diacylglycerol as well as inositoltriphosphate. The latter liberates calcium from intracellular stores and this free calcium then acts in conjunction with diacylglycerol to activate protein kinase C. The involvement of protein kinase C in a reaction can be probed directly by using the phorbol ester TPA which directly activates protein kinase C. Thus far, there has been no convincing evidence that the mitogenic activity of growth factors is mediated by cyclic nucleotides.

Growth factors may be coupling factors. Several osteoblast mitogens can be isolated from bone matrix: TGF-β, PDGF, somatomedin C, IGF-II, and bone-derived growth factor (Mohan *et al.*, 1987). Potentially, matrix resorption could release an active osteoblast mitogen and thereby couple bone formation to resorption. This assumes that the process of bone dissolution by enzymes does not render the growth factors metabolically inactive and that a significant amount is actually released locally.

3.3.1. EGF

EGF was originally isolated by Cohen from mouse submaxillary glands (Cohen and Carpenter, 1975). It is probably the most well-studied growth factor, but its exact physiologic role(s) is still elusive. EGF, which is identical to human urogastrone, has a molecular weight of approximately 6 kDa. It stimulates a wide variety of epithelial and other cell types including osteoblasts, but inhibits collagen synthesis (Canalis and Raisz, 1979). EGF has also been shown to stimulate prostaglandin synthesis in cultured mouse calvaria, to increase bone resorption, and to raise plasma calcium (Tashjian and Levine, 1978; Sammon *et al.*, 1987). The EGF receptor is a protein kinase and the V-erb B oncogene codes for a truncated EGF receptor which may function autonomously.

3.3.2. TGFs and TGF-α

TGFs were discovered by DeLarco and Todaro and originally termed sarcoma growth factor. Further study revealed that this activity was actually composed of two factors (Todaro *et al.*, 1980). They were renamed TGFs (TGF-α and -β) because their primary activity seemed to be the ability to alter the normal phenotype of proliferating cells toward that of neoplastic cells. TGF-α binds to the EGF receptor (Marquardt *et al.*, 1984) but is encoded by a different gene from EGF and the two peptides do not cross-react immunologically. TGF-α may be one of the factors produced by tumors that stimulate bone resorption and cause hypercalcemia of malignancy (Ibbotson *et al.*, 1983).

3.3.3. TGF-β

TGF-β does not bind to the EGF receptor. Susceptible cells cultured in the presence of TGF-β and TGF-α (or EGF) will grow in soft agar, a characteristic called anchorage independence, i.e., the hallmark of transformation. TGF-β can also transform nonneoplastic cells (Frolik *et al.*, 1983). Bone is one of the richest sources of TGF-β per gram of protein (Pfeilschifter and Mundy, 1987). Hormones such as PTH, interleukin 1 (IL-1), and calcitriol, which stimulate bone resorption, release TGF-β activity; calcitonin, which inhibits resorption, decreases it. The source of TGF-β in matrix is probably osteoblasts since cultured neonatal calvaria release TGF-β (Pfeilschifter and Mundy, 1987). Since osteoblasts produce and also respond to increased proliferation to TGF-β it is an autocrine factor for osteoblasts. Preosteoblasts are also stimulated by TGF-β, suggesting that it may have a role in differentiation. It is unlikely that TGF-β can be used to increase bone formation because it stimulates bone resorption and lacks any effect on collagen synthesis. However, the activities of TGF-β do make it a very likely candidate for the putative coupling factor. Dissolution of matrix by such hormones as PTH or calcitriol would release TGF-β. This could couple bone formation to this resorptive activity by stimulating osteoblasts to proliferate and to produce more TGF-β which would continue to act in an autocrine fashion. How this process would be shut off is obviously a subject for investigation, but presumably when the dissolution of matrix stops, there would be insufficient stimulation of osteoblasts for the autocrine production of TGF-β to maintain proliferation. Clearly, other mechanisms could be involved.

3.3.4. IL-1

IL-1 is a protein of approximately 17.5 kDa. It is a monocrine secreted by monocytes/macrophages, bone cells, and many other cells at low levels. There are two IL-1 molecules, α and β, which are the products of distantly related genes and do not always have identical action. IL-1β was found to be one of the osteoclast-activating factors (OAFs). The primary action of IL-1 appears to be to stimulate T and B cell proliferation and function like other cytokines except interferon-γ. IL-1 is a potent stimulator of the resorption of bone and may be a moderator of bone loss in several conditions (Gowen et al., 1985; Sammon et al., 1987). Stimulation of prostaglandins may be only partially responsible for the resorptive actions of IL-1 (Sato et al., 1987). Interferon-γ inhibits the resorptive action of IL-1.

The actions of IL-1 on osteoclasts require that osteoblasts be present (Thompson and Chambers, 1985). Other bone-resorbing hormones such as PTH and calcitriol also have this requirement. The actual mechanism of this requirement may have been elucidated by Elford et al. (1987). They showed that osteoblasts produce M-CSF, which is capable of stimulating bone resorption (Burger et al., 1982). This hypothesis was tested by looking at other bone-resorbing agents that require osteoblasts. IL-1 and calcitriol stimulated osteoblasts to produce M-CSF. However, PTH, TGF, and EGF did not stimulate M-CSF activity under the conditions of study. Therefore, more than one mechanism might be responsible for osteoblast-mediated bone resorption.

The effect of IL-1 on collagen synthesis and bone formation is complicated. Collagen synthesis is inhibited by a prostaglandin mechanism but stimulated by a direct effect on osteoblasts, probably involving procollagen gene transcription since synthesis parallels changes in procollagen mRNA. IL-1 has also been noted to stimulate collagen synthesis in chrondrocytes (Goldring and Krane, 1987). DNA synthesis by osteoblasts may be directly stimulated by or occur secondary to prostaglandin production.

3.3.5. FGF

There are two FGFs, basic and acidic, with molecular weights of approximately 14 kDa. FGF is extremely potent, being active in the picomolar range. A physiologic role for FGF has never been defined. The effects of FGF do not appear to be cell specific since it can stimulate a wide variety of cells, including bone and cartilage cells, to divide (Gospodarowicz and Moran, 1975; Canalis and Raisz, 1980). FGF does not have any appreciable effect on protein synthesis and, as EGF, inhibits osteoblast synthesis of collagen and alkaline phosphatase activity. No effects on bone mineralization have been reported.

FGF is translated in the cell without a signal peptide for export. This means that the cells have to be ruptured to release FGF. Acidic FGF is present in platelets and could be released by a fracture. This could result in early stimulation of osteoblast proliferation during the repair process. In addition, FGF's chemotactic property for fibroblasts and angiogenic activity could be very important.

3.3.6. PDGF

PDGF is a dimeric protein of approximately 28 kDa. Three forms can exist depending on the combination of A and B chains (AA, BB, and AB). Perhaps the most interesting aspect of PDGF is a structural homology to the simian sarcoma virus oncogene (Deuel *et al.*, 1983), one of several findings relating growth factors and oncogenes. PDGF receptors, like those of other growth factors (and some oncogene products), have protein kinase activity. PDGF is found principally in platelet granules but endothelial, malignant, and transformed cells can synthesize it. PDGF stimulates only mesenchymal cells, including fibroblasts, chondrocytes, arterial smooth muscle cells, and osteoblasts. In calvarial bone systems, PDGF stimulates DNA and type I collagen synthesis. The latter is not specific since protein synthesis in general is increased. In addition to its anabolic effect on bone formation, PDGF has been shown to stimulate bone resorption by a prostaglandin-mediated mechanism (Tashjian *et al.*, 1982).

An important aspect of PDGF's mechanism of action is how it can cooperate with other growth factors to stimulate proliferation (Stiles *et al.*, 1979). Fibroblasts arrested in the G_0 state require PDGF to become competent to replicate their DNA. However, after PDGF stimulation, cells are arrested at a restriction point in the S phase. They do not proceed further through the cell cycle unless progression factors such as somatomedin or EGF are present, in which case DNA synthesis is completed and the cells divide. PDGF was shown to increase the somatomedin type 1 receptor level during the critical period in the cell cycle. It is not known whether these growth factors cooperate in the same way in osteoblasts, but evidence is emerging in different cell systems that growth factors (and hormones) interact with each other by affecting receptor populations.

PDGF may play a role in fracture healing. PDGF, released from platelets on clotting in the early stages, could participate in the removal of devitalized bone and the initial stimulation of osteoblasts. PDGF is a chemoattractant for fibroblasts, but whether this activity extends to preosteoblasts or osteoblasts has not been tested. Since platelets participate in the earliest phase of fractures, the source of PDGF for subsequent action could be proliferating endothelial cells, but this has not been demonstrated. Endothelial cell proliferation at the distal end of the growth plate in the zone of calcification could stimulate osteoblast proliferation through a paracrine action of PDGF.

3.3.7. Other Growth Factors

Bone-derived growth factor (β_2-microglobulin) and bone morphogenic protein are examples of other growth factors probably produced by osteoblasts, but their roles have not been clarified (Canalis, 1983; Mohan *et al.*, 1987). Bone morphogenic protein induces mesenchymal cells to form chondrocytes which secrete matrix that becomes mineralized into bone. It may also stimulate preosteoblast differentiation, but does not stimulate collagen synthesis (Urist *et al.*, 1984). Tumor necrosis factor (TNF) is an OAF produced by myeloma cells and is a potent stimulator of bone resorption (Klein *et al.*, 1987).

Skeletal growth factor was originally termed coupling factor (Howard *et al.*, 1981). Successive attempts to purify it from bone matrix have led to a variety of claims concerning its molecular weight and properties. Most recently it has been claimed to be IGF-II. However, IGF-II does not have the properties originally attributed to coupling factor. Because matrix contains several growth factors, subtle changes in extraction procedure will yield different products if a nonspecific mitogenic assay is used.

3.3.8. Cartilage Growth Factors

Chondrocytes are sensitive to some of the same growth factors as osteoblasts including somatomedins, FGF, PDGF, TGF-β, and EGF. Chondrocytes also produce some of the same factors as osteoblasts. Chondrocyte growth factor is probably the same as somatomedin C. Cartilage-derived growth factor (CDGF) has been isolated from cartilage. It is a basic protein of approximately 20 kDa that may function to regulate linear growth (Sullivan and Klagsbrun, 1985). When its sequence is completed, CDGF may be found to be identical to another factor. Other factors are found in chondrocyte conditioned media but they are incompletely categorized.

References

Ameenuddin, S., M. Sunde, H. F. DeLuca, N. Ikekawa and Y. Kobayashi. 1982. 24-Hydroxylation of 25-hydroxyvitamin D$_3$: is it required for embryonic development in chicks? Science 217:451.

Askenasi, R. and N. Demeester-Mirkinen. 1975. Urinary excretion of hydroxylysyl glycosides and thyroid function. J. Clin. Endocrinol. Metab. 40:342.

Au, W. Y. 1976. Cortisol stimulation of parathyroid hormone secretion by rat parathyroid glands in organ culture. Science 193:1015.

Bennett, A., T. L. Chen, D. Feldman, R. L. Hintz and R. G. Rosenfeld. 1982. Characterization of somatomedin C/insulin-like growth factor-I receptors on cultured bone cells: regulation of receptor concentration by glucocorticoids. Calcif. Tissue Int. 34:S2.

Bennett, A. E., H. W. Wahner, B. L. Riggs and R. L. Hintz. 1984. Insulin-like growth factors I and II: aging and bone density. J. Clin. Endocrinol. Metab. 59:701.

Blundell, T. L., S. Bedarkar, E. Rinderknecht, and R. E. Humbel. 1978. Insulin-like growth factor: a model for tertiary structure accounting for immunoreactivity and receptor binding. Proc. Natl. Acad. Sci. USA 75:180.

Bringhurst, R. and J. T. Potts. 1981. Bone collagen synthesis in vitro: structure activity relations among parathyroid hormone fragments and analogs. Endocrinology 108:103.

Burch, W. M. and H. E. Lebovitz. 1981. Hormonal activation of ornithine decarboxylase in embryonic chick pelvic cartilage. J. Physiol. (London) 241:E454.

Burch, W. M. and H. E. Lebovitz. 1982. Triiodothyronine stimulates maturation of porcine growth-plate cartilage in vitro. J. Clin. Invest. 70:496.

Burger, E. H., J. W. M. Van der Meer, J. S. Van de Gevel, J. C. Gribnau, C. W. Thesingh and R. Van Furth. 1982. In vitro formation of osteoclasts from long-term cultures of bone marrow mononuclear phagocytes. J. Exp. Med. 156:1604.

Canalis, E. 1980. Effect of insulin-like growth factor I on DNA and protein synthesis in cultured rat calvaria. J. Clin. Invest. 66:709.

Canalis, E. 1982. Cortisol effects on calvarial collagen and DNA synthesis depend upon the presence of periosteal tissue. Calcif. Tissue Int. 34:S3.

Canalis, E. 1983. The hormonal and local regulation of bone formation. Endocrine Rev. 4:62.

Canalis, E. M., J. W. Dietrich, D. M. Maina and L. G. Raisz. 1977. Hormonal control of bone collagen synthesis in vitro. Effects of insulin and glucagon. Endocrinology 100:668.

Canalis, E. M., W. A. Peck and L. G. Raisz. 1980. Stimulation of DNA and collagen synthesis by autologous growth factor in cultured fetal rat calvaria. Science 210:1021.

Canalis, E. and L. G. Raisz. 1978. Effect of sex steroids on bone collagen synthesis in vitro. Calcif. Tissue Res. 25:105.

Canalis, E. and L. G. Raisz. 1979. Effect of epidermal growth factor on bone formation in vitro. Endocrinology 104:862.

Canalis, E. and L. G. Raisz. 1980. Effect of fibroblast growth factor on cultured fetal rat calvaria. Metabolism 29:108.

Chen, T. L., C. M. Cone, E. Morey-Holton and D. Feldman. 1982. Glucocorticoid regulation of 1,25 $(OH)_2$-vitamin D_3 receptors in cultured mouse bone cells. J. Biol. Chem. 257:13564.

Clemmons, D. R. and D. S. Shaw. 1983. Variables controlling somatomedin production by cultured human fibroblasts. J. Cell. Physiol. 115:137.

Clemmons, D. R., L. E. Underwood, E. C. Ridgway, B. Kliman, R. N. Kjellberg and J. J. Van Wyk. 1980. Estradiol treatment of acromegaly. Amer. J. Med. 69:571.

Cohen, S. and G. Carpenter. 1975. Human epidermal growth factor: isolation and chemical and biological properties. Proc. Natl. Acad. Sci. USA 72:1317.

Copeland, K. C., D. M. Johnson, T. J. Kuehl and V. D. Castracane. 1984. Estrogen stimulates growth hormone and somatomedin-C in castrate and intact female baboons. J. Clin. Endocrinol. Metab. 58:698.

Corvol, M. T., M. F. Dumontier, M. Garabedian and R. Rappaport. 1978. Vitamin D and cartilage. II. Biological activity of 25-hydroxycholecalciferol and 24,25 and 1,25-dihydroxycholecalciferols on cultured growth plate chondrocytes. Endocrinology 102:1269.

Daughaday, W. H., A. C. Herington and L. S. Phillips. 1975. The regulation of growth by endocrines. Annu. Rev. Physiol. 37:211.

Deuel, T. F., J. S. Huang, S. S. Huang, P. Stroobant and M. D. Waterfield. 1983. Expression of a platelet-derived growth factor-like protein in simian sarcoma virus transformed cells. Science 221:1348.

Dietrich, J. W., E. Canalis, D. M. Maina and L. G. Raisz. 1976. Hormonal control of bone collagen synthesis in vitro: effects of parathyroid hormone and calcitonin. Endocrinology 98:943.

Elford, P. R., R. Felix, M. Cecchini and H. Fleisch. 1987. Osteoblast-like cells produce macrophage colony-stimulating factor which is enhanced by various bone-resorbing agents. In: D. V.

Cohn, T. J. Martin and P. J. Meunier (Ed.) Calcium Regulation and Bone Metabolism: Basic and Clinical Aspects. Vol. 9, pp 279–282. Elsevier Science Publ., New York.

Elgin, R. G., W. H. Busby, Jr., and D. R. Clemmons. 1987. An insulin-like growth factor (GF) binding protein enhances the biologic response to IGF-I. Proc. Natl. Acad. Sci. USA 84:3254–3258.

Eriksen, E. F., N. J. Berg, M. L. Graham, K. G. Mann, T. C. Spelsberg and B. L. Riggs. 1987. Evidence of estrogen receptors in human bone cells. J. Bone Miner. Res. 2:238 (Abstr.).

Foley, T. P., S. P. Nissley, R. L. Stevens, C. G. King, V. C. Hascall, R. E. Humbel, P. A. Short and M. M. Rechler. 1982. Demonstration of receptors for insulin and insulin-like growth factors on swarm rat chondrosarcoma chondrocytes. J. Biol. Chem. 257:663.

Froesch, E. R., J. Zapf, T. K. Audhya, E. Ben-Porath, B. J. Segen and K. D. Gibson. 1976. Nonsuppressible insulin-like activity and thyroid hormones: major pituitary-dependent sulfation factors for chick embryo cartilage. Proc. Natl. Acad. Sci. USA 73:2904.

Frolik, C. A., L. L. Dart, C. A. Meyers, D. M. Smith and M. B. Sporn. 1983. Purification and initial characterization of a type B transforming growth factor from human placenta. Proc. Natl. Acad. Sci. USA 80:3676.

Gallagher, J. C., B. L. Riggs and H. F. DeLuca. 1980. Effect of estrogens on calcium absorption and serum vitamin D metabolites in postmenopausal osteoporosis. J. Clin. Endocrinol. Metab. 51:1359.

Gee, I. and T. R. Preston. 1957. The effect of hexoestrol implantation on carcass composition and efficiency of food utilization in fattening lambs. Brit. J. Nutr. 11:329.

Gennant, H. K., J. M. Baron, E. Paloyan and J. Jowsey. 1975. Osteosclerosis in primary hyperparathyroidism. Amer. J. Med. 59:104.

Goldring, M. B. and S. Krane. 1987. Modulation of collagen synthesis and procollagen mRNA levels in human chondrocyte, synovial cell and fibroblast cultures by interleukin 1. In: D. V. Cohn, T. J. Martin and P. J. Meunier (Ed.) Calcium Regulation and Bone Metabolism: Basic and Clinical Aspects. Vol. 9, pp 444–449. Elsevier Science Publ., New York.

Gospodarowicz, D. and J. S. Moran. 1975. Mitogenic effect of fibroblast growth factor on early passage cultures of human and murine fibroblasts. J. Cell Biol. 66:451.

Gourmelen, M., F. Girard and M. Binoux. 1982. Serum somatomedin/insulin-like growth factor (IGF) and IGF carrier levels in patients with Cushing's syndrome or receiving glucocorticoid therapy. J. Clin. Endocrinol. Metab. 54:885.

Gowen, M., D. Wood and R. G. G. Russell. 1985. Stimulation of the proliferation of human bone cells in vitro by human monocyte products with interleukin-1 activity. J. Clin. Invest. 75:1223.

Green, H., M. Minoru and T. Nixon. 1985. A dual effector theory of growth-hormone action. Differentiation 29:195.

Hahn, T. J., S. J. Downing and J. M. Phang. 1971. Insulin effect on amino acid transport in bone: dependence on protein synthesis and Na^+. Amer. J. Physiol. 220:1717.

Hahn, T. J., L. R. Halsteal, S. L. Teitelbaum and B. H. Hahn. 1979. Altered mineral metabolism in glucocorticoid-induced osteopenia effect of 25-hydroxyvitamin D administration. J. Clin. Invest. 64:655.

Halloran, B. P., H. F. DeLuca, E. Barthell, S. Yamada, M. Ohmori and H. Takayama. 1981. An examination of the importance of 24-hydroxylation to the function of vitamin D during early development. Endocrinology 108:2067.

Halme, J., J. Vitto, K. I. Kivirikko and L. Saxen. 1972. Effect of triiodothyronine on the metabolism of collagen in cultured embryonic bones. Endocrinology 90:1476.

Heath, H., P. W. Lambert, J. W. Service and S. Arnaud. 1979. Calcium homeostasis in diabetes mellitus. J. Clin. Endocrinol. Metab. 49:462.

Henry, H. L. and A. W. Norman. 1978. Vitamin D: two dihydroxylated metabolites are required for normal chicken egg hatchability. Science 201:835.

Hill, D. J. 1979. Stimulation of cartilage zones of the calf costochondral growth plate in vitro by growth hormone dependent rat plasma somatomedin activity. J. Endocrinol. 83:219.

Hintz, R. L. and F. Liu. 1977. Demonstration of specific plasma protein binding sites for somatomedin. J. Clin. Endocrinol. Metab. 45:988.

Holder, A. T., M. A. Preece and E. M. Spencer. 1981. Effect of bovine growth hormone and a partially pure preparation of somatomedin on various growth parameters in hypopituitary dwarf mice. J. Endocrinol. 89:275.

Howard, G. A., B. L. Bottemiller, R. T. Tirner, J. L. Rader and D. J. Baylink. 1981. Parathyroid hormone stimulates bone formation and resorption in organ culture: evidence for a coupling mechanism. Proc. Natl. Acad. Sci. USA 78:3204.

Howard, G. A. and E. M. Spencer. 1984. Somatomedins A and C directly stimulate bone cell proliferation in vitro. In: D. V. Cohn, T. Fujita, J. T. Potts and R. V. Talmage (Ed.) Endocrine Control of Bone and Calcium Metabolism. pp 86–89. Elsevier Science Publ., New York.

Ibbotson, K. J., S. M. D'Souze, K. W. Ng, C. K. Osborne, M. Niall, T. J. Martin and G. R. Mundy. 1983. Tumor-derived growth factor increases bone resorption in a tumor associated with humoral hypercalcemia of malignancy. Science 221:1292.

Isaksson, O. G. P., J.-O. Jansson and I. A. M. Gause. 1982. Growth hormone stimulates longitudinal bone growth directly. Science 216:1237.

Jansson, J.-O., S. Eden and O. Isaksson. 1983. Sites of action of testosterone and estradiol on longitudinal bone growth. Amer. J. Physiol. 244:E135.

Kan, K. W., R. L. Cruess, B. I. Posner, H. J. Guyda and S. Solomon. 1984. Hormone receptors in the epiphyseal cartilage. J. Endocrinol. 103:125.

Klein, B., M. Jourdan, B. Durie, J. F. Rossi and R. Bataille. 1987. Spontaneous production of bone cytokines by human myeloma cell lines. J. Bone Miner. Res. 2:42 (Abstr.).

Kragstrup, J., F. Melsen and L. Mosekilde. 1981. Effects of thyroid hormone(s) on mean wall thickness of trabecular bone packets. Metab. Bone Dis. Relat. Res. 3:181.

Levin, M. E., V. C. Boisseau and L. V. Avioli. 1976. Effects of diabetes mellitus on bone mass in juvenile and adult-onset diabetes. N. Engl. J. Med. 294:241.

McNair, P., S. Madsbad, M. S. Christensen, C. Christiansen, O. K. Faber, C. Binder, and I. Transbøl. 1979. Bone mineral loss in insulin-treated diabetes mellitus: studies on pathogenesis. Acta Endocrinol. (Copenhagen) 90:463.

Manologas, S. C., D. C. Anderson and G. A. Lumb. 1979. Glucocorticoids regulate the concentrations of 1,25 dihydroxycholecalciferol receptors in bone. Nature 277:314.

Marquardt, H., M. W. Hunkapiller, L. E. Hood and G. J. Todaro. 1984. Rat transforming growth factor type 1: Structure and relation to epidermal growth factor. Science 223:1079.

Miller, S. C. 1978. Rapid activation of the medullary bone osteoclast cell surface by parathyroid hormone. J. Cell Biol. 76:615.

Miller, S. C., B. P. Halloran, H. F. DeLuca, S. Yamada, H. Takayama and W. S. Jee. 1981. Studies on the role of 24-hydroxylation of vitamin D in the mineralization of cartilage and bone of vitamin D-deficient rats. Calcif. Tissue Int. 33:489.

Mohan, S., T. A. Linkhart, J. C. Jennings and D. J. Baylink. 1987. Identification and quantification of four distinct growth factors stored in human bone matrix. J. Bone Miner. Res. 2:44 (Abstr.).

Mundy, G. R., J. G. Shapiro, E. M. Canalis and L. G. Raisz. 1976. Direct stimulation of bone resorption by thyroid hormones. J. Clin. Invest. 58:529.

Parfitt, A. M. 1976. The actions of parathyroid hormone on bone: relation to bone remodeling and turnover, calcium homeostasis and metabolic bone disease. Metabolism 25:1033.

Peck, W. A., J. Brandt and I. Miller. 1967. Hydrocortisone-induced inhibition of protein synthesis and uridine incorporation in isolated bone cells in vitro. Proc. Natl. Acad. Sci. USA 57:1599.

Peck, W. A. and K. Messinger. 1970. Nucleoside and ribonucleic acid metabolism in isolated bone cells. J. Biol. Chem. 245:2722.

Pfeilschifter, J., and G. R. Mundy. 1987. TGFβ stimulates osteoblast activity and is released during the bone resorption process. In: D. V. Cohn, T. J. Martin and P. J. Meunier (Ed.) Calcium Regulation and Bone Metabolism: Basic and Clinical Aspects. Vol. 9, pp 405–454. Elsevier Science Publ., New York.

Poser, J. W. and P. A. Price. 1979. A method for decarboxylation of gamma-carboxyglutamic acid in proteins. Properties of the decarboxylated gamma-carboxyglutamic acid protein from calf bones. J. Biol. Chem. 254:431.

Raisz, L. G. and B. E. Kream. 1981. Hormonal control of skeletal growth. Annu. Rev. Physiol. 43:225.

Raisz, L. G., C. L. Trummel, M. F. Holick and H. F. DeLuca. 1972a. 1,25-dihydroxycholecalciferol, a potent stimulator of bone resorption in tissue culture. Science 175:768.

Raisz, L. G., C. L. Trummel, J. A. Wener and H. Simmons. 1972b. Effect of glucocorticoids on bone resorption in tissue cultures. Endocrinology 90:961.

Raisz, L. G., J. Y. Vanderhoek, H. A. Simmons, B. E. Kream and K. C. Nicolaou. 1979. Prostaglandin synthesis by fetal rat bone in vitro: evidence for a role of prostacyclin. Prostaglandins 17:905.

Reeve, J., M. Arlot, M. Bernat, S. Charhon, C. Edouard, D. Slovik, F. J. Vismans, and P. J. Meunier. 1981. Calcium-47 kinetic measurements of bone turnover compared to bone histomorphometry in osteoporosis: the influence of human parathyroid fragment (hPTH 1–34) therapy. Metab. Bone Dis. Relat. Res. 3:23.

Reeve, J., P. J. Meunier, J. A. Parsons, M. Bernat, O. L. M. Bijvoet, P. Courpron, C. Edouard, L. Klenerman, R. M. Neer, J. C. Renier, D. Slovik, F. J. F. E. Vismans and J. T. Potts. 1980. Anabolic effect of human parathyroid hormone fragment on trabecular bone in involutional osteoporosis—a multicentre trial. Brit. Med. J. 280:1340.

Rinderknecht, E. and R. E. Humbel. 1978a. The amino acid sequence of human insulin-like growth factor I and its structural homology with proinsulin. J. Biol. Chem. 253:2769.

Rinderknecht, E. and R. E. Humbel. 1978b. Primary structure of human insulin-like growth factor II. FEBS Lett. 89:283.

Rodan, S. B., G. A. Rodan, H. A. Simmons, R. W. Walenga, M. B. Feinstein and L. G. Raisz. 1981. Bone resorptive factor produced by osteosarcoma cells with osteoblastic features is PGE₂. Biochem. Biophys. Res. Commun. 102:1358.

Rowe, D. W. and B. E. Kream. 1982. Regulation of collagen synthesis in fetal rat calvaria by 1,25-dihydroxyvitamin D_3. J. Biol. Chem. 257:8009.

Russell, S. M. and E. M. Spencer. 1985. Local injections of human or rat growth hormone or of purified human somatomedin-C stimulate unilateral tibial epiphyseal growth in hypophysectomized rats. Endocrinology 116:2563.

Salmon, W. D., Jr. and W. H. Daughaday. 1957. A hormonally controlled serum factor which stimulates sulfate incorporation by cartilage in vitro. J. Lab. Clin. Med. 49:825.

Sammon, P. J., T. J. Wronski, J. A. Flueck and D. A. Cohen. 1987. Humoral hypercalcemia of malignancy: Evidence for interleukin-I activity as a bone resorbing factor released by human transitional-cell carcinoma cells. In: D. V. Cohn, T. J. Martin and P. J. Meunier (Ed.) Calcium Regulation and Bone Metabolism: Basic and Clinical Aspects. Vol. 9, pp 383–390. Elsevier Science Publ., New York.

Sato, K., Y. Fujii, T. Kachiuchi, K. Kasono and K. Shizume. 1987. Production of interleukin 1 alpha (IL-1α)-like activity and colony stimulating activity by clonal squamous cell carcinomas derived from patients with hypercalcemia and leukocytosis. In: D. V. Cohn, T. J. Martin and P. J. Meunier (Ed.) Calcium Regulation and Bone Metabolism: Basic and Clinical Aspects. Vol. 9, pp 149–154. Elsevier Science Publ., New York.

Schlechter, N. L., S. M. Russell, E. M. Spencer and C. S. Nicoll. 1986. Evidence suggesting that the direct growth-promoting effect of growth hormone on cartilage *in vivo* is mediated by local production of somatomedin. Proc. Natl. Acad. Sci. USA 83:7932.

Schmid, C., T. Steiner and E. R. Froesch. 1983. Insulin-like growth factors stimulate synthesis of nucleic acids and glycogen in cultured calvaria cells. Calcif. Tissue Int. 35:578.

Schneider, L. E., L. M. Nowosielski and H. P. Schedl. 1977. Insulin-treatment of diabetic rats: effects of duodenal calcium absorption. Endocrinology 100:67.

Schoenle, E., J. Zapf, R. E. Humbel and E. R. Froesch. 1982. Insulin-like growth factor I stimulates growth in hypophysectomized rats. Nature 206:252.

Seeman, E., H. W. Wahner, K. P. Offord, R. Kuman, W. J. Johnson and B. L. Riggs. 1982. Differential effects of endocrine dysfunction on the axial and the appendicular skeleton. J. Clin. Invest. 69:1302.

Shore, R. M., R. W. Chesney, R. B. Mazess, P. G. Rose and G. J. Bargman. 1981. Osteopenia in juvenile diabetes. Calcif. Tissue Int. 33:455.

Si, E. C. C., E. M. Spencer, C. C. Liu and G. A. Howard. 1988. Synergism between insulin-like growth factor-I: a parathyroid hormone in bone metabolism. J. of Bone and Mineral Research 3 (Suppl. 1) 113S.

Silberberg, M. and R. Silberberg. 1971. Steroid hormones and bone. In: G. H. Bourne (Ed.) The Biochemistry and Physiology of Bone. Vol. 3, pp 401–484. Academic Press, New York.

Silbermann, M. 1983. Hormones and cartilage. In: B. K. Hall (Ed.) Cartilage. Vol. 2, pp 327–368. Academic Press, New York.

Simmons, D. J. 1966. Collagen formation and endochondral ossification in estrogen treated mice. Proc. Soc. Exp. Biol. Med. 121:1165.

Simpson, M. E., C. Asling and H. M. Evans. 1950. Some endocrine influences on skeletal growth and differentiation. Yale J. Biol. Med. 23:1.

Slovik, D. M., M. A. Daly, S. H. Doppelt, J. T. Potts, Jr., D. I. Rosenthal and R. M. Neer. 1987. Increases in vertebral bone density of postmenopausal osteoporotics after treatment with hPTH-(1–34) fragment plus 1,25 (OH_2) D_3: an interim report. In: D. V. Cohn, T. J. Martin and P. J. Meunier (Ed.) Calcium Regulation and Bone Metabolism: Basic and Clinical Aspects. Vol. 9, pp 119–122. Elsevier Science Publ., New York.

Smeets, T. and S. van Buul-Offers. 1983. The influence of growth hormone, somatomedins, prolactin and thyroxine on the morphology of the proximal tibial epiphysis and growth plate of Snell dwarf mice. Growth 47:160.

Somjen, D., G. J. Somjen, Y. Weisman and I. Binderman. 1982. Evidence for 24,25-dihydroxycholecalciferol receptors in long bones of newborn rats. Biochem J. 204:31.

Spencer, E. M., M. Khalil and O. Tobiassen. 1980. Experimental diabetes in the rat causes an insulin-reversible decrease in renal 25-hydroxyvitamin D3-1α-hydroxylase activity. Endocrinology 107:300.

Spencer, E. M., M. Ross and B. Smith. 1983. The identity of human insulin-like growth factors I and II with somatomedins C and A with rat IGF I and II. In: E. M. Spencer (Ed.) Insulin-like Growth Factors/Somatomedins. pp 81–96. Walter de Gruyter, Berlin.

Stern, P. H. 1969. Inhibition by steroids of parathyroid hormone-induced Ca^{45} release from embryonic rat bone in vitro. J. Pharmacol. Exp. Ther. 168:211.

Stiles, C. D., G. T. Capone, C. D. Scher, H. N. Antoniades, J. J. Van Wyk and W. J. Pledger. 1979. Dual control of cell growth by somatomedins and platelet-derived growth factor. Proc. Natl. Acad. Sci. USA 76:1279.

Stuart, C. A., R. W. Furlanetto and H. E. Lebovitz. 1979. The insulin receptor of embryonic chicken cartilage. Endocrinology 105:1293.

Sullivan, R. and M. Klagsbrun. 1985. Purification of cartilage-derived growth factor by heparin affinity chromatography. J. Biol. Chem. 260 (4) 2399–403.

Tashjian, A. H., Jr., E. L. Hohmann, H. N. Antoniades and L. Levine. 1982. Platelet-derived growth factor stimulates bone resorption via a prostaglandin-mediated mechanism. Endocrinology 111:118.

Tashjian, A. H., Jr., and L. Levine. 1978. Epidermal growth factor stimulates prostaglandin production and bone resorption in cultured mouse calvaria. Biochem. Biophys. Res. Commun. 85:966.

Tessler, R. H. and W. D. Salmon. 1975. Glucocorticoid inhibition of sulfate incorporation by cartilage of normal rats. Endocrinology 96:898.

Thompson, B. M. and T. J. Chambers. 1985. Lipoxygenase inhibitors abolish responsiveness of osteoblast–osteoclast co-cultures to PTH and IL-1. Calcif. Tissue Int. 38:S14.

Todaro, G. J., C. Fryling and J. E. DeLarco. 1980. Transforming growth factors produced by certain human tumor cells: Polypeptides that interact with epidermal growth factor receptors. Proc. Natl. Acad. Sci. USA 77:5258.

Turner, R. T., J. E. Puzas, M. D. Forte, G. E. Lester and T. K. Gray. 1980. In vitro synthesis of a 1α,25-dihydroxycholecalciferol and 24,25-dihydroxycholecalciferol by isolated calvarial cells. Proc. Natl. Acad. Sci. USA 77:5720.

Urist, M. R., Y. K. Kuo, A. G. Brownell, W. M. Hohl, J. Buyske, A. Lietze, P. Tempst, M. Hunkapiller and R. J. DeLange. 1984. Purification of bovine bone morphogenic protein by hydroxyapatite chromatography. Proc. Natl. Acad. Sci. USA 81:371.

Valentin-Opran, A., P. D. Delmas, P. M. Chavassieux, C. Chenu, S. Saez and P. J. Meunier. 1986. 1,25-dihydroxyvitamin D_3 stimulates the somatomedin-C secretion by human bone cells in vitro. J. Bone Miner. Res. 1 (Suppl. I), (Abstract 139).

Van Sickle, D. C. 1985. Control of bone growth. J. Anim. Sci. 61 (Suppl. 2)76.

Voelkel, E. F., A. H. Tashjian, Jr., and L. Levine. 1980. Cyclooxygenase products of arachidonic acid metabolism by mouse bone in organ culture. Biochim. Biophys. Acta 620:418.

Weiss, R. E. and A. H. Reddi. 1980. Influence of experimental diabetes and insulin on matrix-induced cartilage and bone differentiation. Amer. J. Physiol. 238:E200.

Wiedemann, E., E. Schwartz and A. G. Frantz. 1976. Acute and chronic estrogen effects upon serum somatomedin activity, growth hormone, and prolactin in man. J. Clin. Endocrinol. Metab. 42:942.

Wiske, P. S., S. M. Wentworth, J. A. Norton, Jr., S. Epstein and C. C. Johnston, Jr. 1982. Evaluation of bone mass and growth in young diabetics. Metabolism 31:848.

Endocrine Regulation of Adipogenesis

G. J. HAUSMAN, D. E. JEWELL, AND E. J. HENTGES

1. Introduction

Insight into the regulation of adipose tissue development has been gained by many studies utilizing cell culture techniques. There are several published reviews that appraise the hormonal aspects of adipocyte development *in vitro* (Hausman *et al.*, 1980; Hiragun, 1985; Van, 1985; Vannier *et al.*, 1985) and *in vivo* (Allen, 1976; Hausman, 1985; Martin *et al.*, 1985b; York, 1985). The establishment of several "preadipocyte" cell lines has led to the propagation of a vast amount of literature concerned with various aspects of adipocyte differentiation (Hiragun, 1985; Vannier *et al.*, 1985).

Several recent developments have particular relevance to the endocrinology of adipogenesis. The establishment of serum-free conditions for *in vitro* growth of preadipocyte cell lines (Serrero and Khoo, 1982; Gaillard *et al.*, 1984; Harrison *et al.*, 1985; Gamou and Shimizu, 1986) and for rat and sheep preadipocytes (adipose tissue stromal-vascular cells) in primary culture (Broad and Hamm, 1983; Deslex *et al.*, 1987a) has shown that preadipocytes in primary culture require fewer media components for serum-free growth than do established cell lines. Furthermore, fibroblast growth factor (FGF) was shown to be an essential component for serum-free growth of the ob/17 cell line (Gaillard *et al.*, 1984) whereas FGF inhibited the serum-free growth and differentiation of rat preadipocytes in primary cultures of stromal-vascular (S-V) cells

G. J. HAUSMAN • United States Department of Agriculture, Agricultural Research Service, Richard B. Russell Agricultural Research Center, Athens, Georgia 30613. *D. E. JEWELL* • Purina Mills, St. Louis, Missouri 63131. *E. J. HENTGES* • National Livestock and Meat Board, Chicago, Illinois 60070.

(Deslex *et al.*, 1987a). It is clear that extrapolations (to the *in vivo* situation) from *in vitro* results on established cell lines should be viewed with caution.

An inherent problem with *in vivo* studies is the difficulty in determining "*de novo*" preadipocyte replication and differentiation (Hausman *et al.*, 1980). Therefore, it is difficult to draw meaningful information on endocrine control of adipogenesis from *in vivo* studies. Therefore, this review will cover the endocrine control of adipogenesis *in vitro* with special reference to the type of cell and the conditions of culture media.

2. Insulin

The role of insulin in growth and development has been reviewed (Martin *et al.*, 1984). Gaben-Cogneville *et al.* (1983) were the first to demonstrate that rat preadipocyte (S-V cells) differentiation is dependent on physiological levels of insulin.

The conversion of preadipocytes to adipocytes (fat-filled cells) *in vitro* is

Table I
The Effect of Insulin on Preadipocytes *in Vitro*

Cells	Insulin concentration	Conditions	Effects
3T3-L1[a]	1 μg/ml	6-hr incubation, no serum	↓ mRNA content for glutamine synthetase
3T3-L1[b]	10 μg/ml	Serum-free media; induction with dexamethasone	↑ replication of preadipocytes
3T3-L1[c] and 3T3-F442A	0.24 nM	Media containing serum and no other hormones	↑ mRNA content for glyceraldehyde-3-phosphate dehydrogenase
CHEF/18[d]	10 ng/ml	Serum-free	↑ adipocyte conversion
ob/17[e]	nM range	Media containing serum and physiological levels of T₃	↑ lipogenic enzyme activity and content
3T3-L1[f]	nM range	Media containing serum	↑ conversion rate of fat-free to fat-filled cells
Rat preadipocytes in primary culture[g]	Pharmacological	Media containing serum	↑ adipocyte formation and ↑ LPL activity

Table I (continued)

Cells	Insulin concentration	Conditions	Effects
Rat preadipocytes in secondary culture[h]	Pharmacological	Media containing serum	↑ lipogenic enzyme activity
Rat preadipocytes in primary culture[i]	10^{-13}–10^{-9} M	Media containing serum	↑ adipocyte conversion rate and ↑ lipogenic enzyme activity
Pig preadipocytes in primary culture[j]	10^{-6} M	Media containing 2.5 to 5% serum	0 on lipogenic enzyme activity and cytodifferentiation
ob/17[k]	nM range	Media containing serum and 2 nM T_3	↑ mRNA content for phosphoenolpyruvate carboxykinase (PEPCK) and glycerol-3-phosphate dehydrogenase
ob/17[k] and 3T3-F442A	nM range	Media containing serum and 2 nM T_3	↑ transcription of PEPCK mRNA
Rat preadipocytes in primary culture[l]	nM range	Serum-free with 200 pM T_3	↑ lipogenic enzyme activity and ↑ replication of preadipocytes
ST 13[m]	5 μg/ml	Media containing serum	↑ adipocyte conversion

[a] Miller et al. (1987).
[b] Gamou and Shimizu (1986).
[c] Alexander et al. (1985).
[d] Harrison et al. (1985), Sager and Kovac (1982).
[e] Grimaldi et al. (1983a), Vannier et al. (1985).
[f] Green and Kehinde (1975).
[g] Bjorntorp et al. (1980).
[h] Roncari et al. (1979).
[i] Hausman (1987).
[j] Hentges and Hausman (1987).
[k] Dani et al. (1986).
[l] Deslex et al. (1987a).
[m] Hiragun (1985).

accelerated by insulin in serum-free and serum-supplemented media at physiological and pharmacological concentrations (Table I). Lipogenic enzyme activities are markedly increased by insulin as are the mRNA contents for several lipogenic enzymes. Recent studies of two preadipocyte cell lines have demonstrated that physiological levels of insulin in the presence of triiodothyronine (T_3) (2 nM) increased the transcription of phosphoenolpyruvate carboxykinase (PEPCK) mRNA (Table I).

Physiological levels of insulin in serum-free media stimulated the morphological differentiation of preadipocytes (CHEF/18 cells) and the lipogenic enzyme activity of primary cultures (S-V cells) of rat preadipocytes (Table I). In these studies insulin was the only hormone in the growth media and was present at pharmacological levels (5 μ/ml) in the cultures of rat preadipocytes (S-V cells) and at physiological levels (10 ng/ml) in cultures of CHEF/18 cells (Table I). Insulin in combination with dexamethasone (DEX) in serum-free media stimulated the adipocyte conversion of sheep fibroblasts in primary culture (Broad and Hamm, 1983) and 1246 preadipocytes (Serrero and Khoo, 1982; Serrero, 1986). Several of the other components commonly used in these serum-free media include FGF, epithelial growth factor, transferrin, and thrombin (Broad and Hamm, 1983; Harrison et al., 1985; Gamou and Shimizu, 1986; Serrero, 1986; Deslex et al., 1987a).

The role of insulin as a stimulus of preadipocyte replication is dependent on culture conditions and the type of cell studied. In the ob/17 preadipocyte cell line, insulin stimulates mitotic activity at pharmacological levels with maximal activity at a concentration of 1 μM (reviewed by Vannier et al., 1985). There was no insulin requirement for growth when ob/17 cells were grown in medium supplemented with insulin-depleted serum (Vannier et al., 1985). Autoradiography of tritiated thymidine-labeled cells was used to demonstrate that pharmacological levels of insulin stimulated replication of 3T3-L1 preadipocytes in serum-free medium (Gamou and Shimizu, 1986). A number of studies with 3T3-L1 preadipocytes show that the combination of insulin, DEX, and methyl isobutyl xanthine (IBMX) added to confluent cultures stimulates one round of cell division followed by differentiation-associated events (reviewed by Hiragun, 1985; Gratzner et al., 1985). A clonal analysis of 3T3-L1 preadipocytes indicates that insulin is not the inducer of commitment but merely enhances lipid synthesis in previously committed cells (Steinberg and Brownstein, 1982).

Recent studies by Green and co-workers showed that 3T3 preadipocytes differentiated with growth hormone (GH) were more sensitive to the mitogenic effect of physiological levels of insulin [and insulinlike growth factor I (IGF-I)] than were undifferentiated cells (Zezulak and Green, 1986). Autoradiography after labeling with tritiated thymidine demonstrated that the result of insulin (and IGF-I) action is a selective multiplication of newly differentiated cells (clonal expansion; Zezulak and Green, 1986). Studies of 1246 preadipocytes also showed that GH (10 ng/ml) and insulin had a synergistic effect on adipocyte differentiation in serum-free medium (Serrero, 1986). Unlike in the studies of Zezulak and Green (1986), GH alone did not promote (1246) preadipocyte differentiation whereas GH plus insulin stimulated greater differentiation than did insulin alone (Serrero, 1986).

The effect of insulin has been studied in primary cultures of rat preadipocytes (S-V cells) supplemented with fetal pig serum (FPS) (Hausman, 1987). FPS has a very low insulin titer but contains a high (> 100 ng/ml) level of GH

(Martin *et al.*, 1985a). Therefore, media supplemented with 10% FPS would have no insulin but would have a significant GH titer ($\cong$ 10 ng/ml). Supplementation of physiological levels of insulin to preadipocyte cultures (primary) increased the adipocyte conversion rate and lipogenic enzyme activities (Hausman, 1987). As in the 3T3 and 1246 studies, GH may increase the efficacy of low insulin levels (Hausman, 1987) possibly by stimulating insulin or IGF-I sensitivity. However, the effectiveness of insulin on rat preadipocytes (primary culture) was not decreased when insulin was added to cultures fed pig serum (PS) instead of FPS (Hausman, 1987). Replacing FPS with PS effectively lowered the GH concentration in the culture medium by a factor of ten since the PS contained less than 10 ng GH/ml (D. E. Jewell, unpublished observations). The lower GH level (PS supplementation) did not decrease insulin sensitivity of cultured preadipocytes (Hausman, 1987).

There are several possibilities that could explain the stimulation of cell proliferation in response to pharmacological levels of insulin. One possibility is that insulin binds to the high-affinity receptor of IGF-I. Affinity cross-linking studies showed that there were two IGF receptors (IGF-I, IGF-II) and that the IGF-I receptor had the highest affinity for the insulin molecule (Chernausek *et al.*, 1981; Massague and Czech, 1982). When the receptor for IGF-I was blocked with a monoclonal antibody that did not neutralize the insulin receptor, the mitogenic effect of insulin in cultured human fibroblasts was abolished (Van Wyk *et al.*, 1985).

The interaction of insulin and the IGF-I receptor and its role in the mitogenic response of insulin is still controversial. In another study using a monoclonal antibody to the IGF-I receptor, insulin maintained mitogenic effects on human fibroblasts (Flier *et al.*, 1986). Insulin was shown to stimulate mitosis in rat hepatoma cells, which do not have IGF-I receptors (Koontz and Iwahashi, 1981). Therefore, in some instances insulin may stimulate cell division by binding the high-affinity insulin receptor.

Insulin may mediate a mitogenic effect by stimulating the release of a somatomedinlike peptide (Clemmons, 1985), which in turn may stimulate DNA synthesis (Clemmons and Van Wyk, 1985). Thus, what appears as a direct effect of insulin could be an indirect effect mediated through somatomedinlike molecules.

3. Growth Hormone

The injection of GH into hypophysectomized rats stimulated DNA synthesis of S-V cells (adipose tissue) but did not affect DNA synthesis of preadipocytes or adipose cellularity (Hollenberg and Vost, 1968). Serum from hypophysectomized animals stimulated fibroblast proliferation *"in vitro,"* suggesting that hypophysectomy removes a specific inhibitor of fibroblast growth (Olsen *et al.*, 1979; Murphy and Lazarus, 1984). Supplementing primary cultures of rat S-V cells with sera from hypophysectomized pigs stimulated total cell proliferation but significantly reduced preadipocyte replication (Ramsay *et al.*, 1987a). The response to different doses (1 to 10% of medium) of sera from control and hypophysectomized pigs indicated that hypophysectomy produces an absolute deficiency of adipogenic factors since maximal proliferation was reached at the 5% level for hypophysectomized serum but was not reached at the 10% level for control serum (Ramsay *et al.*, 1987a).

GH promotes the conversion [increased specific activity of α-glycerol phosphate dehydrogenase (α-GPDH) and increased percentage of fat cells] of 3T3 and 1246 preadipocytes to adipocytes (Morikawa *et al.*, 1982; Table II). Furthermore, GH was responsible for one-half of the adipogenic activity (for 3T3 preadipocytes) present in fetal bovine serum (Morikawa *et al.*, 1984). The responsiveness of confluent 3T3 cultures to GH is transient and cannot be recovered by refeeding the cells (Morikawa, 1986).

Recent studies on ob/1771 preadipocytes (a subclone of ob/17 cells) also demonstrate a positive role for GH in adipocyte differentiation (Table II). The mRNA contents for lipogenic enzymes were elevated (ob/1771 cells) by low

Table II
The Effect of GH and IGF-I on Preadipocytes *in Vitro*

Cells	Concentration	Conditions	Effects
	GH		
3T3-F442A [a]	25 ng/ml	Media containing serum	↑ differentiation of preadipocytes
Ob 1771 [b]	1.2 nM	Media containing serum	↑ lipogenic enzyme activity and ↑ intraellular concentration of spermidine
Ob 1771 [c]	1.2 nM	Media containing serum and IBMX	↑ lipogenic enzyme activity; ↑ mRNA content for lipogenic enzymes ↑ transcription of the gene for a structural protein (myelin P_2-like)
1246 [d]	10 ng/ml	Serum-free media with insulin; induction with dexamethasone and IBMX	↑ GPDH activity

Table II *(continued)*

Cells	Concentration	Conditions	Effects
Rat preadipocytes in primary culture[e]	100 ng/ml	Media containing serum from hypophysectomized pigs	↑ preadipocyte proliferation
Rat preadipocytes in primary culture[f]	50 ng/ml	Media containing serum from hypophysectomized pig fetuses	↓ preadipocyte differentiation and 0 on preadipocyte proliferation
	IGF-I		
3T3-F442A[g]	75–150 ng/ml	Media containing serum; differentiation induced with GH	↑ replication of differentiated preadipocytes
Rat preadipocytes in primary culture[h]	nM range	Serum-free media with 200 pM T_3	↑ lipogenic enzyme activity ↑ replication of preadipocytes
Rat preadipocytes in primary culture[i]	nM range	Media containing serum	↑ adipocyte conversion

[a] Morikawa *et al.* (1984), Zezulak and Green (1986).
[b] Amri *et al.* (1986a).
[c] Doglio *et al.* (1986).
[d] Serrero (1986).
[e] Ramsay *et al.* (1987a).
[f] Ramsay *et al.* (1987b).
[g] Zezulak and Green (1986).
[h] Deslex *et al.* (1987a).
[i] Jewell and Hausman (1986).

levels of GH as was the transcription of a gene coding for a structural protein (Doglio *et al.*, 1986). In contrast, an earlier study of ob/17 cells failed to show an effect of GH (wide range of concentrations) on adipose conversion in serum-free conditions (Gaillard *et al.*, 1984). Serum-free conditions support growth of ob/17 cells but do not promote differentiation even when supplemented with T_3, known to be essential for the conversion of ob/17 cells (Vannier *et al.*, 1985).

The effect of GH on ob/1771 preadipocyte differentiation may be mediated by an increase in the intracellular concentration of a polyamine, i.e., spermidine (Amri *et al.*, 1986a). Intracellular concentrations of polyamines were involved in the differentiation of 3T3-L1 (Erwin *et al.*, 1984), 3T3-F442A (Amri *et al.*, 1986a), and ob/1771 preadipocytes (Amri *et al.*, 1986a). Other studies are needed to determine the role of polyamines in mediating actions of other hormones.

Addition of GH to primary cultures of rat S-V cells increases preadipocyte proliferation in the presence of serum from hypophysectomized pigs but does not influence proliferation in the presence of serum from hypophysectomized (decapitated) pig fetuses (Table II). Differentiation of preadipocytes was not augmented by addition of GH to rat S-V cultures in the presence of sera from hypophysectomized fetuses or pigs (Table II). The positive influence of GH on preadipocyte growth in serum from hypophysectomized pigs could be attributed to an indirect effect of prostaglandins of the E type (PGE). As pointed out by Ramsay *et al.* (1987a), GH stimulates PGE release by mammary preadipocyte cell lines (Rudland *et al.*, 1984) and PGEs stimulate 3T3 preadipocyte proliferation (Hiragun, 1985). Furthermore, glucocorticoids can inhibit PGE synthesis by human preadipocytes *in vitro* (Mitchell *et al.*, 1983) and hypophysectomy significantly reduces the titer of serum glucocorticoids. Therefore, hypophysectomy may allow a GH mitogenic response by removing systemic factors (such as glucocorticoids) that regulate PGE synthesis in preadipocytes. Glucocorticoid levels are higher in serum from hypophysectomized fetuses than in serum from hypophysectomized pigs (Martin *et al.*, 1984). Therefore, glucocorticoids may have prevented a GH mitogenic response of rat preadipocytes in serum from hypophysectomized fetuses (Table II).

The role of GH in regulating adipogenesis is clearly unresolved. Studies of preadipocyte cell lines (ob/17, 1246, 3T3) and studies of primary cultures of preadipocytes have produced contradictory results with regard to GH action (Table II). This controversy could be partially resolved by using serum-free cultures for GH studies. GH (1–10 nM) inhibits porcine preadipocyte differentiation in serum-free primary cultures (G. J. Hausman, unpublished observations). This effect is consistent with the observation that GH antagonizes the action of insulin in cultures of porcine adipose tissue (Walton *et al.*, 1986). An indirect effect of GH mediated by the secretion of IGF-I-like molecules is also possible.

4. Insulinlike Growth Factors

In the postnatal animal, GH binds to its receptors in the liver and elicits the release of IGFs which have growth-promoting potential (for review see Chapter 12, this volume). The fibroblast cells TA1 have IGF-I and IGF-II receptors and are responsive to both peptides (increased glucose and amino acid uptake; Shimizu *et al.*, 1986). However, after differentiation into adipocytes, TA1 cells have higher numbers of insulin receptors (than IGF-I receptors) and are more sensitive to insulin than IGF-I (Shimizu *et al.*, 1986). In contrast, differentiated 3T3 preadipocytes are responsive to IGF-I (Table II) whereas undifferentiated 3T3 cells are not (Morikawa *et al.*, 1984; Table II). Replication of 3T3-F442A cells was increased by IGF-I with concurrent or prior exposure to GH (Table II). Physiological levels of IGF-I also increased proliferation of ob/17 preadipocytes (Grimaldi *et al.*, 1983b) and rat preadipocytes (S-V cells) in primary culture (Table II).

The differentiation of rat preadipocytes (S-V cells) in serum-free primary cultures was markedly stimulated by less than 20 nM IGF-I (Table II). Insulin and IGF-I were equipotent in stimulating GPDH activity but insulin increased lipoprotein lipase (LPL) activity more than did IGF-I in primary S-V cultures (Deslex *et al.*, 1987a). The stimulation of GPDH and LPL activities by physiological levels of IGF-I was accentuated by the presence of low levels of insulin (Deslex *et al.*, 1987a). These authors reasoned that low concentrations of IGF-I play a permissive role for adipose cell differentiation when S-V cells are exposed to physiological levels of insulin (Deslex *et al.*, 1987a).

The synthesis and secretion of LPL by mature adipocytes in primary cultures was stimulated by IGF-I at concentrations 40-fold lower than effective insulin concentrations (Kern *et al.*, 1985). This observation indicates that IGF-I could stimulate adipocyte differentiation independent of an effect on cell proliferation (i.e., mature cells would not proliferate).

The role of IGF-I in adipogenesis may be as a mitogen for the proliferation of differentiated adipose precursor cells (Zezulak and Green, 1986; Deslex *et al.*, 1987a). An unanswered question relevant to studies of primary S-V cultures is whether prior exposure to GH is necessary *(in vivo)* for activation of IGF-I receptors (as in 3T3 cell development). Further studies combining *in vivo* and *in vitro* approaches are necessary to address this question.

5. Glucocorticoids

Several combinations of hormones and IBMX have been routinely used to "trigger" adipocyte differentiation in confluent cultures of 3T3-L1 and other cell lines: insulin (1 μg/ml) + hydrocortisone (0.1 μg/ml); DEX (0.25 μM) + IBMX (0.5 mM) ± insulin (1 μg/ml) (Table III). The latter combination requires incubation of confluent cultures for 48 hr with medium containing 10% fetal calf serum (FCS) and DEX + IBMX (± insulin). Subsequently, cultures are maintained an additional 4 to 6 days in medium containing 10% FCS with no added hormone. The combination of DEX + IBMX produces rapid and homogeneous (> 85% of cells) differentiation of 3T3-L1 cells (serum-supplemented medium) and 1246 cells (serum-free medium) (Table III). This "triggering regime" (DEX + IBMX) enhanced the adipose conversion of SF 13 cells (Hiragun *et al.*, 1981) and pig preadipocytes in primary culture (Hentges and Hausman, 1987) but had no effect on ob/17 preadipocytes (Verrando *et al.*, 1981).

Glucocorticoid hormones represent a significant portion of the adipogenic activity in human serum (Schiwek and Loffler, 1987). Purification of the adipogenic factors (for 3T3-L1 cells) that remained in heat-treated sera (no detectable GH) resulted in two adipogenic fractions, i.e., cortisol and cortisone (Schiwek and Loffler, 1987).

The potency of hydrocortisone or DEX alone as inducers of adipose differentiation has been demonstrated in numerous cell culture studies (Table III; Elks and Manganiello, 1985). Hydrocortisone or DEX can induce the differentiation of rat and pig preadipocytes (S-V cells) in primary culture and TA1 preadipocytes (Table III). Furthermore, 1 nM DEX significantly enhanced 3T3-L1 differentiation in the presence of insulin and IBMX (Elks and Manganiello, 1985).

DEX accelerates the differentiation of the adipogenic cell line TA1 by increasing the content of specific mRNAs and by increasing transcriptional activation of adipose-inducible genes (Table III; Ringold et al., 1986). These studies indicate that DEX may affect the triggering mechanisms that orchestrate the transcriptional induction of adipose genes rather than a direct regulation of the genes (Ringold et al., 1986).

Glucocorticoid-treated 3T3-L1 cells undergo one round of cell division followed by de novo synthesis of lipogenic enzymes (Gratzner et al., 1985; Gamou and Shimizu, 1986). However, similar treatment of primary cultures of preadipocytes yielded no increase in cell number (Table III; Hentges and Hausman, 1987). The mitogenic effect of DEX on 3T3-L1 preadipocytes is probably indirect since the serum source (fetal versus calf serum) dictates the mitogenic response to DEX (Miller and Carrino, 1981).

6. Indomethacin, Prostaglandins, and AMP

Since DEX is known to antagonize the production of arachidonic acid metabolites such as prostaglandins (Ringold et al., 1986), indomethacin, an inhibitor of prostaglandin synthesis, has been tested as an inducer of adipose differentiation (Williams and Polakis, 1977; Verrando et al., 1981; Ringold et al., 1986). Indomethacin enhances the differentiation of several preadipocyte cell lines, and in studies of TA1 cells indomethacin was more effective at the molecular level than was DEX (Williams and Polakis, 1977; Verrando et al., 1981; Ringold et al., 1986). These studies suggest an inhibitory role for endogenously synthesized prostaglandins, but two lines of evidence indicate a more complicated situation. For instance, the effect of exogenous prostaglandins is dependent on the stage of differentiation, i.e., stimulatory in the early phase and inhibitory in a later phase (reviewed by Hiragun, 1985). Furthermore, another approach utilized other potent inhibitors of cyclooxygenase (inhibited by indomethacin) on (TA1) preadipocytes (Ringold et al., 1986). Inhibition of cyclooxygenase did inhibit prostaglandin production by TA1 cells but had no

Table III

The Effect of Dexamethasone and Hydrocortisone on Preadipocytea *in Vitro*

Cells	Concentration	Conditions	Effects
	Dexamethasone		
3T3-L1[a]	3 μM	6-hr incubation, no serum	↑ mRNA content for glutamine synthetase
TA1[b]	1 μM	Media containing serum	↑ accumulation of specific mRNAs and ↑ transcription of mRNA for a differentiation-dependent gene
3T3-L1[c]	2.5 nM	Media containing serum and differentiation induced with IBMX and insulin	Caused complete switch from β_1- to β_2-adrenergic subtype
1246[d]	0.25 μM	Serum-free media with IBMX	↑ adipocyte conversion and ↑ lipogenic and lipolytic enzyme activity
3T3-L1[e]	0.25 μM	Media containing serum and IBMX	↑ adipocyte conversion
Pig preadipocytes in primary culture[f]	0.25 μM	Media comtaining serum	↑ GPDH activity ↑ cytodifferentiation
	Hydrocortisone		
Rat preadipocytes in primary culture[g]	50 ng/ml	Media containing 2.5% sera from hypophysectomized pigs	↑ GPDH activity ↓ preadipocyte proliferation
Pig preadipocytes in primary culture[f]	0.1 μg/ml	Media containing serum and insulin	↑ GPDH activity ↑ cytodifferentiation

[a]Miller *et al.* (1987).　[e]Rubin *et al.* (1978).
[b]Ringold *et al.* (1986).　[f]Hentges and Hausman (1987).
[c]Lai *et al.* (1982).　[g]Ramsay *et al.* (1987a).
[d]Serrero (1986).

effect on differentiation (Ringold *et al.*, 1986). Therefore, other aspects of indomethacin and glucocorticoid function may be involved in triggering adipogenesis (Ringold *et al.*, 1986).

Perturbations in intracellular levels of cAMP may be involved in adipocyte differentiation since prostaglandins stimulate cAMP accumulation in 3T3-L1 cells (Hopkins and Gorman, 1981). However, the addition of dibutyryl cAMP during the differentiation of 3T3 preadipocytes reduced the rates of synthesis of several lipogenic enzymes (Spiegelman and Green, 1981). Furthermore, the cellular contents of mRNAs for lipogenic enzymes (e.g., GPDH) were reduced (3T3 cells) by dibutyryl cAMP (Spiegelman and Green, 1981; Bhandari and Miller, 1985; Dobson et al., 1987). Even though lipid accumulation was negligible, the morphological development of 3T3 cells (rounding up) was not inhibited by dibutyryl cAMP (Spiegelman and Green, 1981).

The response of preadipocytes to agents that elevate cAMP may be dependent on the stage of differentiation and their response to hormones (Hiragun, 1985). For example, dibutyryl cAMP inhibited differentiation of rat preadipocytes in primary culture when added after confluence but was stimulatory when added at confluence (Bjorntorp et al., 1979, 1980). Similar observations were reported in studies of 3T3-L1 cells (Williams and Polakis, 1977).

Alterations in cAMP induced by inhibition of soluble cAMP phosphodiesterase may play an important role in IBMX-enhanced differentiation of 3T3 cells (Elks and Manganiello, 1985). However, several studies indicated that IBMX may act via mechanisms independent of cyclic nucleotide metabolism (reviewed by Hiragun, 1985).

7. Thyroid Hormones

T_3 exerts positive effects on preadipocyte cell lines and on primary cultures of S-V cells (Table IV; Zezulak and Green, 1986; Deslex et al., 1987a,b). It is an essential component of serum-free medium for primary cultures of S-V cells from humans (Deslex et al., 1987b) and rats (Deslex et al., 1987a). Serum-containing medium for 3T3-L1, 3T3-F442A, and ob/17 preadipocytes also contains low levels of T_3 (Vannier et al., 1985; Zezulak and Green, 1986).

The synthesis, activity, and mRNA content of fatty acid synthetase (FAS) in differentiating ob/17 preadipocytes were enhanced by T_3 supplementation (Gharbi-Chini et al., 1981, 1984). The positive response of T_3 on lipogenic enzymes in ob/17 cells was associated with an increase in T_3 receptor concentration without any change in receptor affinity (Anselmet et al., 1984).

The differentiation and proliferation of rat preadipocytes (S-V cells) in primary culture were markedly increased by T_3 (Table IV). These results are consistent with those of studies on cell lines except that T_3 had no effect on proliferation of HGFu (Forest et al., 1983) and ob/17 preadipocytes (Grimaldi et al., 1982).

Several lines of evidence suggest that thyroid hormones may have an indirect effect on adipocyte development. For instance, autophosphorylation of the insulin receptor of rat adipocytes is modulated by thyroid hormone status (Correze et al., 1985). Therefore, interaction of insulin and T_3 could take place

Table IV
The Effect of Triiodothyronine on Preadipocytes *in Vitro*

Cells	Triiodothyronine concentration	Conditions	Effects
ob/17[a]	2 nM	Media containing insulin-depleted serum	↑ lipogenic enzyme activity
ob/17[b]	1.5 nM	Media containing serum and insulin	↑ fatty acid synthetase (FAS) synthesis ↑ FAS mRNA content
Rat preadipocytes in primary culture[c]	1.5 nM	Media containing serum from hypophysectomized pigs and fetal pigs	↑ preadipocyte proliferation ↑ preadipocyte differentiation

[a] Vannier *et al.* (1985).
[b] Gharbi-Chihi *et al.* (1984).
[c] Ramsay *et al.* (1987a,b).

at the level of the insulin receptor-linked protein kinase. Additional evidence suggests that proper coupling of adipocyte beta receptors to the adenylate cyclase cascade is also modulated by thyroid hormones (Fain, 1983). Therefore, thyroid hormones may couple growth factors and/or hormones to second messenger systems.

8. Cachectin

Cachectin (tumor necrosis factor) is an endotoxin-induced macrophage mediator that adversely affects lipogenic enzyme activities of adipocytes and other cells. Early work indicated that lipid metabolism was altered in tumor-bearing rats (Lanza-Jacoby *et al.*, 1982). LPL activity was increased in heart tissue but decreased in muscle and adipose tissue (Lanza-Jacoby *et al.*, 1982). Cachectin selectively inhibited the synthesis of enzymes used for *de novo* fatty acid synthesis in 3T3-L1 preadipocytes despite no major disturbances to protein synthesis in general (Pekala *et al.*, 1983). Furthermore, there was a decrease in the ability of insulin to promote glucose uptake and to stimulate anabolic processes in the presence of cachectin (Pekala *et al.*, 1983). Specific receptors for cachectin were demonstrated on differentiated 3T3-L1 preadipocytes and the degree of receptor occupancy was a function of cachectin concentration (Beutler *et al.*, 1985a). The majority (70–80%) of cachectin's anti-LPL activity was achieved with as few as 5% of the receptors occupied (Beutler *et al.*, 1985).

Strong amino acid sequence homology between cachectin and tumor necrosis factor (TNF) has been reported (Beutler *et al.*, 1985b). The gene coding for cachectin has been sequenced and cloned (Beutler *et al.*, 1985b). Four exons code for a 233-amino-acid precursor product, which is then converted to a 157-amino-acid active hormone.

In studies of cachectin, cDNA probes were used to show that levels of mRNAs for lipogenic enzymes in preadipocytes were reduced without affecting levels of other mRNAs in response to cachectin (Torti *et al.*, 1985). The level of mRNA for *sn*-GPDH was reduced in parallel to a decrease in enzyme activity. The response to cachectin was reversible (Torti *et al.*, 1985). The effect of TNF is not tissue specific as the expression of GDPH mRNA was decreased in myoblast and preadipocyte cultures exposed to TNF (Dobson *et al.*, 1987).

Tumor promoters can adversely affect preadipocyte differentiation. For instance, Hiragun *et al.*, (1981) reported that differentiation of a clonal cell line of murine fibroblasts (ST 13) was inhibited by the tumor promoter 12-0-tetradecanoyl phorbol 13-acetate. Furthermore, the tumor promoter dihydroteleocidin B inhibited 3T3-L1 preadipocyte differentiation in serum-free cultures (Gamou and Shimizu, 1986).

9. Autocrine Control

A variety of cell types can secrete biologically active somatomedinlike proteins (reviewed by Underwood *et al.*, 1986). Various hormones (thyroid hormones, insulin) and growth factors stimulate the endogenous secretion of somatomedinlike molecules by cultured fibroblasts (Clemmons and Shaw, 1983) and cultured smooth muscle cells (Clemmons, 1985). Extensive studies on autocrine control of the adipogenic cell line 1246 have been reported by Serrero and co-workers (reviewed by Serrero, 1986). Besides the studies on prostaglandins previously discussed, there are only several isolated reports of the secretion of biologically active molecules by other preadipocytes (Bjorntorp, 1983; Hsu *et al.*, 1984). Extensive studies of other preadipocyte cell lines and S-V cells are needed to further clarify the autocrine control of adipogenesis.

10. Perspectives on Integrated Endocrine Control

An elegant series of studies on ob/17 cells and related subclones led Ailhaud and co-workers to propose that adipocyte differentiation involves the expression of two sets of markers (early and late) that are differentially regulated by hormones (Amri *et al.*, 1984, 1986a,b; Vannier *et al.*, 1985; Dani *et al.*, 1986). Insulin, T_3, and GH are not required for the emergence of early markers of differentiation, i.e., p Ob 24 mRNA and LPL (Amri *et al.*, 1984;

Vannier *et al.*, 1985; Doglio *et al.*, 1986). The early markers appear after cells become confluent but before a period of limited (postconfluent) mitosis (Amri *et al.*, 1986b). The appearance of late markers of adipose conversion, i.e., GPDH and acyl-CoA synthetase as well as several specific mRNAs, takes place after this period of "post-confluent mitosis" (Amri *et al.*, 1986b; Doglio *et al.*, 1986). GH, insulin, and T_3 influenced the appearance of the late markers (Doglio *et al.*, 1986). Although not stated as such, "postconfluent mitosis" is involved in the dual effector theory proposed by Green *et al.* (1985). According to this theory, GH first causes adipocyte differentiation, after which IGF-I (or insulin) serves as a mitotic signal for newly differentiated cells in a postconfluent state (Zezulak and Green, 1986). A common feature of these two theories (Ailhaud, Green) and of results of recent studies of serum-free cultures (primary) of rat and human preadipocytes (S-V cells) (Deslex *et al.*, 1987a,b) is that a mitogenic stimulus is required for significant expression of GPDH, a late marker, whereas expression of early markers such as LPL does not require a mitogenic stimulus. Postconfluent mitosis and the subsequent emergence of GPDH is most effectively stimulated by GH in ob/17 (and subclones) preadipocytes whereas IGF-I and/or insulin stimulate these most effectively in 3T3 cells and in preadipocytes (S-V cells) in primary culture. This apparent conflict is unresolved, but it should be pointed out that the ob/17 cell line was established through a dedifferentiation process of mature adipocytes *in vitro* whereas 3T3 cells were derived from mouse embryos. Therefore, ob/17 cells were derived from much older animals than were 3T3 cells.

Serum-free media have been described for the differentiation of preadipocytes (S-V cells) from the rat (Deslex *et al.*, 1987a; Serrero, 1987), pig (G. J. Hausman, unpublished observations), sheep (Broad and Hamm, 1983), and human (Deslex *et al.*, 1987b). Insulin (5–10 μg/ml) and transferrin were common components of these media and T_3 was present in media for pig (G. J. Hausman, unpublished observations), rat (Deslex *et al.*, 1987a), and human cells (Deslex *et al.*, 1987b). The simple hormonal requirements for differentiation of preadipocytes in serum-free cultures suggest that adipocyte differentiation *in vitro* cannot be a regulatory step for control of adipose tissue accretion. The slow replication rates of preadipocytes in serum-free cultures (Deslex *et al.*, 1987a,b; G. J. Hausman, unpublished observations) indicate that stimulation of the proliferation of growth-arrested, dormant precursor cells by specific mitogens may be a key regulatory event operative during adipose tissue expansion (Deslex *et al.*, 1987a). Several studies (Grimaldi *et al.*, 1983; Zezulak and Green, 1986; Deslex *et al.*, 1987a) provide evidence that IGF-I might be such a specific mitogen for preadipocytes.

A role for glucocorticoids is clearly absent from the above-mentioned theories, which is unusual in light of the vast literature (Table III) and several recent studies (Hauner *et al.*, 1987; Wiederer and Loffler, 1987) on glucocorticoid involvement in adipocyte differentiation. The stimulatory effect of glu-cocorticoids on adipocyte differentiation is dependent on the presence of serum (Table III), a lipid source (Broad and Hamm, 1983), and/or insulin (Table III; Hauner *et al.*, 1987; Wiederer and Loffler, 1987) in the medium. These collec-tive observations indicate that the action of glucocorticoids is indirect, which is consistent with the conclusions of Ringold *et al.* (1986) that glucocorticoids do not initiate but do accelerate adipose differentiation as they do for other tissues.

References

Alexander, M., G. Curtis, J. Avruch and H. M. Goodman. 1985. Insulin regulation of protein biosynthesis in differentiated 3T3 adipocytes: regulation of glyceraldehyde-3-phosphate dehy-drogenase. J. Biol. Chem. 160:11978.

Allen, C. E. 1976. Cellularity of adipose tissue in meat animals. Fed. Proc. 35:2302.

Amri, E., R. Barbaras, A. Doglio, C. Dani, P. Grimaldi and G. Ailhaud. 1986a. Role of sper-midine in the expression of late markers of adipose conversions. Effects of growth hormone. Biochem. J. 239:363.

Amri, E., C. Dani, A. Doglio, J. Etienne, P. Grimaldi and G. Ailhaud. 1986b. Adipose cell differentiation: evidence for a two-step process in the polyamine-dependent Ob 1754 cell line. Biochem. J. 238:115.

Amri, E. Z., P. Grimaldi, R. Negrel and G. Ailhaud. 1984. Adipose conversion of Ob17 cells. Exp. Cell Res. 152:368.

Anselmet, A., J. Gharbi-Chini and J. Torresani. 1984. Nuclear triiodothyronine receptor in differ-entiating preadipocytes cloned from obese and lean mice. Endocrinology 114:450.

Beutler, B., D. Greenwald, J. D. Hulmes, M. Chang, Y. C. E. Pan, J. Mathison, R. Ulevith and A. Cerami. 1985a. Identity of tumour necrosis factor and the macrophage-secreted factor cachectin. Nature 316:552.

Beutler, B., J. Mahoney, N. Le Trang, P. Pekala and A. Cerami. 1985b. Purification of cachectin, a lipoprotein lipase-suppressing hormone secreted by endotoxin-induced RAW 264.7 cells. J. Exp. Med. 161:984.

Bhandari, B. and R. E. Miller. 1985. Glycerol-3-phosphate dehydrogenase mRNA content in cul-tured 3T3-L1 adipocytes: regulation by dibutyryl cyclic AMP. Biochem. Biophys. Res. Com-mun. 131:1193.

Bjorntorp, P. 1983. Interactions of adipocytes and their precursor cells with endothelial cells in culture. Exp. Cell Res. 149:277.

Bjorntorp, P., M. Karlsson, L. Gustafson, U. Smith, L. Sjostrom, L. Cigolini, G. Storck and P. Pettersson. 1979. Quantitation of different cells in epididymal fat pad of the rat. J. Lipid Res. 20:97.

Bjorntorp, P., M. Karlsson, P. Pettersson and G. Synpniewska. 1980. Differentiation and function of rat adipocyte precursor cells in primary culture. J. Lipid Res. 21:714.

Broad, T. E. and R. G. Hamm. 1983. Growth and adipose differentiation of sheep preadipocyte fibroblasts in serum free medium. Eur. J. Biochem. 135:33.

Chernausek, S. D., S. Jacobs and J. J. Van Wyk. 1981. Structural similarities between human receptors for somatomedin C and insulin: analysis by affinity labeling. Biochemistry 20:7345.

Clemmons, D. R. 1985. Variables controlling the secretion of a somatomedin-like peptide by cultured porcine smooth muscle cells. Circ. Res. 56:418.

Clemmons, D. R. and D. S. Shaw. 1983. Variables controlling somatomedin production by cultured human fibroblasts. J. Cell. Physiol. 115:137.

Clemmons. D. R. and J. J. Van Wyk. 1985. Evidence for a functional role of endogenously produced somatomedin-like peptides in the regulation of DNA synthesis in cultured human fibroblasts and porcine smooth muscle cells. J. Clin. Invest. 75:1914.

Correze, C., M. Pierre, H. Thibout and D. Toru-Delbauffe. 1985. Autophosphorylation of the insulin receptor in rat adipocytes is modulated by thyroid hormone status. Biochem. Biophys. Res. Commun. 126:1061.

Dani, C., A. Doglio, P. Grimaldi and G. Ailhaud. 1986. Expression of the phosphoenolpyruvate carboxykinase gene and its insulin regulation during differentiation of preadipose cell lines. Biochem. Biophys. Res. Commun. 138:468.

Deslex, S., R. Negrel and G. Ailhaud. 1987a. Development of a chemically defined serum-free medium for differentiation of rat adipose precursor cells. Exp. Cell Res. 168:15.

Deslex, S., R. Negrel, C. Vannier, J. Etienne and G. Ailhaud. 1987b. Differentiation of human adipocyte precursors in a chemically defined serum-free medium. Int. J. Obesity 11:19.

Dobson, D. E., D. L. Groves and B. M. Spiegelman. 1987. Nucleotide sequence and hormonal regulation of mouse glycerophosphate dehydrogenase mRNA during adipocyte and muscle cell differentiation. J. Biol. Chem. 262:1804.

Doglio, A., C. Dani, P. Grimaldi and G. Ailhaud. 1986. Growth hormone regulation of the expression of differentiation-dependent genes in preadipocyte Ob1771 cells. Biochem. J. 238:123.

Elks, M. L. and V. C. Manganiello. 1985. A role for soluble cAMP phosphodiesterases in differentiation of 3T3-L1 adipocytes. J. Cell. Physiol. 124:191.

Erwin, B. G., D. R. Bethell and A. E. Pegg. 1984. Role of polyamines in differentiation of 3T3-L1 fibroblasts into adipocytes. Amer. J. Physiol. 246:C293.

Fain, J. N. and J. A. Garcia-Sainz. 1983. Adrenergic regulation of adipocyte metabolism. J. Lipid. Res. 24:945.

Flier, J. S., P. Usher and A. C. Moses. 1986. Monoclonal antibody to the type I insulin-like growth factor (IGF-I) receptor blocks IGF-I receptor-mediated DNA synthesis: clarification of the mitogenic mechanisms of IGF-I and insulin in human skin fibroblasts. Proc. Natl. Acad. Sci. USA83:664.

Forest, C., P. Grimaldi, D. Czerucka, R. Negrel and G. Ailhaud. 1983. Establishment of a preadipocyte cell line from the epididymal fat pad of the lean C57 BL/6J mouse—long term effects of insulin and triiodothyronine on adipose conversion. In Vitro 19:344.

Gaben-Cogneville, A. M., Y. Aron, G. Idriss, T. Jahchan, J. Y. Pello and E. Swierczewski. 1983. Differentiation under the control of insulin of rat preadipocytes in primary culture. Isolation of homogenous cellular fractions by gradient centrifugation. Biochim. Biophys. Acta 762:437.

Gaillard, D., R. Negrel, G. Serrero-Dave, C. Cermolacce and G. Ailhaud. 1984. Growth of preadipocyte cell lines and cell strains from rodents in serum-free hormone-supplemented medium. In Vitro 20:79.

Gamou, S. and N. Shimizu. 1986. Adipocyte differentiation of 3T3-L1 cells in serum-free hormone-supplemented media: effects of insulin and dihydroteleocidin B. Cell Struct. Funct. 11:21.

Gharbi-Chini, J., O. Chabaud and J. Torresani. 1984. The role of triiodothyronine in the regulation of synthesis rate and translatable messenger RNA level of fatty acid-synthesis in a preadipocyte cell line. Biochim. Biophys. Acta 783:26.

Gharbi-Chini, J., P. Grimaldi, J. Torresani and G. Ailhaud. 1981. Triiodothyronine and adipose conversion of Ob17 preadipocytes: binding to high affinity sites and effects on fatty acid synthesizing and esterifying enzymes. J. Receptor Res. 2:153.

Gratzner, H. G., P. M. Ahmad, J. Stein and F. Ahmad. 1985. Flow cytometric analysis of DNA replication during the differentiation of 3T3-L1 preadipocytes. Cytometry 6:563.

Green, H. and O. Kehinde. 1975. An established preadipose cell line and its differentiation in culture. II. Factors affecting the adipose conversion. Cell 5:19.

Green, H., M. Morikawa and T. Nixon. 1985. A dual effector theory of growth hormone action. Differentiation 293:195.

Grimaldi, P., P. Djian, C. Forest, P. Poli, R. Negrel and G. Ailhaud. 1983a. Lipogenic and mitogenic effects of insulin during conversion of Ob17 cells to adipose-like cells. Mol. Cell. Endocrinol. 29:271.

Grimaldi, P., C. Forest, P. Poli, R. Negrel and G. Ailhaud. 1983b. Modulation of lipid-synthesizing enzymes by insulin in differentiated Ob17 adipose-like cells. Biochem. J. 214:443.

Grimaldi, P., P. Djian, R. Negrel and G. Ailhaud. 1982. Differentiation of Ob17 preadipocytes to adipocytes: requirement of adipose conversion factor(s) for fat cell cluster formation. EMBO J. 1:687.

Harrison, J. J., E. Soudry and R. Sager. 1985. Adipocyte conversion of CHEF cells in serum-free medium. J. Cell Biol. 100:429.

Hauner, H., P. Schmid and E. F. Pfeiffer. 1987. Glucocorticoids and insulin promote the differentiation of human precursor cells into fat cells. J. Clin. Endocrinol. Metab. 64:832.

Hausman, G. J. 1987. The effect of insulin on primary cultures of rat preadipocytes grown in fetal or postnatal pig serum. J. Anim. Sci. (in press).

Hausman, G. J. 1985. The comparative anatomy of adipose tissue. In: A. Cryer and R. L. R. Van (Ed.) New Perspectives in Adipose Tissue: Structure, Function and Development. pp. 1–12. Butterworths, London.

Hausman, G. J., D. R. Campion and R. J. Martin. 1980. Search for the adipocyte precursor cell and factors that promote its differentiation. J. Lipid Res. 21:657.

Hentges, E. J. and G. J. Hausman. 1987. Induced lipid filling and conversion of preadipocytes from porcine stromal-vascular cells in primary culture. Fed. Proc. 46:1155 (Abstr.).

Hiragun, A. 1985. Cell and tissue culture models of adipocyte development. In: A. Cryer and R. L. R. Van (Ed.) New Perspectives in Adipose Tissue: Structure, Function and Development. pp 333–352. Butterworths, London.

Hiragun, A., M. Sato and H. Mitsui. 1981. Prevention of tumor promoter mediated inhibition of preadipocyte differentiation by dexamethasone. Gann 72:891.

Hollenberg, C. H. and A. Vost. 1968. Regulation of DNA synthesis in fat cells and stromal elements from rat adipose tissue. J. Clin. Invest. 47:2485.

Hopkins, N. K. and R. R. Gorman. 1981. Regulation of 3T3-L1 fibroblast differentiation by prostacyclin prostaglandin E-2. Biochim. Biophys. Acta 663:457.

Hsu, Y. M., J. M. Barry and J. L. Wang. 1984. Growth control in cultured 3T3 fibroblasts: neutralization and identification of a growth-inhibitory factor by a monoclonal antibody. Proc. Natl. Acad. Sci. USA 81:2107.

Jewell, D. E. and G. J. Hausman. 1986. Comparative effects of insulin-like growth factor-1 and insulin on the growth and differentiation of cultured rat preadipocytes. J. Anim. Sci. 63 (Suppl. 1):226 (Abstr.).

Kern, P. A., S. Marshall and R. H. Eckel. 1985. Regulation of lipoprotein lipase in primary cultures of isolated human adipocytes. J. Clin. Invest. 75:199.

Koontz, J. W. and M. Iwahashi. 1981. Insulin as a potent specific factor in a rat hepatoma cell line. Science 211:947.

Lai, E., O. M. Rosen and C. S. Rubin. 1982. Dexamethasone regulates the beta adrenergic receptor subtype expressed by 3T3-L1 preadipocytes and adipocytes. J. Biol. Chem. 257:6691.

Lanza-Jacoby, S., E. E. Miller and F. E. Rosato. 1982. Changes in the activities of lipoprotein lipase and the lipogenic enzymes in tumor-bearing rats. Lipids 17:944.

Martin, R. J., D. R. Campion, G. J. Hausman and J. H. Gahagan. 1985a. Serum hormones and metabolites in fetally decapitated pigs. Growth 48:158.

Martin, R. J., T. R. Kasser, T. G. Ramsay and G. J. Hausman. 1985b. Regulation of adipose tissue development in utero. In: A. Cryer and R. L. R. Van (Ed.) New Perspectives in Adipose Tissue: Structure, Function and Development. pp 303–317. Butterworths, London.

Martin, R. J., T. G. Ramsay and R. B. S. Harris. 1984. Central role of insulin in growth and development. Domest. Anim. Endocrinol. 1:89.

Massague, J. and M. P. Czech. 1982. The subunit structures of two distinct receptors for insulinlike growth factors I and II and their relationship to the insulin receptor. J. Biol. Chem. 257:5038.

Miller, R. E., D. M. Burns and B. Bhandari. 1987. Hormonal regulation of glutamine synthetase in cultured 3T3-L1 adipocytes. In: G. J. Hausman and R. J. Martin (Ed.) Biology of the Adipocyte: Research Approaches. Van Nostrand Reinhold Co., New York.

Miller, R. E. and D. A. Carrino. 1981. An association between glutamine synthetase activity and adipocyte differentiation in cultured 3T3-L1 cells. Arch. Biochem. Biophys. 209:486.

Mitchell, M. D., W. H. Cleland, M. E. Smith, E. R. Simpson and C. R. Mendelson. 1983. Inhibition of prostaglandin biosynthesis in human adipose tissue by glucocorticosteroids. J. Clin. Endocrinol. Metab. 57:771.

Morikawa, M. 1986. Sensitivity of preadipose 3T3 cells to growth hormone. J. Cell. Physiol. 128:293.

Morikawa, M., H. Green and U. J. Lewis. 1984. Activity of human growth hormone and related polypeptides on the adipose conversion of 3T3 cells. Mol. Cell. Biol. 4:228.

Morikawa, M., T. Nixon and H. Green. 1982. Growth hormone and the adipose conversion of 3T3 cells. Cell 29:783.

Murphy, L. J. and L. Lazarus. 1984. Effect of hypophysectomy and subsequent growth hormone replacement in the rat on the ability of serum to stimulate proliferation of human and mouse (BALB/c 3T3) fibroblasts. Horm. Metab. Res. 16:631.

Olsen, R. F., W. H. Patton, R. J. Martin and P. J. Wangsness. 1979. Somatomedin and fibroblast proliferative activity with growth hormone injection in hypophysectomized and sham-operated rats. Horm. Metab. Res. 11:645.

Pekala, P., M. Dawadami, W. Vine, M. D. Land and A. Cerami. 1983. Studies of insulin resistance in adipocytes induced by macrophage mediator. J. Exp. Med. 157:1360.

Ramsay, T. G., G. J. Hausman, R. R. Kraeling and R. J. Martin. 1987a. Alterations in adipogenic and mitogenic activity of porcine sera in response to hypophysectomy. Endocrinology (submitted).

Ramsay, T. G., G. J. Hausman and R. J. Martin. 1987b. Preadipocyte proliferation and differentiation in response to hormone supplementation of decapitated fetal pig sera. J. Anim. Sci. 64:735.

Ringold, G. M., A. B. Chapman, D. M. Knight and F. M. Torti. 1986. Hormonal control of adipogenesis. In: A. G. Goodridge and R. W. Hansen (Ed.) Metabolic Regulation: Application of Recombinant DNA Techniques. pp 109–119. Annals of the New York Academy of Sciences, New York.

Roncari, D. A. K., E. Y. W. Mack and D. K. Yip. 1979. Enhancement of microsomal phosphatidate phosphohydrolase and diacylglycerol acyl-transferase activity by insulin during growth of rat adipocyte precursors in culture. Can. J. Biochem. 57:573.

Rubin, C. S., A. Hirsch, C. Fung and O. M. Rosen. 1978. Development of hormone receptors and hormonal responsiveness in vitro. J. Biol. Chem. 253:7570.

Rudland, P. S., A. C. Twiston-Davies and S. W. Tsao. 1984. Rat mammary preadipocytes in culture produce a trophic agent for mammary epithelia—Prostaglandin E_2. J. Cell. Physiol. 120:364–376.

Sager, R. and P. Kovac. 1982. Preadipocyte determination either by insulin or by 5-azacytidine. Proc. Natl. Acad. Sci. USA 79:480.

Schiwek, C. R. and G. Loffler. 1987. Glucocorticoid hormones contribute to the adipogenic activity of human serum. Endocrinology 120:469.

Serrero, G. 1986. Endocrine and autocrine control of growth and differentiation of teratoma-derived cell lines. In: G. Serrero and J. Hayashi (Ed.) Cellular Endocrinology: Hormonal Control of Embryonic and Cellular Differentiation. pp 191–204. Alan R. Liss, Inc., New York.

Serrero, G. 1987. EGF inhibits the differentiation of adipocyte precursors in primary cultures. Biochem. Biophys. Res. Commun. 146:194.

Serrero, G. and J. C. Khoo. 1982. An in vitro model to study adipose differentiation in serum-free medium. Anal. Biochem. 120:351.

Shimizu, M., F. Torti and R. A. Roth. 1986. Characterization of the insulin and insulin-like growth factor receptors and responsivity of a fibroblast–adipocyte cell line before and after differentiation. Biochem. Biophys. Res. Commun. 137:552.

Spiegelman, B. M. and H. Green. 1981. Cyclic AMP-mediated control of lipogenic enzyme synthesis during adipose differentiation of 3T3 cells. Cell 24:503.

Steinberg, M. M. and B. L. Brownstein. 1982. A clonal analysis of the differentiation of 3T3-L1 preadipose cells: role of insulin. J. Cell. Physiol. 113:359.

Torti, F. M., B. Dieckmann, B. Beutler, A. Cerami and G. M. Ringold. 1985. A macrophage factor inhibits adipocyte gene expression: an in vitro model of cachexia. Science 229:867.

Underwood, L. E., A. J. D'Ercole, D. R. Clemmons and J. J. Van Wyk. 1986. Paracrine functions of somatomedins. Clin. Endocrinol. Metab. 15(1):59.

Van, R. L. R. 1985. The adipocyte precursor cell. In: A. Cryer and R. L. R. Van (Ed.) New Perspectives in Adipose Tissue: Structure, Function and Development. pp 353–379. Butterworths, London.

Vannier, C., D. Gaillard, P. Grimaldi, E.-Z. Amri, P. Kjian, C. Cermolacce, C. Forest, J. Etienne, R. Negrel and G. Ailhaud. 1985. Adipose conversion of Ob17 cells and hormone related events. Int. J. Obesity 9:41.

Van Wyk, J. J., D. C. Graves, S. J. Casella and S. Jacobs. 1985. Evidence from monoclonal antibody studies that insulin stimulates DNA synthesis through type 1 somatomedin receptor. J. Clin. Endocrinol. Metab. 61:639.

Verrando, P., R. Negrel, P. Grimaldi, M. Murphy and G. Ailhaud. 1981. Differentiation of Ob17 preadipocytes: triggering effects of clofenapate and indomethacin. Biochim. Biophys. Acta 663:255.

Walton, P. E., T. D. Etherton and C. M. Evock. 1986. Antagonism of insulin action in cultured pig adipose tissue by pituitary and recombinant porcine growth hormone: potentiation by hydrocortisone. Endocrinology 118:2577.

Wiederer, O. and G. Loffler. 1987. Hormonal regulation of the differentiation of rat adipocyte precursor cells in primary culture. J. Lipid Res. 28:649.

Williams, I. H. and E. Polakis. 1977. Differentiation of 3T3-L1 fibroblasts to adipocytes: the effect of indomethacin, prostaglandin E_1 and cyclic AMP on the process of differentiation. Biochem. Biophys. Res. Commun. 77:175.

York, D. A. 1985. The role of hormonal status in the development of excess adiposity in animal models of obesity. In: A. Cryer and R. L. R. Van (Ed.) New Perspectives in Adipose Tissue: Structure, Function and Development. pp 407–455. Butterworths, London.

Zezulak, K. M. and Green, H. 1986. The generation of insulin-like growth factor-1-sensitive cells by growth hormone action. Science 233:551.

CHAPTER 4

Autocrine, Paracrine, and Endocrine Regulation of Myogenesis

WILLIAM R. DAYTON AND MARCIA R. HATHAWAY

1. Introduction

Rapid and efficient deposition of muscle tissue is the primary goal of meat-animal production. Because numerous events that occur during embryonic development influence the number, size, and type of muscle fibers present in a particular muscle, alterations in the timing or the duration of these events could potentially result in changes in muscle mass or efficiency of muscle deposition in the postnatal animal. Thus, an increased knowledge of the mechanism and regulation of myogenesis, the embryonic development of muscle, is potentially of great significance to our ultimate goal of understanding and regulating muscle growth. The aim of this chapter is to briefly describe the events that occur during myogenesis and to discuss the factors that may regulate this process.

2. Muscle Structure

In order to efficiently perform its highly specialized functions, skeletal muscle has developed a unique and highly organized structure. To fully appreciate the complexity of muscle differentiation, it is necessary to have a general understanding of this structure. Individual muscles are composed of a variable number of elongated, multinucleated fibers or cells that run parallel to each other. These fibers are from 10 to 100 μm in diameter and can be as much as

WILLIAM R. DAYTON ● Department of Animal Science, University of Minnesota, St. Paul, Minnesota 55108. **MARCIA R. HATHAWAY** ● Department of Animal Science, University of Minnesota, St. Paul, Minnesota 55108.

several centimeters in length. The variation in size between similar muscles in different species is a function of the number and length of these fibers. Similarly, differences in fiber number, length, and diameter undoubtedly are partially responsible for variations in muscle mass observed between animals of the same species. For reasons that will be discussed later in the chapter, the number of muscle fibers and, to some extent, the distribution of fiber diameters in a muscle are fixed by events that occur during embryonic development.

For the most part, muscle fibers contain the same organelles found in other somatic cells. However, one significant difference in organelle content is the presence of the contractile apparatus or myofibril in the muscle cell. The myofibril is a contractile thread consisting of 12–14 proteins arranged in a highly organized structure. The proteins comprising the myofibril make up as much as 55% of the total muscle protein with the myofibrillar proteins myosin and actin being found in greatest abundance. Myosin is an ATPase and hydrolyzes ATP in order to provide the energy needed to power the contractile process. Several different isozymes of myosin are expressed in muscle fibers during different stages of differentiation (Gauthier *et al.*, 1982; Caplan *et al.*, 1983; Lowey *et al.*, 1983; Miller *et al.*, 1985; Miller and Stockdale, 1986a,b; Weydert *et al.*, 1987). Thus, presence of a particular myosin isoform has been used as a marker for stage of differentiation. Another significant difference between muscle fibers and most other cells is the multinucleated nature of the muscle fiber. A single muscle fiber may contain several hundred nuclei arrayed randomly just beneath the plasma membrane. Numerous studies have shown that the nuclei present in a multinucleated fiber are not capable of division.

Muscle is called upon to function under a wide range of conditions ranging from maintaining posture to providing the rapid, powerful limb movements needed for running or lifting heavy weights. For this reason, specific muscle fiber types have developed to more efficiently perform specific functions. Some fibers are capable of undergoing very rapid, powerful contractions of short duration. In general, these fibers predominate in propulsion muscles found in the limbs. Other fibers contract more slowly but maintain their contraction for a longer period of time. These fibers predominate in the postural muscles. Several classification schemes have been developed to group the different types of muscle fibers. All of these schemes group fibers into three general categories based primarily on their speed of contraction and the energy source utilized for contraction. In this discussion we will employ the classification system developed by Peter *et al.* (1972). In this system, fibers that contract slowly and utilize oxidative pathways to obtain energy are designated as slow-twitch, oxidative fibers (SO). At the other end of the continuum are the fast-twitch, glycolytic fibers (FG), which contract rapidly and primarily utilize anaerobic pathways to produce energy for contraction. Fast-twitch oxidative–glycolytic fibers (FOG) form a broad category that is intermediate between these two extremes.

FOG fibers are fast-twitch but are capable of utilizing either oxidative or glycolytic energy pathways. The morphology, biochemistry, and histochemistry of these fiber types have been studied extensively and significant differences have been noted between fiber types. For example, the fiber types differ in: isoforms of the contractile proteins; degree of development of the sarcoplasmic reticulum and T-tubule systems; number of mitochondria; and amount of myoglobin. More importantly from the standpoint of muscle growth, the various fiber types vary significantly in diameter. In general, FG fibers are the largest in diameter and the SO fibers the smallest. FOG fibers tend to be intermediate in diameter. During development of a muscle, its fiber composition changes continually, although the basis of these changes is not understood. Because fiber types vary in diameter, factors that affect the fiber type distribution within a muscle could alter ultimate muscle size or growth rate.

3. Myogenesis

Myogenesis is the embryonic development of muscle tissue. Observations in cell culture and *in vivo* have shown that myogenesis originates in muscle precursor cells of mesodermal origin. These myogenic cells proliferate, differentiate, and ultimately fuse to form multinucleated myotubes that begin to synthesize muscle-specific contractile proteins and assemble them into myofibrils. Once nuclei have been incorporated into a myotube, they are no longer capable of division. Thus, the fusion of proliferating, mononucleated myoblasts to form the multinucleated, nonproliferating myotube is the terminal step in muscle differentiation. Once this terminal differentiation is completed during embryonic muscle development, the number of muscle fibers is essentially fixed. Therefore, as stated earlier, the number of muscle fibers is fixed at birth in meat animals. Additionally, nuclei present in fibers are not able to divide. Paradoxically, the amount of DNA present in muscle fibers has been reported to increase as much as eightfold during postnatal muscle growth (Winick and Noble, 1966). This DNA accretion is closely associated with, if not causally related to, the rate and extent of muscle growth in chickens (Moss, 1968), pigs (Powell and Aberle, 1975; Harbison et al., 1976; Swatland, 1977), and cattle (Trenkle et al., 1978). The source of this DNA is currently thought to be mononucleated, myogenic cells known as satellite cells that are located between the plasmalemma and basal lamina of each muscle fiber (Mauro, 1961; reviewed by Campion et al., 1984). Numerous studies have suggested that satellite cells proliferate and fuse with existing muscle fibers to provide nuclei needed during postnatal muscle growth (Moss and Leblond, 1970–1971). However, the origin of satellite cells and regulation of their proliferation, differentiation, and fusion are poorly understood.

The mechanism and regulation of the transition from proliferating myoblast to nonproliferating myotube have been studied extensively over the last 20–30 years. Holtzer and co-workers proposed that muscle differentiation occurs through a unique mitotic event termed a "quantal mitosis" (reviewed by Holtzer, 1978). During a quantal mitosis the genetic program of the cell is altered so that resultant daughter cells possess a more differentiated phenotype than the mother cell. The final quantal mitosis results in production of myoblasts that have irreversibly withdrawn from the cell cycle and are capable of fusing with other myoblasts to form a myotube. More recently, Quinn *et al.* (1985) proposed a more restrictive version of this model in which a fixed number of cell divisions occur during the transition from the stem cell to the nonproliferating, fusion-competent myoblast. The key element of the quantal theory is that the final quantal mitosis not only renders the myoblast fusion-competent but also irreversibly removes it from the cell cycle. Therefore, according to this theory, fusion-competent myoblasts cannot be induced to proliferate by mitogenic stimuli. An alternative mechanism of myogenesis has been proposed by Buckley and Konigsberg (1974, 1977). Their theory is based on observations that differentiating muscle cells spend increasingly longer periods of time in the G_1 phase of the cell cycle. Fusion occurs in the G_1 phase and, therefore, the longer time spent in this phase increases the probability of fusion. The probability that myogenic cells will fuse is also influenced by the presence of environmental factors that stimulate proliferation or fusion. In contrast to the quantal mitosis theory, this "probabilistic" model hypothesizes that the withdrawal from mitosis and the initiation of muscle-specific protein synthesis are the result of the fusion process. The third, and currently most widely accepted, theory of myogenesis (Nadal-Ginard, 1978) combines elements of both the quantal and probabilistic theories. This theory hypothesizes that early in the G_1 phase, differentiating myoblasts either withdraw from the cell cycle and become committed to fusion, or become committed to continuing through the cell cycle. This decision is greatly influenced by the presence or absence of mitogenic factors that stimulate the cell to continue to proliferate. Thus, the withdrawal from the cell cycle occurs as a result of lack of mitogenic stimulation and not as a result of fusion.

During *in vivo* embryonic development of muscle, two morphologically distinct populations of fibers appear to form (reviewed by Kelly and Rubinstein, 1986). Initially, myoblasts fuse to form a set of "'primary fibers." Such fibers are surrounded by mononucleated cells that presumably fuse and give rise to a second population of fibers known as "secondary fibers." Formation of secondary fibers continues until most of the mononucleated muscle precursor cells have been incorporated into multinucleated myotubes. Mononucleated muscle cells not incorporated into fibers during embryonic development will presumably form the satellite cells found in postnatal muscle tissue. The origin of primary and secondary fibers as well as satellite cells may be partially explained by the relatively recent discovery that several different types of myogenic cells are actively proliferating and differentiating during different stages of embryonic development of avian muscle (Miller *et al.*, 1985; Miller and Stockdale, 1986a,b; Stockdale *et al.*, 1986; Narusawa *et al.*, 1987). White *et al.* (1975) initially showed that myogenic cells cloned from muscle tissue obtained at various stages of embryonic development demonstrate dramatic differences in morphology and medium requirements. Based on these differences, two general myogenic cell types were identified and designated "early muscle colony-forming (MCF) cells" and "late muscle colony-forming cells" because the former predominate in the early stages of embryonic development and the latter predominate in the later stages of development of embryonic muscle. It should be emphasized that late MCF cells do not appear to have descended from early MCF cells (Rutz and Hauschka, 1982; Seed and Hauschka, 1984). The time of appearance of myoblasts in early MCF and late MCF lineages corresponds to the time of formation of primary and secondary myotubes observed in anatomical studies. Thus, it has been suggested that early MCF cells may form the primary fibers and late MCF cells may form the secondary fibers (Miller and Stockdale, 1986b; Stockdale and Miller, 1987). More detailed studies have shown that early MCF cells can be subdivided into three subtypes and that late MCF cells can be subdivided into two subtypes based on formation of distinct myotubes expressing a particular myosin heavy chain isozyme (Stockdale and Miller, 1987). Although many aspects of mammalian and avian myogenesis are similar, including the formation of primary and secondary fibers (Kelly and Rubinstein, 1986), it is not yet known whether embryonic mammalian myoblasts are committed to distinct myogenic cell lineages.

In addition to enhancing our understanding of embryonic muscle development, more extensive study of myogenic cell lineages may also lead to a better understanding of the origin of satellite cells. Most studies of the properties of satellite cells have been done in cell culture where they behave very much like myogenic cells of embryonic origin (Bischoff, 1974; Allen *et al.*, 1980; Cossu *et al.*, 1980). Cultured satellite cells proliferate and fuse to form myotubes that are capable of synthesizing muscle-specific proteins. Additionally, satellite cells respond to many growth factors in much the same way as do embryonic myoblasts. These observations have led to speculations that satellite cells are simply myoblasts that, for some undefined reason, did not fuse during embryonic development of muscle. However, more recent studies suggest that there are differences between embryonic myoblasts and satellite cells. These differences include the following: (1) myotubes derived from cultured satellite cells synthesize only about one third as much α-actin per nucleus as do myotubes derived from embryonic myoblasts (Allen *et al.*, 1982); (2) phorbol ester tumor promoters reversibly block differentiation of embryonic myoblasts but have no effect upon differentiation of satellite cells (Cossu *et al.*, 1983, 1985, 1986); (3) acetylcholine receptor channels are present in undifferentiated satellite cells in culture but are detectable only after the onset of terminal differentiation of embryonic myoblasts (Cossu *et al.*, 1987). Based on these reported differences between embryonic myoblasts and satellite cells, it is possible that satellite cells represent yet another myogenic cell lineage. Clearly, factors that differentially affect the proliferation of various myogenic cell lineages could have significant impact on the rate and extent of postnatal muscle growth.

As can be seen from the preceding discussion, the process of myogenesis is extremely complex and is not fully understood. Factors that affect proliferation, fusion, fiber number, development of myogenic cell lineages, or development of specific fiber types could affect muscle mass or efficiency of muscle growth. Neither the identity nor the mechanism of action of factors regulating these processes is known. However, several factors capable of influencing proliferation and differentiation of myogenic cells in culture have been identified. The remainder of this chapter will be devoted to a discussion of these factors and their effect on myogenic cells.

4. Factors Affecting Myogenesis

4.1. Insulinlike Growth Factors (Somatomedins)

Insulinlike growth factors (IGFs) or somatomedins are a class of polypeptides, originally isolated from sera, that have a molecular weight of approximately 7500. At concentrations of 10^{-9} to 10^{-10} M, these polypeptides are mitogenic for a variety of cultured cells. IGFs share considerable sequence homology with insulin and possess insulinlike properties *in vitro;* however, they do not cross-react with insulin antibodies (Zapf *et al.*, 1978). Because these factors were originally isolated from different species by different laboratories, a complex nomenclature has developed around them. To date, two classes of IGFs have been characterized: IGF-I (Rinderknecht and Humbel, 1978a), also referred to as basic somatomedin (pI 8.2–8.4) (Bala and Bhaumick, 1979) or somatomedin C (SmC) (Svoboda *et al.*, 1980); and IGF-II (Rinderknecht and Humbel, 1978b) or neutral somatomedin. Additionally, another family of peptides collectively designated *multiplication-stimulating activity* (MSA) has been isolated from media conditioned by a Buffalo rat liver cell line (BRL 3A) (Moses *et al.*, 1980). MSA III appears to be the rat form of IGF-II, since the primary structure of MSA III shows 93% identity with that of human IGF-II (Marquardt *et al.*, 1981). Circulating levels of IGF-I/SmC and to a lesser extent IGF-II appear to be regulated by somatotropin (see Steele and Elsasser, this volume).

IGF-I/SmC has been shown to stimulate proliferation (Ballard *et al.*, 1986; Hill *et al.*, 1986; Ewton *et al.*, 1987), amino acid uptake (Hill *et al.*, 1986; Ewton *et al.*, 1987), and differentiation (Ewton and Florini, 1981; Schmid *et al.*, 1983; Ewton *et al.*, 1987) in cultured myogenic cells. Additionally, IGF-I has been shown to stimulate protein synthesis in L6 myoblasts (Ballard *et al.*, 1986) and ovine myotubes (Harper *et al.*, 1987) and to decrease protein degradation in L6 myotubes (Ewton *et al.*, 1987) and ovine myotubes (Harper *et al.*, 1987). The mechanism by which IGF-I affects these critical cellular processes is not known. However, Ong *et al.* (1987) have recently used Northern analysis and cytoplasmic RNA blot hybridization to show that IGF-I (6.25 ng/ml) can cause induction of the c-fos oncogene in quiescent L6 muscle cells. Peak c-fos induction occurred after L6 cells had been exposed to IGF-I for 30 min and levels remained elevated for 2 hr. This observation raises the possibility that IGF-I action on differentiating myogenic cells is in part mediated by the c-fos oncogene, which previously was hypothesized to play a crucial role in cell differentiation and proliferation. IGF-II/MSA has been shown to stimulate proliferation (Ewton and Florini, 1980, 1981; Florini and Ewton, 1981; Florini *et al.*, 1984; Beguinot *et al.*, 1985; Ballard *et al.*, 1986; Hill *et al.*, 1986; Ewton *et al.*, 1987), differentiation (Ewton and Florini, 1981; Florini and Ewton, 1981; Florini *et al.*, 1984; Ewton *et al.*, 1987), amino acid transport (Merrill *et al.*, 1977; Janeczko and Etlinger, 1984; Hill *et al.*, 1986; Ewton *et al.*, 1987), and rate of protein synthesis in cultured myogenic cells (Janeczko and Etlinger, 1984; Ballard *et al.*, 1986). Additionally, MSA has been shown to decrease the rate of protein degradation in cultured myotubes (Janeczko and Etlinger, 1984; Ballard *et al.*, 1986; Ewton *et al.*, 1987). Ovine somatomedin and rat MSA/IGF-II have also been reported to stimulate proliferation of cultured satellite cells (Dodson *et al.*, 1985).

Binding studies utilizing ^{125}I-labeled IGF-I, IGF-II, and insulin have shown that most cells possess a minimum of three receptor sites which bind one or more of these peptides. Later affinity-labeling studies revealed the existence of two general classes of membrane receptors that bound insulin and/or the IGFs. The receptors for insulin and for IGF-I have been designated *type I insulin* or *type I IGF receptors*, respectively, because of their remarkably similar structure (reviewed by Czech, 1985, 1986; Rechler and Nissley, 1985; Goldfine, 1987). Both receptors consist of a disulfide-linked heterotetrameric subunit configuration containing two α (125 kDa) and two β (90 kDa) subunits. Both the insulin (Ebina *et al.*, 1985; Ullrich *et al.*, 1985) and the type I IGF receptor (Ullrich *et al.*, 1986) have been cloned and this should help to elucidate structural differences between the two receptors. Both receptors exhibit tyrosine kinase activity, which results in the autophosphorylation of the tyrosine residues in the β subunits (Kasuga *et al.*, 1983; Jacobs *et al.*, 1983). This intrinsic tyrosine kinase activity has also been shown to catalyze phosphorylation of exogenous substrates (Rechler and Nissley, 1985). Similar tyrosine kinase activities are associated with receptors for epidermal growth factor (Ushiro and Cohen, 1980) and platelet-derived growth factor (Elk *et al.*, 1982). Additionally, the src family of viral oncogenes encode enzymes that catalyze tyrosine phosphorylation (Hunter and Cooper, 1985). Consequently, although the specific function of the tyrosine kinase activity of the type I receptors is not known, it appears likely that this activity plays an important role in the functionality of the receptor. Both the type I insulin and type I IGF receptors bind insulin, IGF-I, and IGF-II but with varying affinities. The type I insulin receptor binds insulin with greatest affinity followed by IGF-II and then IGF-I. The type I IGF receptor binds IGF-I with greatest affinity followed by IGF-II and then insulin (Czech, 1986). A third class of receptor, the type II IGF receptor, exhibits highest affinity for IGF-II. This receptor is a 250-kDa glycoprotein that has no disulfide-linked subunits. The type II IGF receptor has no intrinsic tyrosine kinase activity. The receptor binds IGF-I but has no detectable affinity for insulin (Czech, 1986). Current data suggest that the type I insulin and IGF receptors mediate rapid biological effects such as activation of nutrient transport and stimulation of cell proliferation *in vitro* (De Vroede *et al.*, 1984; Mottola and Czech, 1984; Yu and Czech, 1984; Ballard *et al.*, 1986; Ewton *et al.*, 1987). The function of the type II IGF receptor is unknown.

Both type I and type II IGF receptors have been identified on the surface of cultured muscle cells (Yu and Czech, 1984; Beguinot *et al.*, 1985; Ballard *et al.*, 1986; Ewton *et al.*, 1987) and current research is aimed at elucidating the biological role of these receptors. Ewton *et al.*, (1987) examined the effects of human IGF-I, human IGF-II, rat IGF-II, and insulin on proliferation, amino acid uptake, protein degradation, and differentiation in L6 myoblasts. Their results showed that in the L6 cell line, all four parameters examined were significantly more sensitive to IGF-I than to IGF-II or insulin. In this study, the activities of the recombinantly produced Thr[59] analogue of IGF-I and highly purified human IGF-I were compared. Both exhibited equal potencies for affecting proliferation, differentiation, protein degradation, and amino acid uptake. These results suggest that human IGF-I and the Thr[59] analogue have equal affinity for the receptor or receptors responsible for their effect on these parameters. Therefore, it is significant that affinity cross-linking of ^{125}I-labeled human IGF-I and Thr[59]-IGF-I to L6 membrane proteins revealed that significant amounts of human IGF-I bound to the type II IGF receptor whereas very little of the Thr[59]-IGF-I bound to this receptor. The fact that the biological potency of the Thr[59]-IGF-I analogue is equal to that of human IGF-I even though the analogue has little if any affinity for the type II IGF receptor suggests that the type II IGF receptor does not regulate the biological processes measured in this study. Similar conclusions have been reached independently in other studies utilizing muscle cells (Yu and Czech, 1984; Ballard *et al.*, 1986) as well as other cell types (Mottola and Czech, 1984). These results are in apparent conflict with those of Beguinot *et al.* (1985) who reported that both IGF-I and IGF-II stimulate 2-deoxy-D-glucose and α-aminoisobutyric acid uptake in L6 myoblasts by binding to their own receptors. The relationship between IGF-I, IGF-II, and their receptors is further complicated by the fact that circulating levels of IGF-I are extremely low during fetal development whereas levels of IGF-II/MSA are high. This has led to speculation that IGF-II/MSA is the biologically active somatomedin in the fetus and that a "switch over" to IGF-I/ SmC occurs postnatally (Adams *et al.*, 1983). In summary, it appears that binding of IGF-I or IGF-II (at higher concentrations) to the type I IGF receptor is correlated with currently recognized effects of the IGFs on proliferation, differentiation, and protein turnover. As is the case with many other cell types, however, the preponderance of recent studies have raised doubts as to whether the type II IGF receptor in muscle cells plays any direct role in regulating these processes. Thus, although it is very likely that IGFs play a role in myogenesis, neither the specific function of individual IGFs nor their mechanism of action is known with certainty at present.

Traditionally, the liver has been considered to be the source of circulating IGFs that act in an endocrine manner to influence growth and differentiation of other tissues such as muscle and bone (Steele and Elsasser, this volume). However, several recent studies have shown that numerous fetal and adult tissues contain and/or synthesize IGF (D'Ercole *et al.*, 1984, 1986; Hill *et al.*, 1985a,b; Clemmons and Shaw, 1986; Han *et al.*, 1987a,b; Jennische *et al.*, 1987; Jennische and Hansson, 1987). Perhaps the most convincing evidence that IGF is synthesized in nonhepatic tissues is a recent report by Han *et al.* (1987a) who utilized *in situ* hybridization histochemistry to investigate the cellular localization of IGF mRNA in human fetal tissues. These workers reported that IGF-I and IGF-II RNAs were localized to connective tissues or cells of mesenchymal origin in 14 organs and tissues. It thus appears that in many tissues, including muscle, locally produced IGFs may function via autocrine and/or paracrine mechanisms. However, the relative importance of autocrine, paracrine, and endocrine mechanisms of IGF action is unknown.

A key unanswered question concerning the potential autocrine/paracrine effects of IGFs on myogenesis is whether myoblasts and myotubes are able to produce IGFs. Studies suggesting that IGFs are synthesized in myogenic cells include the following: (1) Immunocytochemical studies have shown that immunoreactive IGFs can be localized in fetal muscle fibers (Han *et al.*, 1987b) and in satellite cells and muscle fibers of regenerating muscle (Jennische *et al.*, 1987; Jennische and Hansson, 1987). Although these findings may indicate that IGFs are synthesized in myogenic cells, it is also possible that IGFs from nonmyogenic cells are bound and internalized by muscle cells (Han *et al.*, 1987b). Therefore, the presence of immunoreactive IGFs in myogenic cells does not prove that these cells synthesize IGFs. (2) Cultured embryonic muscle cells have been reported to synthesize and secrete approximately equal amounts of IGF-I/SmC and IGF-II/MSA (Hill *et al.*, 1985a,b).

In contrast to these studies indicating that IGFs are synthesized in myogenic cells, *in situ* hybridization techniques have shown that IGF-I and IGF-II mRNA is not detectable in myogenic cells but, rather, is localized in connective tissue and fibroblasts in skeletal muscle from human fetuses (16–20 weeks of gestation) (Han *et al.*, 1987a). Although these studies indicate that myogenic cells do not produce IGFs, it is possible that IGF mRNA levels in myogenic cells were below the detection threshold of the *in situ* hybridization procedures (Han *et al.*, 1987a). However, localization of IGF mRNA in fibroblasts is consistent with numerous reports that cultured fibroblasts secrete IGFs and Han *et al.* (1987a,b) have suggested that fibroblasts may synthesize and secrete IGFs that act on other cells via paracrine mechanisms. In summary, although it appears that IGFs are synthesized in muscle tissue, there is no agreement as to whether myogenic cells are able to synthesize them.

4.2. Insulin

The importance of insulin in regulating general cell metabolism has been recognized for many years; however, the mechanism by which this regulation is accomplished is still not well understood. Additionally, the role of insulin in regulating myogenesis (reviewed by Florini, 1985) is not clear. Several lines of evidence suggest that insulin may have an anabolic effect on postnatal muscle tissue. Studies of a variety of animal models have demonstrated that wasting of skeletal muscle is a prominent feature of diabetes mellitus and that this wasting is reversed by administration of insulin to affected animals (Pain and Garlick, 1974; Flaim *et al.*, 1980). Additionally, ribosomes isolated from muscle of diabetic rats are less active in *in vitro* protein synthesis systems than are ribosomes from nondiabetic controls. *In vitro* studies with isolated muscles (Manchester and Young, 1970; Fulks *et al.*, 1975; Frayan and Maycock, 1979) and the perfused rat hemicorpus (Jefferson *et al.*, 1977) have shown that insulin increases the rate of protein synthesis and decreases the rate of protein degradation in these systems.

In cultured satellite cells and embryonic muscle cells as well as fibroblasts and fibroblastic cell lines, supraphysiological concentrations of insulin, 1 μg/ml or higher, are required to elicit a maximal response. In muscle cell cultures, these high concentrations stimulate both proliferation and differentiation of myogenic cells (Merrill *et al.*, 1977; Ball and Sanwall, 1980; Ewton and Florini, 1981). Additionally, insulin at high concentrations (10^{-6} M) is a component of synthetic media used to support growth and differentiation of myogenic cells in culture (Florini and Roberts, 1979; Dollenmeier *et al.*, 1981). It has been proposed that the stimulation of growth of fibroblasts by insulin is mediated by its weak binding to receptors for IGFs. At high concentrations it is thought that insulin may bind to the type I IGF receptor and in so doing affect cell growth in a manner similar to that observed for much lower concentrations of IGF-I. This hypothesis is based on work by King *et al.* (1980) who showed that blockade of the high-affinity insulin receptor with antireceptor Fab fragments blocked high-affinity insulin binding but did not prevent insulin-induced stimulation of DNA synthesis in cultured fibroblasts. Furthermore, these investigators showed that anti-insulin-receptor IgG, which triggers a number of acute insulinlike metabolic effects, did not stimulate DNA synthesis. They concluded that the growth-promoting effects of insulin on human fibroblasts were due to binding of insulin to the type I IGF receptor. Although similar studies have not been done on cultured muscle cells, it would seem quite likely that the well-documented effects of supraphysiological concentrations of insulin on proliferation and differentiation of cultured muscle cells are the result of this spillover action of insulin through IGF-I receptors. Even though physiological levels of insulin do not appear to affect proliferation or differentiation of cultured myogenic cells, insulin may facilitate muscle cell proliferation and/or differentiation by maintaining cells in a metabolic state that allows them to respond to other hormones and growth factors that directly affect these processes.

4.3. Transforming Growth Factor-β (TGF-β) (Differentiation Inhibitor)

Three forms of TGF-β have been identified (reviewed by Massague, 1987). TGF-β1, a 25-kDa homodimer, has recently (Florini *et al.*, 1986; Massague *et al.*, 1986) been shown to inhibit differentiation of L6 myoblasts. At 60 pg/ml, TGF-β inhibited fusion and elevation of creatine kinase activity in cultured myoblasts (Florini *et al.*, 1986). However, maximal inhibition of differentiation was detected at or above 0.5 ng/ml. The inhibitory effects of TGF-β could be reversed by removing it from the culture media. Recently, TGF-β has also been shown to inhibit proliferation and to suppress differentiation in cultured rat satellite cells (R. E. Allen, personal communication). Florini and co-workers (Evinger-Hodges *et al.*, 1982; Florini *et al.*, 1984) have also reported that Coon's Buffalo rat liver (BRL) cells secrete a protein that is a potent inhibitor of skeletal myoblast differentiation *in vitro*. This "differentiation inhibitor" (DI) exhibits no detectable mitogenic activity; however, in skeletal myoblast cultures it has been shown to reversibly block fusion, elevation of creatine kinase, and increased binding of α-bungarotoxin. This inhibitor has also been isolated from sera of embryonic origin. Florini *et al.* (1986) have shown that the biological activities, physical properties, and antigenic properties of TGF-β and DI are very similar. For example, TGF-β and DI cofractionate in gel permeation, reverse phase, and ion exchange chromatography. Both are stable under acid conditions and are sensitive to proteases and to sulfhydryl reagents. Additionally, antibodies to human TGF-β detect a single band in partially purified DI preparations and this band comigrates with highly purified TGF-β. Based on these observations, Florini *et al.* (1986) concluded that the principal differentiation inhibitor in their partially purified DI preparations is TGF-β or a very closely related molecule.

Many different types of cells synthesize TGF-β and have receptors for it. Moreover, many studies have shown that TGF-β can have profound effects on proliferation, differentiation, and function in numerous cell types. Consequently, it is possible that TGF-β acts as an autocrine or paracrine regulator of these processes. For example, TGF-β might prevent differentiation of proliferating myoblasts to fusion-competent, nonproliferating myoblasts during the initial stages of embryonic muscle development. This could be instrumental in maintaining the extensive myogenic cell proliferation presumably required to provide the basis for later formation of muscle tissue. In this regard, it is significant that there have been numerous reports of TGF-β-like activities in very early development. Although more study is needed to establish the role of TGF-β in myogenesis, this molecule appears to offer promise as a factor regulating this process.

4.4. Transferrin

Transferrin is an iron-binding glycoprotein that is present in serum (Ozawa and Kokama, 1978) and embryo extract (Ii *et al.*, 1981, 1982). Additionally, transferrinlike molecules have been isolated from both nerve and muscle extracts (Matsuda *et al.*, 1984a,b). In muscle cell cultures, iron-saturated transferrin stimulates both proliferation and differentiation and is essential for maintenance of healthy myotubes (Ozawa and Hagiwara, 1982). Therefore, this protein may be a paracrine effector of myogenesis. The effect of transferrin on muscle growth in culture is absolutely dependent on the presence of iron and appears to be class specific (i.e., mammalian transferrins do not affect avian myoblasts nor do avian transferrins affect mammalian myoblasts) (Shimo-Oka *et al.*, 1986).

4.5. Fibroblast Growth Factor (FGF)

Two forms of FGF (acidic and basic) have been isolated and purified to homogeneity. These molecules are closely related and possess similar biological activities in a wide range of cell types (reviewed by Gospodarowicz *et al.*, 1987). Comparison of the amino acid sequence of acidic and basic FGF shows that they are distinct peptides with 55% sequence identity (Gimenez-Gallego *et al.*, 1985; Esch *et al.*, 1985). Partially purified FGF has been shown to stimulate proliferation and inhibit differentiation in bovine myoblasts (Gospodaro-wicz et al., 1976) and a mouse myoblast cell line (Linkhart et al., 1981). Additionally, FGF has been shown to be a potent mitogen for cultured satellite cells (Allen et al., 1984) and to antagonize differentiation of the nonfusing BC$_3$H1 muscle cell line (Lathrop et al., 1985a,b; Spizz et al., 1986; Wice et al., 1987). Studies utilizing highly purified FGFs have shown that both acidic and basic FGFs stimulate proliferation and inhibit differentiation of MM14 myoblasts with half-maximal effects on proliferation occurring at 30 and at 1 pM, respectively (Olwin and Hauschka, 1986). Affinity-labeling studies and kinetic analysis of binding of ^{125}I-labeled acidic and basic FGF to MM14 myoblasts have shown that these cells possess a single FGF receptor with a molecular weight of 165,000 (Olwin and Hauschka, 1986). This receptor binds both acidic and basic FGF.

FGFs are not found in the general circulation; however, basic FGF has been isolated from a number of tissues and/or cultured cells including pituitary, brain, placenta, corpus luteum, macrophages, and cartilage (Gospodarowicz et al., 1987). Additionally, an FGF-like myogenic factor has recently been isolated and purified from chicken skeletal muscle tissue (Kardami et al., 1985a,b). These observations, viewed in light of the effects of FGFs on myogenic cells and the presence of FGF receptors on myoblasts, suggest that FGF may function as a paracrine regulator of myogenesis. However, the mechanism by which FGF functions is not known.

4.6. Somatotropin

The effect of somatotropin deficiency on muscle growth has been well established for many years. Additionally, long-term administration of somatotropin to pituitary-intact animals has been reported to increase muscling, decrease fat content, and improve feed efficiency in swine (Machlin, 1972; Chung et al., 1985); to increase nitrogen retention in steers (Moseley et al., 1982) and sheep (Davis et al., 1969); and to improve milk production in dairy cattle (Peel et al., 1981). However, it appears unlikely that somatotropin has a direct effect on proliferation and protein turnover in myogenic cells. Although there is an increased incorporation of [^{3}H]thymidine into DNA in muscle from somatotropin-treated hypophysectomized rats as compared to untreated controls (Breuer, 1969), these increases may reflect an effect of somatotropin on proliferation of nonmuscle cells or an indirect effect of somatotropin on muscle satellite cell proliferation. It has also been reported that 10^{-8}M somatotropin stimulates amino acid uptake (Kostyo and Engel, 1960; Albertsson-Wikland and Isaksson, 1976) in in vitro incubations of rat diaphragm muscle. However, recent observations that many types of cells, including fibroblasts, can secrete somatomedin raise the possibility that responses seen in the intact diaphragm are the result of locally produced somatomedins. Moreover, several independent studies have shown that somatotropin, even at concentrations obtainable in somatotropin-injected animals, has no detectable effect on proliferation, differentiation, or rate of protein turnover in cultured satellite cells, myotubes, or embryonic myoblasts (Florini et al., 1977; Merrill et al., 1977; Ewton and Florini, 1980; Allen et al., 1983; Harper et al., 1987; Kotts et al., 1987). Therefore, it appears unlikely that somatotropin plays any direct role in regulating myogenesis or postnatal muscle growth. In apparent contrast to these findings, however, daily intramuscular somatotropin injection has been shown to significantly increase efficiency and extent of muscle deposition in swine. These apparently conflicting observations may be reconciled by the fact that treatment of pigs with exogenous somatotropin causes a significant increase in circulating IGF-I, which coincides with an increased ability of their sera to stimulate proliferation (Kotts et al., 1987), increase protein synthesis rate, and decrease protein degradation rate in cultured myogenic cells (W. R. Dayton, unpublished data). Consequently, it is likely that the stimulation of muscle growth in somatotropin-injected pigs is the result of somatotropin-induced increases in IGF-I levels. Additionally, rats bearing somatotropin-secreting tumors have been shown to have elevated levels of circulating somatomedin (Baxter et al., 1982) as well as increased numbers of muscle fiber nuclei and satellite cells as compared to control animals (McCusker and Campion, 1986). Consequently, it is generally believed that much if not all of the effect of somatotropin on muscle growth is mediated through somatotropin-dependent plasma factors—somatomedins—that are produced in response to somatotropin.

4.7. Glucocorticoids

The role of glucocorticoids in regulating myogenesis has not been studied extensively. Perhaps the most compelling evidence that glucocorticoids are involved in myogenesis is the fact that they are an essential component of synthetic media that support proliferation and differentiation of myogenic cells (Florini and Roberts, 1979; Dollenmeier et al., 1981). Glucocorticoids have been reported to stimulate proliferation and differentiation of primary rat myoblasts and to stimulate glucose transport in the L6 rat myoblast cell line (reviewed by Florini, 1985). Additionally, the synthetic glucocorticoid dexamethasone has been reported to enhance the mitogenic effects of ovine somatomedin and rat IGF-II/MSA on cultured satellite cells (Dodson et al., 1985). When added to cultures at pharmacological levels (10^{-4}M), glucocorticoids (dexamethasone) are toxic to myoblasts (Furcht et al., 1977). The physiological significance of these observations is not clear and more research is necessary to establish the role of the glucocorticoids in muscle differentiation.

4.8. Thyroid Hormone

A growing body of evidence suggests that thyroid hormone (triiodothyronine) may play a role in differentiation of myogenic cells. Most of this evidence is based on the ability of thyroid hormone to alter the myosin isoform composition of muscle fibers. As already discussed briefly, different isoforms of myosin heavy chain are encoded by members of a multigene family. Expression of specific myosin heavy chain genes during embryonic development is regulated in a tissue- and developmental-stage specific manner. In addition, environmental stimuli such as innervation patterns and various hormones have been shown to alter myosin heavy chain expression. During the differentiation of skeletal and cardiac muscle, the normal transition of embryonic to adult myosin heavy chain isozymes is accelerated by hyperthyroidism and inhibited by hypothyroidism (Gustafson *et al.*, 1986; Izumo *et al.*, 1986). Additionally, numerous studies have shown that hyper- or hypothyroid conditions in postnatal animals result in alterations in myosin heavy chain isozymes present in specific muscles (Gustafson *et al.*, 1986; Izumo *et al.*, 1986). Therefore, it is possible that thyroid hormone plays a role in the myosin isozyme transitions that occur during myogenesis.

5. Summary

Myogenesis is an extremely complex process that significantly affects the extent and efficiency of muscle growth. Most of our information about factors that might affect myogenesis has been obtained through experiments utilizing cultures of primary myoblasts or established muscle cell lines. Although several factors affect myogenesis *in vitro*, very little is known about their effect *in vivo*. Moreover, almost nothing is known with certainty about the mechanisms by which these factors influence muscle differentiation. Consequently, much more research is necessary to answer these important questions.

References

Adams, S. O., S. P. Nissley, S. Handwerger and M. M. Rechler. 1983. Developmental patterns of insulin-like growth factor-I and -II synthesis and regulation in rat fibroblasts. Nature 302:150.

Albertsson-Wikland, K. and O. Isaksson. 1976. Development of responsiveness of young normal rats to growth hormone. Metabolism 25:747.

Allen, R. E., M. V. Dodson and L. S. Luiten. 1984. Regulation of skeletal muscle satellite cell proliferation by bovine pituitary fibroblast growth factor. Exp. Cell Res. 152:154.

Allen, R. E., P. K. McAllister and K. C. Masak. 1980. Myogenic potential of satellite cells in skeletal muscle of old rats. Mech. Ageing Dev. 13:105.

Allen, R. E., P. K. McAllister, K. C. Masak and G. R. Anderson. 1982. Influence of age on accumulation of α-actin in satellite cell-derived myotubes in vitro. Mech. Ageing Dev. 18:89.

Allen, R., K. C. Masak, P. K. McAllister and R. A. Merkel. 1983. Effects of growth hormone, testosterone and serum concentration on actin synthesis in cultured satellite cells. J. Anim. Sci. 56:833.

Bala, R. M. and B. Bhaumick. 1979. Radioimmunoassay of a basic somatomedin: comparison of various assay techniques and somatomedin levels in various sera. J. Clin. Endocrinol. Metab. 49:770.

Ball, E. H. and B. D. Sanwall. 1980. A synergistic effect of glucocorticoids and insulin on the differentiation of myoblasts. J. Cell. Physiol. 102:27.

Ballard, F. J., L. C. Read, G. L. Francis, C. J. Bagley and J. C. Wallace. 1986. Binding properties and biological potencies of insulin-like growth factors in L6 myoblasts. Biochem. J. 233:223.

Baxter, R. C., Z. Zaltsman and J. R. Turtle. 1982. Induction of somatogenic receptors in livers of hypersomatotropic rats. Endocrinology 111:1020.

Beguinot, F., C. R. Kahn, A. C. Moses and R. J. Smith. 1985. Distinct biologically active receptors for insulin, insulin-like growth factor I, and insulin-like growth factor II in cultured skeletal muscle cells. J. Biol. Chem. 260:15892.

Bischoff, R. 1974. Enzymatic liberation of myogenic cells from adult rat muscle. Anat. Rec. 180:645.

Breuer, C. B. 1969. Stimulation of DNA synthesis in cartilage of hypophysectomized rats by native and modified placental lactogen and anabolic hormones. Endocrinology 85:989.

Buckley, P. A. and I. R. Konigsberg. 1974. Myogenic fusion and the duration of the post-mitotic gap (G_1). Dev. Biol. 37:193.

Buckley, P. A. and I. R. Konigsberg. 1977. Do myoblasts in vivo withdraw from the cell cycle? A reexamination. Proc. Natl. Acad. Sci. USA 74:2031.

Campion, D. E., R. W. Purchas, R. A. Merkel and D. R. Romsos. 1984. Genetic obesity and the muscle satellite cell. Proc. Soc. Exp. Biol. Med. 176:143.

Caplan, A. I., M. Y. Fiszman and H. M. Eppenberger. 1983. Molecular and cell isoforms during development. Science 221:921.

Chung, C. S., Etherton, T. D. and Wiggins, J. P. 1985. Stimulation of swine growth of porcine growth hormone. J. Anim. Sci. 60:118.

Clemmons, D. R. and D. S. Shaw. 1986. Purification and biologic properties of fibroblast somatomedin. J. Biol. Chem. 261:10293.

Cossu, G., P. Cicinelli, C. Fieri, M. Coletta and M. Molinaro. 1985. Emergence of TPA-resistant satellite cells during muscle histogenesis of human limb. Exp. Cell Res. 160:403.

Cossu, G., F. Eusebi, F. Grassi and E. Wanke. 1987. Acetylcholine receptor channels are present in undifferentiated satellite cells but not in embryonic myoblasts in culture. Dev. Biol. 123:43.

Cossu, G., M. Molinaro and M. Pacifici. 1983. Differential response of satellite cells and embryonic myoblasts to a tumor promoter. Dev. Biol. 98:520.

Cossu, G., M. I. Senni, F. Eusebi, D. Giacomoni and M. Molinaro. 1986. Effect of phorbol esters and liposome-delivered phospholipids on the differentiation program of normal and dystrophic satellite cells. Dev. Biol. 118:182.

Cossu, G., B. Zani, M. Coletta, M. Bouche, M. Pacifici and M. Molinaro. 1980. In vitro differentiation of satellite cells isolated from normal and dystrophic mammalian muscle. A comparison with embryonic myogenic cells. Cell Differ. 9:357.

Czech, M. P. 1985. The nature and regulation of the insulin receptor structure and function. Annu. Rev. Physiol. 47:357.

Czech, M. P. 1986. Structure and functions of the receptors for insulin and the insulin-like growth factors. J. Anim. Sci. 63 (Suppl. 2):27.

Davis, S. L., U. S. Garrigus and F. C. Hinds. 1969. Metabolic effects of growth hormone and diethylstilbestrol in lambs. II. Effects of daily ovine growth hormone injections on plasma metabolites and nitrogen-retention in fed lambs. J. Anim. Sci. 30:236.

D'Ercole, A. J., D. J. Hill, A. J. Strain and L. E. Underwood. 1986. Tissue and plasma somatomedin-C/insulin-like growth factor I (SmC/IGF I) concentrations in the human fetus during the first half of gestation. Pediatr. Res. 20:253.

D'Ercole, A. J., A. D. Stiles and L. E. Underwood. 1984. Tissue concentrations of somatomedin C: further evidence for multiple sites of synthesis and paracrine or autocrine mechanisms of action. Proc. Natl. Acad. Sci. USA 81:935.

De Vroede, M. A., J. A. Romanus, M. L. Standaert, R. J. Pollett, S. P. Nissley and M. M. Rechler. 1984. Interaction of insulin-like growth factors with a nonfusing mouse muscle cell line: binding, action, and receptor down-regulation. Endocrinology 114:1917.

Dodson, M. V., R. E. Allen and K. L. Hossner. 1985. Ovine somatomedin, multiplication-stimulating activity, and insulin promote skeletal muscle satellite cell proliferation *in vitro*. Endocrinology 117:2357.

Dollenmeier, P., D. C. Turner and H. M. Eppenberger. 1981. Proliferation of chick skeletal muscle cells cultured in a chemically defined medium. Exp. Cell Res. 135:47.

Ebina, Y., L. Ellis, K. Jarnagin, M. Edery, L. Graf, E. Clauser, J. H. Ou, F. Maslarz, Y. W. Kan, I. D. Goldfine, R. A. Roth and W. J. Rutler. 1985. The human insulin receptor cDNA: the structural basis for hormone-activated transmembrane signaling. Cell 40:747.

Elk, B., B. Westermark, A. Wasteson and C.-H. Heldin. 1982. Stimulation of tyrosine-specific phosphorylation by platelet-derived growth factor. Nature 295:419.

Esch, F., A. Baird, N. Ling, N. Veno, F. Hill, L. Denoroy, R. Klepper, D. Gospodarowicz, P. Bohlen and R. Guillemin. 1985. Primary structure of bovine pituitary fibroblast growth factor (FGF) and comparison with the amino-terminal sequence of bovine brain acidic FGF. Proc. Natl. Acad. Sci. USA 82:6507.

Evinger-Hodges, M. J., D. Ewton, Z. S. C. Seifert and J. R. Florini. 1982. Inhibition of myoblast differentiation *in vitro* by a protein isolated from liver cell medium. J. Cell Biol. 93:395.

Ewton, D. Z., S. L. Falen and J. R. Florini. 1987. The type II insulinlike growth factor (IGF) receptor has low affinity for IGF-I analogs: pleiotypic actions of IGFs on myoblasts are apparently mediated by the type I receptor. Endocrinology 120:115.

Ewton, D. Z. and J. R. Florini. 1980. Relative effects of the somatomedins, MSA and growth hormone on myoblasts and myotubes in culture. Endocrinology 106:577.

Ewton, D. Z. and J. R. Florini. 1981. Effects of the somatomedins and insulin on myoblast differentiation *in vitro*. Dev. Biol. 86:31.

Flaim, K. E., M. E. Copenhaver and L. S. Jefferson. 1980. Effects of diabetes on protein synthesis in fast- and slow-twitch rat skeletal muscle. Amer. J. Physiol. 239:E88.

Florini, J. R. 1985. Hormonal control of muscle cell growth. J. Anim. Sci. 61 (Suppl. 2):21.

Florini, J. R. and D. Z. Ewton. 1981. Insulin acts as a somatomedin analog in stimulating myoblast growth in serum-free medium. In Vitro 17:763.

Florini, J. R., D. Z. Ewton, M. J. Evinger-Hodges, S. L. Fallen, S. L. Lau, J. F. Ragan and B. M. Vertel. 1984. Stimulation and inhibition of myoblast differentiation by hormones. In Vitro 20:942.

Florini, J. R., M. L. Nicholson and N. C. Dulak. 1977. Effects of peptide anabolic hormones on growth of myoblasts in culture. Endocrinology 101:32.

Florini, J. R. and S. B. Roberts. 1979. A serum-free medium for the growth of muscle cells in culture. In Vitro 15:983.

Florini, J. R., A. B. Roberts, D. Z. Ewton, S. L. Falen, K. C. Flanders, and M. B. Sporn. 1986. Transforming growth factor-β: a very potent inhibitor of myoblast differentiation, identical to differentiation inhibitor secreted by buffalo rat liver cells. J. Biol. Chem. 261:16509.

Frayan, K. N. and P. F. Maycock. 1979. Regulation of protein metabolism by a physiological concentration of insulin in mouse soleus and extensor digitorum longus muscles. Biochem. J. 184:323.

Fulks, R. M., J. B. Li and A. L. Goldberg. 1975. Effects of insulin, glucose and amino acids on protein turnover in rat diaphragm. J. Biol. Chem. 250:290.

Furcht, L. T., G. Wendelschafer-Crabb and P. A. Woolbridge. 1977. Cell surface changes accompanying myoblast differentiation. J. Supramol. Struct. 7:307.

Gauthier, G. F., S. Lowey, P. A. Benfield and A. W. Hobbs. 1982. Distribution and properties of myosin isozymes in developing avian and mammalian skeletal muscle fibers. J. Cell Biol. 92:471.

Gimenez-Gallego, G., J. Rodkey, C. Bennett, M. Rios-Candelore, J. DiSulvo and K. Thomas. 1985. Brain-derived acidic fibroblast growth factor: complete amino acid sequence and homologies. Science 230:1385.

Goldfine, I. D. 1987. The insulin receptor: molecular biology and transmembrane signaling. Endocrine Rev. 8:235.

Gospodarowicz, D., N. Ferrara, L. Schweigerer and G. Neufeld. 1987. Structural characterization and biological functions of fibroblast growth factor. Endocrine Rev. 8:95.

Gospodarowicz, D., J. Weseman, J. S. Moran and J. Lindstrom. 1976. Effect of fibroblast growth factor on the division and fusion of bovine myoblasts. J. Cell Biol. 70:395.

Gustafson, T. A., B. E. Markham and E. Morkin. 1986. Effects of thyroid hormone on α-actin and myosin heavy chain gene expression in cardiac and skeletal muscles of the rat: measurement of mRNA content using synthetic oligonucleotide probes. Circ. Res. 59:194.

Han, V. K. M., A. J. D'Ercole and P. K. Lund. 1987a. Cellular localization of somatomedin (insulin-like growth factor) messenger RNA in the human fetus. Science 236:193.

Han, V. K. M., D. J. Hill, A. J. Strain, A. C. Towle, J. M. Lauder, L. E. Underwood and A. J. D'Ercole. 1987b. Identification of somatomedin/insulin-like growth factor immunoreactive cells in the human fetus. Pediatr. Res. 22:245.

Harbison, S. A., D. E. Goll, F. C. Parrish, V. Wang and E. A. Kline. 1976. Muscle growth in two genetically different lines of swine. Growth 40:253.

Harper, J. M. M., J. B. Soar and P. D. Buttery. 1987. Changes in protein metabolism of ovine primary muscle cultures on treatment with growth hormone, insulin, insulin-like growth factor I or epidermal growth factor. J. Endocrinol. 112:87.

Hill, D. J., C. J. Crace and R. D. G. Milner. 1985a. Incorporation of [^{3}H] thymidine by isolated human fetal myoblasts and fibroblasts in response to human placental lactogen (HPL): possible mediation of HPL action by release of immunoreactive somatomedin-C. J. Cell. Physiol. 125:337.

Hill, D. J., C. J. Crace, S. P. Nissley, D. Morrell, A. T. Holder and R. D. G. Milner. 1985b. Fetal rat myoblasts release both rat somatomedin-C (SM-C)/insulin-like growth factor I (IGF I) and multiplication-stimulating activity in vitro: partial characterization and biological activity of myoblast-derived SM-C/IGF I. Endocrinology 117:2061.

Hill, D. J., C. J. Crace, A. J. Strain and R. D. G. Milner. 1986. Regulation of amino acid uptake and deoxyribonucleic acid synthesis in isolated human fetal fibroblasts and myoblasts: effect of human placental lactogen, somatomedin-C, multiplication stimulating activity, and insulin. J. Clin. Endocrinol. Metab. 62:753.

Holtzer, H. 1978. Cell lineages, stem cells, and the "quantal" cell cycle concept. In: B. I. Lord, C. S. Potten and R. D. Cole (Ed.) Stem Cells and Tissue Homeostasis. pp 1–28. Cambridge Univ. Press, London.

Hunter, T. and J. A. Cooper. 1985. Protein-tyrosine kinases. Annu. Rev. Biochem. 54:897.

Ii, I., L. Kimura, T. Hasegawa and E. Ozawa. 1981. Transferrin is an essential component of chick embryo extract for avian myogenic cell growth in vitro. Proc. Jpn. Acad. 57:211.

Ii, I., I. Kimura, and E. Ozawa. 1982. A myotrophic protein from chick embryo extract: its purification, identity to transferrin, and indispensability for avian myogenesis. Dev. Biol. 94:366.

Izumo, S., B. Nadal-Ginard and V. Mahdavi. 1986. All members of the MHC multigene family respond to thyroid hormone in highly tissue-specific manner. Science 231:597.

Jacobs, S., F. C. Kull, Jr., H. S. Earp, M. E. Svoboda, J. J. Van Wyk and P. Cuatrecasas. 1983. Somatomedin-C stimulates the phosphorylation of the β-subunit of its own receptor. J. Biol. Chem. 258:8581.

Janeczko, R. A. and J. D. Etlinger. 1984. Inhibition of intracellular proteolysis in muscle cultures by multiplication-stimulating activity. Comparison of effects of multiplication-stimulating activity and insulin on proteolysis, protein synthesis, amino acid uptake, and sugar transport. J. Biol. Chem. 259:6292.

Jefferson, L. S., J. B. Li and S. R. Rannels. 1977. Regulation by insulin of amino acid release and protein turnover in perfused rat hemicorpus. Proc. Natl. Acad. Sci. USA 74:816.

Jennische, E. and H.-A. Hansson. 1987. Regenerating skeletal muscle cells express insulin-like growth factor I. Acta Physiol. Scand. 130:327.

Jennische, E., A. Skottner and H.-A. Hansson. 1987. Satellite cells express the trophic factor IGF-I in regenerating skeletal muscle. Acta Physiol. Scand. 129:9.

Kardami, E., D. Spector and R. C. Strohman. 1985a. Myogenic growth factor present in skeletal muscle is purified by heparin-affinity chromatography. Proc. Natl. Acad. Sci. USA 82:8044.

Kardami, E., D. Spector and R. C. Strohman. 1985b. Selected muscle and nerve extracts contain an activity which stimulates myoblast proliferation and which is distinct from transferrin. Dev. Biol. 112:353.

Kasuga, M., Y. Fujita-Yamaguchi, D. L. Blithe, M. F. White and C. R. Kahn. 1983. Characterization of the insulin receptor kinase purified from human placental membranes. J. Biol. Chem. 258:10973.

Kelly, A. M. and N. A. Rubinstein. 1986. Muscle histogenesis and muscle diversity. In: C. Emerson, D. Fischman, B. Nadal-Ginard and M. A. Q. Siddiqui (Ed.) Molecular Biology of Muscle Development. Vol. 29, pp 77–84. Alan R. Liss, New York.

King, G. L., C. R. Kahn, M. M. Rechler and S. P. Nissley. 1980. Direct demonstration of separate receptors for growth and metabolic activities of insulin and multiplication-stimulating activity (an insulin-like growth factor) using antibodies to the insulin receptor. J. Clin. Invest. 66:130.

Kostyo, J. L. and F. L. Engel. 1960. In vitro effects of growth hormone and corticotropin preparations on amino acid transport by isolated rat diaphragms. Endocrinology 67:708.

Kotts, C. E., M. E. White, C. E. Allen and W. R. Dayton. 1987. Stimulation of in vitro muscle cell proliferation by sera from swine injected with porcine growth hormone. J. Anim. Sci. 64:623.

Lathrop, B., E. Olson and L. Glaser. 1985a. Control by fibroblast growth factor of differentiation in the BC$_3$H1 muscle cell line. J. Cell Biol. 100:1540.

Lathrop, B., K. Thomas and L. Glaser. 1985b. Control of myogenic differentiation by fibroblast growth factor is mediated by position in the G$_1$ phase of the cell cycle. J. Cell Biol. 101:2194.

Linkhart, T. A., C. H. Clegg and S. D. Hauschka. 1981. Myogenic differentiation in permanent clonal mouse myoblast cell lines: regulation by macromolecular growth factors in the culture medium. Dev. Biol. 86:19.

Lowey, S. P., A. Benfield, D. D. Le Blanc and G. S. Waller. 1983. Myosin isozymes in avian skeletal muscles. I. Sequential expression in developing chicken pectoralis muscles. J. Muscle Res. Cell Motil. 4:695.

McCusker, R. H. and D. R. Campion. 1986. Effect of growth hormone secreting tumors on skeletal muscle cellularity in the rat. J. Endocrinol. 111:279.

Machlin, L. J. 1972. Effect of porcine growth hormone on growth and carcass composition of the pig. J. Anim. Sci. 35:794.

Manchester, K. L. and F. G. Young. 1970. The influence of the induction of alloxan-diabetes on the incorporation of amino acids into protein of rat diaphragm. Biochem. J. 77:386.

Marquardt, H., G. J. Todaro, L. E. Henderson and S. Oroszlan. 1981. Purification and primary structure of a polypeptide with multiplication-stimulating activity from rat liver cell cultures. J. Biol. Chem. 256:6859.

Massague, J. 1987. The TGF-β family of growth and differentiation factors. Cell 49:437.

Massague, J., S. Cheifetz, T. Endo and B. Nadal-Ginard. 1986. Type β transforming growth factor is an inhibitor of myogenic differentiation. Proc. Natl. Acad. Sci. USA 83:8206.

Matsuda, R., D. Spector, J. Micou-Eastwood and R. C. Strohman. 1984a. There is selective accumulation of growth factor in chicken skeletal muscle. II. Transferrin accumulation in dystrophic fast muscle. Dev. Biol. 103:276.

Matsuda, R., D. Spector and R. C. Strohman. 1984b. There is selective accumulation of a growth factor in chicken skeletal muscle. I. Transferrin accumulation in adult anterior latissimus dorsi. Dev. Biol. 103:267.

Mauro, A. 1961. Satellite cell of skeletal muscle fibers. J. Biophys. Biochem. Cytol. 9:493.

Merrill, G. F., J. R. Florini and N. C. Dulak. 1977. Effects of multiplication stimulating activity (MSA) on AIB transport into myoblasts and myotube cultures. J. Cell. Physiol. 93:173.

Miller, J. B., M. T. Crow and F. E. Stockdale. 1985. Slow and fast myosin heavy chain content defines three types of myotubes in early muscle cell cultures. J. Cell Biol. 101:1643.

Miller, J. B. and F. E. Stockdale. 1986a. Developmental origins of skeletal muscle fibers: clonal analysis of myogenic cell lineages based on fast and slow myosin heavy chain expression. Proc. Natl. Acad. Sci. USA 83:3860.

Miller, J. B. and F. E. Stockdale. 1986b. Developmental regulation of the multiple myogenic cell lineages of the avian embryo. J. Cell Biol. 103:2197.

Moseley, W. M., L. F. Krabill and R. F. Olsen. 1982. Effect of bovine growth hormone administered in various patterns on nitrogen metabolism in the Holstein steer. J. Anim. Sci. 55:1062.

Moses, A. C., S. P. Nissley, P. A. Short, M. M. Rechler and M. M. Podskalny. 1980. Purification and characterization of multiplication-stimulating-activity. Eur. J. Biochem. 103:387.

Moss, F. P. 1968. The relationship between the dimensions of fibers and the number of nuclei during normal growth of skeletal muscle in the domestic fowl. Amer. J. Anat. 122:555.

Moss, F. P. and C. P. Leblond. 1970. Nature of dividing nuclei in skeletal muscle of growing rats. J. Cell Biol. 44:459.

Moss, F. P. and C. P. Leblond. 1971. Satellite cells as the source of nuclei in muscles of growing rats. Anat. Rec. 170:421.

Mottola, C. and M. P. Czech. 1984. The type II insulin-like growth factor receptor does not mediate increased DNA synthesis in H-35 hepatoma cells. J. Biol. Chem. 259:12705.

Nadal-Ginard, B. 1978. Commitment, fusion and biochemical differentiation of a myogenic cell line in the absence of DNA synthesis. Cell 15:855.

Narusawa, M., R. B. Fitzsimons, S. Izumo, B. Nadal-Ginard, N. A. Rubinstein and A. M. Kelly. 1987. Slow myosin in developing rat skeletal muscle. J. Cell Biol. 104:447.

Olwin, B. B. and S. D. Hauschka. 1986. Identification of the fibroblast growth factor receptor of Swiss 3T3 cells and mouse skeletal muscle myoblasts. Biochemistry 25:3487.

Ong, J., S. Yamashita and S. Melmed. 1987. Insulin-like growth factor I induces c-fos messenger ribonucleic acid in L6 rat skeletal muscle cells. Endocrinology 120:353.

Ozawa, E. and Y. Hagiwara. 1982. Degeneration of large myotubes following removal of transferrin from culture medium. Biomed. Res. 3:16.

Ozawa, E. and K. Kohama. 1978. Partial purification of a factor promoting chicken myoblast multiplication in vitro. Proc. Jpn. Acad. 49:852.

Pain, V. M. and P. J. Garlick. 1974. Effect of streptozotocin diabetes and insulin treatment on the rate of protein synthesis in tissues of the rat in vivo. J. Biol. Chem. 249:4510.

Peel, C. J., D. E. Bauman, R. C. Gorewit and C. J. Sniffen. 1981. Effect of exogenous growth hormone on lactational performance in high yielding dairy cows. J. Nutr. 111:1662.

Peter, J. B., R. J. Barnard, V. R. Edgerton, C. A. Gillespie and K. W. Stempel. 1972. Metabolic profiles of three fiber types of skeletal muscle in guinea pigs and rabbits. Biochemistry 11:2627.

Powell, S. E. and E. D. Aberle. 1975. Cellular growth of skeletal muscle in swine differing in muscularity. J. Anim. Sci. 40:476.

Quinn, L. S., H. Holtzer and M. Nameroff. 1985. Generation of chick skeletal muscle cells in groups of 16 from stem cells. Nature 313:692.

Rechler, M. M. and S. P. Nissley. 1985. The nature and regulation of the receptors for insulin-like growth factors. Annu. Rev. Physiol. 47:425.

Rinderknecht, E. and R. E. Humbel. 1978a. The amino acid sequence of human insulin-like growth factor I and its structural homology with proinsulin. J. Biol. Chem. 253:2769.

Rinderknecht, E. and R. E. Humbel. 1978b. Primary structure of human insulin-like growth factor II. FEBS Lett. 89:283.

Rutz, R. and S. D. Hauschka. 1982. Clonal analysis of myogenesis. VII. Heritability of muscle colony type through sequential subclonal passages in vitro. Dev. Biol. 91:103.

Schmid, C., T. Steiner and E. R. Froesch. 1983. Preferential enhancement of myoblast differentiation by insulin-like growth factors (IGF I and IGF II) in primary cultures of chicken embryonic cells. FEBS Lett. 161:117.

Seed, J. and S. D. Hauschka. 1984. Temporal segregation of the migration of distinct myogenic precursor population into the developing chick wing bud. Dev. Biol. 106:389.

Shimo-Oka, T., Y. Hagiwara and and E. Ozawa. 1986. Class Stockdale, F. E. and H. Holtzer. 1961. DNA synthesis and myogenesis. Exp. Cell Res. 24:508.

Spizz, G., D. Roman, A. Strauss and E. N. Olson. 1986. Serum and fibroblast growth factor inhibit myogenic differentiation through a mechanism dependent on protein synthesis and independent of cell proliferation. J. Biol. Chem. 261:9483.

Stockdale, F. E. and J. B. Miller. 1987. The cellular basis of myosin heavy chain isoform expression during development of avian skeletal muscles. Dev. Biol. 123:1.

Stockdale, F. E., J. B. Miller, D. A. Schafer and M. T. Crow. 1986. Myosins, myotubes and myoblasts. Origins of fast and slow muscle fibers. In: C. Emerson, D. Fischman, B. Nadal-Ginard and M. A. Q. Siddiqui (Ed.) Molecular Biology of Muscle Development. Vol. 29, pp 213–223. Alan R. Liss, New York.

Svoboda, M. E., J. J. Van Wyk, D. G. Klapper, R. R. Fellows, F. E. Grissom and R. J. Schlueter. 1980. Purification of somatomedin-C from human plasma: chemical and biological properties, partial sequence analysis, and relationship to other somatomedins. Biochemistry 19:790.

Swatland, H. J. 1977. Accumulation of myofiber nuclei in pigs with normal and arrested development. J. Anim. Sci. 44:759.

Trenkle, A., D. L. DeWitt and D. G. Topel. 1978. Influence of age, nutrition and genotype on carcass traits and cellular development of M. longissimus of cattle. J. Anim. Sci. 46:1597.

Ullrich, A., J. R. Bell, E. Y. Chen, R. Herrera, L. M. Petruzzelli, T. J. Dull, A. Gray, L. Coussens, Y. C. Liao, M. Tsubokawa, A. Mason, P. H. Seeburg, C. Grunfeld, O. M. Rosen and J. Ramachandran. 1985. Human insulin receptor and its relationship to the tyrosine kinase family of oncogenes. Nature 313:756.

Ullrich, A., A. Gray, A. W. Tam. T. Yang-Feng, M. Tsubokawa, C. Collins, W. Henzel, T. Le Bon, S. Kathunia, E. Chen, S. Jacobs, U. Franke, J. Ramachandran and Y. Fujita-Yamaguchi. 1986. Insulin-like growth factor I receptor primary structure: comparison with insulin receptor suggests structural determinants that define functional specificity. EMBO J. 5:2503.

Ushiro, H. and S. Cohen. 1980. Identification of phosphotyrosine as a product of epidermal growth factor-activated protein kinase in A-431 cell membranes. J. Biol. Chem. 255:8363.

Weydert, A., P. Barton, A. J. Harris, C. Pinset and M. Buckingham. 1987. Developmental pattern of mouse skeletal myosin heavy chain gene transcripts in vivo and in vitro. Cell 49:121.

White, N. K., P. H. Bonner, D. R. Nelson and S. D. Hauschka. 1975. Clonal analysis of vertebrate myogenesis. IV. Medium-dependent classification of colony-forming cells. Dev. Biol. 44:346.

Wice, B., J. Milbrandt and L. Glaser. 1987. Control of muscle differentiation in BC_3H1 cells by fibroblast growth factor and vanadate. J. Biol. Chem. 262:1810.

Winick, M. and A. Noble. 1966. Cellular response in rats during malnutrition at various ages. J. Nutr. 89:300.

Yu, K.-T. and M. P. Czech. 1984. The type I insulin-like growth factor receptor mediates the rapid effects of multiplication-stimulating activity on membrane transport systems in rat soleus muscle. J. Biol. Chem. 259:3090.

Zapf, J., E. Rinderknecht, R. E. Humbel and E. R. Froesch. 1978. Nonsuppressible insulin-like activity (NSILA) from human serum: recent accomplishments and their physiologic implications. Metab. Clin. Exp. 27:1803.

CHAPTER 5

The Expression of Protooncogenes in Skeletal Muscle

JEFFREY D. TURNER, JAN NOVAKOFSKI, AND PETER J. BECHTEL

1. Introduction

Oncogenes are DNA sequences that when expressed lead to rapid cellular proliferation characteristic of malignant transformation. The nontransforming normal cellular cognates of the oncogenes are protooncogenes. These sequences
are highly conserved and are thought to be involved in the process of normal
cell replication (Bishop, 1983).

Oncogenes were first identified from rapidly transforming retroviruses. These
are RNA viruses and have a replicative cycle that integrates viral DNA into the
host genome. Portions of the host cellular genes are often added to the viral
sequence in this process. Newly made viral RNA will also contain these added
sequences and in subsequently infected cells the viral promoters dramatically
increase the abundance of these gene products. If the host sequence incorporated into the viral RNA codes for a protein involved in regulating normal
replication, overexpression may result in transformation of the infected cell.
This type of viral sequence, containing coding regions for part of a normal
regulatory protein that is improperly expressed as part of the virus, is an oncogene.

One strategy for identifying proteins with key regulatory functions in normal cells has been to examine oncogenes. Oncogenic modification of regulatory

JEFFREY D. TURNER ● Department of Animal Science, MacDonald College of McGill
University, Quebec, Canada H9X-ICO. JAN NOVAKOFSKI AND PETER J.
BECHTEL ● Muscle Biology Laboratory, University of Illinois, Urbana, Illinois 61801.

genes may occur by nonviral mechanisms as well, but the simplicity of the viral genome has permitted identification of a majority of the more than 40 known oncogenes. In fact, the three-letter designation for oncogenes arose from retroviral gene product nomenclature, such as gag for "group-specific antigens" or env for "envelope proteins." Oncogenes were named with respect to the virus in which the insert was originally identified or some trivial mnemonic designation. Viruses encoding two different inserts have an -A or -B suffix added to the three-letter gene. Protein products of oncogenes are designated p for protein or pp for phosphoprotein, the molecular mass in kilodaltons, and the superscript gene onc, i.e., pp60^{src} (Coffin *et al.*, 1981).

2. Categories of Oncogenes

Oncogene sequences have been grouped into four major categories as to their enzymatic function or the subcellular location of their gene products. The groups are: (1) proteins with tyrosine kinase activity, (2) GTP/GDP-binding proteins, (3) nucleus-located gene products, and (4) growth factor-like proteins. A brief description concerning the nature of these various groups is required to understand the relationship between oncogenes and growth factors/growth factor receptors/intracellular mediators.

2.1. Tyrosine Kinases

Most known oncogenes have protein gene products that possess the enzymatic ability to phosphorylate tyrosine amino acid residues of substrate proteins. Oncogenes with such activity include src, yes, fes, fps, fgr, and abl. The tyrosine kinase activity of these oncogenes is located on the cytosolic side of the plasmalemma. Most proteins from this group are membrane-associated by virtue of a hydrophobic membrane-spanning domain (Witte *et al.*, 1979) or through an interaction between the membrane lipids and lipid covalently bound to the kinase proteins. Examples of the latter include pp60^{v-src} and p120$^{gag-abl}$.

Tyrosine phosphorylation is an extremely rare event in the cell, with less than 0.1% of covalently bound protein phosphate on tyrosine residues. Conversely, phosphorylation of serine and threonine in substrate proteins is common. Changes in the tyrosine phosphorylation state is associated with proliferation induced by growth factors. Indeed, the similarities between growth factor receptors and oncogene products with respect to tyrosine phosphorylation indicate a common pathway between growth stimulation and neoplastic transformation. The identification and function of protein substrate(s) that are phosphorylated and permit these gene products to induce such massive proliferation remain unknown. This paucity of information may reflect the number of potential signal proteins present and their low concentration. Synthetic substrates

10–12 amino acids long, having homology to the autophosphorylation site of pp60^{v-src}, are substrates for the EGF (Pike *et al.*, 1982), PDGF (Pike *et al.*, 1983), and insulin receptors (Kasuga *et al.*, 1983) and a variety of oncogene tyrosine kinases (Hunter, 1982). Alternately, orthovanadate, a compound that prevents the removal of phosphate from tyrosine, potentiates hormonal responses and can cause cellular proliferation.

Many peptide growth factors induce myoblasts and satellite cells to proliferate. Signal transduction includes specific high-affinity binding of the hormone ligand to the extracellular portion of the receptor. This signal is passed to the cytosolic portion of the receptor within the same molecule as with PDGF, EGF, or FGF receptors, or through allosteric interaction between subunits in the case of insulin or IGF receptors. The end result of ligand binding is an increase in the receptors' tyrosine kinase activity. The receptors themselves can be phosphorylated by other receptor kinases and protein kinases or through autophosphorylation. The tyrosine kinase activity is reduced when the receptor is internalized. Conversely, in the case of oncogene products, the tyrosine kinase activity is not dependent on ligand binding.

The most conclusive evidence linking growth factor receptors and oncogene products is the EGF receptor and the oncogene product of the erythroblastosis virus, erb-B. Sequence analysis shows a striking homology in the membrane-spanning and tyrosine kinase domains. The extracellular domain of the EGF receptor is larger, glycosylated, and has the EGF binding site, which is absent in the truncated erb-B protein. The tyrosine kinase activity is regulated in the case of the EGF receptor and unregulated with the erb-B product.

2.2. GTP-Binding Proteins

The second group of oncogenes are those of the ras family. The transforming gene product has a high affinity for guanine nucleotide and possesses GTPase enzymatic activity. Since the mode of action of ras gene products resembles the ubiquitous G proteins and transducins, a brief examination of the latter is warranted.

The G proteins are hydrophobic, membrane-associated proteins arranged in an oligomeric structure containing α, β, and γ subunits (Gilman, 1984). The molecular mass of the α, β, and γ subunits is 45, 35, and 10 kDa, respectively. Activation of membrane-bound receptors caused by ligand association, results in subunit dissociation and increased GTP-binding affinity. The Gα–GTP complex then activates adenylate cyclase on the cytosolic surface of the plasmalemma, thereby affecting cyclic nucleotide pools. Transducin (T), a regulatory protein complex in bovine retina, has a similar α subunit composition and molecular mass of 41 kDa. When activated, Tα inhibits phosphodiesterase activity in the retinal rods.

The ras oncogene product is a protein of 21 kDa designated p21^{ras}. It shares many of the characteristics of the "transducing" proteins mentioned above but has only 189 amino acids. Despite this, p21^{ras} shares with Gα and Tα two hydrophobic domains (Tanabe *et al.*, 1985), which are involved with subunit association or membrane interaction. The p21^{ras} undergoes posttranslational lipid attachment, which enhances association with membranes (Sefton *et al.*, 1982). Three other common domains involve nucleotide binding and GTPase regions. Mutations in the GTPase region in p21^{ras} decrease the GTPase activity by tenfold (Sweet *et al.*, 1984). Point mutation of the c-Ha-ras-1 protooncogene at codon 12 or codon 61 leds to malignant activation forming the Ha-ras oncogene (Fujita *et al.*, 1985). The transforming potential caused by mutations at these "hot spots" reflects a reduction in GTPase activity. As with other proteins that use GTP in their regulatory scheme, the GTPase rate is critical as GTP cleavage is associated with turning the signal off (Gibbs *et al.*, 1984). In the case of p21^{ras}, low GTPase activity allows some presently unknown system to remain on and thereby induce malignancy.

2.3. Nuclear Proteins

Nuclear oncogenes are so named because their protein products are thought to function within the nucleus. Nuclear oncogenes include myc, myb, fos, and ski, of which myc has been studied extensively. The c-myc gene has been completely sequenced (Colby *et al.*, 1983; Watt *et al.*, 1983) and consists of three exons. The last two exons give rise to a transcript coding for a 47-kDa peptide, which would contain long stretches of basic residues and single or tandem proline residues (Colby *et al.*, 1983). *In vitro* translation of c-myc mRNA yields a phosphoprotein of 62 kDa (pp62^{c-myc}) (Hann *et al.*, 1983). However, c-myc from quail embryo is a 60-kDa protein that is not phosphorylated. The discrepancy between predicted molecular weight and those determined by SDS–PAGE may reflect conformational changes due to the presence of basic amino acids or proline.

The retroviral v-myc gene codes for a fusion product between the retroviral gag (group-specific antigen) and c-myc with various 5′ deletions. The resulting protein is p110$^{gag-myc}$. In addition, v-myc lacks the AATAAA hexanucleotide putative polyadenylation signal. This allows transcription to continue through to the 3′ long terminal repeat (LTR), thereby permitting retroviral replication.

In both c-myc and v-myc the COOH-terminal amino acids are highly conserved. This region is high in basic amino acids, which is consistent with other known DNA-binding proteins. Binding of p64^{c-myc} (and p110$^{gag-myc}$) to double-stranded DNA has been observed (Donner *et al.*, 1982 Watt *et al.*, 1985).

Neoplastic transformation caused by myc is typically due to an overexpression or continual unmodulated expression of c-myc. This occurs through at least three mechanisms: (1) the acquisition of upstream promoter elements arising from retroviral integration, (2) chromosomal translocation between c-myc and various immunological loci, (3) amplification of c-myc genes. Viral integration upstream of c-myc has been implicated in mouse leukemia from mink cell focus-forming (MCF) retrovirus (Li *et al.*, 1984), mouse lymphomas from murine leukemia virus (MLV) (Steffen, 1984), and avian B cell lymphomas from avian leukosis virus (ALV) (Neel *et al.*, 1981). These viruses lack classical oncogenes, indicating viral gene products are not necessary for tumor induction. This class of virus has a long latent period and transforms cells less successfully than other viral classes (Neel *et al.*, 1981). Viral integration is the sole requirement for this type of transformation. As the retrovirus often integrates 2–10 kb away from the coding region, rather than adjacent to c-myc, an effect on enhancer elements has been postulated.

Translocation of c-myc occurs in several forms of Burkitt's lymphoma (Pelicci *et al.*, 1986). In these cells, reciprocal chromosomal translocation places c-myc with a 5′ deletion into an immunoglobulin locus. Overexpression of myc protein has been attributed to the presence of enhancer elements of the adjacent immunoglobulin gene. An alternate hypothesis is that the 5′ deletion removed some regulatory elements (Pelicci *et al.*, 1986).

Amplification of myc occurs in human neuroblastoma (Kohl *et al.*, 1983) and an amplification of 25- to 30-fold is typical and can reach levels of 700-fold. This amplification appears to occur after translocation. Although the function of myc protein products is not clear, a correlation exists between myc expression and stimulation of cellular proliferation. Such examples are the stimulation of myc expression by PDGF (Kelly *et al.*, 1983) or bombesin (Letterio *et al.*, 1986).

2.4. Peptide Hormones

The fourth major group of oncogenes are those with gene products resembling known peptide hormone mitogens. The structural homology between the transforming gene product of simian sarcoma virus (SSV) and portions of the human PDGF gene (Devare *et al.*, 1983) is the best understood example. The SSV is a combination between a simian sarcoma-associated virus (SSAV) and a cellular sequence (c-sis) from the genome of the woolly monkey. Transforming potential resides in a region denoted v-sis, which encodes a protein of 28 kDa (p28^{sis}). The acquisition of the v-sis sequence occurred at the expense of the retroviral envelope (env) gene, of which only a small portion remains. The p28^{sis} is immunologically related to PDGF (Robbins *et al.*, 1983; Wang and Williams, 1984). PDGF-like mitogenic activity is found in cell lysates from SSV-transformed cells and possesses a 94% amino acid homology to the B chain of PDGF (Robbins *et al.*, 1983). Within SSV-transformed cells, p28^{sis}

forms dimers with a composite mass of 24 kDa. In this respect, p28^{sis} is similar to the pig B–B PDGF homodimer (Stroobant and Waterfield, 1984).

The transforming potential of SSV appears to involve p28^{sis} acting through an extracellular autocrine loop. The v-sis oncogene is transcribed under the control of the viral promoters within the flanking LTR sequences. Translation of p28^{sis} is initiated upstream at the first or second ATG codon within the env gene (King et al., 1985). The v-sis gene lacks the 5' portion of the c-sis gene, which contains the signal peptide. Vectorial translocation of the nascent chain through the membrane during secretion is dependent upon a hydrophobic region of 11 amino acids encoded within the env gene. Deletion of this env signal peptide or point mutations that replace a hydrophobic residue with a charged amino acid abrogates the SSV transforming potential (Hannik and Donoghue, 1984). Thus, secretion of p28^{sis} is essential for conservation of transforming potential.

Once outside the cell, p28^{sis} exerts its mitogenic effects through association with the PDGF receptor. Antibodies directed at PDGF prevent growth of SSV-transformed fibroblasts by immunoprecipitating the secreted p28^{sis} (Johnsson et al., 1985). In fact, the v-sis gene product induces down-regulation of cellular PDGF receptors (Garrett et al., 1984). Suramin, a compound known to block PDGF from binding to its receptor, causes PDGF receptor numbers to attain their pretransformed levels and blocks the transforming capability of the SSV.

Despite the profound mitogenic action of SSV transformation, human fibroblasts so infected are not immortalized (Johnsson et al., 1986). Prolonged serial passage of SSV-treated fibroblasts results in cells that are nonresponsive to PDGF but have PDGF receptors. This suggests a block in some postreceptor pathway.

3. Oncogene Expression in Skeletal Muscle

Since oncogene expression has been well correlated with the proliferative phenotype, their role in the control of normal cell mitosis has been implied. Studies performed in our laboratory (Turner, 1987) examined oncogene expression using viral and cellular oncogene probes from each family hybridized to RNA isolated from two muscle systems. The first system was the in vitro culture of L8E63 myoblasts. This cell line provided a homogeneous cell population that proliferated rapidly, then at confluency differentiated into myotubes. The second system was growing gastrocnemius from normal rats as well as muscles from rats bearing GH-secreting tumors (Turner et al., 1986).

Oncogene transcripts including abl, erb-B, fos, ki-ras, myc, mos, sis, and src were not detected. The internal controls did show adequate hybridization, indicating that the hybridization procedures had functioned properly (Table I).

Table I

Literature Survey of Oncogene Expression in Muscle

RNA source	Oncogene														
	Nuclear			Tyrosine kinase						GTP-binding		Other			
	fos	myb	myc	abl	fes	fps	ros	src	yes	Ha-ras	Ki-ras	erb-A	erb-B	mos	sis
Rat primary myoblast	ND[a]	−	++	++	−	ND	ND	−	ND	ND	++	−	−	−	ND
Rat L6 cell line	ND	−	++	++	−	ND	ND	−	ND	ND	++	−	−	−	ND
Mouse Cl 10 cell line	ND	−	+	+	−	ND	ND	−	ND	ND	++	−	−	−	ND
(Sejersen *et al.*, 1985a)															
Rat L6 myoblast	ND	−	+	++	−	ND	ND	−	ND	ND	++	−	−	−	++
(Sejersen *et al.*, 1985b)															
Chicken muscle	ND	−	−	ND	ND	ND	ND	−	ND	ND	ND	ND	++	ND	ND
(Gonda *et al.*, 1982)															
Chicken (2 weeks)															
Muscle	ND	ND	ND	ND	ND	−	−	−	++	ND	ND	ND	ND	ND	ND
Heart	ND	ND	ND	ND	ND	−	−	−	++	ND	ND	ND	ND	ND	ND
(Shibuya *et al.*, 1982)															
Chicken (2 weeks)															
Muscle	ND	−	++	ND	ND	ND	ND	−	ND	ND	ND	−	−	ND	ND
(Coll *et al.*, 1983)															
Mouse muscle	ND	ND	ND	ND	ND	ND	ND	ND	ND	++	++	ND	ND	ND	ND
Heart	ND	ND	ND	ND	ND	ND	ND	ND	ND	++	++	ND	ND	ND	ND
(Leon *et al.*, 1987)															
Rat L8 cell line	−	ND	−	−	ND	ND	ND	−	ND	++	−	ND	−	−	−
Rat muscle	−	ND	−	−	ND	ND	ND	−	ND	++	−	ND	−	−	−
(Turner, 1987)															

[a] ++, strong signal; +, detectable signal; −, absence of signal; ND, not determined.

However, these oncogene transcripts could have been present at levels below the detection limit of the experiments.

Turner's (1987) study indicated that a transcript similar in size to the large Ha-ras transcript was present in muscle and that this signal became stronger in the presence of GH$_3$ tumors. In this study, a v-Ha-ras-like transcript was expressed; c-ki-ras was not detected. A similar relationship was reported in mouse skeletal muscle where Ha-ras was three to ten times as abundant as ki-ras (Leon *et al.*, 1987). Extrapolation to the present study would push ki-ras below the limits of detection. In mouse skeletal muscle, Ha-ras transcripts of 1.1 and 1.3 kb were observed (Leon *et al.*, 1987), but mouse skeletal muscle and heart show only the 5.0-kb transcript of N-ras (Leon *et al.*, 1987). Variety in transcript size originates through differential transcript termination. This leads to mRNA species of different lengths but encoding identical proteins (Birnstiel *et al.*, 1985). In fact, the long 3' untranslated regions of some 4.5 kb in Ki-ras and N-ras (Guerrero *et al.*, 1985) are among the longest in nature. Muscle systems use transcription termination as a mechanism of regulating gene expression. The use of different polyadenylation sites in muscle occurs with the protein vimentin (Capatanaki *et al.*, 1983). Similarly, tropomyosin isoforms from muscle and nonmuscle sources result from alternate polyadenylation (Helfman *et al.*, 1986). Whether muscle tissue uses transcription termination to control Ha-ras expression is not known.

Primary myoblasts and cultured myoblasts have been shown to express the myc oncogene but never the myb oncogene (Table I). However, the expression of myc is highly variable and depends upon the particular myoblast cell line or clone of that cell line examined. Sejersen *et al.*, 1985b, reported low myc expression in L6J1 cells, but in another study (Sejersen *et al.*, 1985a) found significant myc expression in the same cell line. These authors conceded that large differences in myc expression were possible within a cell line and between muscle cell lines.

The hormonal milieu of the cell culture greatly alters nuclear oncogene expression. Myoblasts (L6E9) treated with IGF-I, insulin, or 20% fetal calf serum had significant myc expression whereas cells exposed to low levels of serum had low myc expression (Endo and Nadal-Ginard, 1986). Similarly, stimulation of L6E9 myoblasts with IGF-I caused c-fos expression to be induced from undertectable levels to significant expression (Ong *et al.*, 1987). Oncogene expression in cultured cells may be transient as well. Primary cultured myocytes stimulated with norepinephrine (0.002–20 μM) increased c-myc expression tenfold over 2 hr and then declined to background by 6 hr (Starksen *et al.*, 1986).

myc expression reflects the developmental maturity of the muscle tissue donor. Adult chicken muscle RNA contains very few myc or myb transcripts, whereas chicken embryos express significant amounts (Gonda *et al.*, 1982). In

2-week-old chickens, myc is present at four copies per cell whereas less than one copy of myb is found (Coll *et al.*, 1983). In light of these factors, difficulties in detection of myc expression are not unexpected.

Oncogenes mos, src, erb-B, and sis are not typically expressed in muscle tissues (Table I).

In summary, the role of oncogenes and protooncogenes in skeletal muscle growth and differentiation is not well understood. This is in part due to the low abundance of their mRNAs in skeletal muscle tissue and cell lines. In many instances the methodology used may not have been sensitive enough to detect oncogene mRNA levels.

Few reports exist concerning the expression of protooncogenes in traditional farm animals. As referenced above, there are several studies on chicken muscle, which is an excellent model for some types of studies involving retroviral systems.

References

Birnstiel, M. L., M. Busslinger and K. Strub. 1985. Transcription termination and 3′ processing: the end is in site! Cell 41:349.

Bishop, J. M. 1983. Cellular oncogenes and retroviruses. Annu. Rev. Biochem. 52:301.

Capatanaki, Y. G., J. Ngai, C. N. Flytzanis and E. Lazarides. 1983. Tissue-specific expression of two mRNA species transcribed from a single vimentin gene. Cell 35:411.

Coffin, J. M., H. E. Varmus, J. M. Bishop, M. Essex, W. D. Hardy Jr, G. S. Martin, N. E. Rosenberg, E. M. Scolnick, R. A. Weinberg and P. K. Vogtt. 1981. Proposal for naming host cell-derived inserts in retrovirus genomes. J. Virol. 40:953.

Colby, W. W., E. Y. Chen, D. H. Smith and A. D. Levinson. 1983. Identification and nucleotide sequence of a human locus homologous to the v-myc oncogene of avian myelocytomatosis virus MC 29. Nature 301:722.

Coll, J., S. Saule, P. Martin, M. B. Raes, C. Lagrou, T. Graf, H. Beug, I. E. Simon and D. Stehelin. 1983. The cellular oncogenes c-myc, c-myb and c-erb are transcribed in defined types of avian hematopoietic cells. Exp. Cell Res. 149:151.

Devare, S. G., E. P. Reddy, J. D. Law, K. C. Robbins and S. A. Aaronson. 1983. Nucleotide sequence of the simian sarcoma virus genome: demonstration that its acquired cellular sequences encode the transforming gene product p28sis. Proc. Natl. Acad. Sci. USA 80:731.

Donner, P., I. Greiser-Wilke and K. Moelling. 1982. Nuclear localization and DNA binding of the transforming gene product of avian myelocytomatosis virus. Nature 296:262.

Endo, T. and B. Nadal-Ginard. 1986. Transcriptional and post-transcriptional control of c-myc during myogenesis: its mRNA remains induced in differentiated cells and does not suppress the differentiated phenotype. Mol. Cell. Biol. 6:1412.

Fujita, J., S. K. Srivastava, M. H. Kraus, J. S. Rhim, S. R. Tronick and S. H. Aaronson. 1985. Frequency of molecular alterations affecting ras protooncogenes in human urinary tract tumors. Proc. Natl. Acad. Sci. USA 82:3849.

Garrett, J. S., S. R. Coughlin, H. L. Niman, P. M. Tremble, G. M. Giels and L. T. Williams. 1984. Blockade of autocrine stimulation in simian sarcoma virus-transformed cells reverses down-regulation of platelet-derived growth factor receptors. Proc. Natl. Acad. Sci. USA 81:7466.

Gibbs, J. B., I. S. Signal, M. Poe and E. M. Scolnick. 1984. Intrinsic GTPase activity distinguishes normal and oncogenic ras p21 molecules. Proc. Natl. Acad. Sci. USA 81:5704.

Gilman, A. G. 1984. G proteins and dual control of adenylate cyclase. Cell 36:577.

Gonda, T. J., D. K. Sheiness and J. M. Bishop. 1982. Transcripts from the cellular homologs of
retroviral oncogenes: distributions among chicken tissues. Mol. Cell. Biol. 2:617.

Guerrero, I., A. Villasante, V. Corces and A. Pellicer. 1985. Loss of the normal N-ras allele in a
mouse thymic lymphoma induced by a chemical carcinogen. Proc. Natl. Acad. Sci. USA
82:7810.

Hann, S. R., H. D. Abrams, L. R. Rohrschneider and R. N. Eisenman. 1983. Proteins encoded
by v-myc and c-myc oncogenes: identification and localization in acute leukemia virus trans-
formants and bursal lymphoma cell lines. Cell 34:789.

Hannik, M. and D. J. Donoghue. 1984. Requirement for a signal sequence in biological expression
of the v-sis oncogene. Science 226:1197.

Helfman, D. M., S. Cheley, E. Kuissmanen, L. A. Finn and Y. Yamawaki-Katoake. 1986. Nonmuscle and muscle tropomyosin isoforms are expressed from a single gene by alternate RNA
splicing and polyadenylation. Mol. Cell. Biol. 6:3582.

Hunter, T. 1982. Synthetic peptide substrates for a tyrosine protein kinase. J. Biol. Chem. 257:4843.

Johnsson, A., C. Betsholtz, C. H. Heldin and B. Westermark. 1985. Antibodies against platelet-
derived growth factor inhibit acute transformation by simian sarcoma virus. Nature 317:438.

Johnsson, A., C. Betsholtz, C. H. Heldin and B. Westermark. 1986. The phenotype characteristics
of simian sarcoma virus-transformed human fibroblasts suggest that the v-sis gene product acts
solely as a PDGF receptor agonist in cell transformation. EMBO J. 5:1535.

Kasuga, M., Y. Fugita-Yamaguchi, D. L. Blithe and C. R. Kahn. 1983. Tyrosine-specific protein
kinase activity is associated with the purified insulin receptor. Proc. Natl. Acad. Sci. USA
80:2137.

Kelly, K., B. H. Cochran, C. D. Stiles and P. Leder. 1983. Cell specific regulation of the c-myc
gene by lymphocyte mitogens and platelet-derived growth factor. Cell 35:603.

King, C. R., N. A. Giese, K. C. Robbins and S. A. Aaronson. 1985. In virto mutagenesis of the
v-sis transforming gene defines functional domains of its growth factor-related product. Proc.
Natl. Acad. Sci. USA 82:5295.

Kohl, N. E., N. Kanda, R. R. Schreck, G. Bruns, S. A. Latt, F. Gilbert and F. W. Alt. 1983.
Transposition and amplification of oncogene-related sequences in human neuroblastomas. Cell
35:359.

Leon, J., I. Guerrero and A. Pellicer. 1987. Differential expression of the ras gene family in mice.
Mol. Cell. Biol 7:1535.

Letterio, J. J., S. R. Coughlin and L. T. Williams. 1986. Pertussis toxin-sensitive pathway in the
stimulation of c-myc expression and DNA synthesis by bombesin. Science 234:117.

Li, Y., C. A. Holland, J. W. Hartley and N. Hopkins. 1984. Viral integration near c-myc in 10–
20% of MCF 247-induced AKR lymphomas. Proc. Natl. Acad. Sci. USA 81:6808.

Neel, B. G., W. S. Hayward, H. L. Robinson, J. Fang and S. M. Astrin. 1981. Avian leukosis
virus-induced tumors have common proviral integration sites and synthesize discrete new RNAs:
oncogenesis by promoter insertion. Cell 23:323.

Ong, J., S. Yamashita and S. Melmed. 1987. Insulin-like growth factor 1 induces c-foc messenger
ribonucleic acid in L6 rat skeletal muscle cells. Endocrinology 120:353.

Pelicci, P. G., D. M. Knowles, I. Magrath and R. Dalla-Favera. 1986. Chromosomal breakpoints
and structural alterations of the c-myc locus differ in endemic and sporadic forms of Burkitt
lymphoma. Proc. Natl. Acad. Sci. USA 83:2984.

Pike, L. J., D. F. Bowen-Pope, R. Ross and E. G. Krebs. 1983. Characterization of platelet-
derived growth factor-stimulated phosphorylation in cell membranes. J. Biol. Chem. 258:9383.

Pike, L. J., B. Gallis, J. E. Casnellie, P. Bornstein and E. G. Krebs. 1982. Epidermal growth factor stimulates the phosphorylation of synthetic tyrosine-containing peptides by A431 cell membranes. Proc. Natl. Acad. Sci. USA 79:1443.

Robbins, K. C., H. N. Antoniades, S. G. Devare, M. W. Hunkapiller and S. A. Aaronson. 1983. Structural and immunological similarities between simian sarcoma virus gene product(s) and human platelet-derived growth factor. Nature 305:605.

Sefton, B. M., I. S. Trowbridge, J. S. Cooper and E. M. Scolnick. 1982. The transforming proteins of Rous sarcoma virus. Harvy sarcoma virus and Abelson virus contain tightly bound lipid. Cell 31:465.

Sejersen, T., J. Sumegi and N. R. Ringertz. 1985a. Density-dependent arrest of DNA replication is accompanied by decreased level of c-myc mRNA in myogenic but not in differentiation-defective myoblasts. J. Cell. Physiol. 125:465.

Sejersen, T., J. P. Wahrmann, J. Sumegi and N. R. Ringertz. 1985b. Change in expression of oncogenes (sis, ras, myc, abl) during in vitro differentiation of L6 rat myoblasts. In: B. Wahren, G. Holm, S. Hammarstrom and P. Perlmann (Eds.) Molecular Biology of Tumor Cells. pp 243–250. Raven Press, New York.

Shibuya, M., H. Hanafusa and P. C. Balduzzi. 1982. Cellular sequences related to three new oncogenes of avian sarcoma virus (fps, yes and ros) and their expression, in normal and transformed cells. J. Virol. 42:143.

Starksen, N. F., P. C. Simpson, N. Bishopric, S. R. Coughlin, W. M. F. Lee, J. A. Escobedo and L. T. Williams. 1986. Cardiac myocyte hypertrophy is associated with c-myc protoon-cogene expression. Proc. Natl. Acad. Sci. USA 83:8348.

Steffen, D. 1984. Proviruses are adjacent to c-myc in some murine leukemia virus-induced lymphomas. Proc. Natl. Acad. Sci. USA 81:2097.

Stroobant, P. and M. D. Waterfield. 1984. Purification and properties of porcine platelet-derived growth factor. EMBO J. 3:2963.

Sweet, R. W., S. Yokoyama, T. Kamata, J. R. Feramisco, M. Rosenberg and M. Gross. 1984. The product of ras is a GTPase and the T24 oncogene mutant is deficient in the activity. Nature 311:273.

Tanabe, T., T. Nukada, Y. Nishikawa, K. Sugimoto, H. Suzuki, H. Takahashi, M. Noda, T. Haga, A. Ichiyama, K. Kangawa, N. Minamino, H. Matsuo, and S. Numa. 1985. Primary structure of the α-subunit of transducin and its relationship to ras proteins. Nature 315:242.

Turner, J. D. 1987. Molecular mechanisms of growth hormone induced muscle hypertrophy. Ph.D. Thesis. Univ. of Illinois, Urbana.

Turner, J. D., P. J. Bechtel and J. E. Novakofski. 1986. Interaction between age and GH₃ tumor implantation in Wistar–Furth rats. Growth 50:402.

Wang, J. Y. and L. T. Williams. 1984. A v-sis oncogene protein produced in bacteria competes for platelet-derived growth factor binding its receptor. J. Biol. Chem. 259:10645.

Watt, R. A., A. R. Shatzman and M. Rosenberg. 1985. Expression and characterization of the human c-myc DNA-binding protein. Mol. Cell. Biol. 5:448.

Watt, R., L. W. Stanton, K. B. Marcu, R. C. Gallo, C. M. Croce and G. Rovera. 1983. Nucleotide sequence of cloned cDNA of human c-myc oncogene. Nature 303:725.

Witte, O. N., N. Rosenberg and D. Baltimore. 1979. Preparation of syngenic tumor regressor serum reactive with the unique determinants of the Abelson murine leukemia virus-encoded P120 protein at the cell surface. J. Virol. 31:776.

CHAPTER 6

Regulation of Myofibrillar Protein Gene Expression

HOLLY E. RICHTER, RONALD B. YOUNG, AND DEBRA M. MORIARITY

1. Introduction

The contraction of skeletal and cardiac muscle is the result of a physiological conversion of chemical energy into mechanical energy, which takes place in a highly ordered three-dimensional matrix of myofibrillar proteins. The basic unit of the contractile process in striated muscle is the sarcomere (Squire, 1981), which is composed of thick and thin filaments tandemly arranged in the myofibril. The sarcomere is composed of 10–15 myofibrillar proteins, but of these only myosin, actin, troponin, and tropomyosin participate directly in the contractile event. Regulation of expression of the myofibrillar protein genes seems to occur at the transcriptional level, with some of their RNA products exhibiting alternative exon splicing in the generation of multiple protein isoforms. Some of these primary RNA transcripts are also generated by differential initiation at alternate promoters (Nabeshima et al., 1984; Periasamy et al., 1984a; Robert et al., 1984). Muscle protein diversity can further be manifested by termination of primary transcripts at alternative 3' untranslated sequences (Basi et al., 1984; Ruiz-Opazo et al., 1985; Bernstein et al., 1986; Rozek and Davidson, 1986). Generation of different protein forms by alternative splicing of identical primary RNA transcripts must involve trans-acting factors, some of which are tissue specific (Nadal-Ginard et al., 1987). A cis-acting factor refers to a DNA locus that affects the activity of DNA sequences on its own molecule

HOLLY E. RICHTER, RONALD B. YOUNG, AND DEBRA M. MORIARITY • Department of Biological Sciences, University of Alabama, Huntsville, Alabama 35899.

of DNA, whereas a *trans*-acting factor refers to a diffusible product able to act on all receptive sites in the cell. RNA transcripts derived from alternate promoter initiation or termination sequences could contain the information to regulate their own splicing mechanism by *cis*-acting factors, although some interaction of *trans*-acting factors cannot be ruled out (Breitbart *et al.*, 1985).

The existence of structurally distinct, developmental stage- and tissue-specific protein isoforms of myofibrillar proteins reflects the heterogeneity in physiological properties of muscle and its ability to respond to a broad continuum of environmental conditions. The existence of alternate processing mechanisms, the organization of the myofibrillar protein genes, and the response to various physiological stimuli and hormones reflect a level of functional efficiency by which these proteins are differentially expressed. Understanding the factors that regulate the deposition of myofibrillar proteins in skeletal muscle of meat animals is a long-term goal; however, virtually nothing is known about the interrelationship between cellular events of myogenesis and the expression of myofibrillar genes in large farm animals.

2. Multigene Families

As vertebrate skeletal muscle matures through the developmental stages from fetal to neonatal to adult, a progression of developmental stage-specific isoforms of muscle proteins also ensues (Nadal-Ginard *et al.*, 1982; Lompre *et al.*, 1984; Buckingham *et al.*, 1986). Two possible molecular mechanisms may explain the diversity of protein isoforms: (1) the selective expression of one of the genes from a multigene family, dependent upon tissue- and developmental stage-specific factors, and (2) the generation of different protein isoforms from a single gene (Nadal-Ginard *et al.*, 1987). Both mechanisms involve specific *cis*- or *trans*-acting elements (or both). The major vertebrate myofibrillar proteins—myosin heavy chain (MHC), myosin light chain (MLC), actin, tropomyosin (TM), and troponin (TN)—are all encoded by multigene families (Nadal-Ginard *et al.*, 1982). MHC is structurally the predominant polypeptide chain of the myosin molecule, as well as the major protein of the sarcomere. Moreover, MHC is directly involved in ATP hydrolysis and force generation during movement (Wagner and Giniger, 1981).

3. Major Myofibrillar Proteins of the Sarcomere

3.1. Actin

The myofibrillar protein, actin, exists in two forms, G-actin and F-actin (where F-actin is a polymer of the globular G-actin), and is the primary myofibrillar protein constituent of the thin filament. Rabbit muscle G-actin has a molecular mass of approximately 43 kDa and is a single polypeptide molecule. In replicating myogenic cells, β-actin is the predominant cytoplasmic isoform. After fusion initiates, synthesis of β-actin decreases and sarcomeric α-actins begin to appear (Seiler-Tuyns et al., 1984). The type of α-actin and the level of expression appear to be species specific (Singer and Kessler-Icekson, 1978). For example, cardiac α-actin appears to be the major sarcomeric form in the chicken embryo, but both skeletal and cardiac α-actins are expressed in the fetal mouse (Minty et al., 1982). In adult humans the skeletal α-actin is the major isoform in striated muscle, and the cardiac α-actin is the major isoform in the heart. However, the α-actin cardiac gene is expressed in abundance in human fetal skeletal muscle (Gunning et al., 1983), and low levels of cardiac α-actin are observed in adult skeletal muscle. The appearance of these protein isoforms is reflected at the mRNA level as well. It has been suggested that one or the other isoform predominates in adult tissue to give a selective functional advantage for that particular tissue; however, such precise fine tuning of gene expression may not be necessary or desirable at crucial stages in early development when it is more important to generate large quantities of muscle proteins (Buckingham et al., 1985).

Regulation of this differential expression process occurs at the transcriptional level in a tissue-specific manner (Vandekerckhove and Weber, 1979). Unlike MHC genes, the different actin genes are not linked in mammalian genomes (Czosnek et al., 1983; Gunning et al., 1984) and are unlinked to other sarcomeric protein genes (Czosnek et al., 1983). Two upstream cis-acting elements that modulate the tissue-specific transcription of the human cardiac actin gene may be involved (Minty and Kedes, 1986; Minty et al., 1986), and these may be further modulated by tissue-specific transcriptional factors (trans-acting elements) in muscle cells.

3.2. Tropomyosin (TM)

TM is one of two major muscle proteins that serve a regulatory function during contraction. Rabbit muscle TM has a molecular mass of approximately 70 kDa and is composed of two different helical polypeptides with the same number of amino acids but apparent molecular masses of 34 and 36 kDa (Sodek et al., 1978; Mak et al., 1979). The degree of amino acid sequence homology among the skeletal TMs is quite high as shown by the isolation of a human cytoskeletal TM clone using a chick cDNA clone (MacLeod et al., 1986). The limited information available on regulation of expression of the TM gene has been derived from studies in rats and chickens (Ruiz-Opazo et al., 1983; MacLeod et al., 1986; Nadal-Ginard et al., 1987). In the rat the α-TM gene produces both smooth and striated muscle isoforms by using one promoter region, two intragenic regions containing alternatively spliced isotype-specific switch exons and two poly(A) addition sites located approximately 6 kb apart (Nadal-Ginard *et al.*, 1986, 1987). The expression of the TM genes differs from that of MLC, TN-T, actin, and MHC in that it is not regulated in a tissue- and developmental stage-specific manner. Analysis of rat genomic clones with fibroblast TM 1 and skeletal α-TM cDNA clones indicates that there are two separate loci in the rat genome that contain sequences complementary to these TM cDNAs. One of these loci is a pseudogene, and the other contains a single gene made up of 11 exons that span approximately 10 kb (Helfman *et al.*, 1986). The isoform-specific regions delineate the TN-T-binding domains of skeletal muscle TM. The α-TM gene expression is not limited to muscle cells, but has been detected in many other tissues, although not in liver. All these isoforms of α-TM are derived from a single promoter by differential splicing.

3.3. Troponin (TN)

TN is the second major regulatory protein present in the thin filaments. It is a globular protein composed of three polypeptide molecules, TN-C, TN-I, and TN-T, and diverse TN-T isoforms have been found in different muscle tissues at various developmental stages (Wilkinson *et al.*, 1984, 1986; Breitbart *et al.*, 1985; Nadal-Ginard *et al.*, 1986, 1987). Three types of splicing from a primary transcript have been reported in the expression of the TN-T isoforms in rats (Breitbart and Nadal-Ginard, 1986; Nadal-Ginard *et al.*, 1987). There are 18 exons in this gene, and some of those exons coding for the 5' end of the gene (exons 4–8) may be included or excluded in the generation of a mature transcript. Exons 16 and 17 are spliced in a mutually exclusive manner to generate different peptides at the COOH termini and the remaining exons are constitutively spliced. Alternative splicing has been demonstrated to occur in a tissue- and developmental stage-specific manner (Wilkinson *et al.*, 1986; Nadal-Ginard *et al.*, 1987). Konieczny and Emerson (1985) have concluded that the quail TN-I gene contains evolutionarily conserved control sequences that activate its transcription in response to differentiation-specific regulatory signals. The chicken TN-I gene has also been sequenced (Nikovits *et al.*, 1986).

3.4. Myosin Light Chains (MLC)

Myosin is composed of two heavy chains of 200 kDa and four light chains ranging in size from 17 to 21 kDa (Sivaramakrishnan and Burke, 1982). The first indication of myosin polymorphism was observed in MLC in fast and slow twitch muscles (Lowey and Risby, 1971). Each myosin head contains an MLC 2 light chain and either an MLC 1 or MLC 3 light chain (Lowey, 1986). The MLC 1 and MLC 3 light chains are considered the "alkali-soluble" light chains because of their resistance to dissociation from MHC in all conditions other than highly basic denaturing conditions. The alkali MLC are identical in sequence at the COOH terminus but are divergent at the NH_2 terminus of the protein (Frank and Weeds, 1974). The MLC 2 or DTNB (5,5'-dithiobis-2-nitrobenzoic acid) light chain is easier to dissociate from the MHC. MLC 2 plays a large role in the regulation of contraction in smooth muscles (Chacko et al., 1977) and invertebrate muscles (Szent-Gyorgyi et al., 1973).

A single gene codes for both the alkali light chains in rat (Periasamy et al., 1984a) and mouse (Robert et al., 1984), and differential RNA splicing is produced through the alternative use of two promoters separated by 10 kb of DNA in the rat (Nadal-Ginard et al., 1986, 1987). The isoform-specific exons are arranged in a manner necessitating alternative splicing, and the large separation of transcription initiation sites gives rise to two pre-mRNAs of different size and sequence (MLC 1 is 20 kb and MLC 3 is 10 kb in length). Some of these sequences may play a role in determining the alternative patterns of splicing of exons in an isoform-specific manner. This could occur in two ways. Sequences internal to the gene could act in a *cis* manner as enhancer sequences under certain conditions, or these sequences may be receptive to developmental stage- and tissue-specific *trans*-acting factors (or a combination of the two).

3.5. Myosin Heavy Chains (MHC)

The remainder of this chapter will focus on MHC gene structure and regulation of its expression, with emphasis on studies conducted in farm animals where possible. The MHC polypeptide consists of an NH_2-terminal globular head attached to the COOH-terminal rodlike tail. It is a large polypeptide, with 1939 amino acid residues in rat embryonic MHC (Strehler et al., 1986). MHC genes have been found in the genome as multigene families in a number of species including human, rabbit, rat, mouse, chicken, and the invertebrates *Dictyostelium discoideum*, *Acanthamoeba castellanii*, and *Caenorhabditis elegans* (Robbins et al., 1982; Leinwand et al., 1983a; Friedman et al., 1984; Periasamy et al., 1984b; Buckingham, 1985; Warrick et al., 1986; Hammer et al., 1986; Miller et al., 1986; Moriarity et al., 1987). Recent evidence also indicates a multigene family in the bovine system for MHC genes (Richter et al., 1987). The multiple molecular forms of myosin have different ATPase activities (Barany, 1967; Whalen, 1980; Chizzonite et al., 1982) and are expressed in a sequential developmental regime and in a tissue-specific manner. Since a relationship exists between myosin ATPase activity and muscle shortening (Barany, 1967), changes in myosin isoform expression reflect a physiological need for changes in the contractile properties of the muscle.

A high degree of amino acid sequence homology exists among the various MHC isoforms, including both intraspecies and interspecies sequence conservation. This is not surprising in view of the integral functional and structural importance of myosin to the contractile machinery. The percent homology ranges from approximately 47% between rabbit and nematode MHCs to greater than 90% among some vertebrate MHCs (Kavinsky *et al.*, 1983; Saez and Leinwand, 1986a,b; Young *et al.*, 1986). It is important to note that this intriguing sequence conservation in the coding regions for MHC does not exist in either the introns or the 3' and 5' untranslated sequences (Strehler *et al.*, 1986). However, there does appear to be a conservation of intron *location* in both the 3' and the 5' end regions of different MHC genes (Strehler *et al.*, 1985, 1986). Conservation of intron position at the 5' end is consistent with the possibility that these introns carry information required for regulated expression of the MHC genes.

Sequence divergence among the last few COOH-terminal amino acids and the 3' untranslated region (Mahdavi *et al.*, 1982; Wydro *et al.*, 1983) is consistent with the observation that these regions have evolved more rapidly than the regions coding for specific protein domains in myosin (Nadal-Ginard *et al.*, 1982). These regions appear to be highly specific for each MHC gene and reside in separate exons of MHC genes analyzed to date (Periasamy *et al.*, 1985; Strehler *et al.*, 1986). Analyses of divergent *3' end* regions of clones that have a high precentage of homology within coding regions have been reported for human sarcomeric genes (Saez and Leinwand, 1986b), chicken sarcomeric genes (Robbins *et al.*, 1982), rabbit cardiac genes (Friedman *et al.*, 1984), mouse sarcomeric and cardiac genes (Weydert *et al.*, 1983; Buckingham, 1985), and several rat genes (Izumo *et al.*, 1986; Mahdavi *et al.*, 1986). As discussed later, Izumo *et al.*, (1986) exploited this sequence divergence surrounding the 3' end and constructed subclones that recognize individual myosin isoforms with a high degree of specificity.

4. Isolation and Characterization of MHC Sequences

The 3' untranslated region sequence divergence has been exploited in the isolation of MHC coding sequences. Screening of genomic libraries or particular clones with a cDNA probe for MHC not necessarily of the same species has been carried out in chicken (Robbins *et al.*, 1982; Moriarity *et al.*, 1987), rat (Wydro *et al.*, 1983; Mahdavi *et al.*, 1984), *Drosophila* (Bernstein *et al.*, 1983), rabbit (Friedman *et al.*, 1984), and human (Leinwand *et al.*, 1983) systems. In the screening of a bovine genomic library, a quail embryonic MHC cDNA probe (Hastings and Emerson, 1982) containing both translated and nontranslated regions surrounding the 3' end of the gene was used (Richter *et al.*, 1987). Seven unique MHC clones containing insert sizes ranging from 11.5 to 17.5 kb were isolated. Purified DNA was digested singly or in combination with several restriction enzymes which recognized six base-pair sequences, and

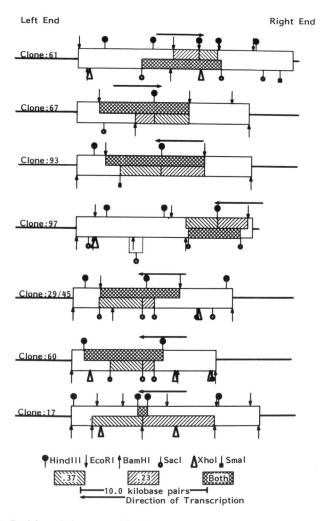

FIGURE 1. Partial restriction maps of bovine MHC genomic clones. Charon 28 lambda DNA is indicated by single lines. The left end refers to the *Bam*HI left arm, which is 23.6 kb; the right end refers to the *Bam*HI right arm, which is 9.2 kb. Insert DNA is indicated by double lines. Symbols for restriction enzymes are shown. The hatched areas indicate hybridization to either the 0.37-kb 3' probe or the 0.23-kb 5' probe, or to both probes. Arrows above the insert indicate the orientation of transcription 5' to 3'.

restriction sites were determined. Partial restriction maps for these seven clones are shown in Fig. 1, where the orientation of the inserted DNA with respect to the 23.6-kb left arm and 9.2-kb right arm of Charon 28 is shown, and the orientation of the direction of transcription is indicated by the arrow. Because the cDNA used in the screening contained sequences at the 3' end of the MHC mRNA as well as the 3' untranslated area, these clones should all contain sequences coding for the rod portion of the molecule. A number of interesting observations can be made with regard to these clones.

The restriction maps showed that two MHC clones, 29 and 45, were identical. None of the clones are large enough to encompass a complete MHC gene, which is approximately 25 kb in size inclusive of intron regions (Strehler *et al.*, 1986). Several features suggest that the clones are related to each other, and yet distinct. For example, MHC clones 17 and 93 do not contain any *Sac*I sites, whereas MHC clone 45/29 has five, 97 contains four, 61 and 60 contain three, and 67 contains only one. Where multiple *Sac*I sites exist, they seem to map very closely to the extreme 3' end of the gene. Every clone has a *Hind*III site just prior to the beginning of the untranslated region. Three clones (17, 45, 97) have a *Bam*HI site 2.5 to 3.5 kb downstream from the *Hind*III site. All clones had numerous *Pst*I sites (> 20 per clone). The recongition sequence for *Pst*I is (CTGCAG), which, if in the reading frame, would encode the dipeptide Leu-Gln, a frequently occurring sequence in the rod portion of MHC (Strehler *et al.*, 1986). *Sma*I restriction sites, which would encode the dipeptide Pro-Gly if in the correct reading frame, are almost entirely absent. This is consistent with the amino acid sequence of rat embryonic MHC, which has a low content of α-helix-breaking Pro and Gly residues in the rod portion of the molecule. In fact, secondary structure calculations predicted that the entire MHC rod, from the last Pro (Pro 838) to the COOH-terminus (Glu 1939), would be an uninterrupted α-helix (Strehler *et al.*, 1986). The *Sma*I sites that do exist map in areas of untranslated or flanking DNA sequence. When one focuses on the area of the clones that should correlate with the 3' untranslated region, clones 45 and 97 are closely related with minor differences in restriction sites, suggesting that these clones represent either similar genes or perhaps allelic variations of the same gene. However, this would still suggest the existence of at least six distinct MHC genes or portions of genes present in the bovine genome. Fragments encompassing the 3' untranslated region of each bovine MHC clone have been isolated and have been shown to be specific for each clone by demonstrating that under stringent conditions each fragment only hybridizes to the clone from which it was isolated (Richter, Hudson, and Young, unpublished data). Furthermore, we have shown that each fragment is unique by performing Southern hybridizations using bovine genomic DNA cut with the restriction enzyme used to isolate the fragment from the clone. Under these conditions, hybridization with the 3' fragment resulted in a single band for each clone.

To characterize definitively each of the bovine clones, DNA sequence analyses must be carried out and the sequences compared with known MHC isoforms from other species (Young *et al.*, 1986). Although our knowledge about the MHC gene family in cattle is clearly limited, detailed characterization of genomic and cDNA MHC clones from other species has been accomplished by hybridizing the MHC genomic or cDNA clones to RNA isolated from different tissues at various developmental stages (Robbins *et al.*, 1982; Mahdavi *et al.*, 1986). Observing the developmental pattern of MHC RNA production provides a first step toward understanding the regulation of MHC gene expression. For example, the most definitive work to date on the characterization of a sarcomeric MHC gene family has been in the rat (Mahdavi *et al.*, 1986), where cDNA and genomic DNA sequences coding for seven MHCs expressed in several striated muscle tissues have been cloned and characterized. These genes encode embryonic skeletal, neonatal skeletal, adult "fast" skeletal types IIB and IIA, ventricular-β, ventricular/atrial-α, and ocular muscle MHC isoforms. The two MHC isoforms in the ventricle, α- and β-MHC (Lompre *et al.*, 1984), are the only MHC isoforms present throughout the development of the heart (Lompre *et al.*, 1984), whereas at least five MHC isoforms are expressed in the leg skeletal muscle (Wydro *et al.*, 1983; Mahdavi *et al.*, 1986). From detailed analyses of the differential expression of MHC isoforms in leg skeletal muscle, it was found that the slow skeletal muscle and cardiac β-MHC isoforms are encoded by the same gene, whereas the cardiac α-MHC isoform is expressed only in the ventricle and atrium. Furthermore, the α-MHC gene is expressed in the atria early in development when there is no evidence for its expression in the ventricle, and the β-MHC gene is expressed at high levels in the slow muscle fibers of both adult and fetal muscle but is expressed at low levels in the adult ventricle. In both rabbits and humans (Lichter *et al.*, 1986), mRNA complementary to β-MHC DNA was detected in both heart muscle and skeletal muscle; however, human skeletal muscles are mixtures of fast and slow fibers, making it difficult to unequivocally demonstrate its presence in slow fibers as in rats and rabbits.

Additional studies have determined that multiple MHC isoforms can be expressed in a single muscle. Using probes specific for rat embryonic, neonatal skeletal, adult skeletal, and cardiac MHC isoforms, Mahdavi *et al.* (1986) showed that rat extraocular muscles expressed all known sarcomeric MHC genes, including the MHC isoform specific for these muscles (Wieczorek *et al.*, 1985). Obviously, the mechanism involved in the expression of MHC isoforms is quite complex as manifested by "incomplete" developmental stage- and tissue-specific expression.

A sequential expression of an embryonic to neonatal to three forms of adult MHC isoforms (two adult fast-oxidative IIA, glycolytic IIB, and slow-oxidative IA, which is identical to cardiac β-MHC) has been observed. This functional expression is also reflected structurally in terms of MHC gene organization in the rat where the skeletal MHC genes are linked in tandem (Mahdavi *et al.*, 1986). Linkage of skeletal MHC genes also exists in humans (Leinwand *et al.*, 1983a) where three MHC genes are located on the short arm of chromosome 17 and in the mouse where skeletal muscle MHC genes are clustered on chromosome 11. Furthermore, the mouse α-cardiac MHC gene has been localized to chromosome 14 (Weydert *et al.*, 1985), and because the α and β cardiac MHC genes are arranged in tandem in the rat (Mahdavi *et al.*, 1984), it is probable that the mouse cardiac β-MHC gene is also located on chromosome 14. Recent evidence suggests that two cardiac MHC genes also exist in humans (Catanzaro and Morris, 1986). These results in rats and mice raise the possibility that human cardiac genes may be located on a chromosome different from that of the skeletal MHC genes. Preliminary results obtained from *in situ* hybridization suggest that some MHC genes may be located on human chromosome 7 (Weydert *et al.*, 1985). The existence of tandemly arranged, sequentially expressed genes may reflect the requirement of a *cis*-acting regulatory mechanism (Weydert *et al.*, 1985). This is in contrast to results observed in the multigene families for actin and MLC where the genes are unlinked but are either coordinately expressed in a given phenotype or coexpressed during sarcomeric muscle development.

5. Effects of Various Stimuli on MHC Gene Expression

In view of the startling advances made in such a short time on the understanding of the molecular aspects of generating structural MHC isoforms, an understanding of the mechanisms that discriminantly signal the turning on or off of MHC genes is probably not as forthcoming. Indeed, no mechanism has yet been clearly identified as to how hormones (e.g., thyroxine) or physiological stimuli (e.g., innervation or exercise) induce or repress genes coding for myofibrillar proteins. Molecular biological techniques combined with classical biochemical, immunocytochemical, and immunochemical analyses will be needed for the complete understanding of the complexities of the differential effects of various stimuli on MHC gene expression.

5.1. Thyroid Hormone

As discussed above, MHC expression undergoes a transition from embryonic to neonatal to adult isoforms during mammalian and avian skeletal muscle development. Developmental transitions in cardiac muscle also occur. Many factors affect MHC isoform expression, including hormonal treatment (Hoh *et al.*, 1977; Lompre *et al.*, 1981), hemodynamic conditions (Mercadier *et al.*, 1981; Izumo *et al.*, 1987), age (Hoh *et al.*, 1977; Fitts *et al.*, 1984), altered contractile ability (Cerny and Bandman, 1986), exercise (Fitts *et al.*, 1984; Gregory *et al.*, 1986), muscle regeneration (Saad *et al.*, 1987), innervation (Whalen, 1985), electrical stimulation (Brown *et al.*, 1983) and pressure overload (Tsuchimochi *et al.*, 1986). Isoform changes appear to be regulated at the transcriptional level (Weydert *et al.*, 1983; Izumo *et al.*, 1986), illustrating the dynamic nature of muscle and its ability to respond to various stimuli. This phenomenon of muscle has been referred to as the "plasticity of the differentiated state" (Blau *et al.*, 1985), implying that although a muscle cell is terminally differentiated and does not undergo further DNA replication, its phenotype is not fixed. For example, skeletal muscle has the remarkable ability to respond to an increased pattern of use when constructed as ventricular pumping chambers (Acker *et al.*, 1987).

The most intensely studied effector of MHC isoform transitions is thyroid hormone and its effect on cardiac muscle. Expression of the various cardiac isomyosins (V1, V2, and V3, composed of MHC α_2, MHC $\alpha\beta$, and MHC β_2, respectively) in rats, mice, and rabbits (Clark *et al.*, 1982; Sinha *et al.*, 1982; Gustafson *et al.*, 1986; Buckingham *et al.*, 1986) has a significant effect on the contractile properties of the tissue. This is because the ATPase activities of the cardiac isomyosins decrease from V1 to V3, and the maximal speed of muscle shortening and the myosin ATPase activity are highly correlated (Barany, 1967). The amounts of the three isomyosins also vary during development (Schwartz *et al.*, 1982; Chizzonite *et al.*, 1982), under dfferent pathological conditions, and during aging, reflecting important adaptation mechanisms of the heart to its environment. Thyroid hormone treatment induces a change from the predominantly MHC-β (V3—low ATPase activity) form to MHC-α (V1—high ATPase activity) form in all species studied to date (Lompre *et al.*, 1984). In contrast, cardiac hypertrophy induced by pressure overload causes isozymic redistribution from MHC-α to MHC-β, and isoform transitions are observed at both the mRNA and protein levels (Tsuchimochi *et al.*, 1986). Interestingly, administration of thyroid hormone results in a reversal of the MHC transition induced by cardiac hypertrophy. In humans, studies on the MHC isoform content as well as the adaptability of the heart to pressure overload have shown that an isozyme redistribution from MHC-α to MHC-β takes place (Tsuchimochi *et al.*, 1986). Additional developmental studies have shown that MHC-α is expressed in the atrium from the early embryonic stage, and only MHC-β is expressed in embryonic ventricular myofibers. However, some of the myofibers replace MHC-β with MHC-α after birth. Curiously, two subtypes of MHC-β have been observed ($\beta1$ and $\beta2$) in the human heart, and it will be interesting to determine if each is coded by its own gene or if they result from differential splicing of the same gene.

Studies on the effect of thyroid hormone on skeletal muscle MHC isoform transitions appear to be more complicated. Early studies (Whalen, 1985) suggested the importance of thyroid hormone in MHC isoform transitions, where it was found that hypothyroid rats went through a normal embryonic to neonatal MHC transition, but transition into an adult fast form was inhibited in fast

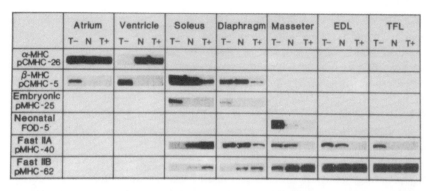

FIGURE 2. S1 nuclease mapping analysis of tissue-specific modulation of MHC genes by thyroid hormone. The density of the band reflects the amount of specific MHC mRNA present in the different tissues. A blank box indicates lack of hybridization to the corresponding MHC mRNA (Izumo *et al.*, 1986). T±, hypothyroid; N, euthyroid or normal thyroid levels; T−, hyperthyroid. Reprinted by permission.

skeletal muscles. Heterogeneity in thyroid responsiveness was indicated by the fact that thyroid hormone levels had little effect in developing slow muscles of hypothyroid rats. Recently, two groups have obtained results illustrating the complexity of effects of thyroid hormone on skeletal muscle MHC isoform expression. Gustafson *et al.* (1986) employed synthetic oligonucleotide probes specific for fast IIa and IIb, MHC-α and MHC-β isoforms, whereas Izumo *et al.* (1986) employed cDNA probes for six different MHC genes to observe effects of thyroid hormone in seven different muscles of adult rats subjected to hypo- or hyperthyroidism. The MHC isoforms studied were embryonic skeletal muscle, neonatal skeletal muscle, fast-twitch oxidative glycolytic (Fast IIa), fast-twitch glycolytic (Fast IIb), and slow-twitch (MHC-β). The results of this study are presented in Fig. 2, and several conclusions are apparent (Izumo *et al.*, 1986): (1) each muscle expresses more than one MHC gene; (2) each MHC gene analyzed is expressed in more than one muscle, but the "set" of MHC genes expressed and levels of expression are characteristic of muscle type; and (3) all muscles responded in some manner to thyroid hormone, but were not consistent for each MHC gene. The expression pattern may depend on tissue-specific factors as well. Clearly, all MHC genes expressed in adult striated muscles are affected by thyroid hormone in a complex tissue-specific manner. The action of thyroid hormone on MHC gene expression may be mediated by interaction with specific nuclear receptors (Oppenheimer and Samuels, 1983) that exert their effect at the transcriptional level. Furthermore, modulation by thyroid hormone of α- and β-MHC mRNAs in the cardiac ventricle seems to occur at the level of transcription rather than by alterations in mRNA stability

(Gustafson *et al.*, 1986). Requirements for tissue-specific *trans*-acting factors, neuronal signaling (Whalen, 1985), second messenger accumulation, or other hormonal and receptor involvement further complicate the production of a model for the effect of thyroid hormone on MHC gene expression in skeletal muscle.

5.2. Exercise/Electrical Stimulation

Of major interest to scientists working with farm animals are the effects of chemical, hormonal, or physical stimuli on the development of skeletal muscle. At the molecular level of MHC gene expression, there appear to be consistent transitions in MHC isoforms induced by electrical stimulation (Brown *et al.*, 1983), hypertrophy, or a combination of hypertrophic stress and exercise (Gregory *et al.*, 1986). A progression takes place from the fast MHC isoforms (Type IIb and Type IIa) to the slow Type I MHC in those muscles composed predominantly of "fast" fibers. This change from "fast" to "slow" MHC in skeletal muscle is consistent with changes in cardiac muscle where pressure overload causes a transition from MHC-α (high ATPase) to MHC-β (low ATPase). Replacement of the fast by the slow MHC reflects the physiological adaptation of a muscle fiber with a reduced intrinsic speed of shortening of the muscle (Barany, 1967). This is supported by the fact that in the slow soleus muscle, changes in response to hypertrophy and exercise were much less pronounced than those in the fast plantaris muscle, suggesting that the MHC isoforms present in the soleus are already suited to that physiological condition. However, an intermediate form (Type IIa) of myosin was found to disappear in the hypertrophied soleus, suggesting the existence of a Type IIb to Type IIa to Type I progression in the MHC isoform transition of fast to slow fibers (Gregory *et al.*, 1986). The fact that a particular muscle fiber type can be formed in response to various stimuli perhaps could help guide animal scientists toward finding modulators of muscle protein synthesis that are economically beneficial.

The amount of time for the induction or repression of a particular MHC isoform also varies, depending on the stimulus and the tissue. For example, administration of thyroid hormone resulted in an increase in α-MHC mRNA in the ventricle within 6–12 hr after treatment of hypothyroid rats (Gustafson *et al.*, 1986); however, at least 6 weeks of thyroid hormone administration was required in the soleus to increase the proportion of fast-oxidative (Type IIa) from 12% to 26%. Perhaps these effects could be explained by the apparent larger number of T_3 nuclear receptors in the ventricle compared to skeletal muscles (Morkin *et al.*, 1983), or to a more elaborate signaling process from T_3 nuclear receptors to the modulation of MHC genes in the soleus. When MHC transitions were induced by hypertrophy and exercise in fast- and slow-twitch muscles of rat hindlimbs (Gregory *et al.*, 1986), a period of 11 weeks elapsed in obtaining a threefold increase in slow myosin in the fast-twitch plantaris muscle. Before a molecular model can be proposed to depict the regulation of MHC gene expression in all the appropriate tissues, further study is needed to determine the *trans*-acting factors that are involved both at the critical initial signaling of the muscle fiber as well as from the nuclear membrane to the gene.

6. Alternative Splicing and MHC Gene Structure

Alternative RNA splicing plays an integral role in the generation of myofibrillar protein diversity that is subject to tissue- and developmental stage-specific regulation (Nadal-Ginard *et al.*, 1986, 1987). As discussed earlier, the genes coding for MLC 1 and MLC 3, α-TM, and TN-T all manifest differential splicing mechanisms, and the structure of these genes is well characterized (Nadal-Ginard *et al.*, 1986). Both cDNA and genomic clones coding for seven different isoforms of MHC in the rat have been isolated (Mahdavi *et al.*, 1986), and one of these genes has been sequenced (Strehler *et al.*, 1986). The existence of differential splicing of the primary transcript has not been observed in the generation of mature transcripts of MHC isoforms isolated from multigene families. In contrast to the MHC multigene families present in numerous species, *Drosophila melanogaster* contains only a single MHC gene (Bernstein *et al.*, 1983), yet two biochemically distinct forms of myosin have been identified (Bernstein *et al.*, 1986), suggesting the existence of more than one form of MHC and the involvement of differential RNA splicing. In fact, four different mature transcripts have been found to be produced from the *Drosophila* MHC gene (Bernstein *et al.*, 1986; Rozek and Davidson, 1986). The transcripts are differentiated by the presence or lack of an exon, and by choice of polyadenylation site that generates an alternative COOH terminus. Transcripts are generated in a tissue- and developmental stage-specific manner, where one form accumulates during both larval and adult muscle differentiation and another isoform accumulates only during adult muscle differentiation in the thoracic region. It was speculated that tissue-specific recognition of splicing signals is involved in this process.

In contrast to the *Drosophila* system, alternative splicing seems not to be involved in the differential expression of the vertebrate MHC gene families studied to date. Several factors suggest that *cis*-acting elements interact with tissue- and developmental stage-specific *trans*-acting factors in the regulation of MHC isoform expression. First, MHC gene structure is split by many introns varying in size and sequence. In MHC genes analyzed to date, intron locations are highly conserved in the 5′ end region, suggesting a functional importance in their regulated expression (Strehler *et al.*, 1986). Second, the 5′ untranslated sequence of vertebrate MHC genes are interrupted by introns as well. The coupling of these two observations with the fact that regulatory sequences can be located in 5' region introns (Gillies *et al.*, 1983; Queen and Baltimore, 1983) strongly suggests some role for these sequences in the regulated and differential expression of MHC genes. Transient gene expression studies were conducted with minigene constructs (Nadal-Ginard *et al.*, 1987) composed of the upstream gene sequences of the embryonic and cardiac α- and β-MHC genes fused to exons 1–3 and 37–41 of the embryonic MHC. CAT gene constructs (Gorman *et al.*, 1982) to these upstream sequences were also employed for transient gene expression assays in cultured HeLa cells, C_2 myoblasts, and myotubes. Expression of these constructs was cell type-specific as well as differentiation stage-specific. Thyroid hormone-induced regulation appears to involve the interaction of two 5' regulating elements with tissue-specific *trans*-acting factors. Surprisingly, these factors are not restricted to muscle cells. Therefore, the regulation of cardiac and skeletal muscle MHC isoforms involves the interaction of *cis*-acting sequences with *trans*-acting factors, some of which are specific to certain muscle cell types and others that are present in both muscle and nonmuscle cells.

7. Conclusions

Recombinant DNA technology has been invaluable in elucidating both the primary structure and arrangement of myofibrillar protein genes. By using probes specific for the MHC isoforms, the differential expression of these genes has been shown to occur in a tissue- and developmental stage-specific manner. The existence of such a large number of isoforms in myofibrillar proteins reflects a high degree of heterogeneity in function during development. Currently, work is under way to gain an understanding of the processes necessary for the coordination of muscle gene expression during differentiation and development. This will involve studying the mode of action of *cis*-acting elements in upstream regions, as well as those that exist intragenically. Isolation and characterization of *trans*-acting molecules and how they interact with the genes are crucial to understanding the mechanisms by which myofibrillar protein gene expression is modulated by hormones and other effector molecules.

The general lack of information on myofibrillar gene structure and function in large farm animals is surprising in view of the importance of this information to the animal scientist and farmer for meat animal production and performance. Hopefully, cooperation between animal scientists and molecular biologists will be fostered as this will benefit the growth and understanding of the regulation of myofibrillar protein gene expression.

References

Acker, M. A., R. L. Hammond, J. D. Mannion, S. Salmons and L. W. Stephenson. 1987. Skeletal muscle as the potential power source for a cardiovascular pump: assessment in vivo. Science 236:324.

Barany, M. 1967. ATPase activity of myosin correlated with speed of muscle shortening. J. Gen. Physiol. 50:197.

Basi, G. S., M. Boardman and R. V. Storti. 1984. Alternative splicing of a Dorsophila tropomyosin gene generates muscle tropomyosin isoforms with different carboxy-terminal ends. Mol. Cell. Biol. 4:2828.

Bernstein, S. I., C. J. Hansen, K. D. Becker, D. R. Wassenberg, II, E. S. Roche, J. J. Donady and C. P. Emerson, Jr. 1986. Alternative RNA splicing generates transcripts encoding a thorax-specific isoform of Drosophila melanogaster myosin heavy chain. Mol. Cell. Biol. 6:2511.

Bernstein, S. I., K. Mogami, J. J. Donady and C. P. Emerson, Jr. 1983. Drosophila muscle myosin heavy chain encoded by a single gene in a cluster of muscle mutations. Nature 302:393.

Blau, H. M., G. K. Pavlath, E. C. Hardeman, C. P. Chiu, L. Silberstein, S. G. Webster, S. C. Miller and C. Webster. 1985. Plasticity of the differentiated state. Science 230:758.

Breitbart, R. E., H. T. Nguyen, R. M. Medford, A. T. Destree, V. Mahdavi and B. Nadal-Ginard. 1985. Intricate combinatorial patterns on exon splicing generate multiple regulated troponin T isoforms from a single gene. Cell 41:67.

Brown, W. E., S. Salmons and R. G. Whalen. 1983. The sequential replacement of myosin subunit isoforms during muscle type transformation induced by long term electrical stimulation. J. Biol. Chem. 258:14686.

Breitbart, R. and B. Nadal-Ginard. 1987. Developmentally induced, muscle specific trans factors control the differential splicing of alternative and constitutive troponin-T exons. Cell 49:793.

Buckingham, M. E. 1985. Actin and myosin multigene families: their expression during the formation of skeletal muscle. Essays Biochem. 20:77.

Buckingham, M., S. Alonso, P. Barton, A. Cohen, P. Daubas, I. Garner, B. Robert and A. Weydert. 1986. Actin and myosin multigene families: their expression during the formation and maturation of striated muscle. Amer. J. Med. Genet. 25:623.

Buckingham, M., S. Alonso, G. Bugaisky, P. Barton, A. Cohen, P. Daubas, A. Minty, B. Robert and A. Weydert. 1985. The actin and myosin multigene families. Adv. Exp. Med. 182:333.

Catanzaro, D. F. and B. J. Morris. 1986. Human cardiac myosin heavy chain genes. Isolation of a genomic clone and its characterization and of a second unique clone also present in the human genome. Circ. Res. 59:655.

Cerny, L. C. and E. Bandman. 1986. Contractile activity is required for the expression of neonatal myosin heavy chain in embryonic chick pectoral muscle cultures. J. Cell Biol. 103:2153.

Chacko, S., M. A. Conti and R. S. Adelstein. 1977. Effect of phosphorylation of smooth muscle myosin on actin-activation and Ca^{2+} regulation. Proc. Natl. Acad. Sci. USA 74:129.

Chizzonite, R. A., A. W. Everett, W. A. Clark, S. Jakovcic, M. Rabinowitz and R. Zak. 1982. Isolation and characterization of two molecular variants of myosin heavy chain from rabbit ventricle. J. Biol. Chem. 257:2056.

Clark, W. A., R. A. Chizzonite, A. W. Everett, M. Rabinowitz and R. Zak. 1982. Species correlations between cardiac isomyosins. J. Biol. Chem. 257:5449.

Czosnek, H., U. Nudel, Y. Mayer, P. E. Barker, D. D. Pravtcheva, F. R. Ruddle and D. Yaffe. 1983. The genes coding for the cardiac muscle actin, the skeletal muscle actin and the cytoplasmic B-actin are located on three different mouse chromosomes. EMBO J. 2:1977.

Fitts, R. H., J. P. Troup, F. A. Witzmann and J. O. Holloszy. 1984. The effect of ageing and exercise on skeletal muscle function. Mech. Ageing Dev. 27:161.

Frank, G. and A. G. Weeds. 1974. The amino acid sequence of the alkali light chains of rabbit skeletal muscle myosin. Eur. J. Biochem. 44:317.

Friedman, D. J., P. K. Umeda, A. M. Sinha, H. J. Hsu, S. Jakovcic and M. Rabinowitz. 1984. Isolation and characterization of genomic clones specifying rabbit α and β ventricular myosin heavy chains. Proc. Natl. Acad. Sci. USA 81:3044.

Gillies, S. D., S. L. Morrison, V. T. Oi and S. Tonegawa. 1983. A tissue-specific transcription enhancer element is located in the major intron of a rearranged immunoglobulin heavy chain gene. Cell 33:717.

Gorman, C. M., L. F. Moffat and B. H. Howard. 1982. Recombinant genomes which express chloramphenicol acetyltransferase in mammalian cells. Mol. Cell. Biol. 2:1044.

Gregory, P., R. B. Low and W. S. Stirewalt. 1986. Changes in skeletal-muscle myosin isoenzymes with hypertrophy and exercise. Biochem. J. 238:55.

Gunning, P., P. Ponte, H. Blau and L. Kedes. 1983. α-Skeletal and cardiac actin genes are coexpressed in adult human skeletal muscle and heart. Mol. Cell. Biol. 3:1985.

Gunning, P., P. Ponte, L. Kedes, R. Eddy and T. Shows. 1984. Chromosomal location of the co-expressed human skeletal and cardiac actin genes. Proc. Natl. Acad. Sci. USA 81:1813.

Gustafson, T. A., B. E. Markham and E. Morkin. 1986. Effects of thyroid hormone on α-actin and myosin heavy chain gene expression in cardiac and skeletal muscles of the rat: measurement of mRNA content using synthetic oligonucleotide probes. Circ. Res. 59:194.

Hammer, J. A., E. D. Korn and B. M. Paterson. 1986. Isolation of a non-muscle myosin heavy chain gene from *Acanthamoeba*. J. Biol. Chem. 261:1949.

Hastings, K. E. M. and C. P. Emerson, Jr. 1982. cDNA clone analysis of six co-regulated mRNAs encoding skeletal muscle contractile proteins. Proc. Natl. Acad. Sci. USA 79:1553.

Helfman, D. M., S. Cheley, E. Kuismanen, L. A. Finn and Y. Yamawaki-Kataoka. 1986. Non-muscle and muscle tropomyosin isoforms are expressed from a single gene by alternative RNA splicing and polyadenylation. Mol. Cell. Biol. 6:3582.

Hoh, J., P. McCrath and P. Hale. 1977. Electrophoretic analysis of multiple forms of rat cardiac myosin: effects of hypophysectomy and thryoxine replacement. J. Mol. Cell. Cardiol. 10:1053.

Izumo, S., A. M. Lompre, R. Matsuoka, G. Koren, K. Schwartz, B. Nadal-Ginard and V. Mahdavi. 1987. Myosin heavy chain messenger RNA and protein isoform transitions during cardiac hypertrophy. J. Clin. Invest. 79:970.

Izumo, S., B. Nadal-Ginard and V. Mahdavi. 1986. All members of the MHC multigene family respond to thyroid hormone in a highly tissue-specific manner. Science 231:597.

Kavinsky, C. J., P. K. Umeda, A. M. Sinha, M. Elzing, S. W. Tong, R. Zak, S. Jackovcic and M. Rabinowitz. 1983. Cloned mRNA sequences for two types of embryonic myosin heavy chains from chick skeletal muscle. J. Biol. Chem. 258:5196.

Konieczny, S. F. and C. P. Emerson, Jr. 1985. Differentiation, not determination, regulates muscle gene activation: transfection of troponin I genes into multipotential and muscle lineages of 10T1/2 cells. Mol. Cell. Biol. 5:2423.

Leinwand, L. A., R. E. K. Fournier, B. Nadal-Ginard and T. Shows. 1983a. Multigene family for sarcomeric myosin heavy chain in mouse and human DNA: localization on a single chromosome. Science 221:766.

Leinwand, L. A., L. Saez, E. McNally and B. Nadal-Ginard. 1983b. Isolation and characterization of human myosin heavy chain genes. Proc. Natl. Acad. Sci. USA 80:3716.

Lichter, P., P. K. Umeda, J. E. Levin and H. P. Vosberg. 1986. Partial characterization of the human β-myosin heavy-chain gene which is expressed in heart and skeletal muscle. Eur. J. Biochem. 160:419.

Lompre, A. M., J. J. Mercadier, C. Wisnewsky, P. Bouveret, C. Pantaloni, A. D'Albis and K. Schwartz. 1981. Species- and age-dependent changes in the relative amounts of cardiac myosin isoenzymes in mammals. Dev. Biol. 84:286.

Lompre, A. M., B. Nadal-Ginard and V. Mahdavi. 1984. Expression of the cardiac ventricular α- and β-myosin heavy chain genes is developmentally and hormonally regulated. J. Biol. Chem. 259:6437.

Lowey, S. 1986. Cardiac and skeletal muscle myosin polymorphism. Med. Sci. Sports Exer. 18:284.

Lowey, S. and D. Risby. 1971. Light chains from fast and slow muscle myosins. Nature 234:81.

MacLeod, A. R., K. Talbot, F. C. Reinach, C. Houlker, L. B. Smillie, C. S. Giometti and N. L. Anderson. 1986. Molecular characterization of human cytoskeletal tropomyosins. In: C. P. Emerson, D. Fischman, B. Nadal-Ginard and M. A. Q. Siddiqui (Ed.) Molecular Biology of Muscle Development. pp 445–455. Alan R. Liss, Inc., New York.

Mahdavi, V., A. P. Chambers and B. Nadal-Ginard. 1984. Cardiac alpha and beta myosin heavy chain genes are organized in tandem. Proc. Natl. Acad. Sci. USA 81:2626.

Mahdavi, V., M. Periasamy and B. Nadal-Ginard. 1982. Molecular characterization of two myosin heavy chain genes expressed in the adult heart. Nature 297:659.

Mahdavi, V., E. E. Strehler, M. Periasamy, D. F. Wieczorek, S. Izumo and B. Nadal-Ginard. 1986. Sarcomeric myosin heavy chain gene family: organization and pattern of expression. Med. Sci. Sports Exer. 18:299.

Mak, A. S., W. G. Lewis and L. B. Smillie. 1979. Amino acid sequence of rabbit skeletal β and cardiac tropomyosins. FEBS Lett. 105:232.

Mercadier, J., A. Lompre, C. Wisnewsky, J. Samuel, J. Bercovice, B. Swynghedauw and K. Schwartz. 1981. Myosin isozyme changes in several models of rat cardiac hypertrophy. Circ. Res. 49:525.

Miller, D. M., F. E. Stockdale and J. Karn. 1986 Immunological identification of the genes encoding the four myosin heavy chain isoforms of *Caenorhabditis elegans*. Proc. Natl. Acad. Sci. USA 83:2305.

Minty, A. J., S. Alonso, M. Caravetti and M. E. Buckingham. 1982. A fetal skeletal muscle actin mRNA in the mouse, and its identity with cardiac actin mRNA. Cell 30:185.

Minty, A., H. Blau and L. Kedes. 1986. Two-level regulation of cardiac actin gene transcription: muscle-specific modulating factors can accumulate before gene activation. Mol. Cell. Biol. 6:2137.

Minty, A. and L. Kedes. 1986. Upstream regions of the human cardiac actin gene that modulate its transcription in muscle cells: presence of an evolutionary conserved repeated motif. Mol. Cell. Biol. 6:2125.

Moriarity, D. M., K. J. Barringer, J. B. Dodgson, H. E. Richter and R. B. Young. 1987. Genomic clones encoding chicken myosin heavy-chain genes. DNA 6:91.

Morkin, E., I. L. Flink and S. Goldman. 1983. Biochemical and physiologic effects of thyroid hormone on cardiac performance. Prog. Cardiovasc. Dis. 25:435.

Nebeshima, Y., Y. Kurijama-Fujii, M. Muramatsu and K. Ogata. 1984. Alternative transcription and two modes of splicing result in two myosin light chains from one gene. Nature 308:333.

Nadal-Ginard, B., R. E. Breitbart, A. Andreadis, M. Gallego, Y. T. Yu, G. Koren, G. White, P. Bouvagnet and V. Mahdavi. 1987 Generation of complex contractile protein phenotypes through promoter selection and alternative pre-mRNA splicing. ICSU Short Rep. 7:62.

Nadal-Ginard, B., R. E. Breitbart, E. E. Strehler, N. Ruiz-Opazo, M. Periasamy and V. Mahdavi. 1986. Alternative splicing: a common mechanism for the generation of contractile protein diversity from single genes. In: C. P. Emerson, D. Fischman, B. Nadal-Ginard and M. A. Q. Siddiqui (Ed.) Molecular Biology of Muscle Development. pp. 387–410. Alan R. Liss, Inc., New York.

Nadal-Ginard, B., R. M. Medford, H. T. Nguyen, M. Periasamy, R. M. Wydro, D. Hornig, R. Gabits, L. I. Garfinkel, D. Weiczorek, E. Bekesi and V. Mahdavi. 1982. Structure and regulation of a mammalian sarcomeric myosin heavy chain gene. In: M. L. Pearson and H. F. Esptein (Ed.) Muscle Development: Molecular and Cellular Control. pp 143–168. Cold Spring Harbor Laboratory, Cold Spring Harbor, NY.

Nikovits, W., Jr., G. Kuncio and C. P. Ordahl. 1986. The chicken fast skeletal troponin I gene: exon organization and sequence. Nucleic Acids Res. 14:3377.

Oppenheimer, J. H. and H. H. Samuels (Ed.). 1983. Molecular Basis of Thyroid Hormone Action. Academic Press, New York.

Periasamy, M., E. E. Strehler, L. I. Garfinkel, R. M. Gubits, N. Ruiz-Opazo and B. Nadal-Ginard. 1984a. Fast skeletal muscle myosin light chains 1 and 3 are produced from a single gene by a combined process of differential RNA transcription and splicing. J. Biol. Chem. 259:13595.

Periasamy, M., D. F. Weiczorek and B. Nadal-Ginard. 1984b. Characterization of a developmentally regulated perinatal myosin heavy-chain gene expressed in skeletal muscle. J. Biol. Chem. 259:13573.

Periasamy, M., R. M. Wydro, M. A. Strehler-Page, E. E. Strehler and B. Nadal-Ginard. 1985. Characterization of cDNA and genomic sequences corresponding to an embryonic myosin heavy chain. J. Biol. Chem. 260:15856.

Queen, C. and D. Baltimore. 1983. Immunoglobulin gene transcription is activated by downstream sequence elements. Cell 33:741.

Richter, H. E., R. B. Young, J. R. Hudson, Jr. and D. M. Moriarity. 1987. Screening of a bovine genomic library for myosin heavy chain genes. J. Anim. Sci. 64:607.

Robbins, J., G. A. Freyer, D. Chisholm and T. C. Gilliam. 1982. Isolation of multiple genomic sequences coding for chicken myosin heavy chain protein. J. Biol. Chem. 257:549.

Robert, B., P. Daubas, M. A. Akimenko, A. Cohen, I. Garner, J. L. Guenet and M. Buckingham. 1984. A single locus in the mouse encodes both myosin light chains 1 and 3, a second locus corresponds to a related pseudogene. Cell 39:129.

Rozek, C. E. and N. Davidson. 1986. Differential processing of RNA transcribed from the single-copy Drosophila myosin heavy chain gene produces four mRNAs that encode two polypeptides. Proc. Natl. Acad. Sci. USA 83:2128.

Ruiz-Opazo, N., J. Weinberger and B. Nadal-Ginard. 1983. Different tissue specific forms of α-tropomyosin encoded by the same gene. J. Cell Biol. 97:329a.

Ruiz-Opazo, N., J. Weinberger and B. Nadal-Ginard. 1985. One smooth and two striated, skeletal and cardiac, α-tropomyosin isoforms are encoded by the same gene. Nature 315:67.

Saad, A. D., T. Obinata and D. A. Fischman. 1987. Immunochemical analysis of protein isoforms in thick myofilaments of regenerating skeletal muscle. Dev. Biol. 119:336.

Saez, L. and L. A. Leinwand. 1986a. Characterization of diverse forms of myosin heavy chain expressed in adult human skeletal muscle. Nucleic Acids Res. 14:2951.

Saez, L. J. and L. A. Leinwand. 1986b. Cloning and characterization of myosin cDNAs in adult human skeletal muscle. In: C. P. Emerson, D. Fischman, B. Nadal-Ginard and M. A. Q. Siddiqui (Ed.) Molecular Biology of Muscle Development. pp 263–272. Alan R. Liss, Inc., New York.

Schwartz, K., A. M. Lompre, P. Bouveret, C. Wisnewsky and R. G. Whalen. 1982. Comparisons of rat cardiac myosins at fetal stages in young animals and in hypothyroid adults. J. Biol. Chem. 257:14412.

Seiler-Tuyns, A., J. D. Eldridge and B. M. Paterson. 1984. Expression and regulation of chicken actin genes introduced into mouse myogenic and nonmyogenic cells. Proc. Natl. Acad. Sci. USA 81:2980.

Singer, R. H. and G. Kessler-Icekson. 1978. Stability of polyadenylated RNA in differentiating myogenic cells. Eur. J. Biochem. 88:395.

Sinha, A. M., P. K. Umeda. C. J. Kavinsky, C. Rajamanickam, H. J. Hsu, S. Jakovcic and M. Rabinowitz. 1982. Molecular cloning of mRNA sequences for cardiac α- and β-form myosin heavy chains: expression in ventricles of normal, hypothyroid and thyrotoxic rabbits. Proc. Natl. Acad. Sci. USA 79:5847.

Sivaramakrishnan, M. and M. Burke. 1982. The free heavy chain of vertebrate skeletal myosin subfragment I shows full enzymatic activity. J. Biol. Chem. 257:1102.

Sodek, J., R. S. Hodges and L. B. Smillie. 1978. Amino acid sequence of rabbit skeletal muscle α-tropomyosin: the COOH terminal half (residue 142–284). J. Biol. Chem. 253:1129.

Squire, J. 1981. The Structural Basis of Muscular Contraction. Plenum Press, New York.

Strehler, E. E., V. Mahdavi, M. Periasamy and B. Nadal-Ginard. 1985. Intron positions are conserved in the 5′ end region of myosin heavy chain genes. J. Biol. Chem. 260:4680.

Strehler, E. E., M. A. Strehler-Page, J. C. Perriard, M. Periasamy and B. Nadal-Ginard. 1986. Complete nucleotide and encoded amino acid sequence of a mammalian myosin heavy chain gene. Evidence against intron-dependent evolution of the rod. J. Mol. Biol. 190:291.

Szent-Gyorgyi, A. G., E. M. Szentkiralyi and J. Kendrick-Jones. 1973. The light chains of scallop myosin as regulatory subunits. J. Mol. Biol. 74:179.

Tsuchimochi, H., M. Kuro-O, F. Takaku, K. Yoshida, M. Kawana, S. I. Kimata and Y. Yazaki. 1986. Expression of myosin isozymes during the developmental stage and their redistribution induced by pressure overload. Jpn. Circ. J. 50:1044.

Vandekerckhove, J. and K. Weber. 1979. The complete amino acid sequence of actins from bovine aorta, bovine heart, bovine fast skeletal muscle, and rabbit slow skeletal muscle. Differentiation 14:123.

Wagner, P. D. and E. Giniger. 1981. Hydrolysis of ATP and reversible binding to F-actin by myosin heavy chains free of all light chains. Nature 292:560.

Warrick, H. M., A. DeLozanne, L. A. Leinwand and J. A. Spudich. 1986. Conserved protein domains in a myosin heavy chain gene from *Dictyostelium discoideum*. Proc. Natl. Acad. Sci. USA 83:9433.

Weydert, A., P. Daubas, M. Caravetti, A. Minty, G. Bugaisky, A. Cohen, B. Robert and M. Buckingham. 1983. Sequential accumulation of mRNAs encoding different myosin heavy chain isoforms during skeletal muscle development in vivo detected with a recombinant plasmid identified as coding for an adult fast myosin heavy chain from mouse skeletal muscle. J. Biol. Chem. 258:13867.

Weydert, A., P. Daubas, I. Lazaridis, P. Barton, I. Garner, D. P. Leader, F. Bonhomme, J. Catalan, D. Simon, J. L. Guenet, F. Gros and M. E. Buckingham. 1985. Genes for skeletal muscle myosin heavy chain are clustered and are not located on the same mouse chromosome as a cardiac myosin heavy chain gene. Proc. Natl. Acad. Sci. USA 82:7183.

Whalen, R. G., G. S. Butler-Browne, S. M. Sell, K. Schwartz, P. Bouveret and I. Pinset-Harstrom. 1980. Transitions in myosin isoforms during muscle development. J. Muscle. Res. Cell Motility 1:473.

Whalen, R. C. 1985. Myosin isoenzymes as molecular markers for muscle physiology. J. Exp. Biol. 115:43.

Wieczorek, D. F., M. Periasamy, G. S. Butler-Browne, R. G. Whalen and B. Nadal-Ginard. 1985. Co-expression of multiple myosin heavy chain genes, in addition to a tissue-specific one, in extraocular musculature. J. Cell Biol. 101:618.

Wilkinson, J. M., I. M. Bird, G. K. Dhoot, B. A. Levine and R. D. Taylor. 1986. Expression of troponin T variants in chicken striated muscle. In: C. P. Emerson, D. Fischman, B. Nadal-Ginard and M. A. Q. Siddiqui (Ed.) Molecular Biology of Muscle Development. pp 423–436. Alan R. Liss, Inc., New York.

Wilkinson, J. M., A. J. G. Moir and M. D. Waterfield. 1984. The expression of multiple forms of troponin T in chicken-fast-skeletal muscle may result from differential splicing of a single gene. Eur. J. Biochem. 143:47.

Wydro, R. M., H. T. Nguyen, R. M. Gubits and B. Nadal-Ginard. 1983. Characterization of sarcomeric myosin heavy chain genes. J. Biol. Chem. 258:670.

Young, R. B., D. M. Moriarity and C. E. McGee. 1986. Structural analysis of myosin genes using recombinant DNA techniques. J. Anim. Sci. 63:259.

CHAPTER 7

Regulation of Growth by Negative Growth Regulators

DENNIS R. CAMPION AND WILLIAM KELLY JONES, JR.

1. Introduction

The concept of growth regulation by endogenous inhibitors was first suggested by studies on wound healing and carcinogenesis (see Wang and Hsu, 1986). This area of research has long suffered from the fact that purification of specific compounds that inhibited cell proliferation (or differentiation) of normal cells proved elusive. Although circumstantial evidence was generated over the last 20 years to indicate their existence, studies involving the inhibition of cell proliferation have been the subject of both criticism and controversy (see Iype and McMahon, 1984: Alison, 1986). With the purification of several inhibitiors of cell proliferation from endogenous sources, a firm experimental basis has now been established and this subject area is coming under more intense investigation. Caution must still be raised with respect to the specific physiological role(s) that any of the inhibitors purified or described to date have in regulating the normal processes of growth and development as this kind of evidence has not been reported.

2. Transforming Growth Factor-β

Transforming growth factor (TGF) activity was originally discovered in conditioned medium from Moloney murine sacroma virus-transformed mouse

DENNIS R. CAMPION • United States Department of Agriculture, Agricultural Research Service, Midwest Area, Peoria, Illinois 61604. *WILLIAM KELLY JONES, JR.* • Nutrition Division, Kraft, Inc., Glenview, Illinois 60025.

cells (De Larco and Todaro, 1978). It was immediately realized that several proteins in conditioned media from neoplastic cells could induce anchorage-independent growth of target cells (see Lawrence, 1985, for a recent review). Subsequently, TGF-β was identified in culture media conditioned by neoplastic cells (Moses *et al.*, 1981; Roberts *et al.*, 1981) and then in a variety of non-neoplastic cells. Although TGF-β was named for its ability to stimulate, under appropriate conditions, anchorage-independent growth of several cell types, it may inhibit, stimulate, or not influence proliferation or differentiation of diverse cell types under anchorage-dependent growth conditions. Thus, TGF is really a misnomer.

TGF-β has been purified from tissues of several species including bovine kidney (Roberts *et al.*, 1983a) and bone (Seyedin *et al.*, 1987), porcine (Cheifetz *et al.*, 1987) and human (Childs *et al.*, 1982; Assoian *et al.*, 1983) platelets, and human placenta (Frolik *et al.*, 1983). The highest concentration is found in the α-granule of the platelet. As determined by cDNA probes for TGF-β mRNA (Derynck *et al.*, 1985), however, TGF-β can be synthesized locally by a variety of cell types. Local production suggests that it may function in an autocrine manner as it has been shown to influence growth of a host of cell types in culture.

The molecule consists of two identical chains cross-linked by disulfide bonds. The molecular mass of the homodimer is 25 kDa; each chain is composed of 112 amino acids (Assoian *et al.*, 1983; Frolik *et al.*, 1983; Roberts *et al.*, 1983a; Derynck *et al.*, 1985, 1986). The gene for TGF-β maps to human chromosome 19 and to mouse chromosome 7 (Fujii *et al.*, 1986). TGF-β has been highly conserved as only one amino acid substitution distinguishes mouse from human TGF-β (Derynck *et al.*, 1985, 1986). In addition, the sequence of the NH_2-terminus residue of bovine TGF-β is identical to the like peptide of human TGF-β (Seyedin *et al.* 1986). Cheifetz *et al.*, (1987) reported the existence of three forms of TGF-β in the pig. TGF-β_1 was analogous to the homodimeric form of TGF-β described in other species. TGF-β_2 was found to be a new homodimer and TGF-$\beta_{1.2}$ was found to be a heterodimer consisting of one TGF-β_1 chain and one TGF-β_2 chain. The three molecules are similar in molecular weight, but the two basic types differ in their NH_2-terminal amino acid sequence.

TGF-β originates from at least one of a family of genes. Some degree of homology has been shown with cartilage-inducing factor-B (Seyedin *et al.*, 1987), with the COOH-terminal domain of Müllerian inhibitory substance (Cate *et al.*, 1986), and with inhibin (Mason *et al.*, 1985, 1986). Cartilage-inducing factors-A and -B have been purified from bovine bone. Cartilage-inducing factor-A is now recognized to be identical to TGF-β (Seyedin *et al.*, 1986). Holley *et al.* (1978) identified a growth inhibitor in conditioned media from confluent cultures of African green monkey kidney cells (BSC-1). Based on similarities in biological activity, molecular characteristics, and cross-reactivity with TGF-β receptor, it has been suggested that this inhibitor is identical, or at least very similar, to TGF-β (Holley et al., 1980; Tucker et al., 1984b; Ristow, 1986).

The cellular actions of TGF-β are presumbly mediated through specific, high-affinity call membrane receptors (Tucker et al., 1984a; Massagué and Like, 1985). Virtually every cell type examined has been shown to possess TGF-β receptors (Sporn et al., 1986). In many cell types, a 560-kDa disulfide-linked glycoprotein is the predominate form (Massagué, 1985; Fanger et al., 1986). This form consists of a 280- to 330-kDa subunit. But other distinct receptors have been described. Cheifetz et al. (1986) described the cellular distribution of a type I (65-kDa) and a type II (85-kDa) receptor. The latter two were not subunits of a larger disulfide-linked receptor complex (Ignotz and Massagué, 1985). Structural similarity exists between the high-molecular-weight receptor and the 65-kDa receptor (Cheifetz et al., 1986). Dissociation coefficients vary from 1 pM (Sporn et al., 1986) to about 1 nM (Ignotz and Massagué, 1985). The three receptor forms may, or may not, be present on each cell type. Based on cross-linking studies, chick embryo myoblasts and rat L6E9 and L8 myoblasts possess only the type I and type II receptors (Massagué et al., 1986). On the other hand, the proadipocyte cell line, 3T3 L1, exhibits predominately the high-molecular-weight species and, to a much lesser extent, the type I and II receptors. In addition, a 135-kDa species was identified in the latter cell type. The relation between heterogeneity of receptor type and cellular response is not known. Indeed, there is no conclusive evidence to suggest whether or not the actions mediated by TGF-β require occupancy of one, all, or a combination of the several apparent receptor types. The 65- and 85-kDa receptors can distinguish between TGF-β_1 and TGF-β_2 as the affinity of these receptors for TGF-β_2 is about one order of magnitude lower than that for TGF-β_1. The high-molecular-weight species could not differentiate between the two homodimers (Cheifetz et al., 1987).

The action of TGF-β is not linked to tyrosine phosphorylation of the cellular 42-kDa proteins (Libby et al., 1986). This suggests that the TGF-β receptor is neither a tyrosine-specific protein kinase nor a stimulator of phosphatidylinositol turnover (Libby et al., 1986). cAMP accumulation is also not associated with the action of TGF-β (Dodson and Schomberg, 1987). Thus, the precise mechanism by which TGF-β exerts its actions is not known. With respect to the mitogenic action of TGF-β, activity may be mediated through induction of c-sis mRNA (Leof et al., 1986), but a cause-and-effect relationship has not been proven.

TGF-β was reported by Massagué et al. (1986) to inhibit terminal differentiation of mammalian (rat L6E9 and L8) myoblasts and chicken skeletal muscle (primary) myoblasts. Expression of differentiation-specific, contractile protein genes was prevented *in vitro* in the presence of picomolar concentrations of TGF-β. Half-maximal inhibition of creatine phosphokinase activity occurred at approximately 10 pM with maximal inhibition at 50 pM. The inhibitory effect on differentiation was dependent on stage of cell commitment to fusion at the time of addition of TGF-β containing medium. Differentiation was blocked only when TGF-β was added to cultures in which the cells had not committed to differentiation. The inhibition was reversed once TGF-β was removed from the medium. The stage dependency of inhibition was not related to any detectable change in receptor number or affinity. Fused cells contained functional receptors as TGF-β was able to stimulate collagen and fibronectin synthesis in differentiated L6E9 myoblasts. Hybridization of cDNA probes to extracted mRNA indicated that regulation was at the level of gene expression. Production of these extracellular matrix proteins was also stimulated in proliferating myoblasts (Ignotz and Massagué, 1986; Massagué *et al.*, 1986; Nusgens *et al.*, 1986). TGF-β also inhibited the differentiation of rat L6 myoblasts (Florini *et al.*, 1986) and of two different mouse myoblast cell lines (C-2 and BC3H1; Olson *et al.*, 1986). Inhibition occurred at the level of muscle-specific mRNA accumulation (Olson *et al.*, 1986). In these studies, TGF-β had no effect of proliferation. TGF-β-like activity was detected in culture medium conditioned by fetal rat myoblasts (Hill *et al.*, 1986b), which suggests that TGF-β could act in an autocrine manner to regulate the differentiation potential of myoblasts.

TGF-β inhibited the developmental process of adipogenesis *in vitro*. Differentiation of the proadipogenic cell line, 3T3 L1 (Ignotz and Massagué, 1985), and BALB/c 3T3 T stem cells (Sparks and Scott, 1986) was inhibited half-maximally at 25 and 3 pM TGF-β, respectively. In the case of 3T3 L1 cells, the inhibitory action was not correlated with receptor characteristics as the number and affinity of receptors were similar for proliferative and differentiated cells. As was the case with myoblasts, TGF-β was without effect when added to cells after commitment to differentiation and did not affect cell proliferation (Ignotz and Massagué 1985).

TGF-β stimulated cell replication and collagen synthesis in cultured fetal rat calvaria (Centrella *et al.*, 1987). But in primary cultures of fetal rat parietal bone cells, the action of TGF-β on proliferation was inhibitory at low cell density and stimulatory at high cell density. Induction of alkaline phosphatase mRNA was associated with TGF-β levels that inhibited DNA synthesis. Collagen synthesis was stimulated at all cell densities with the strongest synthesis when the cells were under reduced proliferation rates. Like the rat myoblasts, rat calvaria were shown to release TGF-β into conditioned culture medium (Centrella and Canalis, 1985). In cultured mouse calvaria, TGF-β induced bone resorption (Tashijian *et al.*, 1985). Therefore, TGF-β may also be an important local regulator of bone growth.

In addition to the action of TGF-β on myogenesis, adipogenesis, and osteogenesis, other studies indicated an important role in cells of the immune system (Kehrl *et al.*, 1986; Ristow, 1986; Rook *et al.*, 1986; Roberts *et al.*, 1986), in connective tissue (Sporn *et al.*, 1983; Roberts *et al.*, 1986; Ignotz and Massagué, 1986), in epithelial cells (Nakamura *et al.*, 1985; Like and Massagué, 1986; Jetten *et al.*, 1986; Masui *et al.*, 1986; Shiota *et al.*, 1986; Shipley *et al.*, 1986; Lin *et al.*, 1987; Kurokawa *et al.*, 1987), in the vasculature (Baird and Durkin, 1986; Frater-Schroder *et al.*, 1986; Heimark *et al.*, 1986; Roberts *et al.*, 1986), and in granulosa cell development (Feng *et al.*, 1986; Dodson and Schomberg, 1987).

Hill *et al.* (1986a) postulated that TGF-β may have a physiological role in the regulation of normal tissue growth and maturation during early mammalian development. In the pig, skeletal muscle fiber number is established by day 70–95 of gestation (Campion *et al.*, 1981). A significant portion of the fat cells are also established fetally (Desnoyers *et al.*, 1980). Therefore, the action of TGF-β on muscle and adipose tissue may be particularly relevant to the fetal animal. The fetal role is supported by the finding that fetal bovine bone, bone marrow, liver, thymus, and kidney possess cells with TGF-β immunoreactivity (Ellingsworth *et al.*, 1986).

Hill *et al.* (1986a) demonstrated that the action of TGF-β on early passage of human fetal fibroblasts was dependent on fetal age/weight, and on the presence of other growth factors. In the presence of FBS, TGF-β stimulated [^{3}H]thymidine incorporation into fibroblasts from human fetuses weighing less than 50 g and potentiated the mitogenic effects of IGF-I, EGF, and PDGF. But in fibroblasts from fetuses weighing more than 100 g, TGF-β inhibited [^{3}H]thymidine incorporation and inhibited the proliferative actions of the other growth factors. In contrast, TGF-β reversibly inhibited DNA synthesis in hepatocytes from human fetuses weighing 34–235 g at a concentration 50-fold lower than that which influenced human fetal fibroblasts (Strain *et al.*, 1986). In these hepatocyte cultures, contaminating nonparenchymal cells did not respond to TGF-β. Neither the mechanisms by which cell sensitivity is imparted nor the factors that regulate the synthesis and release of TGF-β are known.

Regulation of growth of murine mammary gland was recently demonstrated *in vivo* (Silberstein and Daniels, 1987). Slow release of TGF-β into developing mammary gland inhibited mammary growth and morphogenesis. The inhibitory effect was reversible. This study supports the hypothesis that TGF-β is a physiologically active growth-regulatory peptide.

3. Platelet-Derived Inhibitors

TGF-β must also be included in the list of inhibitors isolated from platelets. But unlike the three inhibitors described below, TGF-β is known to be synthesized in other cell types. The ones listed below have not been tested for other potential sites of production.

3.1. 37-kDa Protein

Brown and Clemmons (1986) extracted an inhibitor of proliferation of porcine aortic cells from human platelets. The inhibition was dose dependent, reversible, and not due to cytotoxicity. DNA synthesis was not inhibited in cultures of porcine smooth muscle of BALB/c fibroblasts. Molecular properties were unlike those of TGF-β. The protein was heat sensitive (56°C) and had an approximate size of 37 kDa. Like TGF-β, this inhibitor may be an important regulator of vascular reendothelialization that occurs after injury. Since this is the only report on this substance, there is much to be learned about its mechanism of action as well as its physiological role.

3.2. 27-kDa Protein

A third platelet-derived inhibitor was recently purified to homogeneity (Huang *et al.*, 1986). This protein has a molecular mass of 27 kDa, is stable to heating at 90°C for 3 min, and is inactivated by 2-mercaptoethanol. It inhibited proliferation of several nonmalignant epithelial cell lines, but not of fibroblastic cell lines. Because of the apparent cell type specificity, it was identified as epithelial growth inhibitor (EGI). EGI was active at doses lower than 40 pg/ml in medium containing 10% FBS.

3.3. Proteins Greater than 200 kDa

Nakamura *et al.* (1986) reported that platelets also contained a >200-kDa protein that inhibited proliferation of adult rat hepatocytes in primary culture. Purification was not reported but the inhibitory effect was dose dependent and was maximal at 240 μg/ml. The active fraction was heat and acid labile.

4. Interferons

In addition to the well-recognized role of inhibition of viral replication in infected cells (see Friedman *et al.*, 1986), interferons may encompass a far wider range of functions. For example, γ-interferon reportedly inhibits endothelial cell proliferation (Friesel *et al.*, 1987) and other interferons inhibit proliferation of the Daudi line of human lymphoblastoid cells in culture (McNurlan and Clemens, 1986). β-Interferons are produced in minute amounts following induction of differentiation of hemopoietic cells *in vitro* and may be responsible for maintenance of the differentiated state (Resnitzky *et al.*, 1986).

Of particular interest was the finding by Greene and Tso (1986) that interferons may function to modulate the normal developmental process. These investigators prepared an embryonic and a mature form of interferon from Syrian hamster embryo cells. The embryonic form was prepared from early passage cultures of 9-days gestation embryo cells, and the mature form from 9- or 13-day gestation fetal cells. These two forms differed in their physical characteristics (Greene et al., 1984). In addition, the embryonic interferon preferentially inhibited the proliferation of early gestation embryo cells relative to its effects on fetal cells of late gestation; inhibition by the mature form was similar for the two age groups. The embryonic interferon did not influence the differentiation of a clonal line of preadipocytes, a process that was inhibited to some extent by the presence of the mature interferon. Whether inhibition of these growth functions is mediated through interferon action on c-myc expression (Resnitzky et al., 1986), or on the expression of other (proto) oncogenes (Clemens, 1985) is not known. Regardless, the mature interferon and TGF-β represent two negative regulators of preadipocyte differentiation to adipocytes. As such they may prove valuable tools in determining the mechanism by which adipocyte cell number is regulated.

5. Liver-Derived Inhibitors

Iype and McMahon (1984) recently reviewed the literature on hepatocyte proliferation inhibitors. Progress in purification has been slow. McMahon et al. (1982), however, succeeded in the purification of a rat liver protein of 26 kDa that inhibited the proliferation of nonmalignant liver cells isolated from 12-day old rats. The protein was heat stable, migrated as a single band under denaturing conditions in polyacrylamide gel electrophoresis, and had a pI of 4.65. The inhibition was reversible and the half-maximal dose was 50 ng/ml in the presence of 10% FBS. The inhibitor was localized by immunohistochemical methods to the parenchymal cells, with no staining evident in endothelial or connective tissue cells (McMahon et al., 1984).

An extract of bovine liver was shown by Cook's group (Sekas and Cook, 1976; Sekas et al., 1979; Cook et al., 1982) to inhibit DNA synthesis in, and appeared to be specific for, liver cells. The inhibitor was a polypeptide of < 10 kDa. No further work has been done on this polypeptide.

We recently extracted a protein fraction from bovine liver (Table I) that inhibited proliferation of rat L6 myoblasts in culture. The inhibitor was not associated with albumin, was not an IGF-I-binding protein, and was heat labile at 70°C for 30 min. The inhibitory effect was dose dependent. After elution at low salt concentration from an anion exchange column, the inhibitor was active at 1 μg/ml (Campion, unpublished results). When similar protocols were applied to pig liver (Jones and Campion, unpublished results), an inhibitor with

Table I

Characteristics of Bovine and Porcine Liver Fractions with Inhibitory Activity[a]

Bovine	Porcine
~ 160,000 molecular weight	~ 160,000 molecular weight
Heat labile	Heat labile
Inhibition reversible	Inhibition reversible
Dose dependent	Dose dependent
Not an IGF carrier protein	Not an IGF carrier protein
Active at 1 μg protein/ml medium	Active at 125 μg protein/ml medium
Decreased cartilage sulfate incorporation	Decreased cartilage sulfate incorporation
Not bound to blue sepharose	
Elutes at low salt concentration from	
Mono Q column	

[a]Extract was prepared from liver homogenized in 0.1 M NH$_4$Ac. Supernate was lyophilized, resuspended in 44 mM NaHCO$_3$, and chromatographed on S-300 Sephacryl. Based on the chromatogram, fractions were identified and characterized or subjected to further purification. Inhibition of cell proliferation was determined using rat L6 myoblasts in a culture system described by Kotts *et al.* (1987).

biological characteristics like those from bovine liver was found. The physiological significance of this liver inhibitor has not been ascertained.

The liver appears to be a primary source of serum somatomedin inhibitors as reported by others. Recent studies on this topic are summarized in Table II for liver extracts and in Table III for serum. The somatomedin inhibitor(s) appears to be independent of growth hormone regulation (Bomboy *et al.*, 1983; Vassilopoulou-Sellin *et al.*, 1983, 1984). The inhibitor in the neutral liver extract of rats had a molecular size similar to that of the inhibitor identified in diabetic rat serum. If there is any thread of unity to all the studies on somatomedin inhibitors, it is the fact that nearly all inhibitory fractions are heat labile (see Tables I–III). None of the putative inhibitors of somatomedinlike activity have been purified to date and therefore little can be said about their possible physiological function.

Since the somatomedin carrier proteins produced by the liver can be antagonistic to the actions of somatomedin under certain conditions (Smith, 1984), it has been postulated that the carrier protein may be responsible for the serum inhibitory activity (Binoux *et al.*, 1980). In addition, Kuffer and Herington (1984) identified a human plasma fraction that inhibited the biological expression of somatomedin. The most recent work by Ooi and Herington (1986) suggests that the inhibitory activity of the fraction was due to two IGF-binding proteins of 21.5 and 25.5 kDa. The physical and biological characteristics of these ''inhibitors'' suggest that they are different from the inhibitor isolated in rat sera. Our work with the bovine and porcine liver inhibitory fraction (see Table I) and the rat liver inhibitory fraction (Vassilopoulou-Sellin *et al.*, 1987), on the other hand, failed to detect somatomedin-binding characteristics. Thus, not all of the inhibitory activity can be accounted for by carrier protein.

Table II

Liver Somatomedin Inhibitors

Source	Biological activity	Physical properties	References
Normal rat liver (perfusate, extract)	Inhibition of $^{35}SO_4$ uptake by hypox rat cartilage; inhibition of embryonic chick cartilage growth *in vitro*	Heat labile Acid labile 27–45 kDa	Vassilopoulou-Sellin *et al.* (1980, 1984, 1987)
Diabetic rat liver (perfusate, extract)	Inhibition of $^{35}SO_4$ uptake by hypox rat cartilage	Heat labile Acid labile 27–45 kDa	Vassilopoulou-Sellin *et al.* (1980, 1984)
Malnourished rat liver (perfusate, extract)	Inhibition of $^{35}SO_4$ uptake by hypox rat cartilage	Heat labile Acid labile 27–45 kDa GH independent	Vassilopoulou-Sellin *et al.* (1980, 1984)
Fetal rat liver	Inhibition of [^{3}H]thymidine uptake into DNA of chick embryo fibroblasts	Steroid enhanced	Richman *et al.* (1985)

Table III
Serum Somatomedin Inhibitors

Source	Biological activity	Physical properties	References
Normal rats	Inhibition of $^{35}SO_4$ uptake into cartilage Inhibition of [^{3}H]thymidine uptake into cartilage	Heat labile	Price et al. (1979), Kvinnsland and Kvinnsland (1984)
Diabetic rats	Inhibition of [^{14}C]glucose uptake into adipocytes Inhibition of $^{35}SO_4$ uptake into cartilage Inhibition of [^{14}C]glucose uptake into diaphragm	Heat labile GH independent Noncompetitive ~250,000, ~24,000, ~1000 Da	Vassilopoulou-Sellin et al. (1983), Bomboy et al. (1983), Phillips and Scholz (1982), Phillips et al. (1979, 1983, 1985)
Malnourished rats	Inhibition of [^{3}H]thymidine uptake into cartilage Inhibition of $^{35}SO_4$ uptake into cartilage	Heat labile at neutral pH GH independent Heat stable at acid pH Precipitable with acetone Extractable into acid-ethanol 24,000–40,000 Da	Salmon et al. (1983) Bomboy et al. (1983)
Uremic humans	Inhibition of $^{35}SO_4$ uptake into cartilage Does not inhibit glucose oxidation by adipose tissue	940 Da Noncompetitive Destroyed by proteolytic enzymes	Phillips et al. (1984)
Cushing's disease	Inhibition of $^{35}SO_4$ uptake into cartilage	Two different sizes 0–30,000 Da	Unterman and Phillips (1985)

6. Mammary-Derived Inhibitor

Another bovine growth inhibitor of approximately 13 kDa was partially purified from mammary cells (Bohmer *et al.*, 1984, 1985). This factor inhibited Ehrlich ascites mammary carcinoma cells. As with the other bovine inhibitors, no physiological role has been accepted.

7. Glycopeptide Inhibitors

There is evidence to support the hypothesis that cell surface glycoproteins may interact to prevent cell proliferation. Johnson's group (Kinders and Johnson, 1982; Bascom *et al.*, 1985, 1986; Sharifi *et al.*, 1986) has purified and characterized the biological activity of a sialoglycopeptide inhibitor from bovine cerebral cortex cells. The inhibitor had a molecular mass of 18 kDa and a pI of 3. Saturable and specific binding of the sialoglycopeptide was observed on nontransformed cells, but not on transformed cells. The kinetics of binding correlated with the kinetics of protein synthesis inhibition. At least two other glycoprotein inhibitors of pI 8.1 and 8.3 were identified by this group (Johnson *et al.*, 1985).

Weiser and Oesch (1986) identified the terminal galactose residues of plasma membrane glycoproteins with *N*-glycosidically bound carbohydrates as the antiproliferative agent in human embryonic fibroblasts. Although less well characterized, plasma membranes from bovine aorta endothelial cells inhibited endothelial cell proliferation but did not inhibit proliferation of bovine smooth muscle (Teitel, 1986). Thus, there is some cell specificity.

8. Density-Dependent Inhibitors

The proliferative rate of normal cells decreases *in vitro* as the cell density of the culture increases. This phenomenon was termed density-dependent inhibition (DDI; Stoker and Rubin, 1967). Hsu and Wang (1986) recently succeeded in the purification of a 13-kDa polypeptide (pI 10) from 3T3 cells that inhibited their proliferation. Half-maximal inhibition occurred at 3 ng/ml, but a minimum cell density was required. This polypeptide FGR-s (13K) was acid stable and composed of a single polypeptide chain. The denaturing conditions of SDS-PAGE did not result in loss of bioactivity. Hsu and Wang (1986) noted that the inhibitory factor released into the medium of secondary cultures of mouses embryo fibroblasts (Wells and Malucci, 1983) was similar to their FGR-s (13K) peptide in behavior and polypeptide composition. However, direct comparisons have yet to be made.

9. Summary

Although the number of negative growth regulators that have been isolated from tissues or conditioned culture medium is now gradually increasing, caution must be exercised as to their physiological significance. This is because only with rare exception has the action of any of the purified inhibitors been tested *in vivo*. For these types of studies to be conducted and verified in a number of laboratories will require the availability of larger quantities of pure inhibitors at a lower responsible cost than is now available.

To be sure, the evidence for the existence of negative growth regulators of the normal growth process is becoming more convincing, especially since several polypeptides have been purified and sequenced. Most of these inhibitors are thought to function in an autocrine manner. One notable exception is the inhibitor(s) of somatomedinlike activity, which may act in an endocrine manner, being produced in the liver, released into the circulatory system, and finally acting on target cells. But until these somatomedin inhibitory activities can be assigned to specific molecules, this particular area must also be considered with caution.

References

Alison, M. R. 1986. Regulation of hepatic growth. Physiol. Rev. 66:499.

Assoian, R. K., A. Komoriya, C. A. Meyers, D. M. Miller and M. B. Sporn. 1983. Transforming growth factor-β in human platelets. J. Biol. Chem. 258:7155.

Baird, A. and T. Durkin. 1986. Inhibition of endothelial cell proliferation by type β-transforming growth factor: interactions with acidic and basic fibroblast growth factors. Biochem. Biophys. Res. Commun. 138:468.

Bascom, C. C., B. G. Sharifi and T. C. Johnson. 1986. Receptor occupancy by a bovine sialoglycopeptide inhibitor of protein synthesis. J. Cell. Physiol. 128:202.

Bascom, C. C., B. G. Sharifi, L. J. Melkerson, D. A. Rintoul and T. C. Johnson. 1985. The role of gangliosides in the interaction of a growth inhibitor with mouse LM cells. J. Cell. Physiol. 125:427.

Binoux, M., C. Lassarre and D. Seurin. 1980 Somatomedin production by rat livers in organ culture. II. Studies of cartilage sulfation inhibitors release by the liver and their separation from somatomedins. Acta Endocrinol. 93:83.

Bohmer, F. D., W. Lehmann, F. Noll, R. Samtleben, P. Langen and R. Grosse. 1985. Specific neutralizing antiserum against a polypeptide growth inhibitor for mammary cells purified from bovine mannary gland. Biochim. Biophys. Acta 846:145.

Bohmer, F. D., W. Lehman, H. E. Schmidt, P. Langen and R. Grosse. 1984. Purification of a growth inhibitor for Ehrlich ascites mammary carcinoma cells from bovine mammary gland. Exp. Cell Res. 150:466.

Bomboy, J. D., Jr., V. J. Burkhalter, W. E. Nicholson and W. D. Salmon, Jr. 1983. Similarity of somatomedin inhibitor in sera from starved hypophysectomized, and diabetic rats: distinction from a heat-labile inhibitor of rat cartilage metabolism. Endocrinology 112:371.

Brown, M. T. and D. R. Clemmons. 1986. Platelets contain a peptide inhibitor of endothelial cell replication and growth. Proc. Natl. Acad. Sci. USA 83:3321.

Campion, D. R., S. P. Fowler, G. J. Hausman and J. O. Reagan. 1981. Ultrastructural analysis of skeletal muscle development in the fetal pig. Acta Anat. 110:277.

Cate, R. L., R. J. Mattaliano, C. Hession, R. Tizard, N. M. Farber, A. Cheung, E. G. Ninfa, A. Z. Frey, D. J. Gash, E. P. Chow, R. A. Fisher, J. M. Bertonis, G. Torres, B. P. Wallner, K. L. Ramachandran, R. C. Ragin, T. F. Manganaro, D. T. MacLaughlin and P. K. Donahoe. 1986. Isolation of the bovine and human genes for Mullerian inhibiting substance and expression of the human gene in animal cells. Cell 45:685.

Centrella, M. and E. Canalis. 1985. Transforming and non-transforming growth factors are present in medium conditioned by fetal rat calvariae. Proc. Natl. Acad. Sci. USA 82:7335.

Centrella, M., T. L. McCarthy and E. Canalis. 1987. Transforming growth factor β is a bifunctional regulator of replication and collagen synthesis in osteoblast-enriched cell cultures from fetal rat bone. J. Biol. Chem. 262:2869.

Cheifetz, S., B. Like and J. Massagué. 1986. Cellular distribution of type I and type II receptors for transforming growth factor β. J. Biol. Chem. 261:9972.

Cheifetz, S., J. A. Weatherbee, M. L.-S. Tsang, J. K. Anderson, J. E. Mole, R. Lukas and J. Massagué. 1987. The transforming growth factor-β system, a complex pattern of cross-reactive ligands and receptors. Cell 48:409.

Childs, C. B., R. F. Tucker and H. L. Moses. 1982. Serum contains a platelet-derived transforming growth factor. Proc. Natl. Acad. Sci. USA 79:5312.

Clemens, M. 1985. Interferons and oncogenes. Nature 313:531.

Cook, R. T., K. J. Roetman and G. Steinmetz. 1982. Further purification of an inhibitory factor for DNA synthesis in regenerating rat liver. Cell Biol. Int. Rep. 6:49.

De Larco, J. E. and G. J. Todaro. 1978. Growth factors from murine sarcoma virus-transformed cells. Proc. Natl. Acad. Sci. USA 75:4001.

Derynck, R., J. A. Jarrett, E. Y. Chen, D. H. Eaton, J. R. Bell, R. K. Assoian, A. B. Roberts, M. B. Sporn and P. V. Goeddel. 1985. Human transforming growth factor-β complementary DNA sequence and expression in normal and transformed cells. Nature 316:701.

Derynck, R., J. A. Jarrett, E. Y. Chen and P. V. Goeddel. 1986. The murine transforming growth factor-β precursor. J. Biol. Chem. 261:4377.

Desnoyers, F., G. Pascal, M. Etienne and N. Vodovar. 1980. Cellularity of adipose tissue in the fetal pig. J. Lipid Res. 21:301.

Dodson, W. C. and D. W. Schomberg. 1987. The effects of transforming growth factor-β on follicle-stimulating hormone-induced differentiation of cultured rat granulosa cells. Endocrinology 120:512.

Ellingsworth, L. R., J. E. Brennan, K. Fok, D. M. Rosen, H. Bentz, K. A. Piez and S. M. Seyedin. 1986. Antibodies to the N-terminal portion of cartilage-inducing factor A and transforming growth factor-β. J. Biol. Chem. 261:12362.

Fanger, R. F., L. M. Wakefield and M. B. Sporn. 1986. Structure and properties of the cellular receptor for transforming growth factor-β. Biochemistry 25:3083.

Feng, P., K. J. Catt and M. Kencht. 1986. Transforming growth factor-β regulates the inhibitory actions of epidermal growth factor during granulosa cell differentiation. J. Biol. Chem. 261:14167.

Florini, J. R., A. B. Roberts, D. Z. Ewton, S. L. Falen, K. C. Flanders and M. B. Sporn. 1986. Transforming growth factor-β. A very potent inhibitor of myoblast differentiation, identical to the differentiation inhibitor secreted by Buffalo rat liver cells. J. Biol. Chem. 261:16509.

Frater-Schroder, M., G. Muller, W. Birchmeier and P. Bohlen. 1986. Transforming growth factor-β inhibits endothelial cell proliferation. Biochem. Biophys. Res. Commun. 137:295.

Friedman, R. M., T. Merigan and T. Sreevalsan (Ed). 1986. Interferons as Cell Growth Inhibitors and Antitumor Factors. UCLA Symposia on Molecular and Cellular Biology—New Series. Vol. 50. Alan R. Liss, Inc., New York.

Friesel, R., A. Komoriya and T. Maciag. 1987. Inhibition of endothelial cell proliferation by gamma-interferon. J. Cell Biol. 104:689.

Frolik, C. A., L. L. Dart, C. A. Meyers and D. M. Smith. 1983. Purification and initial characterization of a type β transforming growth factor from human placenta. Proc. Natl. Acad. Sci. USA 80:3673.

Fujii, D., J. E. Brissenden, R. Derynck and U. Francke. 1986. Transforming growth factor beta maps to human chromosome 19 long arm and to mouse chromosome 7. Somatic Cell Mol. Genet. 12:281.

Greene, J. J., R. H. Deyer, L. C. Yang and P. O. P. Tso. 1984. Developmentally regulated expression of the interferon system during syrian hamster embryogenesis. J. Interferon Res. 4:517.

Greene, J. J. and P. O. P. Tso. 1986. Preferential modulation of embryonic cell proliferation and differentiation by embryonic interferon. Exp. Cell Res. 167:400.

Heimark, R. L., D. R. Twardzik and S. M. Schwartz. 1986. Inhibition of endothelial regeneration by type-beta transforming growth factor from platelets. Science 233:1078.

Hill, D. J., A. J. Strain, S. F. Elstwo, I. Swenne and R. D. G. Milner. 1986a. Bi-functional action of transforming growth factor-β on DNA synthesis in early passage human fetal fibroblasts. J. Cell. Physiol. 128:322.

Hill, D. J., A. J. Strain and R. D. G. Milner. 1986b. Presence of transforming growth factor-β-like activity in multiple fetal rat tissues. Cell Biol. Int. Rep. 10:915.

Holley, R. W., R. Armour and J. H. Baldwin. 1978. Density-dependent regulation of BSC-1 cells in culture: growth inhibitors formed by the cells. Proc. Natl. Acad. Sci. USA 75:1864.

Holley, R. W., P. Bohlen, R. Fava, J. H. Baldwin, G. Kleeman and R. Armour. 1980. Purification of kidney epithelial cell growth inhibitors. Proc. Natl. Acad. Sci. USA 77:5989.

Hsu, Y.-M. andf J. L. Wang. 1986. Growth inhibitory control in cultured 3T3 fibroblasts. V. Purification of an Mr 13,000 polypeptide responsible for growth inhibitory activity. J. Cell Biol. 102:370.

Huang, W., T. Timura, K. Mashima, K. Miyazaki, H. Masaki, J. Yamashita and T. Horio. 1986. Purification and properties of epithelial growth inhibitor (EGI) from human platelets: its separation from type β transforming growth factor (TGF-β). J. Biochem. Tokyo 100:687.

Ignotz, R. A. and J. Massagué. 1985. Type β transforming growth factor controls the adipogenic differentiation of 3T3 fibroblasts. Proc. Natl. Acad. Sci. USA 82:8530.

Ignotz, R. A. and J. Massagué. 1986. Transforming growth factor-β stimulates the expression of fibronectin and collagen and their incorporation into the extracellular matrix. J. Biol. Chem. 261:4337.

Iype, P. T. and J. B. McMahon. 1984. Hepatic proliferation inhibitor. Mol. Cell. Biochem. 59:57.

Jetten, A. M., J. E. Shirley and G. Stoner. 1986. Regulation of proliferation and differentiation of respiratory tract epithelial cells by TGF-β. Exp. Cell Res. 167:539.

Johnson, T. C., R. J. Kinders and B. G. Sharifi. 1985. Purification of a cell growth inhibitor from bovine cerebral cortex cells. Life Sci. 37:1117.

Kehrl, J. H., L. M. Wakefield, A. B. Roberts, S. Jakowlew, M. Alvarev-Mon, R. Derynck, M. B. Sporn and A. S. Fauci. 1986. Production of transforming growth factor-β by human T lymphocytes and its potential role in the regulation of T cell growth. J. Exp. Med. 163:1037.

Kinders, R. J. and T. C. Johnson. 1982. Isolation of cell surface glycopeptides from bovine cerebral cortex that inhibit cell growth and protein synthesis in normal but not in transformed cells. Biochem. J. 206:527.

Kotts, C. E., M. E. White, C. E. Allen, F. Martin and W. R. Dayton. 1987. A statistically standardized muscle cell culture bioassay measuring the effect of swine serum on muscle cell proliferation. J. Anim. Sci. 64:615.

Kuffer, A. D. and A. C. Herington. 1984. Partial purification of a specific inhibitor of the insulin-like growth factors by reversed-phase high performance liquid chromatography. J. Chromatogr. 336:87.

Kurokawa, M., K. Lynch and D. K. Podolsky. 1987. Effects of growth factors on an intestinal cell line: transforming growth factor-β inhibits proliferation and stimulates differentiation. Biochem. Biophys. Res. Commun. 142:775.

Kvinnsland, S. and S. Kvinnsland. 1984. Inhibitory activity of cartilage growth in serum of young rats. Acta Anat. 119:94.

Lawrence, D. A. 1985. Transforming growth factors—an overview. Biol. Cell 53:93.

Leof, E. B., J. A. Proper. A. S. Goustin, G. D. Shipley, P. E. Di Corletto and H. L. Moses. 1986. Induction of c-sis mRNA and activity similar to platelet derived growth factor by transforming growth factor-β: a model for indirect mitogenesis involving autocrine activity. Proc. Natl. Acad. Sci. USA. 83:2453.

Libby, J., R. Martinez and M. J. Weber. 1986. Tyrosine phosphorylation in cells treated with transforming growth factor-β. J. Cell. Physiol. 129:159.

Like, B. and J. Massagué. 1986. The antiproliferative effect of type β transforming growth factor occurs at a level distal from receptors for growth-activating factors. J. Biol. Chem. 261:13426.

Lin, P., C. Liu, M.-S. Tsao and J. W. Grisham. 1987. Inhibition of proliferation of cultured rat liver cells at specific cell cycle stages by transforming growth factor-β. Biochem. Biophys. Res. Commun. 143:26.

McMahon, J. B., J. G. Farrelly and P. T. Iype. 1982. Purification and properties of a rat liver protein that specifically inhibits the proliferation of nonmalignant epithelial cells from rat liver. Proc. Natl. Acad. Sci. USA 79:456.

McMahon, J. B., L. Malan-Shibley and P. T. Iype. 1984. Distribution and subcellular localization of a hepatic proliferation inhibitor in rat liver. J. Biol. Chem. 259:1803.

McNurlan, M. A. and M. J. Clemens. 1986. Inhibition of cell proliferation by interferons. Relative contributions of changes in protein synthesis and breakdown to growth control of human lymphoblastoid cells. Biochem. J. 237:871.

Mason, A. J., J. S. Hayflick, N. Ling, F. Esch, N. Ueno, S. Y. Ying, R. Guillemin, H. Niall, and P. H. Seeburg. 1985. Complementary DNA sequences of ovarian follicular fluid inhibin show precursor structure and homology with transforming growth factor-β. Nature 318:659.

Mason, A. J., H. D. Niall and P. H. Seeburg. 1986. Structure of two human ovarian inhibins. Biochem. Biophys. Res. Commun. 135:957.

Massagué, J. 1985. Subunit structure of a high-affinity receptor for type β-transforming growth factor. J. Biol. Chem. 260:7059.

Massagué, J., S. Cheifetz, T. Endo and B. Nadal-Ginard. 1986. Type β transforming growth factor is an inhibitor of myogenic differentiation. Proc. Natl. Acad. Sci. USA 83:8206.

Massagué, J. and B. Like. 1985. Cellular receptors for type β transforming growth factor. J. Biol. Chem. 260:2636.

Masui, T., L. M. Wakefield, J. F. Lechner, M. A. LaVeck, M. B. Sporn and C. C. Harris. 1986. Type β transforming growth factor is the primary differentiation-inducing serum factor for normal human bronchial epithelial cells. Proc. Natl. Acad. Sci. USA 83:2438.

Moses, H. L., E. L. Branum, J. A. Proper and R. A. Robinson. 1981. Transforming growth factor production by chemically-transformed cells. Cancer Res. 41:2842.

Nakamura, T., T. Kitazawa and A. Ichihara. 1986. Partial purification and characterization of masking protein for β-type transforming growth factor from rat platelets. Biochem. Biophys. Res. Commun. 141:176.

Nakamura, T., Y. Tomita, R. Hirai, K. Yamaoka, K. Kaji and A. Ichihara. 1985. Inhibitory effect of transforming factor-β on DNA synthesis of adult rat hepatocytes in primary culture. Biochem. Biophys. Res. Commun. 133:1042.

Nusgens, B., D. Delain, H. Senechal, R. Winard, C. H. M. Lapierre and J. P. Wahrmann. 1986.

Metabolic changes in the extracellular matrix during differentiation of myoblasts of the L_6 line and of a myo-nonfusing mutant. Exp. Cell Res. 162:51.

Olson, E. N., E. Sternberg, J. S. Hu, G. Spizz and C. Wilcox. 1986. Regulation of myogenic differentiation by type β transforming growth factor. J. Cell Biol. 103:1799.

Ooi, G. T. and A. C. Herington. 1986. Covalent cross-linking of insulin-like growth factor 1 to a specific inhibitor from human serum. Biochem. Biophys. Res. Commun. 137:411.

Phillips, L. S., V. R. Bajaj, A. C. Fusco, K. M. Keery and S. Goldstein. 1985. Nutrition and somatomedin—XII. Fractionation of somatomedins and somatomedin inhibitor in normal and diabetic rats. Int. J. Biochem. 17:597.

Phillips, L. S., V. R. Bajaj, A. C. Fusco and C. K. Matheson. 1983. Nutrition and somatomedin. XI. Studies of somatomedin inhibitors in rats with streptozotocin-induced diabetes. Diabetes 32:1117.

Phillips, L. S., A. C. Fusco, T. G. Unterman and F. del Greco. 1984. Somatomedin inhibitor in uremia. J. Clin. Endocrinol. Metab. 59:764.

Phillips, L. S. and T. D. Scholz. 1982. Nutrition and somatomedin. IX. Blunting of insulin-like activity by inhibitor in diabetic rat serum. Diabetes 31:97.

Phillips, L. S., D. C. Young and L. A. Reichard. 1979. Nutrition and somatomedin. VI. Somatomedin activity and somatomedin inhibitory activity in serum from normal and diabetic rats. Endocrinology 104:1519.

Price, D. A., J. M. Wit, S. van Buul-Offers, A. M. Korteland-van Male, A. K. M. van Rooyen-Wehmeijer, C. Hoogerbrugge and J. L. Van den Brande. 1979. Serum somatomedin activity and cartilage metabolism in acutely fasted, chronically malnourished, and refed rats. Endocrinology 105:851.

Resnitzky, D., A. Yarden, D. Zipori and A. Kimchi. 1986. Autocrine β-related interferon controls c-myc suppression and growth arrest during hematopoietic cell differentiation. Cell 46:31.

Richman, R. A., M. R. Benedict, J. R. Florini and B. A. Toly. 1985. Hormonal regulation of somatomedin secretion by fetal rat hepatocytes in primary culture. Endocrinology 116:180.

Ristow, H.-J. 1986. BSC-1 growth inhibitor/type β transforming growth factor is a strong inhibitor of thymocyte proliferation. Proc. Natl. Acad. Sci. USA 83:5531.

Roberts, A. B., M. A. Anzano, L. C. Lamb, J. M. Smith and M. B. Sporn. 1981. New class of transforming growth factors potentiated by epidermal growth factor: isolation from non-neoplastic tissues. Proc. Natl. Acad. Sci. USA 78:5339.

Roberts, A. B., M. A. Anzano, C. A. Meyers, J. Wideman, R. Blacher, E. Yu-Ching, S. Stein, S. R. Lehrman, J. M. Smith, L. C. Lamb and M. B. Sporn. 1983a. Purification and properties of a type β transforming growth factor from bovine kidney. Biochemistry 22:5692.

Roberts, A. B., C. A. Frolik, M. A. Anzano and M. B. Sporn. 1983b. Transforming growth factors from neoplastic and non-neoplastic tissues. Fed. Proc. 42:2621.

Roberts, A. B., M. B. Sporn, R. K. Assoian, J. M. Smith, N. S. Roche, L. M. Wakefield, U. I. Heine, L. A. Liotta, V. Falanga, J. H. Kehrl and A. S. Fauci. 1986. Transforming growth factor-β: rapid induction of fibrosis and angiogenesis in vivo and stimulation of collagen formation in vitro. Proc. Natl. Acad. Sci. USA 83:4167.

Rook, A. H., J. H. Kehrl, L. M. Wakefield, A. B. Roberts, M. B. Sporn, D. B. Burlington, C. H. Lane and A. S. Fauci. 1986. Effects of transforming growth factor-β on the functions of natural killer cells: depressed cytosolic activity and blunting of interferon response. J. Immunol. 136:3916.

Salmon, W. D., Jr., L. A. Holladay and V. J. Burkhalter. 1983. Partial characterization of somatomedin inhibitors in starved rat serum. *Endocrinology.* 112:360.

Sekas, G. and R. T. Cook. 1976. The isolation of a low molecular weight inhibitor of ^3H TdR incorporation into hepatic DNA. Exp. Cell Res. 102:422.

Sekas, G., W. G. Owen and R. T. Cook. 1979. Fractionation and preliminary characterization of a low molecular weight bovine hepatocyte inhibitor of DNA synthesis in regenerating rat liver. Exp. Cell Res. 122:47.

Seyedin, S. M., P. R. Segarini, D. M. Rosen, A. Y. Thompson, H. Bentz and J. Graycar. 1987. Cartilage-induced factor-B is a unique protein structurally and functionally related to transforming growth factor-β. J. Biol. Chem. 262:1946.

Seyedin, S. M., A. Y. Thompson, H. Bentz, D. M. Rosen, J. M. McPherson, A. Conti, N. R. Siegel, G. R. Galluppi and K. A. Piez. 1986. Cartilage-inducing factor-A: apparent identity to transforming growth factor-β. J. Biol. Chem. 261:5693.

Sharifi, B. G., C. C. Bascom, H. Fattaey, S. Nash and T. C. Johnson. 1986. Relationship between protease activity and a sialoglycopeptide inhibitor isolated from bovine brain. J. Cell. Biochem. 31:41.

Shiota, K., T. Nakamura and A. Ichihara. 1986. Distinct effects of transforming growth factor-β on EGF receptors and EGF-induced DNA synthesis in primary cultured rat hepatocytes. Biochem. Int. 13:893.

Shipley, G. D., M. P. Pittlekow, J. J. Willie, Jr., R. E. Scott and H. L. Moses. 1986. Reversible inhibition of normal human prokeratinocyte proliferation by type β transforming growth factor-growth inhibition in serum-free medium. Cancer Res. 46:2068.

Silberstein, G. B. and C. W. Daniels. 1987. Reversible inhibition of mammary gland growth by transforming growth factor-β. Science 237:291.

Smith, G. L. 1984. Somatomedin carrier proteins. Mol. Cell. Endocrinol. 34:83.

Sparks, R. L. and R. E. Scott. 1986. Transforming growth factor-β is a specific inhibitor of 3T3 T mesenchymal stem cell differentiation. Exp. Cell Res. 165:345.

Sporn, M. B., A. B. Roberts, J. H. Shull, J. M. Smith and J. M. Ward. 1983. Polypeptide transforming growth factors isolated from bovine sources and used for wound healing in vivo. Science 219:1329.

Sporn, M. B., A. B. Roberts, L. M. Wakefield and R. K. Assoian. 1986. Transforming growth factor-β: biological function and chemical structure. Science 233:532.

Stoker, M. G. P. and H. Rubin. 1967. Density-dependent inhibition of cell growth in culture. Nature 215:171.

Strain, A. J., D. J. Hill and R. D. G. Milner. 1986. Divergent action of transforming growth factor-β on DNA synthesis in human fetal liver cells. Cell Biol. Int. Rep. 10:855.

Tashijian, A. H., Jr., E. F. Voelkel, M. Lazzaro, F. R. Singer, A. B. Roberts, R. Derynck, M. E. Winkler and L. Levine. 1985. Alpha and beta transforming growth factors stimulate prostaglandin production and bone resorption in cultured mouse calvaria. Proc. Natl. Acad. Sci. USA 82:4535.

Teitel, J. M. 1986. Specific inhibition of endothelial cell proliferation by isolated endothelial plasma membranes. J. Cell. Physiol. 128:329.

Tucker, R. F., E. L. Branum, G. D. Shiplery, R. J. Ryan and H. L. Moses. 1984a. Specific binding to cultured cells of ^{125}I-labeled type β transforming growth factor from human platelets. Proc. Natl. Acad. Sci. USA 81:6757.

Tucker, R. F., G. D. Shipley, H. L. Moses and R. W. Holley. 1984b. Growth inhibitor from BSC-1 cells closely related to platelet type β transforming growth factor. Science 226:705.

Unterman, T. G. and L. S. Phillips. 1985. Glucocorticoid effects on somatomedins and somatomedin inhibitors. J. Clin. Endocrinol. Metab. 61:618.

Vassilopoulou-Sellin, R., R. L. Lock, II, C. O. Oyedeji and N. A. Samaan. 1987. Cartilage sulfation inhibitor from rat liver curtails growth of embryonic chicken cartilage in vitro. Metabolism 36:89.

Vassilopoulou-Sellin, R., C. O. Oyedeji and N. A. Samaan. 1983. Somatomedin inhibitors in serum and liver of growth hormone deficient diabetic rats. Diabetes 32:262.

Vassilopoulou-Sellin, R., C. O. Oyedeji and N. A. Samaan. 1984. Cartilage sulfation inhibitors

(CSI) in serum and liver of growth hormone-deficient starved rats. Horm. Metab. Res. 16:156.

Vassilopoulou-Sellin, R., L. S. Phillips and L. A. Reichard. 1980. Nutrition and somatomedin. VII. Regulation of somatomedin activity by the perfused rat liver. Endocrinology 106:260.

Wang, L. J. and Y. M. Hsu. 1986. Negative regulators of cell growth. Trends Biochem. Sci. 11:24.

Weiser, R. J. and F. Oesch. 1986. Contact inhibition of growth of human diploid fibroblasts by immobilized plasma membrane glycoproteins. J. Cell Biol. 103:361.

Wells, V. and L. Malucci. 1983. Properties of a cell growth inhibitor produced by mouse embryo fibroblasts. J. Cell. Physiol. 117:148.

CHAPTER 8

Skeletal Muscle Proteases and Protein Turnover

DARREL E. GOLL, WILLIAM C. KLEESE, AND ADAM SZPACENKO

1. Introduction

Schoenheimer and Rittenberg's paper published nearly 50 years ago (Schoenheimer and Rittenberg, 1940) established that accumulation of muscle tissue or muscle growth must depend on both the rate of muscle protein synthesis and the rate of muscle protein degradation. Despite this axiom, most of the attention of animal scientists during the period from 1940 to 1980 focused on increasing the rate of muscle growth by increasing the rate of muscle protein synthesis. Because of this, a great deal is known about the mechanism of muscle protein synthesis and how it is controlled. Little is known, however, about the mechanism of muscle protein degradation. It is clear that muscle proteins turn over metabolically with half lives ranging from 2 to 20 days (Low and Goldberg, 1973; Koizumi, 1974; Rubenstein et al., 1976; Martin et al., 1977; Zak et al., 1977; Millward et al., 1978; Martin, 1981; Wolitsky et al., 1984), but the nature of the proteolytic enzymes responsible for this turnover remains unknown. It was learned in 1969 that the rate of muscle protein degradation can vary over a wide range in response to physiological demand (Goldberg, 1969a,b). Therefore, variations in rate of muscle protein degradation could be a primary cause in regulating the rate of muscle growth among domestic animals, and could in some instances be more important than the rate of muscle protein synthesis in such regulation. Indeed, studies on lysine or energy restric-

DARREL E. GOLL, WILLIAM C. KLEESE, AND ADAM SZPACENKO ● Muscle Biology Group, University of Arizona, Tucson, Arizona 85721.

tion in chickens (Maruyama *et al.*, 1978), tenotomy or denervation in rats (Goldspink *et al.*, 1983), trienbolone acetate (Vernon and Buttery, 1976) or clenbuterol (Reeds *et al.*, 1986) administration to rats, rate of muscle growth in layer and broiler chickens (Jones *et al.*, 1986), and rate of myosin heavy chain accumulation in cultures of chicken muscle cells (Orcutt and Young, 1982) have shown that rate of muscle growth in these situations is determined almost entirely by changes in rate of muscle protein degradation with little or no change in rate of muscle protein synthesis. Furthermore, the available evidence indicates that the rapid loss of muscle protein and skeletal muscle mass that occurs during various muscle pathologies, such as the muscular dystrophies (McKeran *et al.*, 1977; Ballard *et al.*, 1979; Li, 1980; Warnes *et al.*, 1981; Yoshikawa and Masaki, 1981), myasthenia gravis (Warnes *et al.*, 1981), denervation (Goldberg, 1969b), phorbol diester treatment (West and Holtzer, 1982), and blockage of muscle membrane Na^+ channels (Crisona and Strohman, 1983), is due primarily to a greatly increased rate of muscle protein degradation with little change or even a slight increase in rate of muscle protein synthesis.

Regulating the rate of muscle protein degradation could cause dramatic changes in the rate of muscle growth. Values reported for fractional rates of muscle protein degradation have ranged from 26.5% per day in breast muscle of young chicks (Maruyama *et al.*, 1978) to 3.4 to 4.4% per day in 22- to 45-kg male pigs (Mulvaney *et al.*, 1985) to 2 to 3.3% per day in growing cattle (McCarthy *et al.*, 1983; Gopinath and Kitts, 1984). Table I shows that if the rate of muscle protein degradation in a 454-kg steer were decreased by only 10% from 3.0% to 2.7% per day, the rate of live weight gain in this animal would more than double from 1.20 to 2.55 kg/day. Moreover, it has been estimated than 15 to 25% of the energy ingested by domestic animals is used

Table I

Potential Effect of Muscle Protein Turnover on Rate of Growth in Domestic Animals

454-kg (1000 lb) bovine animal	Gaining 1.20 kg (2.64 lb)/day
Approximate carcass weight = 272.4 kg (60%)	Gaining 0.72 kg carcass/day
Approximate muscle weight = 163.44 kg (60%)	Gaining 0.432 kg muscle/day
Approximate protein weight = 32.69 kg (20%)	Gaining 0.086 kg protein/day

If this animal turns over muscle protein at a rate of 3% per day, which is normal for bovine animals (McCarthy *et al.*, 1983), then it would degrade $0.03 \times 32.69 = 0.98$ kg of muscle protein per day. It would have to synthesize $0.98 + 0.086 = 1.066$ kg of muscle protein each day to gain 1.20 kg/day.

If the rate of muscle protein degradation were lowered by 10% to 2.7% per day, then only $0.027 \times 32.69 = 0.88$ kg of muscle protein would be degraded each day. If the animal continued to synthesize muscle protein at a rate of 1.066 kg/day, it would be gaining $1.066 - 0.88 = 0.186$ kg/day, which would result in a live weight gain of $0.186/0.2 \times 0.6 = 2.58$ kg (5.69 lb)/day, assuming constant body composition.

to replace muscle protein that is degraded during metabolic turnover (Young *et al.*, 1975). Decreasing the rate of muscle protein degradation, therefore, would decrease the amount of energy needed for protein replacement and should lead directly to increases in efficiency with which animals convert ingested nutrients into edible muscle. It is not certain, on the other hand, that increasing the rate of muscle protein synthesis or number of muscle cells would have any beneficial effect on efficiency of muscle growth. Consequently, it is important to learn what enzymes are involved in intracellular degradation of muscle proteins and how activity of these enzymes is controlled *in vivo*.

2. General Features of Intracellular Protein Degradation

It has generally been assumed that proteolytic enzymes are responsible for degradation of proteins during metabolic turnover (Goldberg and Dice, 1974; Pontremoli and Melloni, 1986), but the nature of these enzymes and the mechanism by which their activity is regulated are still unclear. Proteolytic enzymes or proteases can be divided into two families: endopeptidases or proteinases and exopeptidases. These two families are subdivided into classes having similar functional groups at their active sites (endopeptidases, Table II) or having similar activities (exopeptidases, Table III). Although some intracellular protein degradation clearly is mediated by catheptic proteases in lysosomes (Goldberg

Table II
Classification of Proteolytic Enzymes: Proteinases (Endopeptidases)[a]

Serine proteinases
  Usually optimally active at slightly alkaline pH (7.5–9)
  Inhibited by trypsin inhibitors, diisopropylphosphofluoridate, phenylmethylsulfonyl fluoride
Cysteine proteinases
  Usually optimally active over a wide pH range (4–8)
  Inhibited by leupeptin, iodoacetic acid, epoxysuccinylleucyl-agmatine, *N*-ethylmaleimide,
    p-chloromercuribenzoate
  Include papain, cathepsins B, H, L, N, and S, and the calpains
Aspartic proteinases
  Usually optimally active at acidic pH (2–6)
  Inhibited by pepstatin A, statine
  Include pepsin, cathepsins D and E, renin, and chymosin
Metalloproteinases
  Usually optimally active at neutral or slightly alkaline pH (7–9)
  Inhibited by EDTA, 2,2'-bipyridine, 1,10-phenanthroline, phosphoramidon
  Include collagenases, gelatinases, thermolysin, and angiotensin-converting enzyme
  Usually contain Zn

[a]Nomenclature is taken from that recommended by Barrett and McDonald (1980) and McDonald and Barrett (1986).

Table III
Classification of Proteolytic Enzymes: Exopeptidases[a]

Aminopeptidases
    Cleave the NH_2-terminal amino acid off polypeptides; usually optimally active at pH 7–9; include leucyl aminopeptidase, prolyl aminopeptidase, lysosomal arginyl aminopeptidase

Tripeptidases
    Cleave the NH_2-terminal amino acid off tripeptides only; usually optimally active at pH 7–8; include tripeptide aminopeptidase

Dipeptidyl peptidases
    Cleave dipeptides off the NH_2-terminal region of polypeptides; usually optimally active at pH 5–7, but dipeptidyl peptidase IV is optimally active at pH 7.8–8; include dipeptidyl peptidases I through IV; dipeptidyl peptidases I and II are lysosomal

Tripeptidyl peptidases
    Cleave tripeptides off the NH_2-terminal region of polypeptides; usually optimally active at pH 4–6.5; include tripeptidyl peptidases I (lysosomal) and II

Carboxypeptidases
    Cleave the COOH-terminal amino acid off polypeptides; usually optimally active at pH 4–6 (lysosomal carboxypeptidases) or 7–8; include carboxypeptidases A and B and lysosomal carboxypeptidases A and B

Peptidyl dipeptidases
    Cleave dipeptides off the COOH-terminal region of polypeptides; peptidyl dipeptidase A has a pH optimum of 7–8 and peptidyl dipeptidase B (lysosomal) has a pH optimum of 3.5–6

Dipeptidases
    Cleave dipeptides into two amino acids; usually are optimally active at pH 6.5–8.5 or 4.5–5.5 (lysosomal dipeptidases I and II)

Omega peptidases
    Cleave NH_2-terminal amino acids from polypeptides having blocked NH_2-terminal groups or isopeptide bonds involving γ-glutamyl or β-aspartyl linkages; usually are optimally active at pH 7.3–8.3 except for lysosomal γ-glutamyl carboxypeptidase (pH 4.5)

[a] Nomenclature is taken from that recommended by Barrett and McDonald (1980) and McDonald and Barrett (1986). Most exopeptidases are metalloproteins.

and St. John, 1976; Pösö and Mortimore, 1984), a number of studies have now shown that other nonlysosomal proteases are also involved in metabolic turnover of intracellular proteins (Zeman *et al.*, 1985; Hough *et al.*, 1986; Fagan *et al.*, 1987). The lack of definitive information on the proteases involved and the inability thus far to analyze intracellular protein degradation by monitoring intermediates in the degradation process have hampered studies on the mechanism of intracellular protein degradation. Despite these difficulties, it has been possible by use of different protease inhibitors to determine that several different pathways for protein degradation exist in cells. For convenience, these pathways may be grouped into the three general categories listed in Table IV. The lysosomal system seems to be responsible for the increased intracellular protein degradation that occurs during insulin deprivation. The error-eliminating system has been studied most extensively in reticulocytes and rapidly and selectively degrades denatured or abnormal proteins (Goldberg *et al.*, 1978).

Table IV
Three Types of Proteolytic Systems Involved in Intracellular Protein Turnover

Lysosomal system
  Peptidases are located in lysosomes and function at acidic pH (3–5)
  Includes the catheptic peptidases
  Involved in degradation of endocytosed proteins, including some hormone receptors and some exogenous proteins
  Involved in bulk degradation of some endogenous proteins, but it is unclear how such degradation would produce different half lives for different proteins
Error-eliminating system (cell-sanitation system)
  Peptidases are located in the cell cytoplasm and are optimally active at pH 7.5–8
  Contains serine and cysteine proteinases
  Specific for proteins containing errors of translation
  Requires ATP
Black-box system
  Peptidases are located in the cell cytoplasm and are optimally active at pH 7–8
  Contains serine, cysteine, and metalloproteinase enzymes
  Probably includes many different peptidases, each highly regulated and each with a specific task
  Several of the peptidases in this system may also require ATP

This system requires ATP, which is hydrolyzed during degradation, and involves two to three different enzymes that prepare the protein substrate for degradation. The ubiquitin-requiring pathway (Tanaka et al., 1983, 1984; Pickart and Rose, 1985) seems to function partly in this error-eliminating system. The third cytoplasmic system seems responsible for turnover of short-lived proteins in cultured cells and probably contains many different proteases that are subject to close metabolic control. The ubiquitin-requiring pathway may also function in this third system, as does a second, ATP-requiring, nonubiquitin pathway (Tanaka et al., 1983, 1984; Waxman et al., 1985, 1987). The second ATP-requiring, nonubiquitin pathway involves an ATP-stabilized protease that does not require ATP hydrolysis (called the multicatalytic protease) (Dahlmann et al., 1985a,b; Fagan et al., 1987; Waxman et al., 1987), and a second ATP-requiring protease of very large molecular mass (600 kDa) that requires ATP hydrolysis and that can degrade both ubiquitin-conjugated proteins and proteins that are not so conjugated (Chin et al., 1986; Hough et al., 1986; Fagan et al., 1987; Waxman et al., 1987).

Two recent papers (Bachmair et al., 1986; Rogers et al., 1986) have suggested that the various half lives of different proteins are a direct consequence of their amino acid composition. Rogers et al., (1986) found that proteins having regions rich in proline (P), glutamic acid (E), serine (S), and threonine (T) have very short half lives. Presumably, the PEST regions cause proteins to be more susceptible to proteases responsible for intracellular protein degradation. Bachmair et al., (1986), on the other hand, suggested that half lives of proteins are determined by their NH_2-terminal amino acids. Proteins having Met, Ser,

Ala, Val, Gly, and Thr as NH_2-terminal amino acids have half lives over 20 hr, whereas proteins having Lys or Arg as NH_2-terminal amino acids have half lives of 2–3 min. These hypotheses do not involve a particular role for ubiquitin in "tagging" proteins for intracellular degradation, and they provide no clues about the nature of the proteases that recognize these amino acid sequences or NH_2-terminal amino acids.

3. Intracellular Degradation of Muscle Proteins

Muscle proteins generally are grouped into three categories based on their solubility (Goll *et al.*, 1984): (1) the sarcoplasmic proteins, which are those proteins soluble at ionic strengths of 0.03 or less; this class constitutes 30–35%, by weight, of the proteins in muscle and contains almost all the cytoplasmic proteins in muscle cells: (2) the myofibrillar proteins, which are those proteins that constitute the myofibril in striated muscle cells; high ionic strengths above 0.20 are required to solubilize many of the myofibrillar proteins although some of these proteins are also soluble in water; the myofibrillar proteins constitute 50–55% of total muscle protein, by weight, are exclusively intracellular, and are the major class of muscle proteins both physiologically and in the use of muscle as meat (Goll *et al.*, 1976); and (3) the stroma proteins, which are those proteins insoluble in neutral aqueous solvents; the stroma proteins constitute 15–20% of total muscle protein and are largely extracellular because most of the stroma protein fraction is composed of collagen and elastin. Many studies of muscle protein degradation have measured release of amino acids from muscle strips or have measured bulk muscle weight without attempting to determine which classes of muscle proteins were involved in these changes. Studies of muscle protein degradation in domestic animals, however, should focus on the sarcoplasmic and myofibrillar classes of muscle proteins. Because the myofibrillar protein class is responsible for most of the desirable properties of muscle when used as food, the remainder of this review will emphasize degradation of the myofibrillar or contractile proteins in muscle cells.

Skeletal muscle cells contain far fewer lysosomes and hence less lysosomal proteases than do autophagic cells like macrophages or pinocytosing cells such as hepatocytes (Canonico and Bird, 1970; Iodice *et al.*, 1972; Stauber and Bird, 1974; Kominami *et al.*, 1985). Tables V, VI, and VII summarize some of the properties of the known cathepsins and indicate which ones have been found inside muscle cells. Cathepsins B, D, H, and L seem to have the greatest activities in skeletal muscle cells; immunoassays indicate that skeletal muscle cells contain 92 ng cathepsin B and 52 ng cathepsin H per mg muscle protein (Kominami *et al.*, 1985). It has been estimated that cathepsin L has ten times, cathepsin H five times, and cathepsin D two to three times greater activity per

Table V
Properties of Some Lysosomal Proteases: Cathepsins A–E[a]

Lysosomal protease	Type of protease	Active site	Molecular mass	pH optimum	pI	Found in striated muscle cells?	Other special properties
Lysosomal carboxypeptidase A (cathepsin A, cathepsin I)	Exopeptidase	–OH	35 kDa (100 kDa with 20-, 25-, and 55-kDa subunits in hog kidney)	5.0–6.0	5.0–5.2	Yes	Heat labile, inactivated by 65°C for 40 min; labile above pH 7.0; stabilized by sucrose
Cathepsin B (cathepsin B₁)	Endopeptidase with some exopeptidase activity (dipeptidyl carboxypeptidase)	–SH	27,411 Da	3.5–6.0	4.5–5.5	Yes	Has been cloned and sequenced; widely distributed
Lysosomal carboxypeptidase B (cathepsin B₂)	Exopeptidase	–SH	50 kDa (two subunits of 25 kDa each)	5.0–6.0	5.0	Yes	Unstable above pH 7; widely distributed; contains CHO
Dipeptidyl peptidase I (cathepsin C)	Exopeptidase	–SH	197 kDa (has 24-kDa subunits)	5.0–6.0	5.0–6.0	Yes	Requires halide ions (Cl⁻ or Br⁻) for activity and activated by thiol compounds
Cathepsin D	Endopeptidase	–COOH	36,779 Da	2.5–5.0	5.5–6.5	Yes	A glycoprotein; widely distributed; has been cloned and sequenced
Cathepsin E	Endopeptidase	–COOH	100 kDa (50-kDa subunit?)	Less than 3.0	Very low	No	Still poorly characterized

[a]Most of the information on properties of these proteases was taken from Barrett and McDonald (1980) and McDonald and Barrett (1986).

Table VI
Properties of Some Lysosomal Proteases: Cathepsins G–M[a]

Lysosomal protease	Type of protease	Active site	Molecular mass	pH optimum	pI	Found in striated muscle cells?	Other special properties
Cathepsin G	Endopeptidase	–OH	27.5 kDa	7.0–8.0	10	No	Very similar to mast cell chymase I; specificity like chymotrypsin; leukocytes are a rich source
Cathepsin H (cathepsin B$_3$)	Endopeptidase with aminopeptidase activity	–SH	25,116 Da	5.5–6.5	7.1	Yes	A glycoprotein; very heat stable; loses only 10–20% activity at 50°C for 30 min; has been cloned and sequenced
Cathepsin J	Endopeptidase	–SH	230 kDa	5.5–7.5	5.8	Yes	Specificity is similar to cathepsin L
Cathepsin K	Endopeptidase	–SH	650 kDa	5.0–6.5	5.3	No	Poorly characterized
Cathepsin L	Endopeptidase	–SH	23,720 Da	3.0–6.5	5.8–6.1	Yes	Very labile to autolysis; very active against proteins but less active against peptide substrates; has been cloned and sequenced
Cathepsin M	Exopeptidase (carboxypeptidase)	–SH	30 kDa	4.5–7.0	—	Not tested	Very poorly characterized; only one report

[a]Most of the information on properties of these proteases was taken from Barrett and McDonald (1980) and McDonald and Barrett (1986).

Table VII

Properties of Some Lysosomal Proteases: Cathepsins N–T[a]

Lysosomal protease	Type of protease	Active site	Molecular mass	pH optimum	pI	Found in striated muscle cells?	Other special properties
Cathepsin N	Endopeptidase	–SH	20 kDa	3.0–4.5	6.4	No	A collagenolytic enzyme; human placenta form is 34.5 kDa and pI 5.1
Cathepsin P	Endopeptidase	–SH	31.5 kDa	4.5–6.5	—	Not tested	Converts proinsulin to insulin in pancreatic islet granules
Cathepsin S	Endopeptidase	–SH	25 kDa	3.5–6.0	6.3–6.9	Not tested	Similar to cathepsin L in many of its properties; cleaves proteins but not synthetic substrates
Cathepsin T	Endopeptidase	–SH	35 kDa	6.0–7.5	—	No	A glycoprotein that catalyzes conversion of multiple forms of tyrosine aminotransferase

[a]Most of the information on properties of these proteases was taken from Barrett and McDonald (1980) and McDonald and Barrett (1986).

protease molecule against myosin than does cathepsin B (Schwartz and Bird, 1977; Bird and Carter, 1980). Hence, cathepsin H could cause two to three times greater proteolytic degradation of myosin in muscle cells than does cathepsin B. Schwartz and Bird (1977) found that cathepsins B and D degraded both actin and myosin. Okitani and co-workers (Okitani et al., 1980; Matsukura et al., 1981), however, found that cathepsin B had no effect on any of the myofibrillar proteins and suggested that earlier results showing that cathepsin B could degrade myosin and actin were due to contamination of the cathepsin B preparations with cathepsin L. Cathepsin L degrades myosin heavy chain, actin, α-actinin, troponin T, and troponin I (Matsukura et al., 1981). Okitani et al. (1988) have recently developed a procedure for separating cathepsins B and L from skeletal muscle extracts, so the question of whether cathepsin B can degrade muscle proteins may soon be resolved. In agreement with Schwartz and Bird (1977), Okitani et al. (1981a) found that cathepsin D degraded myosin heavy chain at pH 3 to 4. cDNAs for cathepsins B, D, H, and L have recently been cloned and sequenced (Ishidoh et al., 1987 a,b: Faust et al., 1985). These studies show that all four of these proteases are synthesized in a form that contains a signal peptide of 17 to 21 amino acid residues and a pro-peptide region of 44 (cathepsin D) to 96 amino acid residues. The signal peptide evidently guides the cathepsins into the lysosomes where they are then processed to active proteases by removal of the pro-peptide. All four of these cathepsins may autolyze to produce the heavy and light subunits that were sometimes purified in earlier studies. Although they have not been found inside muscle cells, both cathepsins G and N can degrade collagen. Therefore, depending on the extent to which collagen levels affect the rate of muscle growth, cathepsins G and N may be important in regulating the rate of muscle tissue accumulation.

Although skeletal muscle cells contain sufficient levels of catheptic proteases to degrade all muscle protein in 6–9 days (Schwartz and Bird, 1977), the available evidence indicates that intracellular degradation of myofibrils is not initiated by lysosomal enzymes (Table VIII). Fragments of proteins released from myofibrils by cytoplasmic proteases may ultimately be degraded in lysosomes (Gerard and Schneider, 1979; Table VIII), and lysosomal cathepsins probably have an important role in degradation of the sarcoplasmic proteins (Dayton et al., 1975; Matsukura et al., 1981; Zeman et al., 1985). As stated in Table VIII, however, myofibrils have never been detected inside lysosomes, even in rapidly atrophying muscle, and a large amount of evidence indicates that the lysosomal cathepsins are not active extralysosomally where the pH of the cytoplasm is higher than the pH at which they can degrade myofibrillar proteins. Moreover, cytoplasmic protease inhibitors such as the cystatins (Table XIII) would inhibit many of the lysosomal cathepsins even if they were released into the muscle cell cytoplasm. The rate of muscle protein turnover in cells perfused with L-leucine methyl ester, a compound that disrupts lysosomes,

Table VIII
Evidence Indicating That Lysosomal Cathepsins Are Not Involved in the Initial
Degradation of Myofibrils during Metabolic Turnover

A. Neither intact myofibrils nor entire thick and thin filaments have been observed inside lysosomes, even in muscle cells undergoing rapid atrophy because of denervation (Schiaffino and
Hanzlikova, 1972; Cullen and Pluskal, 1977), dystrophy (Cullen *et al.*, 1979), or metamorpho-
sis (Lockshin and Beaulaton, 1974a,b; Fox, 1975).
  • Products of myofibril degradation or disassembly such as myofilament fragments (Libelius *et
    al.*, 1979) or myofibrillar proteins (Gerard and Schneider, 1979) have been detected in lyso-
    somes, so failure to detect intact myofibrils is not due to inadequate techniques.
  • It seems unlikely that lysosomes could engulf whole myofibrils 10 to 100 μm in diameter
    and 1 to 2 mm long.
  Hence, myofibril degradation is not initiated intralysosomally.

B. Many studies (Bodwell and Pearson, 1964; Martins and Whitaker, 1968; Reville *et al.*, 1976;
Schwartz and Bird, 1977; Bird *et al.*, 1978) have shown that lysosomal cathepsins have little
or no effect on myofibrils or myofibrillar proteins at a pH above 6.0, which would be the
extralysosomal pH in living muscle cells.
  • Lysosomal cathepsins B, D, and L can degrade myofibrils and myofibrillar proteins at a pH
    below 6.0 (Schwartz and Bird, 1977; Bird *et al.*, 1978; Matsukura *et al.*, 1981; Okitani *et
    al.*, 1981a).
  • Lysosomal enzymes released by lysosome disruption do not degrade the proteins in living
    cardiac muscle cells (Reeves *et al.*, 1981).
  Hence, lysosomal cathepsins cannot degrade myofibrils or myofibrillar proteins at the pH values
  that exist outside lysosomes in healthy muscle cells.

C. In several systems, including cardiac (Wildenthal *et al.*, 1980) and insect (Lockshin, 1975)
muscle, rapid muscle atrophy is not prevented by lysosomal enzyme inhibitors (Wildenthal *et
al.*, 1980) or pepstatin (Lockshin, 1975), which is a potent inhibitor of cathepsin D.
  • Although pepstatin inhibits total muscle protein degradation in several mammalian systems
    (McGowan *et al.*, 1976; Stracher *et al.*, 1978, 1979), this inhibition may occur through a
    feedback mechanism from accumulation of fragments of myofibrillar protein in lysosomes.

D. Instances of rapid muscle atrophy have been observed in the absence of detectable lysosomes
(Tweedle *et al.*, 1974) and in starved (Lowell *et al.*, 1986) or streptozotocin-induced diabetic
rats when lysosomal protease content is decreased (Stauber and Fritz, 1985).

E. The myofibrillar proteins have different half lives (Low and Goldberg, 1973; Koizumi, 1974;
Rubenstein *et al.*, 1976; Zak *et al.*, 1977; Wolitsky *et al.*, 1984) and it is unclear how this
would occur if myofibril degradation were initiated by uptake of whole myofibrils into lysosomes.
  • If different proteins have different susceptibilities to proteolytic degradation once inside the
    lysosome, the rate of proteolytic degradation would have to be slower than the rate of uptake
    for this mechanism to produce different half lives (Kay, 1978). If the rate of proteolytic
    degradation were slower than the rate of uptake, however, undegraded proteins would grad-
    ually accumulate inside the lysosome.

decreases only about 30% (Reeves *et al.*, 1981). Also, muscle protein degradation decreases about 25–35% in rats treated with lysosomal protease inhibitors (Janeczko *et al.*, 1985; Lowell *et al.*, 1986). These results suggest that
only 25–35% of total intracellular protein turnover in muscle cells is mediated

by lysosomal enzymes, a level very close to the sarcoplasmic protein content of muscle cells. Consequently, it seems very probable that metabolic turnover of myofibrillar proteins is initiated by cytoplasmic proteases belonging to the third pathway of intracellular protein degradation (Table IV).

4. Neutral and Alkaline Proteolytic Activities

A large number of neutral or alkaline proteolytic activities have been detected in homogenates of skeletal muscle (Tables IX and X), but the significance of many of these proteolytic activities to metabolic turnover of intracellular muscle proteins is unclear. Skeletal muscle tissue contains in addition to muscle cells other cells such as fibroblasts, mast cells, macrophages, and the various cells found in blood. Many of these cells contain large quantities of proteases that can make significant contributions to the total protolytic activity detected in whole muscle homogenates. Purification of the enzymes responsible for some of the proteolytic activities detected in muscle homogenates followed by immunohistochemical localization studies has shown that many of these proteases, such as the alkaline serine protease (Woodbury et al., 1978) and the myosin-cleaving protease (Kuo et al., 1981; Table XI), are located in mast cells and not in muscle cells themselves. Other immunohistochemical studies have shown that the muscle alkaline proteinase and the rat myofibrillar proteinase are identical and cross-react immunologically with mast cell chymase I (Dahlmann et al., 1982). Based on loss of proteolytic activity following injection with "48/80," a compound that disrupts mast cells, it has been suggested that most, if not all, proteolytic activities having an alkaline pH optimum in crude muscle homogenates originate from mast cells in the original tissue and not from striated muscle cells (McKee et al., 1979; Bird and Carter, 1980; Libby and Goldberg, 1980). It is extremely unlikely that proteases in nonmuscle cells are causally involved in protein turnover in muscle cells, and it is important, therefore, to know which proteolytic enzymes are located inside muscle cells when attempting to determine which may be involved in muscle growth.

The subcellular location of the neutral serine proteinase (Table XI) is still somewhat unclear. Immunolocalization studies (Stauber et al., 1983; Kay et al., 1985) have indicated that rat soleus muscle cells have no detectable level of the neutral serine proteinase, whereas rat extensor digitorium longus muscle stained weakly when incubated with antibodies against this proteinase (Kay et al., 1985). Consequently, it seems that some muscle cells may contain the neutral serine proteinase, whereas others do not. The physiological significance of these findings is unclear.

Table IX
Properties of Some Neutral and Alkaline Proteolytic Activities That Have Been Detected in Muscle Homogenates

Protease or proteolytic activity	Cation requirement	Active site	Molecular mass	pH optimum	Location in muscle tissue	Other special properties
μM Ca^{2+}-dependent proteinase (μ-calpain)	2–15 μM Ca^{2+}	–SH	110 kDa (28 and 80 kDa)	7.2–8.0	Inside muscle cells	28-kDa subunit identical with 28-kDa subunit of m-calpain
mM Ca^{2+}-dependent proteinase (m-calpain)	0.6–5 mM Ca^{2+}	–SH	110 kDa (28 and 80 kDa)	7.2–8.0	Inside muscle cells	May relocate in response to physiological stress
Multicatalytic proteinase (Dahlmann et al., 1985a,b Ishiura et al., 1985)	None	–SH or –OH	750 kDa (eight subunits of 19–36 kDa)	8.0–10.0	Inside muscle cells; inter-myofibrillar	Activated 7- to 14-fold by SDS and 50-fold by oleic acid; identical with ingensin(?); pI 5.1–5.2; cylinder shaped
Proteinase I (Dahlmann et al., 1982)	None	–SH	750 kDa	8.0–9.0	Unknown	Stabilized but not activated by ATP; inhibited by leupeptin, antipain, and chymostatin
High-molecular-weight cysteine proteinase (Ismail and Gevers, 1983)	None	–SH	500 kDa	6.5–9.0	Inside muscle cells	Stabilized but not activated by ATP; inhibited by Zn^{2+} but not by leupeptin, antipain, or chymostatin

Table X

Properties of Some Neutral and Alkaline Proteolytic Activities That Have Been Detected in Muscle Homogenates

Protease or proteolytic activity	Cation requirement	Active site	Molecular mass	pH optimum	Location in muscle tissue	Other special properties
Hydrolase H (Okitani et al., 1981b)	None	–SH	340 kDa (51-, 72-, and 92-kDa subunits)	7.0–8.5	Unknown	Has aminoendopeptidase activity in addition to endopeptidase activity; inhibited by Zn^{2+} and leupeptin
ATP-ubiquitin-dependent protease (UCDEN) (Fagan et al., 1987)	Mg^{2+}	–OH and –SH	1500 kDa	7.0–8.5	Inside muscle cells	Cleaves only ubiquitin-conjugated proteins; ubiquitin conjugation requires ATP hydrolysis, which is also required for the proteolysis; very heat-labile; inactivated at 37°C in the absence of ATP
Plasminogen activator (Festoff et al., 1982)	None	–OH	—	7.0–8.0	In cultured cells	Secreted from cultured skeletal muscle cells
Insulin-degrading protease (Shii et al., 1985)	None	–SH	110 kDa	7.0–8.0	Unknown	Most active against insulin but degrades other proteins
Aminopeptidase (Otsuka et al., 1976)	None	–SH	160 kDa	5.5–8.0	Unknown	Removes NH_2-terminal amino acids from several peptides

Table XI

Some Neutral and Alkaline Proteolytic Activities That Have Been Detected in Muscle Homogenates but That Do Not Originate from Inside Muscle Cells

Protease or proteolytic activity	Cation requirement	Active site	Molecular mass	pH optimum	Other special properties
Neutral serine proteinase (Beynon and Kay, 1978)	None	–OH	33 kDa	6.5–8.0	Inactivated by trypsin inhibitors but not chymotrypsin inhibitors; cleaves actin, myosin, α-actinin, tropomyosin, troponin-T, troponin-I, and Z-disks
Group-specific proteinase (Holmes et al., 1971; Katanuma et al., 1975)	None	–OH	24.65 kDa	8.0–9.0	Immunologically identical with chymase II found in mast cells; has been sequenced
Muscle alkaline proteinase (Noguchi and Kandatsu, 1976)	None	–OH	26 kDa	8.0–10.0	Immunologically identical with chymase I found in mast cells; cleaves actin, myosin, tropomyosin, troponin-T, troponin-I but not α-actinin or M-protein
Rat myofibrillar protease (Mayer et al., 1974)	None	–OH	—	8.0–10.0	Probably identical with chymase I; increases during fasting and in rats with genetic muscular dystrophy
Myosin light-chain protease (Bhan et al., 1978)	Ca^{2+} or Mg^{2+}	–SH	14 kDa	8.0–9.0	Specifically cleaves the LC_2 light chain of cardiac myosin; located in mast cells
Myosin-cleaving protease (Murakami and Uchida, 1979; Kuo and Bhan, 1980)	None	–OH	26 kDa	7.5–9.5	Cleaves LC_2 light chain of myosin, myosin heavy chain, M-protein, C-protein, troponin, tropomyosin, and Z-disks; located in mast cells, may be identical with the myosin light-chain protease
Alkaline, cytoplasmic protease (Koszalka and Miller, 1960)	None	–OH	—	7.5–10.5	Probably identical with chymase I and located in mast cells rather than in muscle cells; hydrolyzes endogenous muscle proteins

4.1. Neutral and Alkaline Proteolytic Activities in Muscle Cells

Six neutral proteinases seem to originate from within muscle cells (Tables IX and X). Two of these, the two forms of the Ca^{2+}-dependent proteinase, have clearly been demonstrated by several immunohistochemical studies to be located inside muscle cells (Dayton and Schollmeyer, 1981; Murachi and Yoshimura, 1985; Yoshimura et al., 1986; Kleese et al., 1987). These two proteinases likely have an important role in skeletal muscle growth and will be discussed in more detail subsequently. Three other proteinases are either dependent on ATP hydrolysis or are stimulated by ATP without requiring its hydrolysis. These ATP-related proteinases will be discussed in the next section.

Hydrolase H (Okitani et al., 1981b) is an aminoendopeptidase (Table X) that hydrolyzes the synthetic substrates L-Leu-β-naphthylamide and α-N-benzoyl DL-Arg-β naphthylamide. Although it was purified from rabbit skeletal muscle, which contains few mast cells, the cellular location of this protease has not been determined immunohistochemically, and its role in muscle growth therefore is unclear.

A second protease optimally active at pH 9.0–10.0 and activated by SDS or unsaturated fatty acids such as oleic acid or linoleic acid (Table IX) has been called ingensin (Ishiura et al., 1985) or the multicatalytic proteinase (Dahlmann et al., 1985a, b). This proteinase has a very high molecular mass (Table IX) and seems to have three catalytic sites: one that hydrolyzes benzoyl-Val-Gly-Arg (trypsinlike) with optimal activity at pH 8.5–10.5, one that hydrolyzes succinyl-Ala-Ala-Phe (chymotrypsinlike) with optimal activity at pH 7.0 to 7.5, and one that hydrolyzes benzoyl-Leu-Leu-Glu with optimal activity at pH 8.0–9.5. It has recently been established that ingensin and the multicatalytic protease are the same enzyme (Tsukahara et al., 1988), and the name, multicatalytic protease has been chosen for this enzyme. Proteases similar to the multicatalytic proteinase (Dahlmann et al., 1985a, b; Ishiura et al., 1985) have also been found in bovine pituitary (Wilk and Orlowski, 1983), bovine lens (Ray and Harris, 1985), human lung (Zolfaghari et al., 1987) and erythroleukemia (K562) cells, and rabbit reticulocytes (Ishiura and Sugita, 1986). In addition to their high molecular mass and their activation by SDS or unsaturated fatty acids, all these multicatalytic proteinases dissociate into a group of polypeptides of 19 to 36 kDa in the presence of SDS. Preliminary immunohistochemical results (Dahlmann et al., 1985a; Tanaka et al., 1986) indicate that the multicatalytic proteinase is located inside muscle cells and that after SDS stimulation, the proteinase can degrade sarcoplasmic protein fractions prepared from rat skeletal muscle. The multicatalytic protease has very little proteolytic activity unless it is stimulated by SDS or lipids (Table IX) (Tanaka et al., 1986), and the effect of this proteinase on the myofibrillar proteins is still unknown.

4.2. The ATP-Stimulated and ATP-Dependent Proteinases

It has been known for many years that intracellular proteolysis is greatly stimulated by, and in some instances requires, the presence of ATP (Goldberg and Dice, 1974; Goldberg and St. John, 1976). It was very difficult, however, to purify the proteases responsible for this ATP-related proteolysis and the physiological significance of the ATP-associated proteases remained an enigma. In 1980, it was discovered that proteins whose ϵ-amino groups were conjugated to a small polypeptide called ubiquitin were quickly degraded by an ATP-dependent mechanism involving a neutral protease (Wilkinson et al., 1980; Hershko and Ciechanover, 1982). Conjugation required ATP hydrolysis and seemed to explain the ability of ATP to stimulate intracellular proteolysis. Despite the intriguing qualities of ubiquitin conjugation, however, it was impossible to show that the ubiquitin proteolysis system was involved in anything other than degradation of abnormal proteins in reitculocytes (Table IV; Goldberg and Dice, 1974; Goldberg and St. John, 1976). Consequently, the role of this system in intracellular turnover of normal proteins in mammalian cells was unclear. Moreover, the nature of the proteases involved in this error-eliminating system was unknown.

Recently, it was discovered that the ATP-dependent system in mammalian cells involves at least two high-molecular-mass neutral proteases (Waxman et al., 1985, 1987; Chin et al., 1986; Hough et al., 1986; Fagan et al., 1987). One protease of approximately 1500 kDa requires ATP for activity and degrades only proteins conjugated with ubiquitin (UCDEN, Table X). Hence, this proteolysis pathway requires hydrolysis of a second ATP molecule after ubiquitin conjugation. The second protease has a molecular mass of approximately 600 kDa and requires ATP hydrolysis for degradation of peptide bonds but hydrolyzes proteins that are not conjugated with ubiquitin (Tanaka et al., 1983, 1984; Waxman et al., 1985; Chin et al., 1986). This protease is stimulated weakly by UTP, CTP, and GTP but not by nonhydrolyzable ATP analogues (Waxman et al., 1985). This second protease has been found in reticulocyte lysates and in murine erythroleukemia cells (Tanaka et al., 1983, 1984; Waxman et al., 1985, 1987; Chin et al., 1986). The ubiquitin conjugate degrading enzyme (UCDEN; Table X) has been found in reticulocytes (Waxman et al., 1987) and in liver and skeletal muscle homogenates (Fagan et al., 1987), but has not been localized immunologically in skeletal muscle cells. Also, it is not known whether UCDEN can degrade the myofibrillar proteins, although it can degrade proteins present in crude soluble protein extracts of muscle (Fagan et al., 1987). UCDEN is inhibited by the cystatins, N-ethylmaleimide, decavanadate, and hemin, but not by leupeptin or Ep-475, two potent inhibits of cysteine proteases. The very large size of UCDEN suggests that it is composed of several subunit polypeptides, and it will be important to study the properties of the purified protease.

In addition to the ubiquitin-related proteases, tissues also contain a protease that is stimulated by the presence of ATP but that does not require ATP hydrolysis for this stimulation (Table IX; Dahlman *et al.*, 1982; DeMartino, 1983; Ismail and Gevers, 1983). This ATP-stimulated protease has a molecular mass of 500–750 kDa (Dahlmann *et al.*, 1982; DeMartino, 1983; Ismail and Gevers, 1983; Waxman *et al.*, 1987), but is activated only 2- to 3-fold by ATP, whereas activity of the ATP-dependent proteases is activated 5- to 12-fold by ATP. Moreover, other nucleotides such as ADP, AMP, GTP, GDP, CTP and even pyrophosphate also activate the ATP-stimulated protease but have no effect on the ATP-dependent proteases (DeMartino, 1983). Finally, ATP hydrolysis is not required for activation of the ATP-stimulated protease but is necessary for activation of the ATP-dependent proteases (Fagan *et al.*, 1987; Waxman *et al.*, 1987). Several studies have shown that activity of the ATP-stimulated protease at 37 to 42°C is stabilized by the presence of ATP, and it seems likely that this stabilization causes the ATP stimulation of the protease's activity. The protease has been detected in extracts from isolated cardiac myocytes (DeMartino, 1983) and in extracts from muscle tissue treated with compound 48/80 to remove mast cells (DeMartino, 1983; Ismail and Gevers, 1983), so it is probable that this protease exists inside striated muscle cells. Although the ATP-stimulated protease seems to be a cysteine protease, high levels of leupeptin, antipain, or chymostatin are required to inhibit it. This requirement may account for the conflicting reports as to the ability of leupeptin, antipain, and chymostatin to inhibit this enzyme (Table IX). It recently has been suggested that the ATP-stimulated protease is identical to the multicatalytic protease (Tsukahara *et al.*, 1988), so proteinase I, the high-molecular weight cysteine protease, and the multicatalytic protease (Table IX) may all be the same enzyme. Several studies, however, have failed to find ATP stimulation of multicatalytic protease activity (Ishiura *et al.*, 1985; Ishiura and Sugita, 1986), and the relationship between that protease and the ATP-stimulated protease is still unclear. Because it seems to be located inside muscle cells, it is likely that the ATP-stimulated protease has a role in muscle growth, although the nature of this role is unknown.

4.3. The Ca²⁺-Dependent Proteinases

Busch *et al.* (1972) discovered that skeletal muscle sarcoplasm contained a Ca^{2+}-activated factor that would very quickly remove Z-disks from intact myofibrils without causing other large ultrastructurally detectable changes in myofibrillar structure. This factor was subsequently purified and shown to be a Ca^{2+}-dependent proteinase (CDP) (Dayton, *et al.*, 1976a, b). There has been some confusion regarding nomenclature of the CDPs. In this review, the proteinases will be called μ-calpain (the micromolar Ca^{2+}-dependent proteinase)

FIGURE 1. A schematic diagram showing how the calpains could initiate metabolic turnover of the myofibrillar proteins. (Top) A single sarco-mere of an intact myofibril. The Z-disks are shown as shaded lines in the center of the I-bands, and C-protein bands are shown as open rings encir-cling the thick filament. This schematic is not drawn to scale, and no attempt has been made to include all cross-bridges or C-protein rings on each thick filament. (Middle) The first step proposed for initiating turnover of the myofibrillar proteins. The calpains degrade the Z-disks and remove the C-protein rings from the thick filaments and tro-ponin T, I, and tropomyosin from the thin fila-ments. Degradation of these proteins loosens the myosin molecules on the surface of the thick fil-aments and the actin molecules constituting the thin filaments, so they are released into the sur-rounding sarcoplasm. These filaments are the eas-ily released myofilaments detected by van der Westhuyzen *et al.* (1981). (Bottom) After dissociation into monomers and dimers, the actin and myosin molecules must be degraded to amino acids by other proteases, possibly lysosomal, because the calpains have little or no effect on purified actin and myosin.

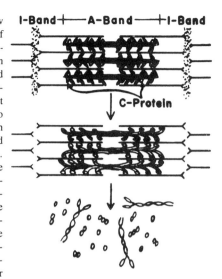

and m-calpain (millimolar Ca^{2+}-dependent proteinase) (see Goll *et al.*, 1985, for a summary of the nomenclature of the CDPs). There was a remarkable similarity between the effects of the calpains on striated muscle myofibrils and the changes occurring in myofibrils during rapidly atrophying muscle, and it was proposed that the proteinases initiated metabolic turnover of the myofibril-lar proteins by releasing thick and thin filaments from the surface of the myofi-bril (Dayton *et al.*, 1975; Fig. 1). Some of the evidence supporting a role for the calpains in initiating myofibrillar protein turnover is summarized in Table XII. The presence of a rapidly labeled group of myofilaments in muscle cells suggests that these filaments originate from the surface of myofibrils and that they are involved in the assembly and disassembly of myofibrils during muscle growth and atrophy (Table XII, A). When isolated myofibrils are treated with m-calpain, the level of the easily released myofilaments increases, suggesting that m-calpain releases myofilaments from the myofibril surface (Fig. 2). The amount of myofilaments released by m-calpain increases when Ca^{2+}, which activates m-calpain, is added to the incubation medium, and decreases when leupeptin, a potent inhibitor of m-calpain is added (Fig. 2). Structurally, Z-disk degradation is the most rapid effect of the calpains on myofibrils (Dayton *et al.*, 1975, 1976b, 1981), and Z-disk alterations, including complete loss of Z-disks, are one of the most consistent features seen in rapidly atrophying muscle afflicted with polymyositis (Cullen and Fulthorpe, 1982; Fig. 3) or various

Table XII
Evidence Supporting a Role for the Ca^{2+}-Dependent Proteinases (Calpains) in Initiating Metabolic Turnover of Myofibrillar Proteins

A. The available evidence indicates that myofibrils grow and turn over metabolically by adding and removing filaments from their surfaces; the calpains have very specific effects on myofibrils that are unique and that would cause disassembly of filaments from the myofibril surface.
   - Myofibrils enlarging in the presence of radioactive amino acids add radioactivity only to their surfaces (Morkin, 1970).
   - Approximately 5% of total myofibrillar protein is in the form of filaments that are easily released from rat myofibrils by treatment with an ATP-relaxing solution (van der Westhuyzen et al., 1981). Pulse-labeling studies with [^{3}H]leucine suggest that these easily released myofilaments originate from the surface of myofibrils and are intermediates in the metabolic turnover of myofibrillar proteins. Treatment with leupeptin, which inhibits the calpains and several other proteases, decreases the size of the releasable myofilament fraction. Treatment with A23187 or treatment of isolated myofibrils with the calpains increases the size of the releasable myofilament fraction.
B. Immunohistochemical studies show that the calpains are located inside muscle cells (Dayton and Schollmeyer, 1981; Goll et al., 1983b, 1985; Yoshimura et al., 1986; Kleese et al., 1987). The calpains are the only proteases known to be located inside muscle cells and to be active at the pH values and ionic strengths that exist inside healthy muscle cells.
C. Calpain activity is elevated during the rapid muscle atrophy associated with denervation (Kohn, 1969), vitamin E deficiency in rabbits (Dayton et al., 1979), and Becker and Duchenne muscular dystrophy in humans (Kar and Pearson, 1976).
D. Rat muscle calpain activity fluctuates during starvation and refeeding (Brooks et al., 1983a) and during diabetes (Brooks et al., 1983b) in a manner consistent with calpain being responsible for initiating metabolic turnover of myofibrillar proteins.
E. Free, intracellular Ca^{2+} concentrations are elevated significantly in dystrophic muscle cells (Bodensteiner and Engel, 1978; Wrogemann et al., 1979; Emery and Burt, 1980; Hudecki et al., 1981; Bertorini et al., 1982), and in infarcted or failing myocardium (Dhalla et al., 1978; Smith, 1978) where myofibrillar degradation is rapid. This elevated Ca^{2+} concentration would be expected to increase activity of the calpains.
F. Rate of total muscle protein turnover increases when muscles are incubated with the Ca^{2+} ionophore A23187 (Publicover et al., 1978; Kameyama and Etlinger, 1979; Sugden, 1980; Lewis et al., 1982; Zeman et al., 1985) or with caffeine, KCl, thymol, procaine, or elevated extracellular Ca^{2+} (Sugden, 1980; Lewis et al., 1982; Zeman et al., 1985), all of which increase intracellular Ca^{2+} concentrations. This increase in muscle protein turnover is inhibited by leupeptin or E-64 (Zeman et al., 1985), and the rate of muscle protein degradation is normal if extracellular Ca^{2+} is absent when A23187 is added (Kameyama and Etlinger, 1979).
G. The most rapid structural alteration of myofibrils caused by the calpains is degradation of Z-disks, and Z-disk alterations, including complete loss of Z-disks, are one of the most consistent structural changes seen in rapidly atrophying muscle afflicted with muscular dystrophy (Cullen et al., 1979; Cullen and Fulthorpe, 1982), after denervation (Cullen et al., 1979), in failing cardiac muscle (Maron et al., 1975, Page and Polimeni, 1977), or in various other muscle pathologies (O'Steen et al., 1975; Afifi et al., 1977; Dayton et al., 1979; Nonaka et al., 1983; Hurst et al., 1984; Ishiura et al., 1984; Otsuka et al., 1985).
H. Several protease inhibitors delay muscle cell degeneration (Libby and Goldberg, 1978; Stracher et al., 1978, 1979) or decrease the rate of muscle protein degradation (Libby and Goldberg, 1980) when administered to dystrophic or denervated muscle or to muscles incubated in vitro. Two of these, leupeptin and chymostatin, are potent inhibitors of the calpains.
I. Increasing intracellular Ca^{2+} concentrations of muscle cells by incubation with cholinesterase

Table XII *(continued)*

agonists causes extensive muscle necrosis at the neuromuscular junctions; this necrosis is characterized by loss of Z-disks and other structural changes that resemble the effects of calpains on muscle cells (Leonard and Salpeter, 1979).

J. SDS–PAGE has shown that the proteolytic changes caused by calpain treatment of myofibrils are very similar to the changes seen in myofibrils prepared from patients with Duchenne muscular dystrophy (Sugita *et al.*, 1980; Obinata *et al.*, 1981).

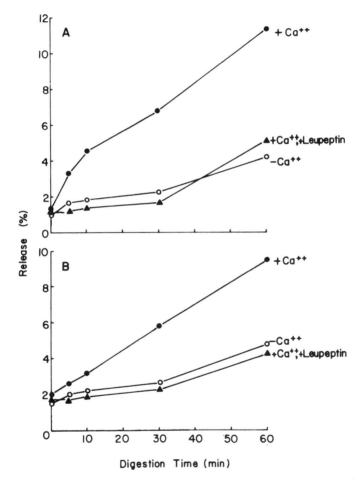

FIGURE 2. Effect of m-calpains on amount of easily released myofilaments obtained from rat skeletal muscle. Myofibrils were treated with m-calpains in the presence or absence of Ca^{2+} and in the presence of Ca^{2+} plus leupeptin, a potent inhibitor of m-calpains. After treatment, the amount of easily released myofilaments that could be obtained was estimated by measuring the amount of myosin (A) or actin (B) present in SDS–polyacrylamide gels of releasable myofilaments. From van der Westhuyzen *et al.* (1981); reprinted by permission of the authors and the American Society of Biological Chemists.

FIGURE 3. Electron micrographs of rat soleus muscle 2 hr after injection of bupivacaine hydrochloride into the muscle. This injection causes rapid atrophy of the treated muscle. The density of Z-disks in hypercontracted myofibrils (right) is markedly decreased. Even in normally arranged sarcomeres (left), Z-disks have selectively disappeared, and the myofibrils resemble those that have been treated with the calpains ×5600; insert ×6500). From Nonaka *et al.* (1983); reproduced by permission of the authors and Springer Verlag.

other muscle pathologies, such as muscle injected with bupivacaine (Nonaka *et al.*, 1983; Fig. 4). These results are consistent with the idea that the calpains degrade Z-disks on the periphery of myofibrils to release filaments from the myofibrillar surface (Dayton *et al.*, 1975; Zeman *et al.*, 1985; Fig. 1). Such a mechanism for metabolic turnover of myofibrils is consistent with the findings that myofibrils in rapidly atrophying muscles typically have diminished diameters rather than being fragmented or having missing sections (Pellegrino and Franzini, 1963) and that myofibrils grow by adding newly synthesized proteins to their surfaces rather than to their interiors (Morkin, 1970). Because the calpains do not degrade myofibrillar proteins to amino acids (Dayton *et al.*, 1975), measuring the release of free amino acids from muscles (Rodemann et al., 1982) does not measure calpain activity, even if it is initiating myofibrillar protein turnover, unless calpain activity is the rate-limiting factor under the conditions chosen and unless degradation of sarcoplasmic and stroma proteins is negligible (Brooks *et al.*, 1983a). The calpains have now been purified from

FIGURE 4. Electron micrograph of muscle from a human afflicted with polymyositis. This patient had muscle weakness and tenderness. Z-disks are completely absent in this sample, and the adjacent thin filaments also are partly destroyed. The thick filaments seem unaffected, and the myofibrils appear structurally similar to myofibrils that have been treated with the calpains. From Cullen and Fulthorpe (1982); reproduced by permission of the authors and John Wiley and Sons, Ltd.

several nonmuscle tissues (Goll *et al.*, 1983b) where their function seems to involve cytoskeletal turnover and release of actin filaments from intracellular attachments. In striated muscle cells, initiating turnover of cytoskeletons likely assumes a more specialized form in which release of actin filaments from Z-disks initiates myofibrillar protein turnover.

4.3.1. Some Properties of m-Calpain

It is now generally agreed that the proteinase originally purified from porcine skeletal muscle by Dayton et al. (1976a, b) is maximally active between pH 6.5 and 8.5; requires 1 to 3 mM Ca^{2+} for maximal activity and has very little activity below 0.1 mM Ca^{2+}; is not activated by Mg^{2+}, Zn^{2+}, Cr^{2+}, or Cd^{2+}; is activated by Mn^{2+}, La^{3+}, and Ba^{2+} to the extent of 10–20% of the maximal activity attained in the presence of Ca^{2+} and is fully activated by high (>10 mM) levels of Sr^{2+} (Tan et al., 1988); requires a reduced sulfhydryl group for activity; and contains two polypeptide chains of 80 and 28 kDa as measured by SDS–PAGE and has a molecular mass of 110 kDa in nondenaturing solvents as measured by sedimentation equilibrium. The cysteine side chain involved in calpain activity is located on the 80-kDa subunit; the role of the 28-kDa subunit is unknown. Care must be taken to prevent proteolysis and loss of the 28-kDa subunit during purification; such loss may account for the occasional reports that the calpains have only an 80-kDa polypeptide.

The available evidence indicates that the calpains have a unique and limited specificity. No synthetic substrate hydrolyzable by the calpains has been found even though m-calpain has been tested with 23 different peptide substrates (Dayton et al., 1976b; Waxman, 1978; Ishiura et al., 1979). The calpains seem to degrade many, but not all, of the contractile and cytoskeletal proteins. In striated and smooth muscle, m-calpain rapidly degrades desmin, vimentin, gelsolin, spectrin, nebulin, vinculin, and troponin T; more slowly degrades titin, troponin I, C-protein, tropomyosin, and tubulin; but does not readily degrade myosin, actin, α-actinin, and troponin C (Dayton et al., 1975; Waxman, 1978; Ishiura et al., 1979; Goll et al., 1983a). The calpains also degrade other cytoskeletal proteins such as the neurofilament proteins (Paggi and Lasek, 1984), ankyrin, Band 4.1 (erythrocytes), microtubule-associated protein (MAP) 2 and more slowly MAP 1 (Sandoval and Weber, 1978), and talin (Fox et al., 1985); degrade the estrogen, progesterone, and EGF receptors; degrade the crystallins in eye lens to produce cataracts (Lorand et al., 1985); and degrade protein kinase C to a Ca^{2+}-independent form. It is critical when studying specificity of the calpains to ensure that the protein substrates are both pure and undenatured because either μ- or m-calpain will rapidly degrade a variety of denatured proteins including actin (Chang and Goll, unpublished observations). We have found that m-calpain causes no bulk degradation of a crude sarcoplasmic protein fraction from rabbit skeletal muscle (Tan et al., 1988) so it seems unlikely that the calpains are causally involved in metabolic turnover of the sarcoplasmic proteins in muscle. In general, the calpains degrade proteins to large fragments rather than to small peptides or free amino acids (Dayton et al., 1975; Goll et al., 1983a).

4.3.2. Autolysis of m-Calpain

Suzuki et al. (1981a) reported that incubation of m-calpain from chicken breast muscle with 0.5 mM Ca^{2+} for 2 min at 0°C reduced the Ca^{2+} required for half-maximal activity from 400 μM to 30 μM. The autolysis that occurred during this period decreased the mass of the large polypeptide subunit, which was the only subunit present in the preparation made by Suzuki and co-workers, from 80 kDa to 79 kDa. Hathaway et al. (1982) subsequently reported that similar autolysis of chicken gizzard m-calpain reduced the Ca^{2+} required for half-maximal activity of this proteinase from 150μM to 5 μM and reduced the mass of the 80-kDa subunit to 76 kDa and the mass of the 28-kDa subunit to 18 kDa. Millimolar Ca^{2+} concentrations are required for autolysis of m-calpain (Suzuki et al., 1981a, b). More recently, Coolican and Hathaway (1984) found that 50 μM phosphatidylinositol or 50 μM dioleoylglycerol lowered the Ca^{2+} required for a half-maximal rate of autolysis from 680 μM to 87 μM. Even 87 μM Ca^{2+}, however, is well above the 0.1 to 10 μM Ca^{2+} concentrations normally found in living, healthy muscle cells.

4.3.3. μ-Calpain

Prompted by an earlier report by Mellgren (1980), Szpacenko et al. (1981), working with bovine cardiac muscle, and Dayton et al. (1981), working with porcine skeletal muscle, found that these tissues contained a second calpain that had subunit polypeptides of 80 and 28 kDa, very similar in mass to those of m-calpain. This second proteinase, μ-calpain, however, had a Ca^{2+} requirement of only 5 to 15 μM for half-maximal activity in contrast to the 150 to 1000 μM required for half-maximal activity of m-calpain (Goll et al., 1986) and possessed a lower net negative charge at pH 7.5 than did the original m-calpain. Otherwise, the two proteinases are very similar: (1) both react with monospecific, polyclonal antibodies against m-calpain (Dayton et al., 1981; Goll et al., 1985; Kleese et al., 1987); (2) both have very similar if not identical effects on myofibrils and myofibrillar proteins (Dayton et al., 1981); (3) SDS–PAGE and NH_2-terminal amino acid analyses indicate that both hydrolyze the same peptide bonds in a given substrate when compared by using the five substrate proteins α-tropomyosin, β-tropomyosin, filamin, vinculin, and gelsolin (Muguruma, Sathe, and Goll, unpublished); (4) both react identically with a variety of inhibitors including the specific calpain inhibitor, calpastatin (Szpacenko et al., 1981); and (5) both autolyze rapidly in the presence of Ca^{2+} and autolysis reduces the Ca^{2+} requirement of both (Goll et al., 1986).

Despite the many similarities between μ- and m-calpain, recent cDNA cloning studies (Emori et al., 1986a, b) have shown that the 80-kDa polypep-

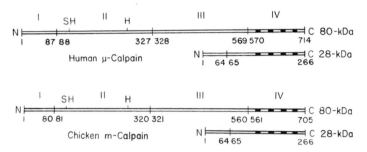

FIGURE 5. Schematic diagram showing the structures of the μ- and m-calpains as indicated by their amino acid sequences. Both proteinases contain 80- and 28-kDa polypeptides. The 28-kDa polypeptides of the two proteinases are identical, and the 80-kDa polypeptides are homologous but not identical. The 28-kDa polypeptide can be divided into two regions: I, the NH_2-terminal region, residues 1–64, which contains 37 Gly residues and which is rich in hydrophobic amino acids; and II, the COOH-terminal region, residues 65–266, which has a sequence homologous to calmodulin and which contains four adjacent sequences homologous to the helix–loop–helix sequences of "E-F calcium-binding hand" structures (four shaded areas). The 80-kDa polypeptides can be divided into four regions numbered from the NH_2 terminus: I, the first 80 or 87 residues (chicken or human) whose sequence is not homologous to any known protein; II, amino acids 88–327 (or 81–320 for chicken) whose sequence is homologous to the sequence of several other cysteine proteinases such as papain and cathepsins B, H, and L; III, amino acids 328–569 (or 321–560 for chicken) whose sequence is not homologous to the sequence of any known protein; and IV, amino acids 570–714 (or 561–705 for chicken) whose amino acid sequence is homologous to the amino acid sequence of several Ca^{2+}-binding proteins such as calmodulin and troponin C and that contains four sequences starting at residues 589, 619, 654, and 684 that are highly homologous to the amino acid sequence of Ca^{2+}-binding "E-F hand" structures. Based on data of Emori *et al.* (1986a, b).

tides of the μ- and m-calpains originate from separate genes and are different polypeptides, although their amino acid sequences are 52% homologous. The domain structures of the polypeptides constituting the μ- and m-calpains are summarized in Fig. 5. The 28-kDa polypeptide originates from a third gene that, in humans at least, is located on a third chromosome (Suzuki, 1987).

It is still unclear whether μ- and m-calpain act in concert on related physiological systems or whether they have completely different functions. Different tissues seem to vary widely in the relative proportions of μ- and m-calpain that they contain (Goll *et al.*, 1986). Platelets and erythrocytes contain less than 10% of their total calpain activity as m-calpain, whereas over 90% of the total calpain activity in chicken skeletal or gizzard muscle is m-calpain (Goll *et al.*, 1986). Skeletal muscle of most mammalian species, including porcine and bovine animals, contains 60–80% of its total calpain activity as m-calpain (Goll *et al.*, 1987). The physiological significance of these differences remains unclear.

Like m-calpain, μ-calpain undergoes autolysis when incubated with Ca^{2+}.

μ-Calpain, however, requires only 50–100 μM Ca^{2+} for autolysis, whereas autolysis of m-calpain is very slow below 300 μM Ca^{2+}. The Ca^{2+} concentration required for half-maximal activity and the polypeptide chains of autolyzed and unautolyzed bovine skeletal muscle calpains are: (1) autolyzed μ-calpain, 0.6 μM Ca^{2+}, 76 and 18 kDa; (2) μ-calpain, 7 μM Ca^{2+}, 80 and 28 kDa; (3) autolyzed m-calpain 180 μM Ca^{2+}, 78 and 18 kDa; and (4) m-calpain, 1000 μM Ca^{2+}, 80 and 28 kDa. The Ca^{2+} requirement of μ-calpain is always lower than the Ca^{2+} requirement of autolyzed m-calpain from the same species (Goll et al., 1986). The physiological significance of the autolyzed calpains is unclear, but the rapid autolysis that occurs in the presence of Ca^{2+} means that Ca^{2+} must be rigorously excluded during purification of the calpains.

4.3.4. Calpain Inhibitor (Calpastatin)

Shortly after m-calpain was purified from porcine skeletal muscle (Dayton et al., 1976a), it was realized that skeletal muscle also contains a protein inhibitor of the calpains (Dayton et al., 1976a; Okitani et al., 1976; Table XIII). It has been very difficult to purify calpastatin without using denaturing solvents and heating steps. These procedures may account for some of the different properties that have been reported for purified calpastatin. Thus, calpastatin from rabbit skeletal muscle has been reported to be 34 kDa (Takahashi-Nakamura et al., 1981), 70 kDa (Cottin et al., 1981), or 110 kDa (Nakamura et al., 1985); that from chicken skeletal muscle as 68 kDa (Ishiura et al., 1982); and that from bovine cardiac muscle as 115 kDa (Otsuka and Goll, 1987) or 145 kDa (Mellgren and Carr, 1983) as measured by SDS–PAGE. Recent cDNA cloning studies show that rabbit calpastatin contains 718 amino acids and has a molecular mass of 76,964 Da (Emori et al., 1987; Table XIII) and that pig heart calpastatin contains 713 residues and has a molecular weight of 77,122 (Takano et al., 1988). These cDNA molecular masses are significantly smaller than the 110 kDa obtained by SDS–PAGE of these same inhibitors and suggest that calpastatin is altered by posttranslational modification or that calpastatin migrates abnormally on SDS–PAGE.

The nature of the undenatured calpastatin molecule and its interaction with the calpains are still unclear. Although the molecular mass of calpastatin in nondenaturing solvents has not yet been determined by sedimentation equilibrium analysis, recent calculations based on Stokes' radius suggest that the undenatured calpastatin molecule is monomeric (Otsuka and Goll, 1987). Various results have shown that one 115-kDa calpastatin molecule inhibits nine or ten calpain molecules (Otsuka and Goll, 1987), that one 125-kDa calpastatin molecule inhibits eight or nine m-calpain molecules (Mellgren and Carr, 1983), that one 64-kDa calpastatin molecule inhibits two calpain molecules (Taka-

Table XIII

Properties of Some Protease Inhibitors That Have Been Detected in Muscle Homogenates: Cysteine Proteinase and Metalloproteinase Inhibitors

Protease inhibitor	Molecular mass	Location in muscle tissue	Specificity and other special properties
Ca^{2+}-dependent proteinase inhibitor (calpastatin) (Otsuka and Goll, 1987)	110 kDa (68,113[a])	Inside muscle cells	Rabbit liver and porcine heart calpastatins have been sequenced and consist of four (liver) or eight (cardiac) repeating domains of 140 amino acids each; these domains presumably are related to the ability of these inhibitors to inhibit four or eight calpain molecules. Calpastatin is specific for the calpains.
High-molecular-weight protease inhibitor (Murakami and Etlinger, 1986)	240 kDa (40-kDa subunits)	Unknown: present in erythrocytes	Only inhibitor of the ATP-stabilized high-molecular-weight proteases detected thus far; ATP seems to abrogate this inhibitor's ability to inhibit the multicatalytic protease; also inhibits the calpains but not trypsin, chymotrypsin, or papain.
Cysteine proteinase inhibitor I-T (Matsuishi et al., 1988)	30,000 (10,000 subunits)	In muscle homogenates	Inhibits cathepsins H and L, moderate inhibition of cathepsin B, inhibits papain only weakly, instability above pH 7 distinguishes it from the known cystatins; not tested against the calpains.
Cystatins, family I (stefins) (Barrett and Salvesen, 1986)	10–14 kDa	Inside muscle and many other cells	The cystatins are a family of protease inhibitors that originate from a common ancestral gene and that inhibit all cysteine proteases except the calpains; all cystatins are stable to extremes in pH and temperature and all sequenced thus far contain a pentapeptide Gln-Val-Val-Ala-Gly that seems to be the "active site"; family I lacks disulfide bonds and contains cystatins A, B, α, and β.
Cystatins, family II (Barrett and Salvesen, 1986)	11–13 kDa	Unknown; but found inside many cells	Contain disulfide bridges and located principally extracellularly, although also found immunologically inside cells; contains cystatins C and S, γ-trace, and chicken cystatin.
Cystatins, family III (kininogens) (Barrett and Salvesen, 1986)	45,723 to 69 kDa	In plasma; not known to be present in muscle cells	Contain low-molecular-weight (L-kininogen; 48 kDa), high-molecular-weight (H-kininogen, 69 kDa), and T-kininogens (45,723 kDa). L-kininogen is identical to α_2-thiol proteinase inhibitor and T-kininogen is identical to α_1-thiol proteinase inhibitor. L- and H-kininogens inhibit the calpains.
α_2 Macroglobulin	740 kDa (four 185-kDa subunits)	In plasma; not known to be present in muscle cells	Inhibits all four types of proteases; is the major inhibitor (95%) of metalloproteinases in the plasma; inhibits proteinases by entrapping them in the center of its four subunits.

[a]Calculated molecular weight of a molecule whose sequence was deduced from DNA sequencing was 68,113 g/mole even though this same molecule had a size of 110,000 by SDS–PAGE (Emori et al., 1987).

hashi-Nakamura *et al.*, 1981), or that one 68-kDa polypeptide inhibits only one m-calpain molecule. Emori *et al.* (1987) found that truncated, mature rabbit calpastatin (68,113 daltons) contains four repeating amino acid sequences of approximately 140 residues each, and that each of these repeats contains a Glu-Lys-Leu-Gly-Glu-Xaa-Glu-Xaa-Ile-Pro-Pro-Xaa-Tyr-Arg sequence that may be the inhibitory site. Subsequent studies have shown that pig heart calpastatin contains a 150 residue N-terminal sequence followed by four repeating sequences of approximately 140 residues each (Takano *et al.*, 1988) and that a cloned 130 amino acid segment of pig heart calpastatin has calpain inhibitory activity (Maki *et al.*, 1987). These studies indicate that calpastatin is a multifunctional inhibitor containing at least four inhibitory sites per molecule. Imajoh *et al.* (1987) recently found that a truncated calpastatin containing two repeating units inhibited two moles of calpain per mole whereas the native rabbit liver inhibitor containing four repeating units inhibited four moles of calpain per mole. Calpastatin is specific for inhibiting the calpains and differs from other protease inhibitors (Tables XIII and XIV) such as the cystatins that inhibit all cysteine proteases except the calpains.

   Ca^{2+} is required for calpastatin to bind the calpains (Ishiura *et al.*, 1982; Mellgren and Carr, 1983; Otsuka and Goll, 1987). The amount of Ca^{2+} required for half-maximal binding of bovine cardiac m-calpain to calpastatin is 0.53 mM and is significantly less than the 0.92 mM Ca^{2+} required for half-maximal activity of the same m-calpain (Otsuka and Goll, 1987). Consequently, if calpastatin and m-calpain are located in the same region of bovine cardiac muscle cells, calpastatin would bind m-calpain and inactivate it even if intracellular Ca^{2+} concentrations reached the millimolar range. The amount of Ca^{2+} required for calpastatin to bind μ-calpain is not known, but it is in the micromolar range (Szpacenko *et al.*, 1981). Clearly, information on the cellular locations of μ-calpain, m-calpain, and calpastatin relative to each other and on possible regulation of the calpain/calpastatin interaction would be important to understanding how activity of the calpains is controlled *in vivo* and how this control might be modified to alter the rate of muscle growth.

4.3.5. Localization of the Calpains and Calpastatin

   Differentiation centrifugation studies (Reville *et al.*, 1976) initially indicated that m-calpain was not located in lysosomes but was free in the cell cytoplasm or the extracellular fluid or was adsorbed loosely to the plasma membrane (Dayton *et al.*, 1976a). Ishiura *et al.* (1980) first suggested on the basis of immunohistochemical studies on isolated myofibrils that part of the calpain in skeletal muscle was located at the level of the Z-disk. It was subsequently shown, however, that Z-disk calpain was probably adventitiously adsorbed to the Z-disk during the homogenization used to prepare myofibrils. Several im-

Table XIV

Properties of Some Protease Inhibitors That Have Been Detected in Muscle Homogenates: Serine Proteinase Inhibitors

Protease inhibitor	Molecular mass	Location in muscle tissue	Specificity and other special properties
Inhibitor of intestinal smooth muscle protease (Carney et al., 1980)	9 kDa	Unknown; found in smooth muscle homogenates	Specific for the neutral serine proteinase found in smooth muscle homogenates (Table XI); has a K_i of 4 M, so it binds weakly to the proteinase.
Protease nexin (Eaton and Baker, 1983)	53 kDa	In cultured skeletal muscle cells	Inhibits urokinase, plasmin, thrombin, and trypsin but not elastase; probably is excreted from cells and functions extracellularly in plasma.
Pancreatic trypsin inhibitor family (Kunitz inhibitors or aprotinin) (Barrett and Salvesen, 1986)	6.512 kDa	Has not been found in muscle cells; rich in lung and pancreas	Has been sequenced and three-dimensional structure determined by X-ray crystallography; has three disulfide bonds; is very stable to heat and pH; inhibits trypsin, chymotrypsin, plasmin, and the kallikreins.
Bowman–Birk family (Barrett and Salvesen, 1986)	8 to 20 kDa	Not in muscle; from leguminous seeds	Includes a large number of inhibitors such as soybean trypsin inhibitor and mung bean inhibitor; most Bowman–Birk inhibitors have two sites: one that inhibits chymotrypsinlike enzymes and another that inhibits trypsin-like enzymes.
Serpins (plasma-serine proteinase inhibitors) (Barrett and Salvesen, 1986)	40 to 60 kDa	Not in muscle; principally in plasma	A large number of inhibitors related by their amino acid sequence; includes α_1-antiproteinase, ovalbumin, antithrombin, α_1-antichymotrypsin, angiotensinogen, α_2-antiplasmin, and CI-inhibitor.
Inter-α-trypsin inhibitor family (Barrett and Salvesen, 1986)	15 to 180 kDa	Not in muscle; found in serum, urine, and nasal secretions	Includes a variety of different proteins that have similar sequences in the first 90 amino acids from the NH_2-terminal region. Different members of this family are related immunologically, but may differ widely in molecular weight.
Muscle serine protease inhibitor (Kuehn et al., 1984)	15 kDa	Found in muscle homogenates	Immunologically different from the serum serpins, α-proteinase inhibitor, rat proteinase inhibitor I, and rat proteinase inhibitor II, but location unknown; inhibits trypsin but not chymotrypsin.

munohistochemical studies at the light (Dayton and Schollmeyer, 1981; Goll *et al.*, 1985; Murachi and Yoshimura, 1985; Kleese *et al.*, 1987) and electron (Yoshimura *et al.*, 1986) microscope levels have now shown that μ- and m-calpain are both located intracellularly and are widely distributed throughout the cell interior. Although Dayton and Schollmeyer (1981) suggested that the calpains were concentrated near the outer cell membrane of muscle cells, no such preferential localization has been observed in subsequent studies (Goll *et al.*, 1983b, 1985; Murachi and Yoshimura, 1985; Yoshimura *et al.*, 1986; Kleese *et al.*, 1987). Schollmeyer (1986a, b) recently suggested that just before cell division of myoblasts, the intracellular location of m-calpain changes from a widely dispersed distribution to a preferential location near the periphery of the cell. During cell division in PtK cells, m-calpain relocated to an association with the mitotic chromosomes, then to a perinuclear location in anaphase, and finally to a midbody location in telophase. It would be important to know whether the intracellular location of m-calpain also varies in mature muscle cells and whether the intracellular distribution of μ-calpain and calpastatin varies in response to different physiological situations.

There have been very few immunohistochemical studies on calpastatin distribution. Lane *et al.* (1985), using several different monoclonal antibodies to bovine cardiac calpastatin in immunofluorescence studies, found that calpastatin seemed preferentially located near the sarcolemma and along the Z-disks of myocytes. Our results indicate that some intracellular calpastatin is located along desmin filaments in mature skeletal muscle cells and in the dense bodies and attachment plaques of chicken gizzard smooth muscle. It seems possible that the location of calpastatin in muscle cells is more restricted than the location of the calpains and that muscle cells contain regions that have the calpains but do not have calpastatin. Such a possibility would have important consequences for regulation of the calpains in skeletal muscle cells.

5. Protease Inhibitors

It has been suggested that a protease inhibitor exists for every protease. Because many of these inhibitors are specific for certain proteases, it is impossible to discover the inhibitors until after the protease has been discovered. Consequently, knowledge of protease inhibitors has lagged behind knowledge of the proteases themselves. The inhibitor of the calpains has been discussed in the preceding section (4.3.4). Recently, an inhibitor of the multicatalytic protease was discovered in reticulocytes (Murakami and Etlinger 1986; Table XIII). It is not known whether such an inhibitor exists in skeletal muscle cells.

Based on the information available, the protease inhibitors have been classified on the basis of the types of proteases that they inhibit (e.g., inhibitors of cysteine proteases, the cystatins; inhibitors of serine proteases, the serpins. The protease inhibitor classes identified thus far are listed in Tables XIII and XIV. In addition to the classes listed, another poorly characterized class called the tissue inhibitors of metalloproteinases has been identified. This class of inhibitors has molecular masses in the range of 30 kDa and has been found in smooth muscle cells but not in skeletal muscle cells. They comprise a group of inhibitors that are resistant to heat (60°C) and to acid pH. They usually are detected by their ability to inhibit collagenases. One of the inhibitors in this class has been sequenced and shown to possess carbohydrate.

6. Summary

Intracellular protein degradation can have important effects on the rate of muscle growth in domestic animals. Despite the potential impact of protein degradation on muscle growth, very little is known about the mechanism of intracellular proteolysis in muscle cells. Although proteolytic enzymes must be involved in muscle protein degradation, the identity and nature of these enzymes are unclear. A substantial amount of circumstantial evidence has accumulated to suggest that lysosomal proteases are not responsible for myofibrillar protein degradation in healthy muscle cells. Despite this circumstantial evidence, however, no unequivocal data exclude lysosomal proteases from some role in myofibrillar protein turnover. Relatively few intracellular proteases have been found that are active at the neutral pH's that exist in muscle cells. Only three of these neutral proteases—the multicatalytic protease, the ATP-stimulated protease, and the calpains—have been demonstrated to exist inside muscle cells. Of these three, only the calpains have been characterized. Although some circumstantial evidence suggests that the calpains could initiate metabolic turnover of the myofibrillar proteins, other proteases must also be involved in this turnover.

In addition to the lack of information on the nature of the proteases involved in intracellular protein degradation, virtually nothing is known about the mechanism by which activity of these proteases is regulated. A protein inhibitor of the calpains has been discovered, but the role this inhibitor has in regulating activity of these proteinases is unclear. It is possible that muscle cells contain many neutral proteases that have remained undiscovered because their activity is suppressed by an endogenous inhibitor. Future studies on the proteases involved in muscle protein degradation should focus on establishing the intracellular location of proteolytic activities detected in muscle homogenates and should involve initial fractionations in an attempt to separate proteases from endogenous inhibitors.

ACKNOWLEDGMENTS. The research that is described in this chapter and that originated from the authors' laboratory was supported by the Arizona Agriculture Experiment Station and by grants from the National Institutes of Health (HL-20984), the Muscular Dystrophy Association, and the United States Department of Agriculture (85-CRCR-1-1818). We thank Donna Sloan for excellent technical assistance and Mavis Collins and Patricia Nagel for their indefatigable help in preparing the manuscript.

References

Afifi, A. K., A. M. Al-Gailany, J. M. Salman and N. B. Bahuth. 1977. Nerve and muscle in steroid-induced weakness in the rabbit. Arch. Phys. Med. Rehabil. 58:143.

Bachmair, A., D. Finley and A. Varshavsky. 1986. In vivo half-life of a protein is a function of its amino-terminal sequence. Science 234:179.

Ballard, F. J., F. M. Thomas and L. M. Stern. 1979. Increased turnover of muscle contractile proteins in Duchenne muscular dystrophy as assessed by 3-methylhistidine and creatinine excretion. Clin. Sci. 56:347.

Barrett, A. J., and J. K. McDonald. 1980. Mammalian Proteases: A Glossary and Bibliography. Vol. I. Academic Press, New York.

Barrett, A. J. and G. Salvesen. 1986. Proteinase Inhibitors. Elsevier, Amsterdam.

Bertorini, T. E., S. K. Bhattacharya, G. M. A. Palmieri, C. M. Chesney, D. Pifer and B. Baker. 1982. Muscle calcium and magnesium content in Duchenne muscular dystrophy. Neurology 32:1088.

Beynon, R. J. and J. Kay. 1978. The inactivation of native enzymes by a neutral proteinase from rat intestinal muscle. Biochem. J. 173:291.

Bhan, A., A. Malhotra and V. B. Hatcher. 1978. Partial characterization of a protease from cardiac myofibrils of dystrophic hamsters. In: H. L. Segal and D. J. Doyle (Ed.) Protein Turnover and Lysosome Function. pp 607–618. Academic Press, New York.

Bird, J. W. C. and J. H. Carter. 1980. Proteolytic enzymes in striated and non-striated muscle. In: K. Wildenthal (Ed.) Degradative Processes in Heart and Skeletal Muscle. pp. 51–85. Elsevier North-Holland Publ. Col, Amsterdam.

Bird, J. W. C., A. M. Spanier and W. N. Schwartz. 1978. Cathepsins B and D: proteolytic activity and ultrastructural localization in skeletal muscle. In: H. L. Segal and D. J. Doyle (Ed.) Protein Turnover and Lysosome Function. pp 589–604. Academic Press, New York.

Bodensteiner, J. B. and A. G. Engel. 1978. Intracellular calcium accumulation in Duchenne dystrophy and other myopathies: a study of 567,000 muscle fibers in 114 biopsies. Neurology 28:439.

Bodwell, C. E. and A. M. Pearson. 1964. The activity of partially purified bovine catheptic enzymes on various natural and synthetic substrates. J. Food Sci. 29:602.

Brooks, B. A., D. E. Goll, Y.-S. Peng, J. A. Greweling and G. Hennecke. 1983a. Effect of starvation and refeeding on activity of a Ca^{2+}-dependent protease in rat skeletal muscle. J. Nutr. 113:145.

Brooks, B. A., D. E. Goll, Y.-S. Peng, J. A. Greweling and G. Hennecke. 1983b. Effect of alloxan diabetes on a Ca^{2+}-activated proteinase in rat skeletal muscle. Amer. J. Physiol. 244:C175.

Busch, W. A., M. H. Stromer, D. E. Goll and A. Suzuki. 1972. Ca^{2+}-specific removal of Z-lines from rabbit skeletal muscle. J. Cell Biol. 52:367.

Canonico, P. G. and J. W. C. Bird. 1970. Lysosomes in skeletal muscle tissue. Zonal centrifugation evidence for multiple cellular sources. J. Cell Biol. 45:321.

Carney, I. T., C. G. Curtis, J. Kay and N. L. Birket. 1980. A low-molecular-weight inhibitor of the neutral proteinase from rat intestinal smooth muscle. Biochem. J. 185:423.

Chin, D. T., N. Carlson, L. Kuehl and M. Rechsteiner. 1986. The degradation of guanidinated lysozyme in reticulocyte lysate. J. Biol. Chem. 261:3883.

Coolican, S. A. and D. R. Hathaway. 1984. Effect of L-α-phosphatidylinositol on a vascular smooth muscle Ca^{2+}-dependent protease. Reduction of the Ca^{2+} requirement for autolysis. J. Biol. Chem. 259:11627.

Cottin, P., P. L. Vidalenc, and A. Ducastaing. 1981. Ca^{2+}-dependent association between a Ca^{2+}-activated neutral protease (CaANP) and its specific inhibitor. FEBS Lett. 136:221.

Crisona, N. J. and R. C. Strohman. 1983. Inhibition of contraction of cultured muscle fibers results in increased turnover of myofibrillar proteins but not of intermediate filament proteins. J. Cell Biol. 96:684.

Cullen, M. J., S. T. Appleyard and L. Bindoff. 1979. Morphological aspects of muscle breakdown and lysosomal activtion. Ann. N.Y. Acad. Sci. 317:440.

Cullen, M. J. and J. J. Fulthorpe. 1982. Phagocytosis of the A-band following Z line and I band loss. Its signficance in skeletal muscle breakdown. J. Pathol. 138:129.

Cullen, M. J. and M. G. Pluskal. 1977. Early changes in the ultrastructure of denervated rat skeletal muscle. Exp. Neurol 56:115.

Dahlmann, B., I. Block, L. Kuehn, M. Rutschmann and H. Reinauer. 1982. Immunological evidence for the identity of three proteinases from rat skeletal muscle. FEBS Lett. 138:88.

Dahlmann, B., L. Kuehn, M. Rutschmann and H. Reinauer. 1985a. Purification and characterization of a multicatalytic high-molecular-mass proteinase from rat skeletal muscle. Biochem. J. 228:161.

Dahlmann, B., M. Rutschmann, L. Kuehn and H. Reinauer. 1985b. Activation of the mutlicatalytic proteinase from rat skeletal muscle by fatty acids or sodium dodecyl sulfate. Biochem. J. 228:171.

Dayton, W. R., D. E. Goll, M. H. Stromer, W. J. Reville, M. G. Zeece and R. M. Robson. 1975. Some properties of a calcium-activated protease that may be involved in myofibrillar protein turnover. In: E. Reich, D. B. Rifkin and E. Shaw (Ed.) Cold Spring Harbor Conferences on Cell Proliferation. Vol. 2, pp. 551–577. Cold Spring Harbor Laboratory, Cold Spring Harbor, NY.

Dayton, W. R., D. E. Goll, M. G. Zeece, R. M. Robson and W. J. Reville, 1976a. A Ca^{2+}-activated protease possibly involved in myofibrillar protein turnover. Purification from porcine muscle. Biochemistry 15:2150.

Dayton, W. R., W. J. Reville, D. E. Goll and M. H. Stromer. 1976b. A Ca^{2+}-activated protease possibly involved in myofibrillar protein turnover. Partial characterization of the purified enzyme. Biochemistry 15:2159.

Dayton, W. R. and J. V. Schollmeyer. 1981. Immunocytochemical localization of a calcium-activated protease in skeletal muscle cells. Exp. Cell Res. 136:423.

Dayton, W. R., J. V. Schollmeyer, A. C. Chan and C. E. Allen. 1979. Elevated levels of a calcium-activated muscle protease in rapidly atrophying muscles from vitamin E-deficient rabbits. Biochim. Biophys. Acta 584:216.

Dayton, W. R., J. V. Schollmeyer, R. A. Lepley and L. R. Cortes. 1981. A calcium-activated protease possibly involved in myofibrillar protein turnover. Isolation of a low-calcium-requiring form of the protease. Biochim. Biophys. Acta 659:48.

DeMartino, G. N. 1983. Identification of a high molecular weight alkaline protease in rat heart. J. Mol. Cell. Cardiol. 15:17.

Dhalla, N. S., P. K. Das and G. P. Sharma. 1978. Subcellular basis of cardiac contractile failure. J. Mol. Cell. Cardiol. 10:363.

Eaton, D. L. and J. B. Baker. 1983. Evidence that a variety of cultured cells secrete protease nexin and produce a distinct cytoplasmic serine protease-binding factor. J. Cell. Physiol. 117:175.

Emery, A. E. H. and D. Burt. 1980. Intracellular calcium and pathogenesis and antenatal diagnosis of Duchenne muscular dystrophy. Brit. Med. J. 280:355.

Emori, Y., H. Kawaskai, S. Imajoh, K. Imahori and K. Suzuki. 1987. Endogenous inhibitor for calcium-dependent cysteine protease contains four internal repeats that could be responsible for its multiple reactive sites. Proc. Natl. Acad. Sci. USA 84:3590.

Emori, Y., H. Kawaskai, S. Imajoh, S. Kawashima and K. Suzuki. 1986a. Isolation and sequence analysis of cDNA clones for the small subunit of rabbit calcium-dependent protease. J. Biol. Chem. 261:9472.

Emori, Y., H. Kawasaki, H. Sugihara, S. Imajoh, S. Kawashima and K. Suzuki. 1986b. Isolation and sequence analyses of cDNA clones for the large subunits of two isozymes of rabbit calcium-dependent protease. J. Biol. Chem. 261:9465.

Fagan, J. M., L. Waxman and A. L. Goldberg. 1987. Skeletal muscle and liver contain a soluble ATP-ubiquitin-dependent proteolytic system. Biochem. J. 243:335.

Faust, P. L., S. Kornfeld, and J. M. Chirgwin. 1985. Cloning and sequence analysis of cDNA for human cathepsin D. Proc. Natl. Acad. Sci. U.S.A. 82:4910.

Festoff, B. M., M. R. Patterson and K. Romstedt. 1982. Plasminogen activator: the major secreted neutral protease of cultured skeletal muscle cells. J. Cell. Physiol. 110:190.

Fox, H. 1975. Aspects of tail muscle ultrastructure and its degeneration in Rana temporaria. J. Embryol. Exp. Morphol. 24:191.

Fox, J. E. B., D. E. Goll, C. C. Reynolds and D. R. Phillips. 1985. Identification of two proteins (actin-binding protein and P235) that are hydrolyzed by endogenous Ca^{2+}-dependent protease during platelet aggregation. J. Biol. Chem. 260:1060.

Gerard, K. W. and D. L. Schneider. 1979. Evidence for degradation of myofibrillar proteins in lysosomes. Myofibrillar proteins derivatized by intramuscular injection of N-ethymaleimide are sequestered in lysosomes. J. Biol. Chem. 254:11798.

Goldberg, A. L. 1969a. Protein turnover in skeletal muscle. I. Protein catabolism during work-induced hypertrophy and growth induced with growth hormone. J. Biol. Chem. 244:3217.

Goldberg, A. L. 1969b. Protein turnover in skeletal muscle. II. Effects of denervation and cortisone on protein catabolism in skeletal muscles. J. Biol. Chem. 244:3223.

Goldberg, A. L. and J. F. Dice. 1974. Intracellular protein degradation in mammalian and bacterial cells. Annu. Rev. Biochem. 43:835.

Goldberg, A. L., J. Kowit, J. Etlinger and Y. Klemes. 1978. Selective degradation of abnormal proteins in animal and bacterial cells. In: H. L. Segal and D. J. Doyle (Ed.) Protein Turnover and Lysosome Function. pp 171–196. Academic Press, New York.

Goldberg, A. L. and A. C. St. John. 1976. Intracellular protein degradation in mammalian and bacterial cells: Part 2. Annu. Rev. Biochem. 45:747.

Goldspink, D. F., P. J. Garlick and M. A. McNurlan. 1983. Protein turnover measured in vivo and in vitro in muscles undergoing compensatory growth and subsequent denervation atrophy. Biochem. J. 210:89.

Goll, D. E., T. Edmunds, W. C. Kleese, S. K. Sathe and J. D. Shannon. 1985. Some properties of the Ca^{2+}-dependent proteinase. In: E. A. Khairallah, J. S. Bond and J. W. C. Bird (Ed.) Intracellular Protein Catabolism. pp 151–164. Alan R. Liss, New York.

Goll, D. E., W. C. Kleese, D. A. Sloan, J. D. Shannon and T. Edmunds. 1986. Properties of the Ca^{2+}-dependent proteinases and their protein inhibitor. Cienc. Biol. (Luanda) 11:75.

Goll, D. E., Y. Otsuka, P. A. Nagainis, J. D. Shannon, S. K. Sathe and M. Muguruma. 1983a. Role of muscle proteinases in maintenance of muscle integrity and mass. J. Food Biochem. 7:137.

Goll, D. E., R. M. Robson and M. H. Stromer. 1976. Muscle proteins. In: J. R. Whitaker and S. Tannenbaum (Ed.) Food Proteins. pp 121–174. AVI Publ. Co., Westport, CT.

Goll, D. E., R. M. Robson and M. H. Stromer. 1984. Skeletal muscle. In: M. J. Swenson (Ed.) Dukes' Physiology of Domestic Animals (10th Ed.). pp 548–580. Cornell Univ. Press, Ithaca, NY.

Goll, D. E., J. D. Shannon, T. Edmunds, S. K. Sathe, W. C. Kleese and P. A. Nagainis. 1983b. Properties and regulation of the Ca^{2+}-dependent proteinase. In: B. de Bernard, G. L. Sottocasa, G. Sandri, E. Carafoli, A. N. Taylor, T. C. Vanaman and R. J. P. Williams (Ed.) Calcium-Binding Proteins. pp. 19–35. Elsevier, Amsterdam.

Gopinath, R. and W. D. Kitts. 1984. Growth, N^r-methylhistidine excretion and muscle protein degradation in growing beef steers. J. Anim. Sci. 59:1262.

Hathaway, D. R., D. K. Werth and J. R. Haeberle. 1982. Limited autolysis reduces the Ca^{2+} requirement of a smooth muscle Ca^{2+}-activated protease. J. Biol. Chem. 257:9072.

Hershko, A. and A. Ciechanover. 1982. Mechanisms of intracellular protein breakdown. Annu. Rev. Biochem. 51:335.

Holmes, D., M. E. Parsons, D. C. Park and R. J. Pennington. 1971. An alkaline proteinase in muscle homogenates. Biochem. J. 125:98p.

Hough, R., G. Pratt and M. Rechsteiner. 1986. Ubiquitin–lysozyme conjugates. Identification and characterization of an ATP-dependent protease from rabbit reticulocyte lysates. J. Biol. Chem. 261:2400.

Hudecki, M. S., C. M. Pollina, R. R. Heffner and A. K. Bhargava. 1981. Enhanced functional ability in drug-treated dystrophic chickens: trial results with indomethacin, diphenylhydantoin, and prednisolone. Exp. Neurol. 73:173.

Hurst, L. C., M. A. Badalamente, J. Ellstein and A. Stracher. 1984. Inhibition of neural and muscle degeneration after epineural neurorrhaphy. J. Hand Surg 9A:564.

Iodice, A. A., J. Chin, S. Perker and I. M. Weinstock. 1972. Cathepsins A, B, C, D, and autolysis during development of breast muscle of normal and dystrophic chickens. Arch. Biochem. Biophys. 152:166.

Imajoh, S., H. Kawashima, Y. Emori, S. Ishiura, Y. Minami, H. Sugita, K. Imahori, and K. Suzuki. 1987. A fragment of an endogenous inhibitor produced in *Escherichia coli* for calcium-activated neutral protease (CANP) retains an inhibitory activity. FEBS Lett. 215:274.

Ishidoh, K., S. Imajoh, Y. Emori, S. Ohno, H. Kawasaki, Y. Minami, E. Kominami, N. Katunuma, and K. Suzuki. 1987. Molecular cloning and sequencing of cDNA for rat cathepsin H. Homology in pro-peptide regions of cysteine proteinases. FEBS Lett. 226:33.

Ishidoh, K., T. Towatari, S. Imajoh, H. Kawasaki, E. Kominami, H. Katunuma, and K. Suzuki. 1987. Molecular cloning and sequencing of cDNA for rat cathepsin L. FEBS Lett. 223:69.

Ishiura, S., I. Nonaka, H. Nakase, A. Tada and H. Sugita. 1984. Two-step mechanism of myofibrillar protein degradation in acute plasmid-induced muscle necrosis. Biochim. Biophys. Acta 798:333.

Ishiura, S., M. Sano, K. Kamakura and H. Sugita. 1985. Isolation of two forms of the high-molecular-mass serine protease, ingensin, from porcine skeletal muscle. FEBS Lett. 189:119.

Ishiura, S. and H. Sugita. 1986. Ingensin, a high-molecular-mass alkaline protease from rabbit reticulocyte. J. Biochem. 100:753.

Ishiura, S., H. Sugita, I. Nonaka and K. Imahori. 1980. Calcium-activated neutral protease. Its localization in the myofibril especially at the Z-band. J. Biochem. 87:343.

Ishiura, S., H. Sugita, K. Suzuki and K. Imahori. 1979. Studies of a calcium-activated neutral protease from chicken skeletal muscle. II. Substrate specificity. J. Biochem. 86:579.

Ishiura, S., S. Tsuji, H. Murofushi and K. Suzuki. 1982. Purification of an endogenous 68,000-dalton inhibitor of Ca^{2+}-activated neutral protease from chicken skeletal muscle. Biochim. Biophys. Acta 701:216.

Ismail, F. and W. Gevers. 1983. A high-molecular-weight cysteine endopeptidase from rat skeletal muscle. Biochim. Biophys. Acta 742:399.

Janeczko, R. A., R. M. Carriere and J. D. Etlinger. 1985. Endocytosis, proteolysis, and exocytosis of exogenous proteins by cultured myotubes. J. Biol. Chem. 260:7051.

Jones, S. J., E. D. Aberle and M. D. Judge. 1986. Skeletal muscle protein turnover in broiler and layer chicks. J. Anim. Sci. 62:1576.

Kameyama, T. and J. D. Etlinger. 1979. Calcium-dependent regulation of protein synthesis and degradation in muscle. Nature 279:344.

Kar, N. C. and C. M. Pearson. 1976. A calcium-activated neutral protease in normal and dystrophic human muscle. Clin. Chim. Acta 73:293.

Katanuma, N., E. Kominami, K. Kobayashi, Y. Banno. K. Suzuki, K. Chichibu, Y. Hamaguchi and T. Katsunuma. 1975. Studies on new intracellular proteases in various organs of rat. I. Purification and comparison of their properties. Eur. J. Biochem. 52:37.

Kay, J. 1978. Intracellular protein degradation. Biochem. Soc. Trans. 6:789.

Kay, J., R. Heath, B. Dahlmann, L. Kuehn and W. T. Stauber. 1985. Serine proteinases and protein breakdown in muscle. In: E. A. Khairallah, J. S. Bond and J. W. C. Bird (Ed.) Intracellular Protein Catabolism. pp 195–205. Alan R. Liss, New York.

Kleese, W. C., D. E. Goll, T. Edmunds and J. D. Shannon. 1987. Immunofluorescent localization of the Ca^{2+}-dependent proteinase and its inhibitor in tissues of Crotalus atrox. J. Exp. Zool. 241:277.

Kohn, R. R. 1969. A proteolytic system involving myofibrils and a soluble factor from normal and atrophying muscle. Lab. Invest. 20:202.

Koizumi, T. 1974. Turnover rates of structural proteins of rabbit skeletal muscle. J. Biochem. 76:431.

Kominami, E., J. Tsukahara, Y. Bando and N. Katunuma. 1985. Distribution of cathepsins B and H in rat tissues and peripheral blood cells. J. Biochem. 98:87.

Koszalka, T. R. and L. L. Miller. 1960. Proteolytic activity of rat skeletal muscle. I. Evidence for the existence of an enzyme active optimally at pH 8.5 to 9.0. J. Biol. Chem. 235:665.

Kuehn, L., B. Dahlmann and H. Reinauer. 1984. Identification of four distinct serine proteinase inhibitors in rat skeletal muscle. Biochem. Biophys. Res. Commun. 120:96.

Kuo, T. and A. Bhan. 1980. Studies of a myosin-cleaving protease from dystrophic hamster heart. Biochem. Biophys. Res. Commun. 92:570.

Kuo, T. H., F. Giacomelli, K. Kithier and A. Malhotra. 1981. Biochemical characterization and cellular localization of serine protease in myopathic hamster. J. Mol. Cell. Cardiol. 13:1035.

Lane, R. D., R. L. Mellgren and M. T. Mericle. 1985. Subcellular localization of bovine heart calcium-dependent protease inhibitor. J. Mol. Cell. Cardiol. 17:863.

Leonard, J. P. and M. M. Salpeter. 1979. Agonist-induced myopathy at the neuromuscular junction is mediated by calcium. J. Cell Biol. 82:811.

Lewis, S. E. M., P. Anderson and D. F. Goldspink. 1982. The effects of calcium on protein turnover in skeletal muscles of the rat. Biochem. J. 204:257.

Li, J. B. 1980. Protein synthesis and degradation in skeletal muscle of normal and dystrophic hamsters. Amer. J. Physiol. 239:E401.

Libby, P. and A. L. Goldberg. 1978. Leupeptin, a protease inhibitor, decreases protein degradation in normal and diseased muscles. Science 199:534.

Libby, P. and A. L. Goldberg. 1980. Effects of chymostatin and other proteinase inhibitors on protein breakdown and proteolytic activities in muscle. Biochem. J. 188:213.

Libelius, R., J. O. Josefsson and I. Lundquist. 1979. Endocytosis in chronically denervated mouse skeletal muscle. A biochemical and ultrastructural study with horseradish peroxidase. Neuroscience 4:283.

Lockshin, R. A. 1975. Failure to prevent degeneration of insect muscles with pepstatin. Life Sci. 17:403.

Lockshin, R. A. and J. Beaulaton. 1974a. Programmed cell death. Cytochemical evidence for lysosomes during the normal breakdown of the intersegmental muscles. J. Ultrastruct. Res. 46:43.

Lockshin, R. A. and J. Beaulaton. 1974b. Programmed cell death. Cytochemical appearance of lysosomes when the death of the intersegmental muscles is prevented. J. Ultrastruct. Res. 46:63.

Lorand, L., S. M. Conrad and P. T. Velasco. 1985. Formation of a 55,000-weight cross-linked β-crystallin dimer in the Ca^{2+}-treated lens. A model for cataract. Biochemistry 24:1525.

Low, R. B. and A. L. Goldberg. 1973. Nonuniform rates of turnover of myofibrillar proteins in rat diaphragm. J. Cell Biol. 56:590.

Lowell, B. B., N. B. Ruderman and M. N. Goodman. 1986. Evidence that lysosomes are not involved in the degradation of myofibrillar proteins in rat skeletal muscle. Biochem. J. 234:237.

McCarthy, F. D., W. G. Bergen and D. R. Hawkins. 1983. Muscle protein turnover in cattle of differing genetic backgrounds as measured by urinary N^τ-methylhistidine excretion. J. Nutr. 113:2455.

McDonald, J. K. and A. J. Barrett. 1986. Mammalian Proteases: A Glossary and Bibliography. Vol. 2. Academic Press, New York.

McGowan, E. B., S. A. Shafiq and A. Stracher. 1976. Delayed degeneration of dystrophic and normal muscle cells cultures treated with pepstatin, leupeptin, and antipain. Exp. Neurol. 50:649.

McKee, E. E., M. G. Clark, C. J. Beinlich, J. A. Lins and H. E. Morgan. 1979. Neutral-alkaline proteases and protein degradation in rat heart. J. Mol. Cell. Cardiol. 11:1033.

McKeran, R. O., D. Halliday and P. Purkiss. 1977. Increased myofibrillar protein catabolism in Duchenne muscular dystrophy measured by 3-methylhistidine excretion in the urine. J. Neurol. Neurosurg. Psychiatry 40:979.

Maki, M., E. Takano, H. Mori, R. Kannagi, T. Murachi and M. Hatanaka. 1987. Repetitive region of calpastatin is a functional unit of the proteinase inhibitor. Biochem. Biophys. Res. Commun. 143:300.

Maron, B. J., V. J. Ferrans and W. C. Roberts. 1975. Ultrastructural features of degenerated cardiac muscle cells in patients with cardiac hypertrophy. Amer. J. Pathol. 79:387.

Martin, A. F., 1981. Turnover of cardiac troponin subunits. Kinetic evidence for precursor pool of troponin I. J. Biol. Chem. 256:964.

Martin, A. F., M. Rabinowitz, R. Blough, G. Prior and R. Zak. 1977. Measurements of half-life of rat cardiac myosin heavy chain with leucyl tRNA used as a precursor pool. J. Biol. Chem. 252:3422.

Martins, C. B. and J. R. Whitaker. 1968. Catheptic enzymes and meat tenderization. I. Purification of cathepsin D and its action on actomyosin. J. Food Sci. 33:59.

Maruyama, K., M. L. Sunde and R. W. Swick. 1978. Growth and muscle protein turnover in the chick. Biochem. J. 176:573.

Matsuishi, M., A. Okitani, Y. Hayakawa, and H. Kato. 1988. Cysteine proteinase inhibitors from rabbit skeletal muscle. Int. J. Biochem. 20:259.

Matsukura, U., A. Okitani, T. Nishimuro and H. Kato. 1981. Mode of degradation of myofibrillar proteins by an endogenous protease, cathepsin L. Biochim. Biophys. Acta 662:41.

Mayer, M., R. Amin and E. Shafrir. 1974. Rat myofibrillar protease: enzyme properties and adaptive changes in conditions of muscle protein degradation. Arch. Biochem. Biophys. 161:20.

Mellgren, R. L. 1980. Canine cardiac calcium-dependent proteases: resolution of two forms with different requirements for calcium. FEBS Lett. 109:129.

Mellgren, R. L. and T. C. Carr. 1983. The protein inhibitor of calcium-dependent proteases: purification from bovine heart and possible mechanisms of regulation. Arch. Biochem. Biophys. 225:779.

Millward, D. J., P. C. Bates, G. J. Laurent and C. C. Lo. 1978. Factors affecting protein breakdown in skeletal muscle. In: H. L. Segal and D. J. Doyle (Ed.) Protein Turnover and Lysosome Function. pp 619–644. Academic Press, New York.

Morkin, E. 1970. Postnatal muscle fiber assembly: localization of newly synthesized myofibrillar proteins. Science 167:1499.

Mulvaney, D. R., R. A. Merkel and W. G. Bergen. 1985. Skeletal muscle protein turnover in young male pigs. J. Nutr. 115:1057.

Murachi, T. and N. Yoshimura. 1985. Intracellular localization of low and high calcium-requiring forms of calpain. In: E. A. Khairallah, J. S. Bond and J. W. C. Bird (Ed.) Intracellular Protein Degradation. pp. 165–174. Alan R. Liss, New York.

Murakami, K. and J. D. Etlinger. 1986. Endogenous inhibitor of nonlysosomal high molecular weight protease and calcium-dependent protease. Proc. Natl. Acad. Sci. USA 83:7588.

Murakami, U. and K. Uchida. 1979. Degradation of rat cardiac myofibrils and myofibrillar proteins by a myosin-cleaving protease. J. Biochem. 86:553.

Nakamura, M., M. Inomata, M. Hayashi, K. Imahori and S. Kawashima. 1985. Purification and characterization of 210,000-dalton inhibitor of calcium-activated neutral protease from rabbit skeletal muscle and its relation to 50,000-dalton inhibitor. J. Biochem. 98:757.

Noguchi, T. and M. Kandatsu. 1976. Some properties of alkaline protease in rat muscle compared with that in peritoneal cavity cells. Agr. Biol. Chem. 40:927.

Nonaka, I., A. Takagi, S. Ishiura, H. Nagase and H. Sugita. 1983. Pathophysiology of muscle fiber necrosis induced by bupivacaine hydrochloride (maraine). Acta Neuropathol. 60:167.

Obinata, T., K. Maruyama, H. Sugita, K. Kohama and S. Ebashi. 1981. Dynamic aspects of structural proteins in vertebrate skeletal muscle. Muscle Nerve 4:456.

Okitani, A., D. E. Goll, M. H. Stromer and R. M. Robson. 1976. Intracellular inhibitor of a Ca^{2+}-activated protease involved in myofribrillar protein turnover. Fed. Proc. 35:1746.

Okitani, A., M. Matsuishi, T. Matsumoto, E. Kamoshida, M. Sato, U. Matsukura, M. Watanabe, H. Kato, and M. Fujimaki. 1988. Purification and some properties of cathepsin B from rabbit skeletal muscle. Eur. J. Biochem. 171:377.

Okitani, A., U. Matsukura, H. Kato, and M. Fujimaki. 1980. Purification and some properties of a myofibrillar protein-degrading protease, cathepsin-L, from rabbit skeletal muscle. J. Biochem. 87:1133.

Okitani, A., T. Matsumoto, Y. Kitamura and H. Kato. 1981a. Purification of cathepsin D from rabbit skeletal muscle and its action towards myofibrils. Biochim. Biophys. Acta 662:202.

Okitani, A., T. Nishimura and H. Kato. 1981b. Characterization of hydrolase H, a new muscle protease possessing aminoendopeptidase activity. Eur. J. Biochem. 115:269.

Orcutt, M. W. and R. B. Young. 1982. Cell differentiation, protein synthesis rate and protein accumulation in muscle cell cultures isolated from embryos of layer and broiler chickens. J. Anim. Sci. 54:769.

O'Steen, W. K., C. R. Shear and K. V. Anderson. 1975. Extraocular muscle degeneration and regeneration after exposure of rats to incandescent radiant energy. J. Cell Sci. 18:157.

Otsuka, Y. and D. E. Goll. 1987. Purification of the Ca^{2+}-dependent proteinase inhibitor from bovine cardiac muscle and its interaction with the millimolar Ca^{2+}-dependent proteinase. J. Biol. Chem. 262:5839.

Otsuka, Y., Y. Kumojima, Y. Ishikawa and E. Kawabara. 1985. Ca^{2+}-activated protease activity in vitamin E-deficient rats. Agr. Biol. Chem. 49:2105.

Otsuka, Y., A. Okitani, R. Katakai and M. Fujimaki. 1976. Purification and properties of an aminopeptidase from rabbit skeletal muscle. Agr. Biol. Chem. 40:2335.

Page, E. and P. I. Polimeni. 1977. Ultrastructural changes in the ischemic zone bordering experimental infarcts in rat left ventricles. Amer. J. Pathol. 86:81.

Paggi, P. and R. J. Lasek. 1984. Degradation of purified neurofilament subunits by calcium-activated neutral protease: characterization of the cleavage products. Neurochem. Int. 6:589.

Pellegrino, C. and C. Franzini. 1963. An electron microscope study of denervation atrophy in red and white skeletal muscle fibers. J. Cell Biol. 17:327.

Pickart, C. M. and I. A. Rose. 1985. Ubiquitin-carboxyl-terminal hydrolase acts on ubiquitin-carboxyl-terminal amides. J. Biol. Chem. 260:7903.

Pontremoli, S. and E. Melloni. 1986. Extralysosomal protein degradation. Annu. Rev. Biochem. 55:455.

Pösö, A. R. and G. E. Mortimore. 1984. Requirement for alanine in the amino acid control of deprivation-induced protein degradation in liver. Proc. Natl. Acad. Sci. USA 81:4270.

Publicover, S. J., C. J. Duncan and J. L. Smith. 1978. The use of A23187 to demonstrate the role of intracellular calcium in causing ultrastructural damage in mammalian muscle. J. Neuropathol. Exp. Neurol. 37:544.

Ray, K. and H. Harris. 1985. Purification of neutral lens endopeptidase: close similarity to a neutral proteinase in pituitary. Proc. Natl. Acad. Sci. USA 82:7545.

Reeds, P. J., S. M. Hay, P. M. Dorwood and R. M. Palmer. 1986. Stimulation of muscle growth by clenbuterol: lack of effect on muscle protein biosynthesis. Brit. J. Nutr. 56:249.

Reeves, J. P., R. S. Decker, J. S. Crie and K. Wildenthal. 1981. Intracellular disruption of rat heart lysosomes by leucine methyl ester: effects on protein degradation. Proc. Natl. Acad. Sci. USA 78:4426.

Reville, W. J., D. E. Goll, M. H. Stromer, R. M. Robson and W. R. Dayton. 1976. A Ca^{2+}-activated protease possibly involved in myofibrillar protein turnover. Subcellular localization of the protease in porcine skeletal muscle. J. Cell Biol. 70:1.

Rodemann, H. P., L. Waxman and A. L. Goldberg. 1982. The stimulation of protein degradation in muscle by Ca^{2+} is mediated by prostaglandin E_2 and does not require the Ca^{2+}-activated protease. J. Biol. Chem. 257:8716.

Rogers, S., R. Wells and M. Rechsteiner. 1986. Amino acid sequences common to rapidly degraded proteins: the PEST hypothesis. Science 234:364.

Rubenstein, N., J. Chi and H. Holtzer. 1976. Coordinated synthesis and degradation of actin and myosin in a variety of myogenic and non-myogenic cells. Exp. Cell Res. 97:387.

Sandoval, I. V. and K. Weber. 1978. Calcium-induced inactivation of microtubule formation in brain extracts. Presence of a calcium-dependent protease acting on polymerization-stimulating microtubule-associated proteins. Eur. J. Biochem. 92:463.

Schiaffino, S. and V. Hanzlikova. 1972. Studies on the effect of denervation in developing muscle. II. The lysosomal system. J. Ultrastruct. Res. 39:1.

Schoenheimer, R. and D. Rittenberg. 1940. The study of intermediary metabolism of animals with the aid of isotopes. Physiol. Rev. 20:218.

Schollmeyer, J. E. 1986a. Role of Ca^{2+} and Ca^{2+}-activated protease in myoblast fusion. Exp. Cell. Res. 162:411.

Schollmeyer, J. E. 1986b. Possible role of calpain I and calpain II in differentiating muscle. Exp. Cell. Res. 163:413.

Schollmeyer, J. E. 1988. Calpain II involvement in mitosis. Science 240:911.

Schwartz, W. N. and J. W. C. Bird. 1977. Degradation of myofibrillar proteins by cathepsins B and D. Biochem. J. 167:811.

Shii, K., S. Baba, K. Yokono and R. A. Roth. 1985. Covalent linkage of ^{125}I-insulin to a cytosolic insulin-degrading enzyme. J. Biol. Chem. 260:6503.

Smith, A. L. N. 1978. Effects of starvation on vacuolar apparatus of cardiac muscle tissue determined by electron microscopy, marker-enzyme assays, and electrolyte studies. Cytobios 18:111.

Stauber, W. T. and J. W. C. Bird. 1974. S-ρ zonal fractionation studies of rat skeletal muscle lysosome-rich fractions. Biochim. Biophys. Acta 338:234.

Stauber, W. T. and V. K. Fritz. 1985. Decreased lysosomal protease content of skeletal muscles from streptozotocin-induced diabetic rats: a biochemical and histochemical study. Histochem. J. 17:613.

Stauber, W. T., V. Fritz, B. Dahlmann and H. Reinauer. 1983. Immunohistochemical localization of two proteinases in skeletal muscle. J. Histochem. Cytochem. 31:827.

Stracher. A., E. B. McGowan, A. Hedrych and S. A. Shafiq. 1979. *In vivo* effect of protease inhibitors in denervation atrophy. Exp. Neurol. 66:611.

Stracher, A., E. B. McGowan and S. A. Shafiq. 1978. Muscular dystrophy: inhibition of degeneration in vivo with protease inhibitors. Science 200:50.

Sugden, P. H. 1980. The effects of calcium ions, ionophore A23187 and inhibition of energy metabolism on protein degradation in the rat diaphragm and epitrochlearis muscles *in vitro*. Biochem. J. 190:593.

Sugita, H., S. Ishiura, K. Suzuki and K. Imahori. 1980. Ca^{2+}-activated neutral protease and its inhibitors: in vitro effect on intact myofibrils. Muscle Nerve 3:335.

Suzuki, K. 1987. Calcium-activated neutral protease: domain structure and activity regulation. Trends Biochem. Sci. 12:103.

Suzuki, K., S. Tsuji, S. Ishiura, Y. Kimura, S. Kubota and K. Imahori. 1981a. Autolysis of calcium-activated neutral and protease of chicken skeletal muscle. J. Biochem. 90:1787.

Suzuki, K., S. Tsuji, S. Kubota, Y. Kimura and K. Imahori. 1981b. Limited autolysis of Ca^{2+}-activated neutral protease (CANP) changes its sensitivity to Ca^{2+} ions. J. Biochem. 90:275.

Szpacenko, A., J. Kay, D. E. Goll and Y. Otsuka. 1981. A different form of the Ca^{2+}-dependent proteinase activated by micromolar levels of Ca^{2+}. In: V. Turk and L.j. Vitale (Ed.) Proteinases and Their Inhibitors: Structure, Function, and Applied Aspects. pp 151–161. Pergamon Press, Elmsford, NY.

Takahashi-Nakamura, M., S. Tsuji, K. Suzuki and K. Imahori. 1981. Purification and characterization of an inhibitor of calcium-activated neutral protease from rabbit skeletal muscle. J. Biochem. 90:1583.

Takano, E., M. Maki, H. Mori, M. Hatanaka, T. Marti, K. Titani, R. Kannagi, T. Oui, and Y. Murachi. 1988. Pig heart calpastatin: identification of repetitive domain structures and anomalous behavior in polyacrylamide gel electrophoresis. Biochemistry 27:1964.

Tan, F. C., D. E. Goll, and Y. Otsuka. 1988. Some properties of the millimolar Ca^{2+}-dependent proteinase from bovine cardiac muscle. J. Mol. Cell Card. (in press).

Tanaka, K., K. Ii, A. Ichihara, L. Waxman, and A. L. Goldberg. 1986. A high molecular weight protease in the cytosol of rat liver. I. Purification, enzymological properties, and tissue distribution. J. Biol. Chem. 261:15197.

Tanaka, K., L. Waxman and A. L. Goldberg. 1983. ATP serves two distinct roles in protein degradation in reticulocytes, one requiring and one independent of ubiquitin. J. Cell Biol. 96:1580.

Tanaka, K., L. Waxman and A. L. Goldberg. 1984. Vanadate inhibits the ATP-dependent degradation of proteins in reticulocytes without affecting ubiquitin conjugation. J. Biol. Chem. 259:2803.

Tsukahara, T., S. Ishiura, and H. Sugita. 1988. The "ATP-dependent protease" in human erythroleukemia (K562) cells is identical to a high-molecular-mass protease, ingensin. Proc. Japan Acad. 64B:72.

Tweedle, C. D., H. Popiela and C. S. Thornton. 1974. Ultrastructure of the development and subsequent breakdown of muscle in aneurogenic limbs *(Ambystoma)*. J. Exp. Zool. 190:155.

van der Westhuyzen, D. R., K. Matsumoto and J. D. Etlinger. 1981. Easily releasable myofilaments from skeletal and cardiac muscles maintained *in vitro*. Role in myofibrillar assembly and turnover. J. Biol. Chem. 256:11791.

Vernon, B. G. and P. J. Buttery. 1976. Protein turnover in rats treated with Trienbolone acetate. Brit. J. Nutr. 36:575.

Warnes, D. M., F. M. Tomas and F. J. Ballard. 1981. Increased rates of myofibrillar protein breakdown in muscle-wasting diseases. Muscle Nerve 4:62.

Waxman, L. 1981. Calcium-activated proteases in mammalian tissues. Methods Enzymol. 80:664.

Waxman, L., J. M. Fagan and A. L. Goldberg. 1987. Demonstration of two distinct high molecular weight proteases in rabbit reticulocytes. one of which degrades ubiquitin conjugates. J. Biol. Chem. 262:2451.

Waxman, L., J. M. Fagan, K. Tanaka and A. L. Goldberg. 1985. A soluble ATP-dependent system for protein degradation from murine erythroleukemia cells. Evidence for a protease which requires ATP hydrolysis but not ubiquitin. J. Biol. Chem. 260:11994.

West, C. M. and H. Holtzer. 1982. Protein synthesis and degradation in cultured muscle is altered by a phorbol diester tumor promoter. Arch. Biochem. Biophys. 219:335.

Wildenthal, K., J. R. Wakeland, J. M. Ord and J. T. Stull. 1980. Interference with lysosomal proteolysis fails to reduce cardiac myosin degradation. Biochem. Biophys. Res. Commun. 96:793.

Wilk, S. and M. Orlowski. 1983. Evidence that pituitary cation-sensitive neutral endopeptidase is a multicatalytic protease complex. J. Neurochem. 40:842.

Wilkinson, K. D., M. K. Urban and A. L. Hass. 1980. Ubiquitin is the ATP-dependent proteolysis factor I of rabbit reticulocytes. J. Biol. Chem. 255:7529.

Wolitsky, B. A., M. S. Hudecki and H. L. Segal. 1984. Turnover of myofibrillar proteins in cultured muscle cells from normal and dystrophic chick embryos. Biochim. Biophys. Acta 803:106.

Woodbury, R. G., G. M. Gruzenski and D. Lagunoff. 1978. Immunofluorescent localization of a serine protease in rat small intestine. Proc. Natl. Acad. Sci. USA 75:2785.

Wrogemann, K., W. A. K. Hayward and M. C. Blanchaer. 1979. Biochemical aspects of muscle necrosis in hamster dystrophy. Ann. N.Y. Acad. Sci. 317:30.

Yoshikawa, A. and T. Masaki. 1981. Increase in protein synthetic activity in chicken muscular dystrophy. J. Biochem. 90:1775.

Yoshimura, N., T. Murachi, R. Heath, J. Kay, B. Jasani and G. R. Newman. 1986. Immunogold electron-microscopic localization of calpain I in skeletal muscle of rats. Cell Tissue Res. 244:265.

Young, V. R., W. P. Steffee, P. B. Pencharz, J. C. Winterer and N. S. Scrimshaw. 1975. Total human body protein synthesis in relation to protein requirements at various ages. Nature 253:192.

Zak, R., A. F. Martin, G. Prior and M. Rabinowitz. 1977. Comparison of turnover of several myofibrillar proteins and critical evaluation of double isotope method. J. Biol. Chem. 252:3430.

Zeman, R. J., T. Kameyama, K. Matsumoto, P. Bernstein and J. D. Etlinger. 1985. Regulation of protein degradation in muscle by calcium. Evidence for enhanced nonlysosomal proteolysis associated with elevated cystolic calcium. J. Biol. Chem. 260:13619.

Zolfaghari, R., C. R. F. Baker, P. C. Canizaro, A. Amirgholami and F. J. Behal. 1987. A high-molecular-mass neutral endopeptidase-24.5 from human lung. Biochem. J. 241:129.

CHAPTER 9

Regulation of Protein Turnover

PETER J. REEDS

1. Introduction

The dynamic nature of cellular proteins was demonstrated over 40 years ago (Schoenheimer *et al.*, 1939). But it was not until 1967–1972 that the measurement of protein turnover *in vivo* received systematic attention, notably by Waterlow's group (Waterlow and Stephen, 1967; Picou and Taylor-Roberts, 1969; Garlick, 1969; Garlick and Marshall, 1972). It is now generally accepted that cellular proteins are always subject to continual degradation even when protein is neither gained nor lost from the body.

The reasons why such a system evolved are by no means certain and a variety of plausible propositions have been made. For example, active turnover of cellular proteins may minimize the accumulation of proteins containing errors of translation. Furthermore, proteins that fail to be incorporated into their specific subcellular locations tend to be rapidly degraded (Wheatley, 1982; Jacobs et al., 1985). Whatever the reason, the phenomenon is critical to the control of tissue protein mass in general and to the levels of specific enzymes in particular. As protein synthesis and degradation are mechanistically distinct (see Goll *et al.*, this volume), both are amenable to separate controls.

The rates of protein synthesis and degradation generally greatly exceed the rate of accretion of protein and between 15 and 22% of total energy expenditure is used in maintaining these processes (Reeds *et al.*, 1985a). Thus, protein turnover can be regarded as a multicomponent "futile" substrate cycle (Newsholme & Start, 1973; Crabtree and Newsholme, 1985; Newsholme, 1986). The rate at which enzyme levels change after the impositon of an external stimulus

PETER J. REEDS ● Children's Nutrition Research Center, Department of Pediatrics, Baylor College of Medicine, Houston, Texas 77030.

is a function of the basal rate of turnover (Schimke and Doyle, 1970) and it is of interest that the cells of the intestinal mucosa (McNurlan *et al.*, 1979), and liver (Garlick *et al.*, 1975; Pain *et al.*, 1978; McNurlan *et al.*, 1979), that are either exposed to the external environment or change their metabolism rapidly in response to specific metabolic demands turn over at high rates. Furthermore, enzymes, such as ornithine decarboxylase (Seely *et al.*, 1983) and cyclooxygenase (Fagan and Goldberg, 1986), that may be involved in the transduction of hormonal signals within cells, have particularly rapid rates of turnover with half lives on the order of 15 min.

2. Factors That Affect Tissue Growth and Protein Turnover

2.1. Protein Turnover in Different Tissues and the Effect of Developmental Age

Interspecific variations in the fractional rate of protein synthesis in immature individuals appear to relate primarily to their differing fractional (i.e., weight-specific) growth rates (Reeds and Harris, 1981; Reeds and Garlick, 1985). The age of the individual, therefore, primarily affects that portion of protein turnover directly related to growth. Although different tissues grow at broadly similar fractional rates, there are inherent differences in the fractional rates of protein synthesis and degradation among tissues (Table II). These differences relate primarily to the number of ribosomes (generally inferred from the RNA concentration) per "cell." In a well-nourished individual the quantity of protein synthesized per unit of RNA is remarkably constant (see Millward *et al.*, 1975; 1981; Lewis *et al.*, 1984). Thus, the contributions of different tissues to total body protein turnover differ from their contributions to body protein mass (Table III) and to some extent, their relative contributions vary with age (Table III; Goldspink *et al.*, 1984; Lewis *et al.*, 1984; Goldspink and Kelly, 1984).

In addition, fundamentally different mechanisms of growth control may operate in the liver and kidney (and perhaps in other rapidly turning over tissues) as opposed to skeletal muscle. Food (Hutson and Mortimore, 1982; Botbol and Scornik, 1985) and protein intake (Conde and Scornik, 1976; Bur and Conde, 1982) stimulate growth of the liver and kidney largely by suppressing protein degradation. The same is true for regrowth after partial hepatectomy (Scornik and Botbol, 1976). This mechanism may apply generally to the control of protein mass in cells that have retained the ability to proliferate (Ballard, 1982). In skeletal muscle, which, once differentiated, has a relatively limited proliferative potential, food intake (Garlick *et al.*, 1983, nutritional rehabilitation (Millward *et al.*, 1975; Millward and Waterlow, 1978), and work-induced hypertrophy (Turner and Garlick, 1974; Laurent *et al.*, 1978; Goldspink *et al.*,

Table I
Total-Body Protein Synthesis in a Variety of Species at Nitrogen Equilibrium: Its Relationship to Metabolic Body Weight ($kg^{0.75}$)

Species	Body weight (kg)	Protein synthesis		Reference
		g/day	$g/kg^{0.75}$ per day	
Rat, immature	0.2	6	18.8[a]	Reeds and Harris (1981)
Rat, mature	0.35	8.8	16.9[a]	Reeds and Harris (1981)
	0.45	9.6	14.4[b]	Goldspink and Kelly (1984)
Rabbit, mature	3.6	33.0	12.6[a]	G. E. Lobley and V. Milne (unpublished study)
Pig, immature	32	268	18.9[a]	Reeds et al. (1980a)
Sheep, mature	56	262	13.1[a]	Pell et al. (1986)
	63	351	15.7[a]	P. J. Reeds and M. I. Chalmers (unpublished study)
Man, immature	5.0	56	16.6[c]	Golden et al. (1977)
Man, mature	71	310	13.4[a]	Reeds and Garlick (1985)
Cattle, immature	400	1806	18.0[a]	Lobley et al. (1987)
Cattle, mature	575	1740	14.8[a]	Lobley et al. (1980)

[a] Values for protein synthesis based on plasma leucine flux.
[b] Value determined by the method of Garlick et al. (1980).
[c] Value determined by the method of Picou and Taylor-Roberts (1969).

1983) uniformly increase protein synthesis and generally increase protein degradation (see Reeds and Palmer, 1986, for summary).

2.2. Nutrient Intake

The ways in which nutrient intake could influence growth and protein turnover are potentially complex. The short-term effects of fasting, which presumably affect the activities of existing populations of ribosomes and proteolytic enzymes, should be separated from longer-term, and perhaps adaptive (Waterlow, 1986), changes that accompany the imposition of persistent changes in nutrient intake. Long-term changes in growth are presumably accompanied by changes in the protein synthetic and degradative capacities of the cells. There is no reason to suppose a priori that the various organic components of the diet (i.e., the relative amounts and composition of the protein and nonprotein nutrients) will necessarily have similar effects on protein metabolism. Furthermore, protein turnover can be regarded as consisting of two portions: an essentially inevitable turnover associated with the maintenance of cell function and a vari-

Table II
Fractional Rates of Protein Synthesis (Percent of Protein Pool Synthesized per Day) in Various Tissues

	Skeletal muscle	Heart muscle	Liver	Kidney	Gut mucosa	Whole gut	Bone	Ref.[a]
Rat	15	12	59	32	120	78		1,2
	(15.8)[b]	(11.5)	(14.9)	(16.1)	(13.5)	(18.5)		
	17[c],19[d]	18	86		123			3
	(16.7)(12.4)	(14.5)	(17.5)		(17.9)			
Pig	17.4		69			67	64	4
	(18.4)		(16.6)			(14.9)		
	5	7	24	20				5,6
Sheep	17		95			87		7
	5		54			79		8
Cattle	2		32			53		9

[a]References: 1, Goldspink and Kelly (1984), Goldspink et al. (1984), Lewis et al., (1984): 8-week-old rats; 2, McNurlan et al. (1979): 6-week-old rats; 3, Millward et al. (1981); 4, Seve et al. (1986): weaning piglets; 5, Garlick et al. (1976a): prepubertal pigs; 6, Simon et al. (1978): prepubertal pigs; 7, Attaix et al. (1986): preweaning lambs; 8, Davis et al. (1981): prepubertal lambs; 9, Lobley et al. (1980): mature cattle.
[b]Where available, the rate per unit of RNA is shown in parentheses.
[c]Gastrocnemius (Mixed Type II).
[d]Soleus (Type I) muscle.

able turnover associated with growth. The latter approaches zero in the mature individual. Thus, in the well-nourished adult, protein deposition appears to be sensitive to nutrient intake only when protein is being lost from the body. When nutrients are supplied in excess of those required to maintain a state of nitrogen equilibrium, there is little or no increase in either the rate of protein

Table III
Contributions of Different Tissues to Body Protein Mass and Synthesis in Rats at Weaning, after Puberty, and at Maturity[a]

Age (weeks)	Tissue	Percentage of body protein	Fractional rate of protein synthesis	Percentage of body protein synthesis
3	Muscle	27.3	15.4	12.3
	Liver	6.5	52.5	10.8
	Intestine	10.8	93.6	32.9
8	Muscle	45.9	12.4	32.1
	Liver	7.0	58.8	25.2
	Intestine	5.9	78.3	25.5
44	Muscle	50.6	7.0	24.9
	Liver	4.3	42.5	15.4
	Intestine	2.6	60.8	14.9

[a]Data from Goldspink and Kelly (1984), Goldspink et al. (1984), and Lewis et al. (1984). Muscle mass measured in the Rowett strain of Hooded Lister rats (S. M. Hay and P. J. Reeds, unpublished results).

deposition or that of body protein turnover (Reeds *et al.*, 1980b; Motil *et al.*, 1981; Reeds and Fuller, 1983), whereas in young animals, increases in protein intake above those required for nitrogen equilibrium clearly accelerate body protein synthesis. Given the importance of these factors, there is remarkably little information regarding the changes in protein dynamics that accompany "normal" variations in nutrient intake, particularly in animals of agricultural interest.

With an increase in total nutrient intake of sufficient duration to enhance growth of the animal, there is a coordinated increase in protein synthesis and degradation (Golden *et al.*, 1977; Reeds *et al.*, 1980a; Lobley *et al.*, 1987) both in the whole body and in skeletal muscle (Millward *et al.*, 1975; El Haj *et al.*, 1986). In the whole body of infants, pigs, and cattle, each additional gram of protein deposited requires the additional synthesis of 1.5–2 g of protein. In skeletal muscle, the increased rate of protein synthesis is accompanied by a rise in the RNA : protein ratio.

Changes in protein intake, dietary protein quality, and nonprotein intake have different effects on protein turnover even though they stimulate protein accretion (Table IV). Thus, the major influence on whole-body (Castellino *et al.*, 1987; Tessari *et al.*, 1987) and skeletal muscle (Garlick, 1988) protein synthesis appears to be protein intake. In addition, Clugston and Garlick (1983) argued that the protein synthetic response to fasting is largely a response to the reduced intake of protein. Other dietary factors, such as increased energy intake (Reeds and Fuller, 1983) and improved dietary protein quality (Fuller *et al,*

Table IV

Changes in Protein Deposition, Body Protein Synthesis, and Degradation in Pigs (30–40 kg) Receiving Different Diets[a]

Dietary supplement	Change in		
	Protein deposition	Protein synthesis	Protein degradation
		g protein/day	
Whole diet	+62	+97	+35
Protein quantity	+25[b]	+154	+129
	+21[c]	+100	+79
Protein quality	+75[b]	+39	−36
	+75[c]	+41	−34
Dietary energy			
Carbohydrate	+43	+20	−23
Fat	+60	+20	−40

[a]Data from Reeds *et al.* (1980a, 1985a) and Fuller *et al.* (1987).
[b]Based on leucine flux and oxidation.
[c]Based on [^{15}N]urea excretion during the constant administration of ^{15}N-labeled yeast protein.

Table V

Changes in Whole-Body and Hindquarter Tissue Protein Synthesis in Young (28–38 kg) Pigs at Different Times after Receiving a Dietary Supplement of Carbohydrate[a]

	Body[b]		Hindquarter[c]	
	Protein synthesis	Protein degradation	Protein synthesis	Protein degradation
		g protein/kg$^{0.75}$ per day		
Basal	32.6 ± 1.7	27.6 ± 2.3	6.60 ± 0.22	5.49 ± 0.24
+ Carbohydrate (6 hr)	30.9 ± 2.2	22.1 ± 2.7	8.47 ± 0.62	6.70 ± 0.60
+ Carbohydrate (3 days)	39.3 ± 2.2	25.2 ± 2.4	8.15 ± 0.28	6.22 ± 0.33

[a]Unpublished data of P. J. Reeds, M. F. Fuller, A. Cadenhead, and S. M. Hay.
[b]Whole-body protein synthesis and degradation estimated from plasma leucine turnover and urea excretion.
[c]Hindquarter protein synthesis and degradation estimated from total leucine label uptake, net balance of leucine, and oxidation of [1-^{14}C]leucine to $^{14}CO_2$ across the whole hindquarter.

1987), appear to reduce body protein degradation with variable effects on protein synthesis. However, these changes may conceal qualitatively different changes in different tissues with a higher proportional increase in protein synthesis in the peripheral tissues (Table V). There is a clear need for more dynamic tissue-directed information on this matter.

2.3. Functional Load

Whereas the metabolic response to nutrients is coordinated in the sense that it involves changes in a number of tissues, it is also clear that the levels and the turnover of individual proteins respond to factors that can be loosely categorized as "functional demand." In the liver, protein accretion is responsive to the balance of protein intake and body protein deposition, reflected in the urea synthetic load (Scornik, 1984). Furthermore, the levels of individual enzymes change in response to specific intermediary metabolic demands via changes in the transcription and translation of specific mRNAs (Rucker and Tinker, 1986) and in the degradative rates of the proteins (Schimke and Doyle, 1970). Compensatory renal growth in unilaterally nephrectomized animals is also a well-established phenomenon, and acute diabetes, which leads to increased urine production, produces a small renal hypertrophy despite the absence of insulin (Albertse et al., 1978). Once again these hepatic and renal growth-responses seem to result primarily from a reduction in protein degradation.

It is equally well established that both skeletal and cardiac muscles, even in adults, increase their growth rate and protein synthetic capacity and rate of degradation in response to changes in tension (Turner and Garlick, 1974; Goldspink *et al.*, 1983) and isometric work-load (Laurent *et al.*, 1978; Goldspink *et al.*, 1983). Furthermore, skeletal muscle responds to changes in activity (Gregory *et al.*, 1986; Morrison *et al.*, 1987) as well as to changes in electrical stimulation by radically altering the pattern of gene expression (see Richter *et al.*, this volume) and protein synthetic capacity (Seedorf *et al.*, 1986).

These of course are organ-specific phenomena and unraveling the mechanism whereby cells sense such demands and transduce this information into appropriate changes in protein accretion and enzyme pattern represents a major challenge to our understanding. Two recent papers (Seedorf *et al.*, 1986; Morrison *et al.*, 1987) serve to illustrate the complexity of the control processes involved. As some aspects of the response can be induced *in vitro* both in the liver (Muira and Fukui, 1976) and in skeletal (Vandenburg and Kaufman, 1979; Palmer *et al.*, 1981) and cardiac muscle (Morgan *et al.*, 1980), presumably the induction of these changes in protein metabolism resides in changes in the nature of the intracellular environment probably involving components of the plasma membrane (Vandenburgh and Kaufman, 1981; Muira *et al.*, 1981; Reeds and Palmer, 1986).

3. Factors That Control Protein Turnover

3.1. Substrates and Hormones and Their Relation to Nutritional Control

3.1.1. Substrates

From studies conducted *in vivo* there is little evidence that protein turnover in skeletal muscle is sensitive to physiological changes in the concentrations of free amino acids in general, of leucine in particular (McNurlan *et al.*, 1982), or of glucose (Garlick *et al.*, 1983). There is likely a "preferred range" of amino acid concentrations within which the rates of protein synthesis and degradation are optimized. This range can be broadly defined as concentrations above the point at which amino acid activation and amino acyl-tRNA synthesis are dependent on amino acid concentrations (i.e., at saturation concentrations for the appropriate enzymes, hence the apparent "zero-order" kinetics of protein kinetics), above the point at which autophagy is induced by amino acid deprivation (Schworer *et al.*, 1981), and below the point at which the amino acids are toxic. In fact, as the free (i.e., nonprotein bound) pool of some amino acids (e.g., leucine and phenylalanine) may be only 1% of the daily deposition of these amino acids in protein (Waterlow *et al.*, 1978), rather than controlling protein turnover, the concentrations of free amino acids reflect changes in the relative rates of protein synthesis and degradation on the one hand and those of amino acid catabolism and absorption on the other. In fact, only small changes in arterial amino acid concentrations occur even when substantial increases in protein accretion are induced (Davis *et al.*, 1981; Reeds and Fuller, 1983; Pell *et al.*, 1986). The maintenance of such free amino acid homeostasis probably requires a coordinated change in all aspects of protein turnover and amino acid catabolism. One important implication of this is that the changes in protein metabolism measured in an animal that has adapted to a new steady state may be qualitatively different from those that induced the change in state (Reeds and Fuller, 1983; Waterlow, 1986; Reeds *et al.*, 1987).

These comments do not deny a central role for substrate absorption in the control of tissue protein turnover, but this role is probably indirect and linked to protein metabolism via changes in the endocrine status of the animal. Nevertheless, despite the clear effects of aberrations in the secretion of four main hormones—growth hormone (probably acting via insulinlike growth factor-I) insulin, cortisol/corticosterone, and the thyroid hormones—on growth and protein metabolism, the precise role of each hormone in the link between nutrient intake, protein metabolism, and growth remains controversial (e.g., see Millward, 1987, and Sharpe *et al.*, 1987).

3.1.2. Insulin

The reduced protein synthetic activity in the muscle of diabetic animals (Forker *et al.*, 1951) and the ability of insulin to stimulate muscle protein biosynthesis *in vitro* have been known for 35 years. Yet, the specific and direct role insulin has in mediating all the protein metabolic effects of experimental diabetes has been challenged (Brown *et al.*, 1981; Scheiwiller *et al.*, 1986) and not all the metabolic changes that accompany diabetes are simple reflections of a lack of insulin. Nevertheless, Garlick *et al.* (1983) showed that insulin infusion alone restored, within 20 min, the rate of muscle protein synthesis to a "fed" value in young rats that were fasted overnight. The rapid response supports the conclusion (Jefferson, 1980) that insulin acutely stimulates the activity of the existing population of ribosomes. In addition, the acute effect of food intake as well as that of the intragastric infusion of glucose and amino acids on muscle protein synthesis is greatly reduced by the administration of anit-insulin antibodies (Preedy and Garlick, 1986).

An important new development is the emergence of strong evidence for an interaction between amino acid supply and the sensitivity of protein metabolism in both the liver and skeletal muscle to insulin. Thus Mortimore's group (Mortimore et al., 1987; 1988) have demonstrated in the perfused liver than insulin potentiates the effects of amino acids on proteolysis when the amino acid concentrations approximate to those found in fed rats. Analagous results from Garlick's group (Garlick and Grant, 1988) have shown that the infusion of amino acids *in vivo* reduce the threshold concentration of insulin necessary to stimulate muscle protein synthesis by nearly a factor of three from 60 μU to 24μU/ml. It is noteworthy that once again this interaction was evident in the regulation of protein degradation in the liver and of protein synthesis in skeletal muscle, confirming the conclusion that these are the primary sites of control of hepatic and muscle protein mass respectively.

Although insulin concentrations within the physiological range stimulate muscle protein synthesis *in vitro* and *in vivo*, a physiological role for insulin in controlling muscle protein degradation is more tenuous. Clear-cut effects on degradation have been shown only in hypercatabolic isolated muscles at insulin concentrations greater than 500 μU/ml (see Palmer *et al.*, 1985, for discussion) or in diabetic animals *in vivo* (Pain *et al.*, 1983). However, there is evidence for major tissue differences in the effect of insulin. Insulin concentrations below 1mU/ml have neither an acute (Mortimore and Mondon, 1970) nor chronic (Millward *et al.*, 1982) effect on protein synthesis in the liver, but changes in insulin concentrations within the physiological range reduce the rate of hepatic protein degradation (Mortimore and Mondon, 1970).

Although there have been repeated demonstrations of correlations between insulin concentrations and muscle protein deposition under extreme conditions of undernutrition (Robinson *et al.*, 1974), the evidence implicating persistent changes in insulin concentrations in longer-term effects of "normal" changes in dietary intake, especially carbohydrate intake, is less strong. What seems likely is that these adaptive responses are related more closely to what might be termed the "anabolic balance," i.e., the relative concentrations of catabolic and anabolic influences (Millward *et al.*, 1981).

3.1.3. Adrenal Glucocorticoids

Positive correlations have been found among the insulin : cortisol ratio, weight gain, and nitrogen retention in a variety of studies. The effects of dietary energy supply in pigs (Reeds *et al.*, 1987), age and feed intake in cattle (Trenkle and Topel, 1978), and the anabolic steroid trenbolone acetate in sheep (Sharpe *et al.*, 1986) correlated with these traits. Generally, these correlations were associated with an increase in plasma cortisol at low feed intakes and suggest that the glucocorticoid hormones may play a crucial long-term role in controlling protein metabolism, particularly at intakes close to energy maintenance. The glucocorticoids have long been associated with the induction of muscle wasting (Goldberg, 1969; Odedra *et al.*, 1983), but differ from insulin inasmuch as their effects in suppressing muscle protein synthesis are slow in onset, both *in vitro* (McGrath and Goldspink, 1982; Reeds and Palmer, 1984) and *in vivo* (P. J. Garlick, personal communication). This is in keeping with a transcriptional mode of action and certainly prolonged corticosterone treatment reduces the RNA : protein ratio muscle. The glucocorticoids, moreover, are strictly antagonistic to insulin in that they not only reduce muscle protein synthesis but also reduce the effect of insulin *in vitro* (Palmer, 1987) and *in vivo* (Odedra and Millward, 1982; Millward *et al.*, 1985). Perhaps both hormones act through the same control system (see below). Interestingly, as with insulin the effect of glucocorticoids on muscle degradation is dose dependent and different muscle types may vary in their responses. Low doses of dexamethasone *in vitro* (Reeds and Palmer, 1986) and of cortisol *in vivo* (Odedra *et al.*, 1983) slightly suppress protein degradation (while consistently reducing protein synthesis) but not all are agreed that high doses stimulate muscle protein degradation (cf. Odedra *et al.*, 1983; Tomas *et al.*, 1984).

3.1.4. Thyroid Hormones

An effect of thyroid hormones on whole-body protein turnover was an early observation (Crispel *et al.*, 1956). Although the thyroid status of the animal undoubtedly affects the rate of muscle growth and protein turnover (Brown *et al.*, 1981; Brown and Millward, 1983), the effects of thyroidectomy and thyroid hormone administration differ from the effects of insulin or corticosterone in three important respects. First, there is a clear positive relationship between the level of free triiodothyronine (T_3) and muscle protein degradation (Millward *et al.*, 1982), especially that portion that involves calcium-stimulated proteases (Zeman *et al.*, 1986). Second, they are much slower in onset. Although there is a difference in the speed of response of muscle protein synthesis to insulin and to the glucocorticoids, the difference is only about 3 hr (cf. McGrath and Goldspink, 1982; Garlick *et al.*, 1983; Reeds and Palmer, 1984). The effects of thyroidectomy (Brown *et al.*, 1981), on the other hand, take up to 5 days to become manifest *in vivo*. Also, significant effects of thyroid hormone replacement on the rate of protein synthesis in the perfused hemicorpus are seen only after 5 days (Flaim *et al.*, 1978).

The third important difference relates to the basis of the changes in protein synthesis. Insulin and the glucocorticoids appear to have a dual effect. They have a rapid effect on the ability of existing ribosomes to synthesize protein, and they have a longer-term effect on the RNA, and hence on the ribosome content of muscle (Rannels and Jefferson, 1980; Pain *et al.*, 1983). In contrast, the thyroid hormones only affect RNA content with, at best, a small effect on the ability of the ribosomes to enter into translation (Brown *et al.*, 1981; Millward *et al.*, 1982; Siehl *et al.*, 1985). This situation is compatible with the suggestion that the primary site of action of the thyroid hormones is at the level of gene transcription, an effect that includes a stimulation of rRNA synthesis (Siehl *et al.*, 1985).

Millward's group proposed that changes in thyroid status underlie the changes in muscle ribosome number both in diabetes (Brown *et al.*, 1981) and during long periods of nutritional deprivation (Millward *et al.*, 1982, 1985). A general mechanism for hormonal control of muscle ribosome number was suggested. This proposal rests on a "normalization" of RNA concentration in T_3-treated diabetic animals and on the correlation between the serum levels of free T_3 and the rates of muscle protein synthesis and the RNA : protein ratio in animals subjected to varying degrees of protein and energy malnutrition. An argument can be made against the proposal, however. Although the muscle RNA : protein ratio was undoubtedly improved by treatment of diabetic animals with T_3, the change in this ratio occurred because the hormone increased the rate of muscle protein loss with a much smaller effect on muscle RNA loss, rather than because it stimulated the rate of RNA accretion. In addition, in undernourished animals, free T_3 is only one of a number of other hormones that consistently change. Nevertheless, Millward's proposition is intriguing and one that should be investigated in greater detail.

3.2. The Role of Initiation and Ribosomal Accretion in the Control of Translation

Control of the relative levels of different proteins involves a complex interplay of changes in transcription, pre- and posttranslational controls as well as changes in degradation that is only now being identified (Rucker and Tinker, 1986) and is beyond the scope of this review. Separate from, but related to, these functions is the control of total protein synthesis. The minute-to-minute control of the efficiency of protein synthesis (i.e., the protein synthesized per ribosome) is probably a function of the metabolic signals generated as an immediate result of the interaction of a given factor (e.g., stretch or insulin in muscle) with the cell surface, whereas the long-term control of ribosome number (the synthetic capacity) is exerted over the relative rates of rRNA and ribosomal (r-)protein synthesis and degradation. As with any other "enzyme," the rate at which ribosome number changes is in itself a function of the rate of turnover of the ribosomes. For example, in fasted or diabetic rat muscle, RNA concentrations do not attain their new steady state for some days (Pain *et al.*, 1983; Ashford and Pain, 1986a) whereas in the liver this is reached by about 48 hr (McNurlan *et al.*, 1979).

There are a number of excellent discussions of translation (e.g., Chapter 3 in Waterlow *et al.*, 1978; Pain and Clemens, 1980; Moldave, 1985; Pain, 1986; see Figure I). Studies of the short-term effects of insulin and glucocorticoids in the perfused hemicorpus (Jefferson, 1980; Rannels and Jefferson, 1980),

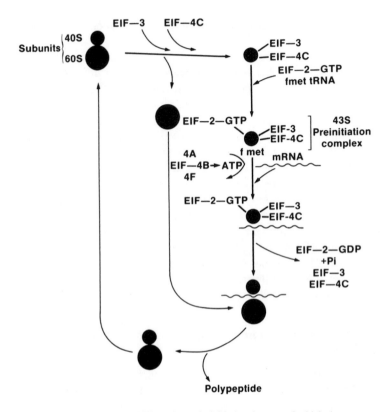

FIGURE 1. The ribosome cycle. EIF = eukaryotic initiation factors, of which there are several.

of the effects of feeding and fasting on muscle protein synthesis (Rannels *et al.*, 1978; Harmon *et al.*, 1984) as well as studies of insulin's effects in cultured cells (Sato *et al.*, 1981; Hesketh *et al.*, 1986) showed that rapid increases in translation rates are generally associated with reciprocal changes in the levels of ribosomal subunits (and nontranslating monomers) and polyribosomal aggregates. This is most likely to occur if the rate of initiation is limiting in the "stepped-down" condition and is preferentially stimulated when growth is resumed. Furthermore, cell-free preparations obtained from cells or tissues in which protein synthesis was inhibited showed a reduced ability to form the 43 S preinitiation complex (i.e., the complex of the 40 S ribosome with EIF_2–GTP and f-Met-tRNA$_f$; London *et al.*, 1976; Harmon *et al.*, 1984; Kelly and Jefferson, 1985).

Although the other initiation factors are clearly required for optimum protein synthesis, especially the EIF_4 factors that are involved in the eventual binding to the initiator codon AUG on 5' capped mRNA, the critical factor appears to be the presence of active EIF_2 (i.e., EIF_2–GTP) available to bind to f-MET-tRNA. This in turn is related to the recharging of EIF–GDP released when the 43 S preinitiation complex is joined with mRNA and the 60 S subunit at the first committed step of translation (i.e., the incorporation of the NH_2-terminal N-formyl methionine). In reticulocytes at least, the recycling of EIF_2-GDP to active EIF_2–GTP is catalyzed by a trimeric translation factor, variously termed reversing factor (RF) or guanine nucleotide exchange factor (GEF), which catalyzes the GTP–GDP exchange on EIF_2 (Matts et al., 1986).

In heme-depleted reticulocyte lysates (London et al., 1976) and in cells starved of amino acids (Pain and Clemens, 1980), the block in initiation can be relieved by the addition of EIF_2 and appears to be due to the reversible phosphorylation of the α subunit of EIF_2 (Ernst et al., 1979). This phosphorylation stabilizes the EIF–RF complex in such a way that the subsequent RF–GTP exchange is blocked. As the level of RF appears to be much lower than that of EIF_2, phosphorylation of the $EIF_2\alpha$ subunit effectively sequesters the pool of RF in an inactive form (Matts and London, 1984; Matts et al., 1986).

The significance of this finding on the activation of growth and nutrient storage by insulin may be great. Insulin stimulation of carbohydrate and fat storage involves the dephosphorylation of key enzymes (summarized by Larner et al., 1982; Ingebritsen and Cohen, 1983). These enzymes include the amino acyl tRNA synthase (Damuni et al., 1982) and branched-chain keto acid dehydrogenase (Harris et al., 1985) complexes. It remains to be proven whether the control of EIF_2-α phosphorylation involves primarily the reduction of the activity of a protein kinase, as suggested by the results of Buse et al., (1984), or the activation of a counteracting phosphatase (Foulkes et al., 1980; Ernst et al., 1982; Macauley and Jarett, 1985). At the present time, there is little direct evidence for a role for EIF_2 phosphorylation in cells other than the reticulocyte (but see Towle et al., 1984), although it is tempting to speculate that precisely the same "factor" is involved in the coordinate stimulation of muscle protein synthesis, fat deposition, glucose oxidtion, and storage induced by insulin (Table VI).

Phosphorylation/dephosphorylation reactions may be involved at another level in the control of polypeptide translation. Insulin leads to the phosphorylation of acetyl CoA carboxylase (Denton et al., 1981; Brownsey et al., 1984), and under a number of circumstances when the growth of cells is stimulated, a specific protein on the 40 S subunit (S6) is progressively phosphorylated on up to five tyrosine residues (Thomas et al., 1979, 1982; Perisic and Traugh, 1983, 1985). Recent evidence has shown that this phosphorylation of S6 can be stimulated by microinjection of the insulin-receptor protein kinase (Maller et al., 1986). The role of this phosphorylation step is controversial and, unlike EIF_2 phosphorylation, there is little evidence to suggest that it is involved in initia-

Table VI
Phosphorylation of Key Enzymes Involved in Carbohydrate Storage and Protein Synthesis

	Proposed change with insulin stimulation	Stimulation (+)/ inhibition (−)
Glycogen synthase	−P	+
Pyruvate dehydrogenase	−P	+
Branched chain keto-acid dehydrogenase	−P	−
EIF-2 subunit	−P	+
Amino acyl-tRNA synthase	−P	+
RNA polymerase I	−P	(+)
Acetyl CoA carboxylase	+P	+
ATP citrate lyase	+P	(+ ?)
Ribosome protein S6	+P	(?)
Lipomodulin	+P	−
Receptor tyrosine kinase	+P	(?)

tion, and *in vivo* the tissues of rapidly growing animals do not necessarily show an enhancement in the degree of S6 phosphorylation (Nielson *et al.*, 1982).

The critical factor in resolving the controversy (see Pain, 1986) may be that the evidence that favors the involvement of the phosphorylation of S6 in the control of translation has all been derived from cells that retain the ability to proliferate. One part of the growth response is stimulation of the accretion of ribosomes and there is evidence from primitive eukaryotes that suggests that the phosphorylation of S6 may be involved in the control of r-protein degradation (Kristiansen and Kruger, 1979).

The fact that proliferative responses involve an early and specific increase in the ribosome number applies also to muscle. In the rat (but perhaps not in cattle; DiMarco *et al.*, 1987), differentiated muscle grows largely by hypertrophy, but the activation of muscle growth is followed by some of the features of G_0 to G_1 progression including an increase in the RNA : protein ratio (Millward *et al.*, 1975; Laurent *et al.*, 1978; Pain *et al.*, 1983; Goldspink *et al.*, 1983) and presumably in ribosome number.

The central problem in understanding the mechanisms that underlie these changes is that the synthesis of the ribosomal subunits involves not only the synthesis of at least two specific RNA species (the 45 S precursor rRNA, which ultimately matures into 28 S, 18 S, and 5.8 S RNA, and 5 S RNA) but also the synthesis of at least 70 r-proteins (Nomura *et al.*, 1984). In many cells the rates of these three processes are closely coordinated (Nomura *et al.*, 1984; Schmidt *et al.*, 1985) in part by a series of feedback controls. Thus, the levels of some free r-proteins as well as the subunits themselves, appear to exert an influence not only on the synthesis of other r-protein mRNAs but also on their translation. Thus, a rapid rise in translation, by lowering the levels of free ribosomal (i.e., 40 S and 60 S) subunits, could activate ribosome synthesis and a stimulation of either rRNA or r-protein synthesis could automatically stimulate the synthesis of all the other components of the ribosome. Similarly, a decreased rate of r-protein translation, occurring as part of a more general change in protein synthesis, would eventually lead to an adaptive fall in rRNA transcription.

Despite the possibility of "passive" feedback control, it appears from recent work on the growth of skeletal and cardiac muscle that the accumulation, synthesis, and degradation of RNAs and protein species is under a specific "active" control (Jacobs et al., 1985). During the stimulation of muscle growth by insulin therapy in diabetic rats (Ashford and Pain, 1986b) and during the induction of cardiac hypertrophy by thyroid hormone administration (Siehl et al., 1985), the synthesis of ribosomes temporarily exceeds that of total muscle protein. In addition, during the recovery from diabetes, the rate of degradation of r-proteins appears to be specifically depressed. Furthermore, an increase in ribosomal content appears to precede the generalized increase in protein synthesis (Siehl et al., 1985; Ashford and Pain, 1986b; Seedorf et al., 1986; McMillan, 1987). Presumably, further maturational changes (perhaps involving increased levels of the soluble factors involved in the translation) may also have to occur before a fully active ribosomal system is achieved. Interestingly, the new ribosomes apparently appear as 80 S monomers (Hollta, 1985; Seedorf et al., 1986), suggesting that immediately after assembly, they underwent at least one round of protein synthesis and then failed to dissociate. This suggests that EIF$_3$ might play an important role in the final production of an active ribosome population.

As regards the "active" control of ribosome number, there is support for a role for the polyamines (whose production is controlled in part by ornithine decarboxylase, an enzyme that is induced rapidly following the activation of cell growth; Russel, 1981) in controlling rRNA synthesis (Russel, 1981; Tabor and Tabor, 1984). It has even been suggested that ornithine decarboxylase directly interacts with RNA polymerase I to stimulate its activity, although this view has been strongly challenged. It has also been argued that the involvement of this system in the control of ribosomal accretion is not at the level of ribosome synthesis but on some process associated with the ultimate formation of active ribosomes (Hollta, 1985; Hollta and Hovi, 1985). Nevertheless, it is noteworthy that recent accounts (e.g., Tabor and Tabor, 1984) have stressed the indirect nature of the evidence linking the polyamine system with ribosomal accretion and activation and it should be emphasized that mechanistic studies are hampered by the extremely rapid rate of ornithine decarboxylase turnover. Understanding the mechanisms that underlie the modulation of this complex but critical process of control of protein synthesis is a major priority in future research on translational control.

3.3. Comments on the Link between Receptor Activation and Protein Synthesis

Although activated cytoplasmic thyroid/glucocorticoid receptors interact directly with DNA, the cell surface receptors of the peptide hormones and growth factors that may also control protein turnover, probably do not bind to the ribosomes, to RNA polymerases I and III, or to the proteolytic systems within cells. Perhaps even more strikingly, mechanical forces stimulate protein synthesis in isolated muscles (Reeds et al., 1980b; Palmer et al., 1981; Smith et al., 1983) and in embryonic myotubes in culture (Vandenburg and Kaufman, 1979) and here it is difficult to argue that the link between the receptor and the ribosomes is anything other than indirect. The key appears to be chemical events that follow the binding of hormones to specific receptors either on the cell surface (peptide factors) or that follow the structural changes in the plasma membrane that are forced by the mechanical distortion.

Three intimately related factors seem crucial to the link between cell surface receptor activation and the stimulation of cell growth (Fig. 2). First, stimulation of the hydrolysis of phospholipids leads to the release of a series of phosphorylated inositols (Berridge, 1984), 1,2-diacylglycerol (Kishimoto et al., 1980), and arachidonic acid (Hong et al., 1976; Jiminez de Asua et al., 1977; Rodemann and Goldberg, 1982; Smith et al., 1983) (Fig. 2A). Second, membrane proteins are phosphorylated either by the tyrosine kinases that appear to be intimately associated with the plasma membrane receptors (Rosen et al., 1983; Nemenoff et al., 1984; Downward et al., 1985) or by the Ca^{2+}/phospholipid-stimulated protein kinase C (Nishizuka, 1984), which is itself stimulated by 1,2-diacylglycerol (Fig. 2B). The precise nature and the functions of the phosphorylated membrane proteins are not known with any certainty. However, they include lipocortin (lipomodulin; Hirate et al., 1984; Pepinsky and Sinclair, 1986), a protein that inhibits phospholipase A_2, which catalyzes arachidonic acid release (Hirata et al., 1980). One end result of these events appears to be the release into the intracellular phase of a number of "mediators." These may include both peptides (Larner et al., 1982; Buse et al., 1984) and glycoinositides (Saltiel and Cuatrecasas, 1986). These mediators then presumably act by altering the activities of the cytoplasmic, mitochondrial, and nuclear protein kinases and phosphatases. At least one protein kinase is released from the plasma membrane and this phosphorylates protein S6. It is of interest that this enzyme seems to be derived by proteolysis of protein kinase C (Perisic and Traugh, 1983; Hashimoto et al., 1985).

Arachidonic acid metabolites appear also to be involved in the stimulation of cell proliferation by serum (Jiminez de Asua et al., 1977), in the induction of ornithine decarboxylase by tumor-promoting phorbol esters (Jetten and Shirley, 1985), and in hepatic regeneration (Muira et al., 1981). They may also be involved in the control of ribosome accretion in work-induced hypertrophy (McMillan, 1987). In the limited context of the short-term control of muscle metabolism, the prostaglandin products of membrane arachidonic acid metabolism seem to be of specific importance in stretch (Smith et al., 1983) and insulin (Reeds and Palmer, 1983; Reeds et al., 1985b; Palmer et al., 1986) stimulation and in steroid inhibition (Reeds and Palmer, 1984) of protein turnover. Muscle protein synthesis and degradation are stimulated in vitro by very low concentrations of arachidonic acid (Rodemann and Goldberg, 1982; Smith et al., 1983), and prostaglandin ($PGF_2\alpha$) appears to have a specific effect on protein synthesis (Rodemann et al., 1982; Smith et al., 1983; Palmer and Wahle, 1987), whereas the synthetic glucocorticoid dexamethasone reduces $PGF_2\alpha$ production with the same time course as its effects on protein synthesis (Reeds and Palmer, 1984) suggesting that the effect of ths hormone on phospholipase A_2 (Hong et al., 1976; Hirata et al., 1981) is mechanistically involved in its effects on protein synthesis. In view of the frequent observation of a coordinate stimulation of protein synthesis and degradation in rapidly growing muscles, it is of extreme interest that Goldberg's group (Rodemann and Goldberg, 1982) showed that PGE_2, another major product of arachidonic acid metabolism in muscle, stimulates protein degradation, by an as yet unknown mechanism. Our group has found consistent positive correlations between the release of these two prostaglandins under a wide variety of circumstances (Palmer et al., 1983; Reeds and Palmer, 1986). Evidence is also now appearing that the rates of protein turnover and growth of muscle can be manipulated by altering the synthesis of prostaglandins in septic animals (Ruff and Secrist, 1984; Wan and Grimble, 1985) and during stretch-induced hypertrophy (McMillan, 1987).

The prostaglandins, especially PGE_2, may be involved in the modulation of pyruvate dehydrogenase activity by insulin and glucocorticoids (Begum et al., 1983, 1985) and in the antilipolytic activity of insulin (Lambert and Jacquemin, 1980). It appears that the twin prostaglandins PGF_2 and PGE_2 may be involved in insulin action in a way that is analogous to the common involvement of protein phosphorylation and dephosphorylation. It is tempting to speculate that the two mechanisms are related and that activation of arachidonic acid release is involved in the initiation of the other changes in the plasma membrane that ultimately serve to modulate protein turnover in general and protein synthesis in particular.

4. Conclusion

Protein accretion in general, and the levels of specific enzymes in particular, are potentially subject to control by changes in both protein synthesis and

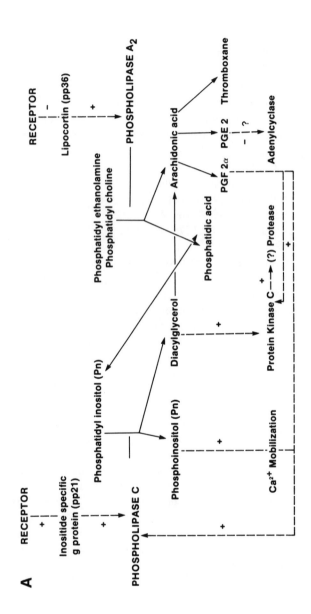

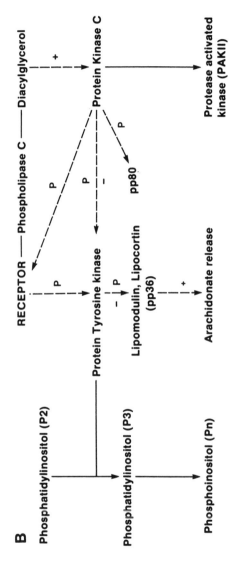

FIGURE 2. Summary of changes in the plasma membrane related to activation of cell surface receptors. (A) Phospholipid pathways; (B) potential protein kinase links. Solid arrows indicate a reaction; dashed arrows indicate a control function, with "+" for a stimulatory effect and "−" for an inhibitory effect. "p" signifies that the protein is phosphorylated via that pathway. "(Pn)" signifies that multiple phosphorylated forms are present. "(?)" signifies a proposed, but unproven effect. pp^{21}, pp^{36}, pp^{80} refer to specific proteins of apparent molecular mass of 21, 36, and 80 kDa, respectively; pp^{21} is possibly the product of the ras protooncogene and pp^{36} the phospholipase A_2 inhibitory protein lipocortin (lipomodulin). PGF 2α and PGE 2 are prostaglandin $F_2\alpha$ and prostaglandin E_2. Phosphatidylinositol (P2) represents phosphatidylinositol 4-monophosphate and phosphatidylinositol (P3) represents phosphatidylinositol 4,5-bisphosphate.

degradation. Much of the research in the last 15 years centered on the development of appropriate techniques for the quantitative measurement of these processes and on the definition of the changes that initiate alterations in growth. Fundamental tissue differences in the relative importance of protein synthesis and degradation to the control of cell protein mass have been identified and these may well be related to the retention or loss of cellular proliferative ability as postnatal development proceeds. Diverse factors such as insulin, functional load, and nutrition, which can cause changes in the growth of different tissues, do so in association with qualitatively different changes in protein synthesis and degradation. In the liver, and perhaps in other proliferative tissues, accelerated growth is achieved primarily by a reduction in protein degradation, whereas in the skeletal musculature, increased rates of protein accretion appear to involve a coordinate increase in both protein synthesis and degradation.

Mechanistic research has made inroads into understanding the factors that link development and the environment and the growth of the organism and into the mechanisms, especially those associated with cell membranes, whereby these factors produce their ultimate effects. Although the metabolic changes that induce accelerated growth may be cell-specific, the control pathways may be common.

Research involving quantitative measurements of these metabolic processes, the identification of the factors that control them, and the study of the molecular mechanisms responsible for linking changes in the extracellular environment to changes in protein and nucleic acid metabolism is now becoming more closely coordinated. This coordination of effort offers hope that mechanistically based approaches to the manipulation of growth may be developed in the near future.

References

Albertse, E. C., V. M. Pain and P. J. Garlick, 1978. Protein synthesis and breakdown in muscle and kidney of diabetic and insulin-treated rats. Proc. Nutr. Soc. 39:19A (Abstr.).

Ashford, A. J. and V. M. Pain. 1986a. Effect of diabetes on the rates of synthesis and degradation of ribosomes in rat muscle and liver in vivo. J. Biol. Chem. 261:4059.

Ashford, A. J. and V. M. Pain. 1986b. Insulin stimulation of growth in diabetic rats. Synthesis and degradation of ribosomal and total tissue protein in skeletal muscle and heart. J. Biol. Chem. 261:4066.

Attaix, D., A. Manghebati, J. Grizard and M. Arnal. 1986. Assessment of in vivo protein synthesis in lamb tissues with [³H] valine flooding doses. Biochim. Biophys. Acta. 882:387.

Ballard, F. J. 1982. Regulation of protein accumulation in cultured cells. Biochem. J. 208:275.

Begum, N., H. M. Tepperman and J. Tepperman. 1983. Effect of dexamethasone on adipose tissue and liver pyruvate dehydrogenase and its stimulation by insulin-generated mediator. Endocrinology 114:99.

Begum, N., H. M. Tepperman and J. Tepperman. 1985. Studies on the possible involvement of prostaglandins in insulin generation of pyruvate dehydrogenase activator. Diabetes 34:29.

Berridge, M. J. 1984. Inositol trisphosphate and diacylglycerol as second messengers. Biochem. J. 220:345.

Botbol, V. and O. A. Scornik. 1985. Peptide intermediates in the degradation of cellular proteins. In: E. A. Khairallah, J. S. Bond and J. W. C. Bird (Ed.) Intracellular Protein Catabolism. pp. 573–583. Alan R. Liss, New York.

Brown, J. G., P. C. Bates, M. A. Holliday and D. J. Millward. 1983. Thyroid hormones and muscle protein turnover. Biochem. J. 194:771.

Brown, J. G. and D. J. Millward. 1983. Dose response of protein turnover in rat skeletal muscle to triiodothyronine treatment. Biochim. Biophys. Acta 757:182.

Brownsey, R. W., N. J. Edgell, T. J. Hopkirk and R. M. Denton. 1984. Studies on insulin-stimulated phosphorylation of acetyl-CoA carboxylase, ATP citrate lyase and other proteins in rat epididymal adipose tissues. Biochem. J. 218:733.

Bur, J. A. and R. D. Conde. 1982. Decreased rate of protein breakdown in nutritional recovery of mouse kidney. Amer. J. Physiol. 243:E360.

Buse, M. G., I. R. Cheema, M. Owens, B. E. Ledford and R. A. Galbraith. 1984. Muscle protein synthesis: regulation of a translational inhibitor. Amer. J. Physiol. 246:E510.

Castellino, P., L. Luzi, D. C. Simonson, M. Haymond and R. A. deFronzo. 1987. Effect of insulin and plasma amino acid concentrations on leucine metabolism in man. J. Clin. Invest. 80:1784.

Clugston, G. A. and P. J. Garlick. 1983. The response of whole-body protein synthesis to feeding in obese subjects given a protein-free low-energy diet for 3 weeks. Hum. Nutr. Clin. Nutr. 36C:391.

Conde, R. D. and O. A. Scornik. 1976. Role of protein degradation in the growth of livers after a nutritional shift. Biochem. J. 158:385.

Crabtree, B. and E. A. Newsholme. 1985. A quantitative approach to metabolic control. Current Topics in Cellular Regulation 25:21.

Crispel, K. R., W. Parson, G. Hollifield and S. Brent. 1956. A study of the rate of protein synthesis before and during the administration of L-triiodothyronine to patients with mxyodaema and healthy volunteers using ^{15}N-glycine. J. Clin. Invest. 35:164.

Damuni, Z., B. Caudwell and P. Cohen. 1982. Regulation of the amino acyl-tRNA synthetase complex of rat liver by phosphorylation/dephosphorylation in vitro and in vivo. Eur. J. Biochem. 129:57.

Denton, R. M., R. W. Brownsey and G. J. Belsham. 1981. A partial view of insulin action. Diabetologia 21:347.

Davis, S. R., T. N. Barry and G. A. Hughson. 1981. Protein synthesis in tissues of growing lambs. Brit. J. Nutr. 46:409.

DiMarco, O. N., R. L. Baldwin and C. C. Calvert. 1987. Relative contributions of hyperplasia and hypertrophy to growth in cattle. J. Anim. Sci. 65:150.

Downward, J., M. D. Waterfield and P. J. Parker. 1985. Autophosphorylation and protein kinase C phosphorylation of the epidermal growth factor receptor. J. Biol. Chem. 260:14538.

El Haj, A. J., S. E. M. Lewis, D. F. Goldspink, B. J. Merry and A. M. Holehan. 1986. The effect of chronic and acute dietary restriction on the growth and protein turnover of fast and slow types of rat skeletal muscle. Comp. Biochem. Physiol. 85A:281.

Ernst, V., D. H. Levin, J. G. Foulkes and I. M. London. 1982. Effects of skeletal muscle protein phosphatase inhibitor-2 on protein synthesis and protein phosphorylation in rabbit reticulocyte lysates. Proc. Natl. Acad. Sci. USA 79:7092.

Ernst, V., D. H. Levin and I. M. London. 1979. In situ phosphorylation of the α-subunit of eukaryotic initiation factor 2 in reticulocyte lysates inhibited by heme deficiency, double-stranded RNA, oxidized glutathione or the heme regulated kinase. Proc. Natl. Acad. Sci. USA 76:2118.

Fagan, J. M. and A. L. Goldberg. 1986. Inhibitors of protein and RNA synthesis cause a rapid block of prostaglandin production at the prostaglandin synthase step. Proc. Natl. Acad. Sci. USA 83:2771.

Flaim, K. E., J. B. Li and L. S. Jefferson. 1978. Effect of thyroxine on protein turnover in rat skeletal muscle. Amer. J. Physiol. 235:E231.

Forker, L. L., I. L. Chaikoff, C. Entenman and H. Tarver. 1951. Formation of muscle protein in diabetic dogs studied with ^{35}S-methionine. J. Biol. Chem. 188:37.

Foulkes, J. G., L. S. Jefferson and P. Cohen. 1980. The hormonal control of glycogen metabolism: dephosphorylation of protein phsophatase inhibitor-I in vivo in response to insulin. FEBS Lett. 112:21.

Fuller, M. F., P. J. Reeds, A. Cadenhead, B. Seve and T. Preston, 1987. Effects of quantity and quality of dietary protein on nitrogen metabolism and protein turnover in growing pigs. Brit. J. Nutr. 57:287.

Garlick, P. J. 1969. Turnover rate of muscle protein measured by constant intravenous infusion of ^{14}C-glycine. Nature 223:61.

Garlick, P. J., T. L. Burk and R. W. Swick. 1976a. Protein synthesis and RNA in the tissues of the pig. Amer. J. Physiol. 230:1108.

Garlick, P. J., M. Fern and V. R. Preedy. 1983. The effect of insulin infusion and food intake on muscle protein synthesis in post-absorptive rats. Biochem. J. 210:669.

Garlick, P. J. and I. Grant. 1988. Amino acid infusion increases the sensitivity of muscle protein synthesis in vivo to insulin. Effect of branched-chain amino acids. Biochem. J. 253:In press.

Garlick, P. J., M. A. McNurlan and V. R. Preedy. 1980. A rapid and convenient technique for measuring the rate of protein synthesis in tissue by injection of [^{2}H]-phenylalanine. Biochem. J. 192:719.

Garlick, P. J. and I. Marshall. 1972. A technique for measuring brain protein synthesis. J. Neurochem. 19:577.

Garlick, P. J., D. J. Millward, W. P. T. James and J. C. Waterlow. 1975. The effect of protein deprivation and starvation on the rate of protein synthesis in tissues of the rat. Biochim. Biophys. Acta 414:71.

Goldberg, A. L. 1969. Protein turnover in skeletal muscle. II. Effects of denervation and corticosterone on protein catabolism in skeletal muscle. J. Biol. Chem. 244:3223.

Golden, M., J. C. Waterlow and D. Picou. 1977. The relationship between dietary intake, weight change, nitrogen excretion and protein turnover in man. Amer. J. Clin. Nutr. 30:1345.

Goldspink, D. F., P. J. Garlick and M. A. McNurlan. 1983. Protein turnover measured in vivo and in vitro in muscle undergoing compensatory growth and subsequent denervation atrophy. Biochem. J. 210:89.

Goldspink, D. F. and F. J. Kelly. 1984. Protein turnover and growth in the whole-body, liver and kidney of the rat from the fetus to senility. Biochem. J. 217:507.

Goldspink, D. F., S. E. M. Lewis and F. J. Kelly. 1984. Protein synthesis during the development growth of the small and large intestine of the rat. Biochem. J. 217:527.

Gregory, P. W., R. B. Low and W. S. Stirewalt, 1986. Changes in skeletal-muscle myosin isozymes with hypertrophy and exercise. Biochem. J. 238:55.

Harmon, C. S., C. G. Proud and V. M. Pain. 1984. Effects of starvation, diabetes and acute insulin treatment on the regulation of polypeptide initiation in rat skeletal muscle. Biochem. J. 223:687.

Harris, R. A., S. M. Powell, R. Paxton, G. E. Gillim and H. Nagae. 1985. Physiological covalent regulation of rat liver branched-chain keto acid dehydrogenase. Arch. Biochem. Biophys. 243:542.

Hashimoto, E., K. Mizuta and H. Yamazura. 1985. Protease activated protein kinase in rat liver plasma membrane. Biochem. Biophy. Res. Commun. 131:246.

Hesketh, J. E., G. Campbell and P. J. Reeds. 1986. Rapid response of protein synthesis to insulin in 3T3 cells. Effect of protein kinase-C depletion and difference from the response to serum. Bioscience Reports 6:797.

Hirata, F., E. Schiffman, K. Venkatasubramania, D. Saloman, and J. Axelrod. 1980. A phospholipase A$_2$ inhibitory protein in rabbit neutrophils induced by glucocorticoids. Biochemistry 77:2533.

Hirata, F. 1981. The regulation of lipomodulin, a phospholipase inhibitory protein, in rabbit neutrophils by phosphorylation. J. Biol. Chem. 256:7730.

Hirata, F., K. Matsuda, Y. Notsu, T. Hattori and R. del Carmine. 1984. Phosphorylation at a tyrosine residue of lipomodulin in mitogen-stimulated murine thymocytes. Proceedings of the National Academy of Sciences, USA 81:4717.

Hollta, E. 1985. Polyamine requirement for polysome formation and protein synthesis in human lymphocytes. In: L. Selmeci, M. E. Brosnan and N. Seiler (Ed.) Recent Progress in Polyamine Research, pp 137–150, Akademiai Kiado, Budapest.

Hollta, E. and T. Hovi, 1985. Polyamine depletion results in an impairment of polyribosome formation and protein synthesis before onset of DNA synthesis in mitogen-activated human lymphocytes, FEBS Lett, 152:229.

Hong, S-CL, R. Polsky-Cynkin and L. Levine. 1976. Stimulation of prostaglandin biosynthesis by vasoactive substances in methylcholanthrene-transformed mouse BALB/3T3. J. Biol. Chem. 251:776.

Hutson, S. N. and G. E. Mortimore. 1982. Suppression of cytoplasmic protein uptake by lysosomes as the mechanism of protein regain in livers of starved-refed mice. J. Biol. Chem. 257:9548.

Ingebritsen, T. S. and P. Cohen. 1983. Protein phosphatases: properties and role in cellular regulation. Science 221:331.

Jacobs, F. A., R. C. Bird and B. H. Sells. 1985. Differentiation of rat myoblasts. Regulation of turnover of ribosomal proteins and their mRNA's. Eur. J. Biochem. 150:255.

Jefferson, L. S. 1980. The role of insulin in the regulation of protein synthesis. Diabetes 29:487.

Jetten, A. M. and J. E. Shirley. 1985. Retinoids antagonise the modulation of ornithine decarboxylase activity by phorbol esters and phospholipase-C in rat tracheal epithelial cells. J. Cell. Physiol. 123:386.

Jiminez de Asua, L., D. Clingan and P. S. Rudland. 1975. Initiation of cell proliferation in cultured mouse fibroblasts by PGF$_2$. Proc. Natl. Acad. Sci. USA 72:2724.

Jiminez de Asua, L., M. O'Farrell, D. Bennett, D. Clingan and P. Rudland. 1977. Interaction between two hormones and their effects on observed rate of initiation of DNA synthesis in 3T3 cells. Nature 265:151.

Kelly, F. J. and L. S. Jefferson. 1985. Control of peptide chain initiation in rat skeletal muscle. Development of methods for preparation of native ribosomal subunits and analysis of the effect of insulin on formation of 40S initiation complexes. J. Biol. Chem. 260:6677.

Kishimoto, A., Y. Takai, U. Kikkawa and Y. Nishizuka. 1980. Activation of phospholipid-dependent kinase by diaclyglycerol. Its possible relationship tp phosphatidylinositol turnover. J. Biol. Chem. 255:2273.

Kristiansen, K. and A. Kruger. 1979. Phosphorylation and degradation of ribosomes in starved Tetrahymena pyriformis. Exp. Cell Res. 118:159.

Lambert, B. and C. Jacquemin. 1980. Synergic effect of insulin and protaglandin E on stimulated lipolysis. Prostaglandin Med. 5:375.

Larner, J., K. Cheng, C. Schwartz, K. Kikuchi, S. Tamura, S. Creacy, R. R. Dubler, R. Galasko, C. Pullin and M. Katz. 1982. Insulin mediators and their control of metabolism through protein phosphorylation. Recent Prog. Horm. Res. 38:511.

Laurent, G. J., M. P. Sparrow and D. J. Millward. 1978. Turnover of muscle protein in the fowl.

Changes in protein synthesis and breakdown during hypertrophy of the anterior and posterior latissimus dorsi muscles. Biochem. J. 176:407.

Lewis, S. E. M., F. J. Kelly and D. F. Goldspink. 1984. Pre- and post-natal growth and protein turnover in smooth muscle, heart and slow- and fast-twitch skeletal muscles of the rat. Biochem. J. 217:517.

Lobley, G. E., A. Connel and V. Buchan. 1987. Effect of food intake on protein and energy metabolism in beef steers. Brit. J. Nutr. 57:457.

Lobley, G. E., V. Milne, J. Lovie, P. J. Reeds, and K. Pennie. 1980. Whole body and tissue protein synthesis in cattle. British Journal of Nutrition 43:491.

London, I. M., M. J. Clemens, R. S. Ranu, D. H. Levin, L. F. Cherbas and V. Ernst. 1976. The role of hemin in the regulation of protein synthesis in erythroid cells. Fed. Proc. 35:2218.

Macauley, G. L. and R. Jarett. 1985. Insulin mediator causes the dephosphorylation of the α-subunit of pyruvate dehydeogenase by stimulated phosphatase activity. Arch. Biochem. Biophys. 237:142.

McGrath, J. A. and D. F. Goldspink. 1982. Glucocorticoid action on protein synthesis and protein breakdown in isolated skeletal muscles. Biochem. J. 206:641.

McMillan, D. N. 1987. The role of prostaglandins in the control of protein turnover in the tissues of the rat and the rainbow trout (Salmo gairdneri Richardson). Ph.D. Thesis. Univ. of Aberdeen.

McNurlan, M. A., E. B. Fern and P. J. Garlick. 1982. Failure of leucine to stimulate protein synthesis *in vivo*. Biochem. J. 204:831.

McNurlan, M. A., A. M. Tomkins and P. J. Garlick. 1979. The effect of starvation on the rate of protein synthesis in rat liver and small intestine. Biochem. J. 178:373.

Maller, J. L., L. J. Pike, G. R. Freidenberg, R. Cordera, B. J. Stith, J. M. Olefsky and E. G. Krebs. 1986. Increased phosphorylation of the ribosomal protein S6 following microinjection of insulin receptor-kinase into Xenopus oocytes. Nature 320:459.

Matts, R. L., D. H. Levin and I. M. London. 1986. Fate of reversing factor during restoration of protein synthesis by hemin or GTP in heme-deficient reticulocyte lysates. Proc. Natl. Acad. Sci. USA 83:1217.

Matts, R. L. and I. M. London. 1984. The regulation of initiation of protein synthesis by phosphorylation of eIF-2(α) and the role of reversing factor in the recycling of eIF-2. J. Biol. Chem. 259:6708.

Millward, D. J. 1987. Cortisol and growth in sheep. Letter to the Editor. Brit. J. Nutr. 57:157.

Millward, D. J., P. C. Bates, J. G. Brown and M. Cox. 1981. Protein turnover and the regulation of growth. In: J. C. Waterlow and J. M. L. Stephen (Ed.) Nitrogen Metabolism in Man. pp 409–418. Applied Science Publ., London.

Millward, D. J., P. C. Bates, J. G. Brown, M. Cox, R. Guigliano, M. Jepson and J. Pell. 1985. Role of thyroid, insulin and corticosteroid hormones in the physiological regulation of proteolysis in muscle. In: E. A. Khairallah, J. S. Bond and J. W. C. Bird (Ed.) Intracellular Protein Catabolism. pp 531–542. Alan R. Liss, New York.

Millward, D. J., B. de Benoist, J. G. Brown, M. Cox, D. Halliday, B. Odedra and M. J. Rennie. 1982. Protein turnover: The nature of the phenomenon and its physiological regulation. In: M. Arnal, R. Pion and D. Bonin (Ed.) Protein Metabolism and Nutrition. pp 69–96. INRA, Paris.

Millward, D. J., P. J. Garlick, R. J. C. Stewart, D. O. Nnanyelugo and J. C. Waterlow. 1975. Skeletal muscle growth and protein turnover. Biochem. J. 150:235.

Millward, D. J. and J. C. Waterlow. 1978. Effect of nutrition on protein turnover in skeletal muscle. Fed. Proc. 37:2283.

Moldave, K. 1985. Eukaryotic protein synthesis. Annu. Rev. Biochem. 54:1109.

Morgan, H. E., B. H. L. Chua, E. O. Fuller and D. H. Siehl. 1980. Regulation of protein synthesis and breakdown during in vitro cardiac work. Amer. J. Physiol. 238:E431.

Morrison, P. R., J. A. Montgomery, T. S. Wong and F. W. Booth, 1987. Cytochrome protein-synthesis rates and mRNA contents during atrophy and recovery in skeletal muscle. Biochem. J. 241:257.

Mortimore, G. E., B. R. Lardeaux and C. E. Adams. 1988. Regulation of microautophagy and basal protein turnover in rat liver. J. Biol. Chem. 263:2506.

Mortimore, G. E. and C. E. Mondon. 1970. Inhibition by insulin of valine turnover in liver. J. Biol. Chem. 245:2375.

Mortimore, G. E., A. R. Poso, M. Kadowaki and J. J. Wert. 1987. Multiphasic control of hepatic protein degradation by regulatory amino acids. J. Biol. Chem. 262:16322.

Motil, K. J., D. Matthews, D. M. Bier, J. F. Burke and V. R. Young. 1981. Whole-body leucine and lysine metabolic response to dietary protein intake in young men. Amer. J. Physiol. 240:E712.

Muira, Y. and N. Fukui, 1976. Pleiotypic responses in regenerating liver. In: G. Weber (Ed.) Adv. Enz. Regulation. 14:393.

Muira, Y., I. Mahmud, F. Karimi-Tari and N. Fukui. 1981. The arachidonate cascade in normal liver and in hepatoma cells. Adv. Enzyme Regul. 19:27.

Nemenoff, R. A., Y. C. Kwok, G. I. Shulman, P. J. Blackshear, R. Osathanondh and J. Avruch. 1984. Insulin-stimulated tyrosine protein kinase. J. Biol. Chem. 259:5058.

Newsholme, E. A. 1986. Substrate cycles and energy metabolism. Their biochemical, biological, physiological, and pathological importance. In: P. W. Moe, H. F. Tyrell and P. J. Reynolds (Ed.) Energy Metabolism of Farm Animals. pp 174–187. Rowman and Littlefield, Totowa, New Jersey.

Newsholme, E. A. and C. Start. 1973. Regulation in Metabolism. John Wiley, New York.

Nielson, P. J., K. L. Manchester, H. Towbin, J. Gordon and G. Thomas. 1982. The phosphory-lation of ribosomal protein S6 in rat tissues following cycloheximide injection, in diabetes and after denervation of the diaghragm. J. Biol. Chem. 257:12316.

Nishizuka, Y. 1984. The role of protein kinase C in cell surface signal transduction and tumor promotion. Nature 308:693.

Nomura, M., R. Gourse and G. Baughman. 1984. Regulation of the synthesis of ribosomes and ribosomal components. Annu. Rev. Biochem. 53:75.

Odedra, B. R., P. C. Bates and D. J. Millward. 1983. Time course of the effect of catabolic doses of corticosterone on protein turnover in rat skeletal muscle and liver. Biochem. J. 214:616.

Odedra, B. R. and D. J. Millward. 1982. Effect of corticosterone treatment on muscle protein turnover in adrenalectomised rats and diabetic rats maintained on insulin. Biochem. J. 204:663.

Pain, V. M. 1986. Initiation of protein synthesis. Biochem. J. 235:625.

Pain. V. M., E. C. Albertse and P. J. Garlick. 1983. Protein metabolism in skeletal muscle, diaphragm and heart of diabetic rates. Amer. J. Physiol. 245:E604.

Pain, V. M. and M. J. Clemens. 1980. Protein synthesis in mammalian systems. In: M. Florkin, A. Neuberger and L. L. M. Van Deenen (Ed.) Comprehensive Biochemistry. Vol. 19B, pp 1–76. Elsevier, Amsterdam.

Pain, V. M., M. J. Clemens and P. J. Garlick. 1978. The effect of dietary protein deficiency on albumin synthesis and on concentrations of active albumin ribonucleic acid in rat liver. Biochem. J. 172:129.

Palmer, R. M. 1987. The role of prostaglandins in the hormonal control of protein synthesis. Ph.D. Thesis. Univ. of Aberdeen.

Palmer, R. M., P. A. Bain and P. J. Reeds. 1985. The effect of insulin and intermittent mechanical stretching on rates of protein synthesis and degradation in isolated rabbit muscle. Biochem. J. 230:117.

Palmer, R. M., P. A. Bain and P. J. Reeds. 1986. Time dependent effect of indomethacin on insulin stimulation of protein synthesis in vitro. Biosci. Rep. 6:157.

Palmer, R. M., P. J. Reeds, T. Atkinson and R. H. Smith. 1983. The influence of changes in tension on protein synthesis and prostaglandin release in isolated rabbit muscles. Biochem. J. 214:1011.

Palmer, R. M., P. J. Reeds, G. E. Lobley and R. H. Smith. 1981. The effect of intermittent changes in tension on protein and collagen synthesis in isolated rabbit muscles. Biochem J. 198:491.

Palmer, R. M. and K. W. J. Wahle. 1987. Protein synthesis and degradation in isolated muscle: effect of ω-3 and ω-6 fatty acids. Biochem. J. 242:615.

Pell, J. M., E. M. Caldarone and E. N. Bergman. 1986. Leucine and -keto isocaproate metabolism and interconversions in fed and fasted sheep. Metabolism 35:1005.

Pepinsky, R. B. and L. K. Sinclair. 1986. Epidermal growth factor-dependent phosphorylation of lipocortin. Nature 321:81.

Perisic, O. and J. A. Traugh. 1983. Protease-activated Kinase II as the potential mediator of insulin stimulated phosphorylation of ribosomal protein S6. J. Biol. Chem. 258:9589.

Perisic, O. and J. A. Traugh. 1985. Protein activated kinase II as the modulator of epidermal growth factor stimulated phosphorylation of ribosomal protein S6. FEBS Lett. 183:215.

Picou, D. I. M. and T. Taylor-Roberts. 1969. The measurement of total protein synthesis and catabolism and nitrogen turnover in infants of different nutritional states and receiving different amounts of dietary protein. Clin. Sci. 38:283.

Preedy, V. R. and P. J. Garlick. 1986. The response of muscle protein synthesis to nutrient intake in postabsorptive rats. The role of insulin and amino acids. Biosci. Rep. 6:177.

Rannels, S. R. and L. S. Jefferson. 1980. Effects of glucocorticoids on muscle protein turnover in perfused rat hemicorpus. Amer. J. Physiol. 238:E564.

Rannels, S. R., D. E. Rannels, A. E. Pegg and L. S. Jefferson. 1978. Effect of starvation on initiation of protein synthesis in skeletal muscle and heart. Amer. J. Physiol. 235:E126.

Reeds, P. J., A. Cadenhead, M. F. Fuller, G. E. Lobley and J. D. McDonald. 1980a. Protein turnover in growing pigs. The effects of age and food intake. Brit. J. Nutr. 43:445.

Reeds, P. J. and M. F. Fuller. 1983. Nutrient intake and protein turnover. Proc. Nutr. Soc. 42:463.

Reeds, P. J., M. F. Fuller, A. Cadenhead, G. E. Lobley and J. D. McDonald. 1981. Effect of changes in the intake of protein and non-protein energy on whole-body turnover in growing pigs. Brit. J. Nutr. 45:539.

Reeds, P. J., M. F. Fuller and B. A. Nicholson. 1985a. Metabolic basis of energy expenditure with particular reference to protein. In: J. Garrow and D. Halliday (Ed.) Substrate and Energy Metabolism in Man. pp 46–57. John Libbey, London.

Reeds, P. J. and P. J. Garlick. 1985. Nutrition and protein turnover. In: H. H. Draper (Ed.) Advances in Nutrition Research. Vol 6, pp 93–126. Plenum Press, New York.

Reeds, P. J. and C. I. Harris. 1981. Protein and synthesis in animals. Man in his context. In: J. C. Waterlow and J. M. L. Stephen (Ed.) Nitrogen Metabolism in Man. pp 391–407. Applied Science Publishers, London.

Reeds, P. J., S. M. Hay, A. Cadenhead and M. F. Fuller. 1987. Urea synthesis and leucine turnover in pigs. Brit. J. Nutr. 58:301.

Reeds, P. J., S. M. Hay, R. T. Glennie, W. S. Mackie and P. J. Garlick. 1985b. The effect on indomenthacin on the stimulation of protein synthesis by insulin in young postabsorptive rats. Biochem. J. 227:225.

Reeds, P. J., E. R. Orskov and N. A. MacLeod. 1980b. Whole body protein synthesis in cattle sustained by infusions of volatile fatty acids and casein. Proc. Nutr. Soc. 40:50A (Abstr.)

Reeds, P. J. and R. M. Palmer. 1983. The possible involvement of $PGF_2\alpha$ in the stimulation of muscle protein synthesis by insulin. Biochem. Biophys. Res. Commun. 116:1084.

Reeds, P. J. and R. M. Palmer. 1984. Changes in prostaglandin release associated with inhibition of muscle protein synthesis by dexamethasone. Biochem. J. 219:953.

Reeds, P. J. and R. M. Palmer. 1986. The role of prostaglandins in the control of muscle protein turnover. In: P. J. Buttery, N. B. Haynes and D. B. Lindsay (Ed.) Control and Manipulation of Animal Growth. pp 161–185. Butterworths, London.

Reeds, P. J., R. M. Palmer and R. H. Smith. 1980b. Protein and collagen synthesis in rat diaphragm incubated *in vitro*. The effects of alterations in tension produced by electrical or mechanical means. Int. J. Biochem. 11:7.

Robinson, H., T. Cocks, D. Kerr and D. Picou. 1974. Fasting and post-prandial levels of plasma insulin and growth hormone in malnourished children, during catch-up growth and after recovery. In: L. I. Gardner and P. Amacher (Ed.) Endocrine Aspects of Malnutrition. pp 45–72. The Kroc Foundation, Santa Ynez.

Rodemann, H.-P. and A. L. Goldberg. 1982. Arachidonic acid, prostaglandins E_2 and F_2 influence rates of protein turnover in skeletal and cardiac muscle. J. Biol. Chem. 257:1632.

Rodemann, H. P., L. Waxman and A. L. Goldberg. 1982. The stimulation of protein degradation in muscle by Ca^{2+} is mediated by prostaglandin E_2 and does not require the calcium-activated protease. J. Biol. Chem. 257:8716.

Rosen, O. M., R. Harrera, Y. Olowe, L. M. Petruzelli and M. H. Cobb. 1983. Phorphorylation activates the insulin-receptor tyrosine protein kinase. Proc. Natl. Acad. Sci. USA 80:3237.

Rucker, R. and D. Tinker. 1986. The role of nutrition in gene expression. A fertile field for the application of molecular biology. J. Nutr. 116:177.

Ruff, R. and D. Secrist. 1984. Inhibitors of prostaglandin synthesis or cathepsin D prevent muscle wasting due to sepsis in the rat. J. Clin. Invest. 73:1483.

Russel, H. 1981. Ornithine decarboxylase: transcriptional induction by trophic hormones via a cAMP and cAMP-dependent protein kinase pathway. In: D. R. Morris and L. J. Marton (Ed.) Polyamines in Biology and Medicine. pp 109–125. Marcel Dekker, New York.

Saltiel, A. R. and P. Cuatrecasas. 1986. Insulin stimulates the generation from hepatic plasma membranes of modulators derived from an inositol glycolipid. Proc. Natl. Acad. Sci. USA 83:5793.

Sato, F., G. G. Ignotz, R. A. Ignotz, T. Gansler, K. Tsukada and I. Lieberman. 1981. On the mechanism by which insulin stimulates protein synthesis in chick embryo fibroblasts. Biochemistry 20:5550.

Scheiwiller, E., H.-P. Guler, J. Merryweather, C. Scandella, W. Maerki, J. Zapf and E. R. Froesch. 1986. Growth restoration of insulin-deficient diabetic rats by recombinant insulin-like growth factor I. Nature 323:169.

Schimke, R. T. and D. Doyle. 1970. Control of enzyme levels in animal tissues. Annu. Rev. Biochem. 39:929.

Schmidt, T., P. S. Chen and M. Pellegrini. 1985. The induction of ribosome synthesis in a nonmitotic secretory tissue. J. Biol. Chem. 260: 7675.

Schoenheimer, R., S. Rattner and D. Rittenberg. 1939. Studies in protein metabolism. X. The metabolic activity of body proteins investigated with L(-) leucine containing two isotopes. J. Biol. Chem. 130:709.

Schworer, C. M., K. A. Schiffer and G. E. Mortimore. 1981. Quantitative relationship between autophagy and proteolysis during graded amino acid deprivation in perfused rat liver. J. Biol. Chem. 256:7652.

Scornik, O. A. 1984. Role of protein degradation in the regulation of cellular protein content and amino acid pools. Fed. Proc. 43:1283.

Scornik, O. A. and V. Botbol. 1976. Role of changes in protein degradation in the growth of regnerating livers. J. Biol. Chem. 251:2891.

Seedorf, V., E. Leberer, B. J. Kirschbaum and D. Pette. 1986. Neural control of gene expression in skeletal muscle. Biochem. J. 239:115.

Seely, J. E., H. Poso and A. E. Pegg. 1983. Effect of androgens on turnover of ornithine decarboxylase in mouse kidney. J. Biol. Chem. 258:7549.

Seve, B., P. J. Reeds, M. F. Fuller, A. Cadenhead, and M. Hay. 1986. Protein synthesis and retention in some tissues of the young pig as influenced by dietary protein intake after early-weaning. Possible connection to the energy metabolism. Reproduction Nutrition and Development 26:849.

Sharpe, P. M., P. J. Buttery and N. B. Haynes. 1986. The effect of manipulating growth by diet or anabolic agents on plasma cortisol and muscle glucocorticoid receptors. Brit. J. Nutr. 56:289.

Sharpe, P. M., P. J. Buttery and N. B. Haynes. 1987. Cortisol and growth in sheep. Letter to the Editor. Brit. J. Nutr. 57:158.

Siehl, D., B. H. L. Chua, N. Lautensack-Belser and H. E. Morgan. 1985. Faster protein and ribosome synthesis in thyroxine-indued hypertrophy of rat heart. Amer. J. Physiol. 248:C309.

Simon, O., R. Munchmeyer, H. Bergner, T. Zebrowska and L. Buraczewska. 1978. Estimation of rate of protein synthesis by constant infusion of labelled amino acids in pigs. Brit. J. Nutr. 40:243.

Smith, R. H., R. M. Palmer and P. J. Reeds. 1983. Protein synthesis in isolated rabbit forelimb muscles. The possible role of arachidonic acid metabolites in the response to intermittent stretching. Biochem. J. 214:153.

Tabor, C. W. and H. Tabor. 1984. Polyamines. Annu. Rev. Biochem. 53:749.

Tessari, P., Inchiostro, S., Biolo, G., Trevisan, R., Fantin, G., Marescotti, M. C., Iori, E., Tiengo, A. and G. Crepaldi. 1987. Differential effects of hyperinsulinaemia and hyperaminoacidemia on leucine-carbon metabolism in vivo. J. Clin. Invest. 79:1062.

Thomas, G., J. Martin-Perez, M. Siegmann and A. M. Otto. 1982. The effect of serum, E.G.F., PGF_2 and insulin on S6 phosphorylation and the initiation of protein and DNA synthesis. Cell 30:235.

Thomas, G., M. Siegmann and J. Gordon. 1979. Multiple phosphorylation of ribosomal protein S6 during transition of quiescent 3T3 cells to early G1 and cellular compartmentalization of the phosphate donor. Proc. Natl. Acad. Sci. USA 76:3952.

Tomas, F. M., A. J. Murray and L. M. Jones. 1984. Modification of glucocorticoid induced changes in myofibrillar protein turnover in rats by protein and energy deficiency as assessed by urinary excretion of N^r-methyl histidine. Brit. J. Nutr. 51:323.

Towle, C. A., H. J. Mankin, J. Avruch and B. V. Treadwell. 1984. Insulin promoted decrease in the phosphorylation of protein synthesis initiaiton factor EIF-2. Biochem. Biophys. Res. Commun. 121:134.

Trenkle, A. and D. G. Topel. 1978. Relationships between some endocrine measurements to growth and carcass composition of cattle. J. Anim. Sci. 46:1604.

Turner, L. V. and P. J. Garlick. 1974. The effect of unilateral phrenicectomy on the rate of protein synthesis in rat diaphragm in vivo. Biochim. Biophys. Acta 349:109.

Vandenburg, H. and S. Kaufman. 1979. In vitro model for stretch induced hypertrophy of skeletal muscle. Science 203:265.

Vandenburg, H. H. and S. Kaufman. 1981. Stretch-induced growth of skeletal myotubes correlates with activation of the sodium pump. J. Cell. Physiol. 109:205.

Wan, J. and R. F. Grimble. 1985. Inhibitory effects of indomethacin on some features of the metabolic response to Escherichia coli endotoxin in rats. Proc. Nutr. Soc. 45:51A (Abstr.)

Waterlow, J. C. 1986. Metabolic adaptation to low intakes of energy and protein. Annu. Rev. Nutr. 6:495.

Waterlow, J. C., P. J. Garlick and D. J. Millward. 1978. Protein Turnover in Mammalian Tissues and in the Whole Body. pp 15–54. North-Holland, Amsterdam.

Waterlow, J. C. and J. M. L. Stephen. 1967. Lysine turnover in man measured by intravenous infusion of L-[^{14}C]-lysine. Clin. Sci. 33:507.

Wheatley, D. N. 1982. Significance of the rapid degradation of newly synthesized proteins in mammalian cells: a working hypothesis. J. Theor. Biol. 98:283.

Zeman, R. J., P. L. Bernstein, R. Ludemann and J. D. Etlinger. 1986. Regulation of Ca^{2+}-dependent protein turnover in skeletal muscle by thyroxine. Biochem. J. 240:269.

Energy Balance Regulation

ROY J. MARTIN, J. LEE BEVERLY, AND GARY E. TRUETT

1. Introduction

Energy balance regulation is a complex process involving several controlling systems and is similar in complexity to the regulation of body temperature and blood glucose. In fact, some of the mechanisms for energy balance regulation overlap with these two systems. The hierarchy of the systems is based on the highest priority of the organism for survival. This chapter will focus on signals of energy balance status from peripheral sources as well as the role of the central nervous system in integration of these signals and active control of either food intake or metabolism (Fig. 1).

2. Evidence for Energy Balance Regulation in Farm Animals

Energy balance regulation occurs in farm species as well as in the much studied laboratory rodent. There are some differences, which will be pointed out where information is available. However, some of the basic concepts of energy balance regulation are applicable to all species. There are reviews of this subject that focus on ruminants (Della-Ferra and Baile, 1984; Baile *et al.*, 1987), pigs (Houpt, 1984, 1985), and poultry (Forbes, 1985).

ROY J. MARTIN, J. LEE BEVERLY, AND GARY E. TRUETT ● Department of Foods and Nutrition, University of Georgia, Athens, Georgia 30602.

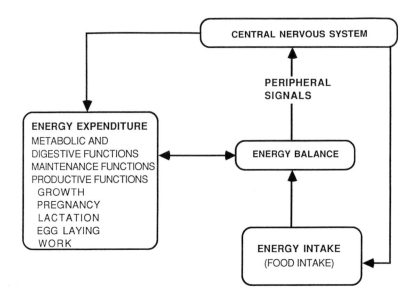

FIGURE 1. General scheme of central regulation of energy balance involving production functions found in farm animal species.

2.1. Lactation

The energy demand for **lactation** is a good example of energy balance regulation in action. Food intake is stimulated in the lactating cow when energy reserves, primarily fat stores, are depleted, much like in the lactating rat. In the lactating rodent, food intake is stimulated threefold to maintain energy balance (Wang, 1925; Truett, 1985). In the high-producing dairy cow, food intake is stimulated two- to threefold (Forbes, 1985). The mechanism by which this change in food intake occurs to maintain energy balance is not known. The energy demand for lactation results in a decrease in body fat (Flatt *et al.,* 1969). It may be that a signal is generated from the lipid stores that results in the stimulation of food intake. The lipostatic theory of food intake regulation is discussed later.

An alternate mechanism stimulated by lactation may involve the high demand of glucose for lactose synthesis. A deficiency of glucose caused by elevated rates of lactose synthesis may stimulate glucose-sensitive neurons in the hypothalamus (Oomura, 1980) and food intake in simple-stomach animals. In ruminant animals, the role of glucose in food intake control has been questioned since the major supply of energy in the ruminant is volatile fatty acids. Insulin-induced hypoglycemia, which normally increases food intake in simple-stomach animals, does not stimulate food intake in ruminants (Baile and Forbes,

1974). Volatile fatty acids have been shown to influence food intake and may be an important factor in the control of feeding in ruminant animals. The glucostatic theory of food intake control will be discussed later.

2.2. Growing Animals

In **growing animals** where there is a high demand for nutrients for tissue deposition, food intake is increased. During the weaning period, food intake per body weight is as high as in animals with brain lesions that cause hyperphagia (Kennedy, 1957). It was thought that there were no satiety factors or energy balance regulation operating in animals during the most rapid phase of weight gain and that food intake was maximally stimulated. However, recent studies indicate that growing animals that are force-fed will grow at a faster rate and when released from force feeding will reduce their intake and return to the body weight of control animals (Drewry, 1986). Growing pigs (Wangsness and Soroka, 1978) and sheep (Baumgardt and Peterson, 1971) respond to dietary caloric dilution by increasing feed intake. It is evident that energy balance mechanisms are operating in young rapidly growing animals.

2.3. Compensatory Growth

Compensatory growth is another example of energy balance regulation in animals. If feed-restricted for a short period of time and then returned to ad libitum intake, animals demonstrate a compensatory increase in food intake until the control body weight is recovered (Wilson and Osbourn, 1960; Harris and Martin, 1984a). During compensatory growth in young lambs, food intake per body weight and protein utilization rates were increased (Asplund *et al.*, 1975). These observations indicate that long-term mechanisms of energy balance are operating during compensatory growth.

During a rapid growth phase, protein deposition may be a major factor in food intake control. Amino acid requirements for protein synthesis are likely to be a factor in determining food intake level. When lean tissue growth is stimulated by a growth factor, the dietary amino acid requirements for optimal performance are increased (Anderson *et al.*, 1987) and when growth hormone is administered to animals there is an increase in growth and an increase in feed intake (Baile *et al.*, 1987). Young growing pigs are able to select an adequate protein intake from a choice of protein-free and protein-rich diets (Robinson, 1974). Compared to mature slow-growing animals, young growing animals select a higher content of protein when given a choice between a high- and a low-protein diet (Anderson, 1979). Taken together, these observations indicate that the requirements for growth play a role in energy balance regulation, perhaps through signals indicating amino acid status. The aminostatic theory of food intake regulation will be discussed later.

2.4. Egg Production

Egg production is another physiological situation that increases the demand for nutrient intake. The mechanisms that induce this rapid change in food intake are unknown. The laying hen undergoes significant changes in body composition as well as endocrine status in the production of eggs. Lipectomy of leghorn chicks did not result in a compensatory increase in feed intake (Maurice *et al.*, 1983). That the laying hen does not recognize the loss of body fat may signify that the endocrine changes occurring as a consequence of egg production are the means by which food intake is increased. The increase in feeding behavior of migratory birds is another example of spontaneous change in energy balance in avian species. This photostimulated increase in feeding behavior (Meier and Martin, 1971) and lipid synthesis (Wheeland *et al.*, 1976) can be mimicked by a unique pattern of injections of prolactin and glucocorticoids. It is possible that the endocrine changes occurring during egg production and photoperiod are responsible for the increase in feeding. More information on the basic mechanisms of energy balance regulation in domestic birds is needed.

3. Mechanisms of Energy Balance Regulation

To maintain energy balance, the rate of energy expenditure and energy intake have to be equal. The relative importance of the control of energy expenditure in the energy balance is probably lower in farm species that are in a state of high production than in laboratory animals. The examples given above for lactation and egg production indicate that when the energy demand for production functions is increased, energy balance is maintained primarily by the stimulation of food intake. However, there are some examples that indicate that energy released in the form of heat is altered in different production states.

3.1. Lactation

During **lactation,** part of the increase in energetic efficiency may be achieved by a reduction in energy expenditure. Energy expenditure in response to a meal is decreased in lactating women (Illingworth *et al.*, 1986). Similar data are not available in lactating cows; however, the efficiency of lipid deposition is higher in lactating than in nonlactating animals (Moe *et al.*, 1971). This increase in efficiency does not appear to be due to a change in energy required for maintenance. When estimated by regression of total energy balance on metabolizable intake, lactating cows have a higher maintenance requirement than nonlactating cows (Moe *et al.*, 1971). In addition, the metabolic activity of tissue from a lactating cow is higher in support of increased demand for milk synthesis (Baldwin and Bywater, 1984).

3.2. Brown Adipose Tissue Metabolism

Brown adipose tissue metabolism has been suggested to be an important factor in energy balance regulation (Himms-Hagen, 1985). The metabolism of this tissue is under CNS regulation (Shimazu, 1986). Specifically, the ventromedial hypothalamus–sympathetic nervous system is stimulated in states where heat production is needed (i.e., exposure to cold). The increased NE release at nerve terminals in brown adipose tissue results in a stimulation of brown fat metabolism and uncoupling of oxidative phosphorylation through the stimulation of thermogen or GDP-binding protein (Himms-Hagen, 1985). In high-producing lactating cows where there is a heat load caused by the heat increment associated with milk production, there may be a reduction in normal mechanisms for maintenance of heat production. In support of this suggestion is the observation that brown adipose tissue thermogenesis is suppressed during lactation in mice (Trayhurn *et al.*, 1982). Other mechanisms for temperature regulation may be operative in lactating cattle.

3.3. Compensatory Growth

During **compensatory growth,** the increase in overall energetic efficiency is caused partially by the increase in food intake above maintenance energy requirements and by a reduction of the energy required for basal metabolic rate. During dietary energy restriction, there is a reduction in basal metabolic rate in man and in animals (Blaxter, 1971; DeBoer *et al.*, 1986). There are several mechanisms by which energy expenditure may be decreased during diet restriction. Chronic food restriction reduces thermogenic capacity and brown fat metabolic activity (Rothwell and Stock, 1982). However, since mature farm animals do not have readily detectable brown fat, heat production by some other tissues may contribute more substantially to whole-body thermogenesis. Energy expenditure associated with Na–K transport mechanisms has been shown to vary with physiological state of the animal and appears to be a substantial component of maintenance of energy expenditure of ruminant animals (Milligan and McBride, 1985). Feed-restricted or fasted sheep have decreased Na,K-ATPase activity in liver and intestinal epithelium (McBride and Milligan, 1985). The activity of Na,K-ATPase is under the control of thyroid hormones. In sheep, the induction of Na,K-ATPase activity in skeletal muscle (Milligan and McBride, 1985) follows exposure to cold. Na,K-ATPase activity and heat production are decreased during feed restriction presumably because there is a reduction in thyroid hormone levels. Decreased thyroid function during feed restriction has been demonstrated in cattle (Fox *et al.*, 1974) and pigs (Kasser *et al.*, 1981).

In addition to reduce thyroid function, a decrease in sympathetic nervous system activity also results in lower energy expenditure during periods of energy deficit (Young *et al.*, 1982). The control of metabolism by hypothalamic areas is discussed later.

4. CNS Control of Food Intake

Feeding, like all other behaviors, is under the control of the CNS. To govern food intake, the perception of when nutrients are needed, when they are consumed, and how they are utilized must all be accounted for by the regulating system. A focal point in the integration of brain pathways dealing with all three of these aspects is the hypothalamus. This brain area contains numerous discrete nuclei of neurons that are highly interrelated with each other and with a much broader network of neural processes from brain areas outside the hypothalamus.

The hypothalamus does not operate autonomously in this role but, nevertheless, has a major controlling involvement. Quantitative and qualitative adjustments in food intake (e.g., appetite for specific nutrients) are derived from sensory and metabolic inputs to the hypothalamus (Leibowitz, 1986a).

4.1. Brain Areas Involved

Lesions or stimulation of specific hypothalamic areas lead to acute or chronic changes in food intake and/or body composition. Feeding is increased by lesions in the ventromedial hypothalamus (VMH) or stimulation of the ventrolateral hypothalamus (VLH). Conversely, lesions in the VLH or stimulation of the VMH results in aphagia. These results led to the formulation of the "dual center hypothesis" (Anand and Brobeck, 1951; Stellar, 1954) in which feeding and satiation are controlled by two separate centers. These centers were inferred to be functional centers, not necessarily anatomical centers. Several distant brain sites may actually, via interconnecting fiber pathways, form one functional entity.

Lesions to the paraventricular nucleus (PVN) and the dorsomedial nucleus (DMN) also induce hyperphagia, demonstrating that hypothalamic nuclei other than the VMH and VLH influence feeding behavior (Leibowitz, 1986a; Bernardis and Bellinger, 1987). Within the hypothalamus, a general medial stimulatory (via α-adrenergic receptors), and a lateral inhibitory (via β-adrenergic receptors) arrangement has been proposed to explain the functional order between hypothalamic areas (Leibowitz, 1970). Long-term infusions of NE into the VMN result in increased food intake and eventual obesity (Lichtenstein *et al.*, 1984; Shimazu *et al.*, 1986).

Integration of sensory and effector information by the hypothalamus allows mediation of the feeding process. Inputs to the VMH and VLH via the limbic system from visual, olfactory, and gustatory centers are supplemented by inputs from hepatic and intestinal glucoreceptors and gastric mechanoreceptors (Oomura, 1980). The VLH especially seems to be involved in monitoring internal conditions (Kandel and Schwartz, 1985).

Hypothalamic input into control of behavior is provided by reciprocal fiber connections with the prefrontal and motor cortex (Ono *et al.*, 1981). Connections between the hypothalamus and amygdala nuclei, which exert both facilitatory and inhibitory influences on hypothalamic function, allow expression of hunger-motivated food acquisition (Aou *et al.*, 1983). That the impetus for feeding arises from the hypothalamus, and not cortical areas, is demonstrated by the application of endogenous neurotransmitters to areas within both structures. Specific alterations in food intake are observed only following application into the hypothalamus (Leibowitz, 1986a).

4.2. Mechanism of Action

How each site functions to regulate feeding behavior is in part a function of the nutritional status (Myers *et al.*, 1986). Within each nucleus of cells are a number of different neuroactive compounds. Four classes of these compounds are involved in the modulation of feeding: monoamines, amino acids, prostaglandins, and peptides (Morley and Mitchell, 1986). Interactions among these compounds may be synergistic or each may act independently with the existing profile, at a given time, being the mediator of feeding behavior (Leibowitz, 1986b; Morley and Mitchell, 1986). The majority of these compounds inhibit food intake; only a few stimulate feeding. This provides a system described as one primarily of food seeking/ingestion that is held in check by inhibitory inputs, mainly from a "peripheral satiety system" (Morley and Levine, 1985).

Changes in food intake are often accompanied by nutrient-specific appetites demonstrated by a change in macronutrient selection. Carbohydrate intakes are increased following administration of NE or neuropeptide Y to the medial hypothalamus (Stanley and Leibowitz, 1985). Microinjection of opioids to the medial areas seems to increase the preference for fat intake and to a smaller extent protein intake (Marks-Kaufman, 1982). Protein intake can also be modified by inhibiting the dopaminergic system or by stimulating the serotonegeric system (Leibowitz, 1986b).

A physiological function of the medial hypothalamic noradrenergic system may be to rapidly replenish carbohydrate stores as this system is more active during periods of food deprivation and at the beginning of the dark period in rats (Leibowitz, 1986b). Release of NE in the medial hypothalamus increases during feeding (McCaleb et al., 1979) and is influenced by the nutritional state (Myers et al., 1986). Modulating effects of neuropeptides on NE release also vary with the nutritional status (McCaleb and Myers, 1982; Myers et al., 1986).

5. CNS Control of Peripheral Metabolism

A change in nutrient level is compensated for by altering feeding patterns or by changing metabolic use of the nutrient in various tissues. Nutrient metabolism is controlled by the composite of autonomic and hormonal inputs to the cells of a given tissue. The autonomic and hormonal profile, as well as feeding behavior, is mediated directly and indirectly by the hypothalamus (Swanson, 1986). Although all nutrients are influenced, glucose, the obligatory fuel for the CNS, displays small variations in circulating levels, reflecting a tight regulation by a glucoregulatory system (Woods et al., 1986).

5.1. Brain Areas Involved

The VMH and VLH are involved in the hypothalamic regulation of fuel substrate availability, matching utilization with food intake (Steffens et al., 1985). Modulation of tissue metabolism is performed by nerve fibers arising from the ventral hypothalamus (Shimazu, 1981). Regulation of peripheral metabolism occurs due to the functionally reciprocal relationship between the sympathetic and parasympathetic branches of the autonomic nervous system (Bray, 1984).

In general, the VMH initiates responses that are manifested by the sympathetic nervous system (SNS) and the lateral hypothalamic area (LHA) utilizes the parasympathetic nervous system (PNS) (Ban, 1975). Interaction between the two branches is implied by electrical stimulation and lesioning data. Electrical stimulation of the VMH decreases the firing rate of the pancreatic vagus nerve and increases activity in the pancreatic splanchnic nerve (Shimazu, 1981; Oomura and Kita, 1981). Opposite responses are generated by stimulation of the VLH (Oomura and Kita, 1981) and lesions to the VMH, PVN, or DMN (Yoshimatsu et al., 1984). Connections between the VMH, VLH, PVN, and DMN with areas in the brain stem relay fibers of the PNS to the nucleus tractus solitarius and dorsal motor nucleus of the vagal nerve in the brain stem. Other connections are to areas associated with fibers of the SNS, specifically the intermediolateral cell column in the spinal cord (Oomura, 1980).

Specific neurotransmitters in central sites are involved in eliciting changes in peripheral metabolism. For example, NE-sensitive neurons in the VMH and cholinergic neurons in the VLH are involved in regulation of hepatic glycogen metabolism (Shimazu, 1981). Sympathetic nervous tone is decreased by β-adrenergic receptors in the LHA and increased by α-adrenergic receptors in the VMH (Bray, 1984).

5.2. Direct Regulation

Control of nutrient metabolism is exerted directly by innervation of hepatic, muscle, and adipose tissues. Innervation of the pancreas and adrenal glands also influences nutrient utilization by regulating the release of hormones having primarily, but not exclusively, glucoregulatory effects.

5.2.1. Pancreas

Innervation of the pancreas by both the PNS and SNS provides for the control of insulin and glucagon release. Neural influences, along with substrate and peptidergic input, provide for a tight regulation that allows for matching of nutrient availability with adequate amounts of the appropriate hormone. Stimulation of the LHA increases both vagal tone and serum insulin levels (Woods and Porte, 1974). Vagal activity increases during feeding, starting during the cephalic phase of ingestion, to increase insulin levels and prevent a rapid elevation in blood glucose (Powley, 1977). Stimulation of the VHM or splanchnic nerve depresses insulin secretion and elevates glucagon levels (Oomura and Niijima, 1983).

Inhibition of insulin release by the splanchnic nerve is removed by VMH lesions resulting in hyperinsulinemia (Inoue *et al.*, 1978). Ectopic positioning of pancreatic islets under the kidney capsules, with no innervation, prevented the hyperinsulinemia associated with VMH lesioning (Inoue *et al.*, 1978).

5.2.2. Liver

Hepatic innervation by the autonomic nervous system includes afferents and efferents associated primarily with the vasculature but also with hepatocytes (Tanikawa, 1968). Glycogenolysis is enhanced as a result of electrical or chemical (NE) stimulation in the VMH, primarily due to an increase in glycogen phosphorylase activity (Shimazu, 1981). This response is insulin and glucagon independent (Woods and Porte, 1974) but is not mediated by locally released NE, which implies mediation by other effectors (Woods *et al.*, 1986). Vagal stimulation, or cholinergic stimulation of the LHA, increases glycogenesis by increasing the rate of glycogen synthetase activity (Shimazu, 1981).

5.2.3. Adrenals

As a functional unit, the sympathoadrenal system is comprised of the adrenal medulla and the SNS (Landsberg and Young, 1984). In response to NE released by sympathetic nerve endings, epinephrine is released into the general circulation where it has several physiological roles. A glucoregulatory role in stimulating hepatic gluconeogenesis and glycogenolysis to elevate blood glucose is under the control of the CNS (Niijima, 1977).

5.2.4. Adipose Tissue

An increase in circulating free fatty acids and glycerol observed following electrical stimulation or administration of NE centrally (Barbosa and Miglorini, 1982) and to the VMH (Steffens *et al.*, 1972) reflects a role of CNS control of lipid metabolism. Sympathetic arousal by β-adrenergic systems in the VMH increases lipolysis, whereas the opposite effect is observed following arousal of the PNS (Steffens *et al.*, 1972).

The release of free fatty acids following stimulation of nerves innervating white adipose tissue is abolished by sympathectomy or β-adrenergic blockage (Weiss and Maickel, 1968). Peripheral chemical sympathectomy or denervation of fat pads also attenuates the loss of adipose tissue during starvation (Bray and Nishizawa, 1978). Lesions to the VMH produce similar results (Bray and Nishizawa, 1978).

5.3. Indirect Regulation

The interface between the CNS and the endocrine system resides in the hypothalamus where sensory information on the internal environment influences the endocrine profile (Swanson, 1986). Besides being the origin of nerves directly innervating the endocrine pancreas and adrenals, the hypothalamus also contains neurosecretory releasing factors that modulate the secretion of pituitary hormones. Five releasing factors, all peptides, have been shown to be synthesized and secreted in the hypothalamus and to regulate the release of anterior pituitary hormones (Swanson, 1986). Several of these influence feeding behavior (Morley and Levine, 1985) as well as have a direct metabolic effect on other organs. A summary of central regulation of metabolism and the proposed pathways is presented in Fig. 2.

6. Theories of Signals Regulating Food Intake

A common feature of theories of food intake regulation is the presence or absence of circulating factors, including nutrients that signal energy status. The

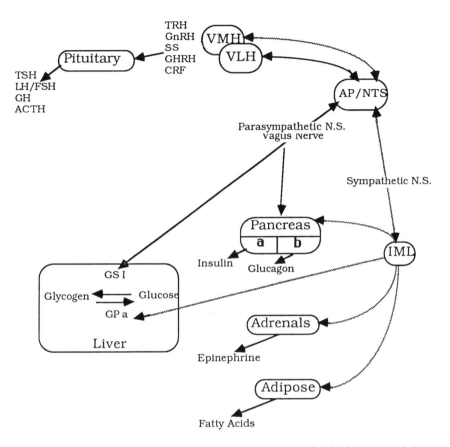

FIGURE 2. The role of the central nervous system in the control of endocrine status and tissue metabolism.

recognition and response to a particular signal may have short-term influence (i.e., meals) or have consequences that can only be measured on a long-term basis (i.e., several days). The latter has particular relevance to maintenance of energy balance regulation. The complexity with which these various inputs are processed and effect changes in intake implies that no one signal has dominant control over feeding. A multiplicity of inputs (signals) with central integration is proposed.

6.1. Short-Term Mechanisms

Those macronutrients that are thought to act as signals for short-term energy status include glucose, volatile fatty acids, glycerol, and amino acids. The following discussion covers some of the classic theories of food intake control involving macronutrients and some recent information that either supports or disputes these theories.

6.1.1. Glucostatic Theory

The basis of the glucostatic theory (Mayer, 1963) is that decreased glucose utilization by neurons in the VMH stimulates feeding. Alterations in serum glucose levels influence food intake in most mammals. Circulating glucose is decreased in rats approximately 10% preceding meals (Louis-Sylvestre and Le Magnen, 1980; Campfield et al., 1985) and serum glucose levels rise after a meal (Le Magnen, 1981). The rate of change in serum glucose concentration may initiate a meal, or determine the interval between meals (Campfield and Smith, 1985, 1987). Inhibiting the slow decline in blood glucose occurring prior to a meal increases the time before the onset of the next meal (Campfield and Smith, 1985). It is the temporal pattern of the decline, however, and not the nadir level that precipitates meal initiation (Campfield and Smith, 1987).

Large fluctuations in serum glucose, such as following insulin injection, also modify food intake in rats (Booth and Pitts, 1968), pigs (Houpt et al., 1979), and ruminants (Anil and Forbes, 1980) but not chickens (Lepkovsky et al., 1967). In avian species, insulin and glucoprivation appear to have effects opposite those seen in mammals (Forbes, 1986).

Changes in circulating glucose levels may have less impact on feeding in ruminants, reflecting the difference in glucose metabolism in these animals (Baile and Forbes, 1974). However, what should have been an increased food intake by sheep and goats following intravenous injections of 2-deoxy-D-glucose was blocked by prior administration of glucose (Houpt, 1974), indicating that glucose utilization in ruminants does exert some influence on feeding behavior.

The stimulus for feeding in mammals may be the availability of glucose to the hypothalamus (Epstein et al., 1975). Nonmetabolizable glucose analogues, such as 2-deoxy-D-glucose, also initiate feeding as a result of neuronal glucoprivation (Miselis and Epstein, 1975). Specific neuronal cell damage in the hypothalamus occurs following gold thioglucose injection into normal mice and results in changes in central catecholamine levels (Lorden et al., 1979), hyperphagia, and obesity (Marshall and Mayer, 1954; Martin and Lamprey, 1974). The ability to cause these specific lesions and hyperphagia is influenced by adrenal gland glucocorticoids (Debons et al., 1986) and insulin. The lesions are absent in diabetic mice (Debons et al., 1968), indicating glucose uptake, or utilization by the hypothalamus has an insulin-sensitive component (Le Magnen, 1985).

Glucoreceptor neurons in the VMH alter firing rate in response to iontophoretically applied glucose. Firing rates are increased by glucose, further elevated by insulin and glucose, but inhibited by insulin alone (Oomura and Kita, 1981). In contrast, the firing rate of glucosensitive neurons in the VLH responds in the opposite manner. The reciprocal relationship between lateral and medial hypothalamus is discussed later.

A "memory" aspect of glucoprivic-induced feeding was indicated by increased food intake occurring several hours after the return of serum glucose to normal levels, if the animals were not fed (Ritter *et al.*, 1978; Engeset and Ritter, 1980). Theories to explain this phenomenon include: altered NE turnover in the VMH (Ritter *et al.*, 1981); reduced glycogen content in the caudal brain stem (Nzuzi and Ritter, 1985); hepatic control of satiety (Friedman and Granneman, 1983); and changes in GABA shunt activity in the VLH (Kasser *et al.*, 1985) and VMH (Beverly and Martin, 1987).

Glucose receptors outside the hypothalamus may also participate in glucostatic regulation of feeding. Reduced glucose availability to brain stem areas (Ritter *et al.*, 1981) and to hepatocytes (Russek, 1963; Novin *et al.*, 1973) influences feeding. Alterations in the firing rate of vagal afferents (Niijima, 1977) and glucoreceptors in the VMH (Shimazu, 1981) occur in response to portal infusion of glucose. The importance of hepatic glucoreceptors in food intake control has been diminished by the observations that hepatic vagotomy does not interfere with normal responses to peripheral glucose deprivation (Bellinger *et al.*, 1984) or with long-term regulation of body weight (Egli *et al.*, 1986).

6.1.2. Aminostatic Theory

Under normal dietary conditions, the intake of a constant ratio of protein energy to total energy is maintained (Musten *et al.*, 1974). The relative proportion of plasma essential amino acids, especially the precursors of neurotransmitters, is instrumental in regulating protein intake. Brain levels of an amino acid are not only reflective of plasma levels but are also proportional to the plasma levels of all other amino acids competing for the same endothelial transport carrier (Pardridge, 1979). Considerable influence on transport mechanisms at the blood–brain barrier is exerted by various hormones, usually by altering relative plasma levels of the amino acids (Pardridge, 1979).

Tryptophan, the limiting amino acid in most diets, it the precursor for the neurotransmitter serotonin and may be a focal point in protein-regulating mechanisms (Anderson, 1979). Similar correlations between tyrosine, catecholamine neurotransmitters, and carbohydrate intake have also been reported (Wurtman and Fernstrom, 1976).

A decrease in food intake and growth rate follows the feeding of diets containing disproportionate levels of amino acids (Harper *et al.*, 1970). This response involves a mechanism different from that regulating total protein intake and is sensitive to the level of any dietary-limiting amino acid (Rogers and Leung, 1973). Protein synthesis in all tissues, combined with increased competition for brain uptake, affects the availability of a limiting amino acid to the brain. Intake of an amino acid-limiting diet will return to normal if the limiting amino acid is injected into the carotid artery but not if applied in the jugular vein (Leung and Rogers, 1969).

An aminostatic mechanism regulating protein/food intake is well established but when diets of normal composition are available, the aminostatic mechanism is subservient to those mechanisms involved in energy balance maintenance (Rogers and Leung, 1977).

6.1.3. Integrating Theories

Several theories have been forwarded to explain and incorporate some of the disparate empirical evidence generated by "single-nutrient" hypotheses. Generally, these theories state that it is the sum nutrient availability for energy synthesis that is monitored and corrected for.

The "ischymetric hypothesis" accounts for the integration of energy substrates of all origins and the effect of endocrine and humoral factors on energy processing (Even and Nicolaidis, 1986). The metabolic rate ["Metabolism de forn" (MF)] provides a signal as to the state of depletion or repletion of energy. Changes in cellular "power" production, evidenced by changes in the MF, initiate or terminate meals. In the "energostatic theory" (Booth, 1972), energy intake is matched to the immediate and/or past energy needs and this "matched" intake then influences meal size and the intervals between meals. Either the total amount of energy supplied by metabolism of various substrates, or a common product of glycolysis, amino acid catabolism, and fatty acid oxidation is monitored in the liver.

The liver also is a monitor of nutrient availability and is the basis of the "utilizable fuel theory" (Friedman and Stricker, 1976). When hepatic oxidation of glucose or fatty acids is depressed, an increase in food intake is observed (Friedman et al., 1986). A synergistic response occurs if both glucose and fatty acid metabolism in the liver are impaired (Friedman and Tordoff, 1986). A vagally mediated satiety signal may be generated by hepatic fatty acid oxidation (Langhans et al., 1985). The effect on satiety is greater when high-fat diets are fed (Langhans and Scharrer, 1987). The interaction between long-term and short-term food intake regulatory mechanisms may be via hepatic glucose and fatty acid oxidation (Le Magnen, 1981).

6.1.4. Anatomical and Humoral Factors

Innervation of the viscera exerts an influence on food intake such that stretching of the gastric or intestinal epithelium (Kraly and Smith, 1978) or ruminal distention (Baile and Della-Ferra, 1981) can influence food intake. Monitoring of intraluminal contents is suggested by observations that following the introduction of calories into the stomach a constant caloric intake is maintained (McHugh and Muran, 1978). Regulation was equally effective for calories in the form of protein, fats, or carbohydrate. In ruminants, infusion of volatile fatty acids into the rumen regulated the amount of food consumed (Baile and Forbes, 1974). Additionally, the rate of gastric emptying is regulated and supplies a constant rate of delivery of calories to the small intestine (McHugh and Muran, 1979).

Of greater importance to the initiation of sataiety are the myriad of humoral factors produced or released in association with digestion. Peptide hormones released by the gut mucosa affect enzyme secretion into the lumen, blood flow in the intestinal epithelium, and influence satiety through the liver or directly in the brain (Woods et al., 1986).

Pancreatic and adrenal hormones released after stimulation by the products of digestion and by direct neural influence also have demonstrable roles in food intake control. Generally, in addition to influencing circulating nutrient levels, these hormones are involved in the maintenance of tissue structure and the normal biochemical profile, which in turn influence energy balance (Bray, 1984).

6.2. Long-Term Mechanisms

It is more difficult to determine long-term mechanisms of intake regulation (Nicolaidis, 1981). Interaction with or modulation by mechanisms involved in body weight regulation tends to make determination of specific signals difficult to interpret. The evolution of the ''set point'' theory of body weight regulation arises from data in which compensatory changes in food intake are manifested following concerted efforts to induce changes in normal body weight or composition. Following prolonged periods of restriction or overfeeding, there is a time when food intake is adjusted until body weight returns to the level preceding the initial change. The dominant theory maintains that what is being regulated is body fat, though lean tissue (i.e., protein) may also exert control on this process (Harris and Martin, 1984a; Harris et al., 1986).

Lipostatic (Set Point) Theory

A constant degree of body lipid is maintained in adult animals (Porte and Woods, 1981) by alterations in food intake or metabolic efficiency (Bray, 1984). Long-term food intake is involved in maintaining adipose tissue mass but its control is secondary to the regulation of body energy stores (Kennedy, 1953). This suggestion by Kennedy was the foundation of the ''lipostatic hypothesis,'' the basis of which is that food intake is regulated by the ability of circulating substrates to be incorporated into adipose tissue. A liporegulatory mechanism residing in the VMH modifies long-term food intake as a function of glucose availability (Le Magnen, 1981). Glucose is utilized by both adipose tissue and nervous tissue and represents a common substrate whose availability to central structures may reflect the level of peripheral adipose mass. An insulin-sensitive glucoreceptor network in the VMH is proposed to serve as the monitor (Le Magnen, 1983, 1985). Fatty acids are another adipose tissue metabolite that have been suggested to fulfill a long-term regulatory function in relating adipose tissue mass and the control of food intake. Lipolytic rate is increased in large fat cells compared to small fat cells. Therefore, when cells become large near the "set point," they release more fatty acids into the circulation, which is detected by the brain. Evidence in support of this suggestion is as follows: continuous infusion of long-chain fatty acids in sheep resulted in a decrease in food intake (Vandermeerschen-Doize and Paquay, 1984). When fatty acid oxidation is inhibited, food intake is stimulated (Langhans and Scharrer, 1987).

Maintenance of adipose tissue levels was demonstrated by lipectomy studies. Removal of adipose tissue from various depots resulted in an increased hyperplastic and hypertrophic growth until body adiposity returned to normal levels (Larson and Anderson, 1978; Bailey and Anderson, 1980). Additionally, forced changes in body fat are spontaneously corrected (Woods and Porte, 1974).

Experiments in which the blood from hyperphagic rats was exchanged with that from normal rats suggest the existence of a circulating humoral factor associated with adiposity (Harris and Martin, 1984b). The circulating factor acts as a feedback signal to centers regulating energy balance as to the state of peripheral energy stores. Parabiosis experiments in which one partner was made obese by VMH lesion (Hervey, 1959), increased food intake by lateral hypothalamic stimulation (Paraneswaren *et al.*, 1977), or by overfeeding (Harris and Martin, 1986) resulted in decreased food intake by the nontreated partner. The blood-borne factor involved in the control of feeding in this model has yet to be identified.

Some blood-borne factors that influence feeding have been investigated. A satiety factor ("satietin") has been identified in human, ruminant, and avian blood which causes a reduction in food intake in rats (Knoll, 1985). A low-molecular-weight fraction of plasma from free-feeding chickens has been shown to decrease food intake when administered centrally to domestic chickens (Skewes *et al.*, 1986). Although these factors have been shown to suppress feeding behaviors, it is not known if they are involved in normal food intake control.

7. Role of Neurotransmitters and Neuropeptides

Although more than 20 neurotransmitters have been implicated in food intake regulation (Hoebel, 1985; Morley *et al.*, 1985; Leibowitz, 1986b) we will focus on those that have been investigated in agricultural animals.

NE stimulates feeding when injected into the PVN (Leibowitz, 1978), and chronic infusion of NE or the α_2-adrenergic agonist clonidine results in hyperphagia and obesity (Lichtenstein *et al.*, 1984). High concentrations of α_2 receptors are found in the PVN (Young and Kuhar, 1980) and vary with the diurnal rhythym of food intake and corticosterone (Jhanwar-Uniyal *et al.*, 1986). The feeding response to NE is dependent on corticosterone (Leibowitz *et al.*, 1984).

GABA agonists stimulate food intake when injected in medial hypothalamic sites of rats (Grandison and Guidotti, 1977) and sheep (Girard *et al.*, 1985). In addition, GABA agonists block NE-induced feeding. GABA injections have uncertain effects, probably because of the rapid breakdown and reuptake of injected GABA (Girard *et al.*, 1985).

The most potent feeding stimulators are the homologous peptides neuropeptide Y (NPY) and peptide YY (PYY), both of which stimulate feeding when injected intracerebroventricularly (i.c.v.) in the rat (Levine and Morley, 1984; Morley *et al.*, 1985); NPY also stimulates food intake in the pig (Parrott *et al.*, 1986). Although PYY produces a greater feeding response, it is a gut endocrine peptide (Tatemoto, 1982; O'Donohue *et al.*, 1985), whereas NPY is found in especially high concentrations in feeding-associated areas of the brain (Allen *et al.*, 1983). Because of this compartmentalization, NPY is thought to be the peptide that acts physiologically in food intake regulation. Other notable characteristics of NPY are that it preferentially stimulates carbohydrate intake (Stanley *et al.*, 1985) and it is colocalized with NE in many neurons (Everitt *et al.*, 1984). Although NPY and NE have similar properties, NE antagonists do not affect NPY-stimulated feeding (Levine and Morley, 1984).

Endogenous opioid peptides are a diverse group of peptide products. In rats, much work has been directed toward dissecting the activities of the various opioid receptors. The κ receptor appears to be responsible for the increase in food intake seen in rats and sheep after central injection of dynorphin, methionine-enkephalin, and β-endorphin. The μ and δ receptors probably mediate a reduction in food intake (Baile *et al.*, 1987).

Recent reports suggest an uncertain role for growth hormone-releasing factor (GRF) in the regulation of food intake. GRF has been shown to both stimulate (Vaccarino *et al.*, 1985) and inhibit food intake in the rat (Imaki *et al.*, 1985). The actions of GRF on feeding appear to be centrally mediated, dose-dependent, and independent of GH. Hay-fed sheep also increase their intake when injected i.c.v. or intravenously (i.v.) with GRF. When fed concentrate, sheep increase their intake for the first hour, but total intake per day is the same as for control sheep. Because i.c.v. or i.v. insulin blocks the increase in intake by hay-fed sheep, insulin appears to modulate the effectiveness of GRF and may be responsible for the diet effect (Riviere and Bueno, 1987).

Cholecystokinin (CCK) is an example of a peptide that shows dissimilar effects on feeding in various species. Central injection of CCK reduces food intake in pigs (Parrott and Baldwin, 1981), sheep (Della-Ferra and Baile, 1980), and chicks (Denbow and Myers, 1982); however, it appears to be less important in central regulation of feeding in rats (Della-Ferra and Baile, 1979). Della-Ferra *et al.* (1981) provided convincing evidence for the central effects of CCK by infusing antibodies to CCK and producing an increase in food intake; they proposed that satiety was delayed in these animals by the removal of CCK by the antibodies. Furthermore, the effective dose of CCK increases with increasing time of fasting (Baile and Della-Ferra, 1984).

Other neurotransmitters that have been shown to affect food intake in sheep are thyrotropin-releasing hormone and corticotropin-releasing factor, both of which suppress food intake (Ruckebusch and Malbert, 1986), and calcitonin gene-related peptide (CGRP), which stimulates total daily food intake in sheep (Bueno *et al.*, 1986), but suppresses food intake in rats (Krahn *et al.*, 1984). Calcitonin is not synthesized in brain, but suppresses food intake in rats (Levine and Morley, 1981) and sheep (Bueno *et al.*, 1986).

8. Summary

The regulation of energy balance in farm species is a unique process that has some basic similarities to the laboratory rodent model. However, it is evident that each farm animal will have to be studied to fully understand the control mechanisms involved.In addition, the genetic selection for enhanced productive traits (lactation, growth, and egg production) has probably altered the range in which energy balance can be regulated and the mechanisms involved.

References

Allen, Y. S., T. E. Adrian, J. M. Allen, K. Tatemoto, T. J. Crow, S. R. Bloom, and J. M. Polak. 1983. Neuropeptide Y distribution in the rat brain. Science 221:877.

Anand, B. K. and J. R. Brobeck. 1951. Hypothalamic control of food intake in rats and cats. Yale J. Biol. Med. 24:123.

Anderson, D. B., E. L. Veenhuisen, W. P. Waitt, R. E. Paxton and S. S. Young. 1987. The effect of dietary protein on nitrogen metabolism, growth performance and carcass composition for finishing pigs fed Ractopamine. Fed. Proc. 46:1021 (Abstr.).

Anderson, G. H. 1979. Control of protein and energy intake: role of plasma amino acids and brain neurotransmitters. Can. J. Physiol. Pharmacol. 57:1043.

Anil, M. H. and J. M. Forbes, 1980. Effect of insulin and gastro-intestinal hormones on feeding and plasma insulin in sheep. Horm. Metab. Res. 12:407.

Aou, S., Y. Oomura, H. Nishino, A. Inokuchi and Y. Mizuno. 1983. Influence of catecholamines on reward related neuronal activity in monkey orbitofrontal cortex. Brain Res. 267:165.

Asplund, J. M., H. B. Hendrick and C. D. Haugebak. 1975. Performance, digestibility and ^{40}K levels in lambs during compensation for feed restriction. J. Anim. Sci. 40:138.

Baile, C. A. and M. A. Della-Ferra. 1981. Nature of hunger and satiety control systems in ruminants. J. Dairy Sci. 64:1140.

Baile, C. A. and M. Della-Ferra. 1984. Peptidergic control of food intake in food-producing animals. Fed. Proc. 43:2898.

Baile, C. A., and M. J. Forbes. 1974. Control of feed intake and regulation of energy balance in ruminants. Physiol. Rev. 54:160.

Baile, C. A., C. L. McLaughlin, F. C. Buonomo, T. J. Lauterio, L. Marson and M. A. Della-Ferra. 1987. Opioid peptides and the control of feeding in sheep. Fed. Proc. 46:173.

Bailey, J. W. and D. B. Anderson. 1980. Rate of fat compensation and growth efficiency of lipectomized Sprague–Dawley rats. J. Nutr. 110:1785.

Baldwin, R. L. and A. C. Bywater. 1984. Nutritional energetics of animals. Annu. Rev. Nutr. 4:101.

Ban, T. 1975. Fiber connections in the hypothalamus and some autonomic functions. Pharmacol. Biochem. Behav. 3 (Suppl. 1):3.

Barbosa, M. C. and R. H. Miglorini. 1982. Free fatty acid mobilization in rats following intracerebroventricular norepinephrine. Amer. J. Physiol. 242:E248.

Baumgardt, B. R. and A. D. Peterson. 1971. Regulation of food intake in ruminants. 8. Caloric density of diets for young growing lambs. J. Anim. Sci. 54:1191.

Bellinger, L. L., V. E. Mendel, F. E. Williams and T. W. Castonguay. 1984. The effect of liver denervation on meal patterns, body weight and body composition of rats. Physiol. Behav. 33:661.

Bernardis, L. L. and L. L. Bellinger. 1987. The dorsomedial hypothalamic nucleus revisited: 1986 update. Brain Res. Rev. 12:321.

Beverly, J. L. and R. J. Martin. 1987. Glucose contribution to GABA shunt activity in the rat ventral hypothalamus: comparison of hyperphagic models. Fed. Proc. 46:1482 (Abstr.).

Blaxter, K. L. 1971. Methods of measuring the energy metabolism of animals and interpretation of results obtained. Fed. Proc. 30:1436.

Booth, D. A. 1972. Postabsorptively induced suppression of appetite and the energostatic control of feeding. Physiol. Behav. 9:199.

Booth, D. A. and M. E. Pitts. 1968. The role of glucose in insulin-induced feeding and drinking. Physiol. Behav. 3:447.

Bray, B. A. 1984. Hypothalamic and genetic obesity: an appraisal of the autonomic hypothesis and the endocrine hypothesis. Int. J. Obesity 8 (suppl.):119.

Bray, G. A. and Y. Nishizawa. 1978. Ventromedial hypothalamus modulates fat mobilization during fasting. Nature 274:900.

Bueno, L., M. J. Fargeas and P. Julie. 1986. Effects of calcitonin and CGRP alone in or in combination of food intake and forestomach (reticulum) motility in sheep. Physiol. Behav. 36:907.

Campfield, L. A., P. Brandon and F. J. Smith. 1985. On-line continuous measurement of blood glucose and meal pattern in free-feeding rats: the role of glucose in meal initiation. Brain Res. Bull. 14:605.

Campfield, L. A. and F. J. Smith. 1985. Functional coupling between transient declines in blood glucose and feeding behavior: temporal relationships. Brain Res. Bull. 17:427.

Campfield, L. A. and F. J. Smith. 1987. Glucose dynamics during feeding predict frequency of ingestion. Fed. Proc. 46:1482.

DeBoer, J. O., L. A. Roovers, J. M. A. van Raaij and J. G. Hautvast. 1986. Adaptation of energy metabolism of overweight women to low energy intake, studied with whole body calorimeters. Amer. J. Clin. Nutr. 44:585.

Debons, A. F., I. Krimsky, H. J. Likuski, A. From and R. J. Cloutier. 1968. Gold thioglucose damage to the satiety center: inhibition in diabetes. Amer. J. Physiol. 214:652.

Debons, A. F., L. D. Zurek, C. S. Tse and S. Abrahasen. 1986. Central nervous system control of hyperphagia in hypothalamic obesity: dependence on adrenal glucocorticoids. Endocrinology 118:1678.

Della-Ferra, M. A. and C. A. Baile. 1979. Cholecystokinin octapeptide: continuous picomole injections into the cerebral ventricles of sheep suppress feeding. Science 206:471.

Della-Ferra, M. A. and C. A. Baile. 1980. CCK octapeptide injected in CSF decreases meal size and daily food intake in sheep. Peptides (Fayetteville) 1:51.

Della-Ferra, M. A. and C. A. Baile. 1984. Control of food intake in sheep. J. Anim. Sci. 59:1362.

Della-Ferra, M. A., C. A. Baile, B. S. Schneider and J. Grinker. 1981. Cholecystokinin antibody injected in cerebral ventricles stimulates feeding in sheep. Science 212:687.

Denbow, D. M. and R. D. Myers. 1982. Eating, drinking and temperature responses to intracerebroventricular cholecystokinin in the chick. Peptides (Fayetteville) 3:739.

Drewry, M. M. 1986. Developmental changes in response to overfeeding. M. S. Thesis. Univ. of Georgia, Athens.

Egli, G., W. Langhans and E. Scharrer. 1986. Selective hepatic vagotomy does not prevent compensatory feeding in response to body weight changes J. Auton. Nerv. Syst. 15:45.

Engeset, R. M. and R. C. Ritter. 1980. Intracerebroventricular 2-DG causes feeding in the absence of other signs of glucoprivation. Brian Res. 202:229.

Epstein, A. N., S. Nicolaidis and R. Miselis. 1975. The glucoprivic control of food intake and the glucostatic theory of feeding behavior. In: G. E. Magenson and F. R. Calarescu (Ed.) Neural Integration of Physiological Mechanisms and Behavior. p 148. Univ. of Toronto Press, Toronto.

Even, P. and S. Nicolaidis. 1986. Short-term control of feeding: limitation of the glucostatic theory. Brain Res. Bull. 17:621.

Everitt, B. J., T. Hokfelt, L. Terenius, K. Tatemoto, V. Mutt and M. Goldstein. 1984. Differential co-existence of neuropeptide Y (NPY)-like immunoreactivity with catecholamines in the central nervous system of the rat. Neuroscience 11:443.

Flatt, W. P., P. W. Moe, A. W. Munson and T. Cooper. 1969. Energy utilization in high producing cows. Proceedings 4th Symposium on Energy Metabolism. European Assoc. Animal Production Publ., Warsaw, Poland. Oriel Press, Newcastle upon Tyne, England.

Forbes, J. M. 1985. Similarities and differences between intake control mechanisms in pigs, chickens and ruminants. Proc. Nutr. Soc. 44:331.

Forbes, J. M. 1986. Review of theories of food intake control. In: The Voluntary Food Intake of Farm Animals. Butterworths, London, pp. 15–34.

Fox, D. G., R. L. Preston, B. Senft and R. R. Johnson. 1974. Plasma growth hormone levels and thyroid secretion rates during compensatory growth in cattle. J. Anim. Sci. 38:437.

Friedman, M. I. and J. Granneman. 1983. Food intake and peripheral factors after a recovery from insulin-induced hypoglycemia. Amer. J. Physiol. 244:R374.

Friedman, M. and E. M. Stricker. 1976. The physiological psychology of hunger: a physiological perspective. Psychol. Rev. 83:409.

Friedman, M. I. and M. G. Tordoff. 1986. Fatty acid oxidation and glucose utilization interact to control food intake in rats. Amer. J. Physiol. 251:R840.

Friedman, M. I., M. G. Tordoff and I. Ramirez. 1986. Integrated metabolic control of food intake. Brain Res. Bull. 17:855.

Girard, C. L., J. R. Seoane and J. J. Matte. 1985. Studies of the role of gamma-aminobutyric acid in the hypothalamic control of feed intake in sheep. Can. J. Physiol. Pharmacol. 63:1297.

Grandison, L. and A. Guidotti. 1977. Stimulation of food intake by muscimol and beta-endorphin. Neuropharmacology 16:533.

Harper, A. E., N. J. Benevenga and R. M. Wohlueter. 1970. Effects of ingestion of disproportionate amounts of amino acids. Physiol. Rev. 50:428.

Harris, R. B. S., T. R. Kasser and R. J. Martin. 1986. Dynamics of recovery of body composition after overfeeding, food restriction or starvation of mature female rats. J. Nutr. 116:2536.

Harris, R. B. S. and R. J. Martin. 1984a. Recovery of body weight from below "set point" in mature female rats. J. Nutr. 114:1143.

Harris, R. B. S. and R. J. Martin. 1984b. Lipostatic theory of energy balance: concepts and signals. Nutr. Behav. 1:253.

Harris, R. B. S. and R. J. Martin. 1986. Metabolic response to a specific lipid-depleting factor in parabiotic rats. Amer. J. Physiol. 250:R276.

Hervey, G. R. 1959. The effects of lesions in the hypothalamus in parabiotic rats. J. Physiol. (London) 145:336.

Himms-Hagen, J. 1985. Brown adipose tissue metabolism and thermogenesis. Rev. Nutr. 5:69.

Hoebel, B. G. 1985. Brain neurotransmitters in food and drug reward. Amer. J. Clin. Nutr. 42:113.

Houpt, T. R. 1974. Stimulation of food intake in ruminants by 2-deoxy-D-glucose and insulin. Amer. J. Physiol. 227:161.

Houpt, K. A., T. R. Houpt, and W. G. Pond 1979. The pig as a model for the study of obesity and control of food intake: a review. Yale J. Biol. Med. 53:307.

Houpt, T. R. 1984. Control of feeding in pigs. J. Anim. Sci. 59:1345.

Houpt, T. R. 1985. The physiological determinants of meal size in pigs. Proc. Nutr. Soc. 44:323.

Illingworth, P. J., R. T. Jung, P. W. Howie, P. Leslie and T. E. Isles. 1986. Diminution in energy expenditure during lactation. Brit. Med. J. 292:437.

Imaki, T., T. Shibasaki, M. Hotta, A. Masuda, H. Demura, K. Shizume and N. Ling. 1985. The satiety effect of growth hormone-releasing factor in rats. Brain Res. 340:186.

Inoue, S., G. A. Bray and Y. S. Mullen. 1978. Transplantation of pancreatic B-cells prevents development of hypothalamic obesity in rats. Amer. J. Physiol. 235:E266.

Jhanwar-Uniyal, M., C. R. Roland, and S. F. Leibowitz. 1986. Diurnal rhythm of alpha$_2$-noradrenergic receptors in the paraventricular nucleus and other brian areas: relation to circulating corticosterone and feeding behavior. Life Sci. 38:473.

Kandel, E. R. and J. W. Schwartz. 1985. Hypothalamus and limbic system II: motivation. In: E. R. Kandel and J. W. Schwartz (Ed.) Principles of Neural Science (2nd Ed.). Elsevier, Amsterdam, pp 450–460.

Kasser, T. R., R. B. S. Harris and R. J. Martin. 1985. Level of satiety: GABA and pentose shunt activities in three brain sites associated with feeding. Amer. J. Physiol. 248:R453.

Kasser, T. R., R. J. Martin, J. H. Gahagan and P. J. Wangsness. 1981. Fasting plasma hormones and metabolites in feral and domestic newborn pigs. J. Anim. Sci. 53:420.

Kennedy, G. C. 1953. The role of depot fat in the hypothalamic control of food intake in rats. Proc. R. Soc. London Ser. B 140:578.

Kennedy, G. C. 1957. The development with age of hypothalamic restraint upon appetite of the rat. J. Endocrinol. 16:9.

Knoll, J. 1985. Satietin, a blood-borne, highly selective and potent anorectic glycoprotein. Biomed. Biochim. Acta 44:317.

Krahn, D. D., A. Gosnell, A. S. Levine and J. E. Morley. 1984. Effects of calcitonin gene-related peptide on food intake. Peptides 5:861.

Kralym, F. S. and G. P. Smith. 1978. Combined pregastric and gastric stimulation by food is sufficient for normal meal size. Physiol. Behav. 21:405.

Landsberg, L. and J. B. Young. 1984. The role of the sympathoadrenal system in modulating energy expenditure. Clin. Endocrinol. Metab. 13:475.

Langhans, W., G. Egli and E. Scharrer. 1985. Regulation of food intake by hepatic oxidative metabolism. Brain Res. Bull. 15:425.

Langhans, W. and E. Scharrer. 1987. Evidence for a vagally mediated satiety signal derived from hepatic fatty acid oxidation. J. Auton Nerv. Syst. 18:13.

Larson, K. A. and D. B. Anderson. 1978. The effects of lipectomy on remaining adipose tissue depots in the Sprague–Dawley rat. Growth 42:469.

Leibowitz, S. F. 1970. Reciprocal hunger-regulating circuits involving alpha- and beta-adrenergic receptors located, respectively, in the ventromedial and lateral hypothalamus. Proc. Natl. Acad. Sci. USA 67:1063.

Leibowitz, S. F. 1978. Paraventricular nucleus: a primary site mediating adrenergic stimulation of feeding and drinking. Pharmacol. Biochem. Behav. 8:163.

Leibowitz, S. F. 1986a. Brain neurochemistry and eating behavior. In: E. Ferrari and F. Brambilla (Ed.) Disorders of Eating Behavior: A Psychoneuroendocrine Approach. pp. 65–72. Pergamon Press, Elmsford, NY.

Leibowitz, S. F. 1986b. Brain monoamines and peptides: role in the control of eating behavior. Fed. Proc. 46:1396.

Leibowitz, S. F., C. R. Rowland, L. Hor and V. Squillary. 1984. Noradrenergic feeding elicited via the paraventricular nucleus is dependent upon circulating corticosterone. Physiol. Behav. 32:857.

Le Magnen, J. 1981. The metabolic basis of dual periodicity of feeding in rats. Behav. Brain Sci. 4:561.

Le Magnen, J. 1983. Body energy balance and food intake: a neuroendocrine regulatory mechanism. Physiol. Rev. 63:314.

Le Magnen, J. 1985. Hunger. Cambridge Univ. Press, London.

Lepkovsky, S., M. K. Demick, F. Furuta, N. Snapir, R. Park, N. Narita and K. Komatsu. 1967. Response of blood glucose and plasma free fatty acids to fasting and injection of insulin and testosterone in chickens. Endocrinology 81:1001.

Leung, P. M. B. and Q. R. Rogers. 1969. Food intake: regulation by plasma amino acid pattern. Life Sci. 8:1.

Levine, A. S. and J. E. Morley. 1981. Reduction in feeding in rats by calcitonin. Brain Res. 222:187.

Levine, A. S. and J. E. Morley. 1984. Neuropeptide Y: a potent inducer of consummatory behavior in rats. Peptides 5:1025.

Lichtenstein, S. S., C. Marinescu and S. F. Leibowitz. 1984. Chronic infusion of norepinephrine and clonidine into the hypothalamic paraventricular nucleus. Brain Res. Bull. 13:591.

Lorden, J. F., R. Dawson and M. Callahan. 1979. Effect of gold thioglucose lesions on central catecholamine levels in the mouse. Pharmacol. Biochem. Behav. 10:165.

Louis-Sylvestre, J. and J. Le Magnen. 1980. A fall in blood glucose precedes meal onset in free feeding rats. Neurosci. Biobehav. Res. 4 (Suppl. 1):13.

McBride, B. W. and L. P. Milligan. 1985. Influence of feed intake and starvation on the magnitude of Na K-ATPase dependent respiration in duodenal mucosa of sheep. Brit. J. Nutr. 53:605.

McCaleb, M. L. and R. D. Myers. 1982. 2-Deoxy-D-glucose and insulin modify release of norepinephrine from rat hypothalamus. Amer. J. Physiol. 242:R596.

McCaleb, M. L., R. D. Myers, G. Singer and G. Willis. 1979. Hypothalamic norepinephrine in the rat during feeding and push–pull perfusion with glucose, 2-DG, or insulin. Amer. J. Physiol. 236:R313.

McHugh, P. R. and T. H. Muran. 1978. The accuracy of the regulation of caloric ingestion in the rhesus monkey. Amer. J. Physiol. 235:R29.

McHugh, P. R. and T. H. Muran. 1979. Calories and gastric emptying: a regulatory capacity with implications for feeding. Amer. J. Physiol. 236:R254.

Marks-Kaufman, R. 1982. Increased fat consumption induced by morphine administration in rats. Pharmacol. Biochem. Behav. 16:949.

Marshall, N. B. and J. Mayer. 1954. Energy balance in gold thioglucose obesity. Amer. J. Physiol. 178:271.

Martin, R. J. and P. Lamprey. 1974. Changes in liver and adipose tissue enzymes and lipogenic activities during the onset of hypothalamic obesity in mice. Life Sci. 14:1121.

Maurice, D. V., J. E. Whisenhunt, J. E. Jones and K. D. Smoak. 1983. Effect of lipectomy on control of feed intake and homeostasis of adipose tissue in chickens. Poult. Sci. 62:1466.

Mayer, J. 1953. Glucostatic mechanism of regulation of food intake. N. Engl. J. Med. 249:13.

Meier, A. H. and D. D. Martin. 1971. Temporal synergism of corticosterone and proclactin controlling fat storage in white throated sparrow, Zonotrichia albicollis. Gen. Comp. Biochem. 17:311.

Milligan, L. P. and B. W. McBride. 1985. Shifts in animal requirements across physiological and alimentation states. J. Nutr. 115:1374.

Miselis, R. R. and A. N. Epstein. 1975. Feeding induced by intracerebroventricular 2-deoxy-D-glucose in the rat. Amer. J. Physiol. 229:1438.

Moe, P. W., H. F. Tyrrell and W. P. Flatt. 1971. Energetics of body tissue mobilization. J. Dairy Sci. 54:548.

Morley, J. E., T. J. Bartness, B. A. Gosnell and A. S. Levine. 1985. Peptidergic regulation of feeding. Int. Rev. Neurobiol. 27:207.

Morley, J. E. and A. S. Levine. 1985. The pharmacology of eating behavior. Annu. Rev. Pharmacol. Toxicol. 25:127.

Morley, J. E., A. S. Levine, M. Grace and J. Kneip. 1985. Peptide YY (PYY), a potent orexigenic agent. Brain Res. 341:200.

Morley, J. E. and J. E. Mitchell. 1986. Neurotransmitter/neuromodulator influences on eating. In: E. Ferrari and F. Brambilla (Ed.) Disorders of Eating Behavior: A Psychoneuroendocrine Approach. pp 11–20. Pergamon Press, Elmsford, NY.

Musten, B., D. Peace and G. H. Anderson. 1974. Food intake regulation in the weanling rat: self-selection of protein and energy. J. Nutr. 104:563.

Myers, R. D., H. S. Swortzwelder, J. M. Peinado, T. F. Lee, J. R. Helper, D. M. Denbow and J. M. R. Ferrer. 1986. CCK and other peptides modulate hypothalamic norepinephrine release in the rat: dependence on hunger and satiety. Brain Res. Bull. 17:583.

Nicolaidis, S. 1981. Lateral hypothalamic control of metabolic factors related to feeding. Diabetologia 20:426.

Niijima, A. 1977. Nervous regulatory mechanism of blood glucose levels. In: Y. Katsuki, M. Sato, S. F. Takagi and Y. Oomura (Ed.) Food Intake and Chemical Senses. pp 413–426. Japan Sci. Soc. Press, Tokyo.

Novin, D., D. A. VanderWeele and M. Rezek. 1973. Infusion of 2-deoxy-D-glucose into the hepatic portal system causes eating: evidence for peripheral glucoreceptors. Science 181:858.

Nzuzi, L. and S. Ritter. 1985. Persistence of glycogen depletion in caudate and hindbrain parallels presistence of feeding after insulin-induced glucoprivation. Soc. Neurosci. Abstr. 11:343.

O'Donohue, T. L., B. M. Chronwall, R. M. Pruss, E. Mezey, J. Z. Kiss, L. E. Eiden, V. J. Nassari, R. E. Tessel, V. M. Pickel, D. A. Dimaggio, A. J. Hotchkiss, W. R. Crowley and Z. Zukowska-Grojec. 1985. Neuropeptide Y and peptide YY neuronal and endocrine systems. Peptides 6:755.

Ono, T., Y. Oomura, H. Nishino, K. Sasaki, M. Fukuda and K. Muramoto. 1981. Neural mechanisms of feeding behavior. In: Y Katsuki, R. Norgren and M. Sato (Ed.) Brain Mechanisms of Sensation. pp 271–286. John Wiley and Sons, New York.

Oomura, Y. 1980. Input–output organization in the hypothalamus relating to food intake behavior. In: P. J. Morgane and J. Panksepp (Ed.) Handbook of the Hypothalamus. Vol. II, p 557. Marcel Dekker, New York.

Oomura, Y. and H. Kita. 1981. Insulin acting as a modulator of feeding through the hypothalamus. Diabetologia 20:290.

Oomura, Y. and A.Niijima. 1983. Chemosensitive neurons and neural control of pancreatic secretion. In: E. N. Mngola (Ed.) Diabetes 1982, Proceedings of the 11th Congress of the International Diabetes Foundation. pp 201–211. Excerpta Medica, Amsterdam.

Paraneswaran, S. V., A. B. Steffens, G. R. Hervey and L. deRuiter. 1977. Involvement of a humoral factor in regulation of body weight in parabiotic rats. Amer. J. Physiol. 232:R150.

Pardridge, W. M. 1979. The role of blood–brain barrier transport of tryptophan and other neutral amino acids in the regulation of substrate-limited pathways of brain amino acid metabolism. J. Neural Transm. Suppl. 15:43.

Parrott, R. F. and B. A. Baldwin. 1981. Operant feeding and drinking in pigs following intracerebroventricular injection of synthetic cholecystokinin octapeptide. Physiol. Behav. 26:419.

Parrott, R. F., R. P. Heavens and B. A. Baldwin. 1986. Stimulation of feeding in the satiated pig by intracerebroventricular injection of neuropeptide Y. Physiol. Behav. 36:523.

Porte, D., Jr. and S. C. Woods. 1981. Regulation of food intake and body weight by insulin. Diabetologia 20:274.

Powley, T. L. 1977. The ventromedial hypothalamic syndrome, satiety, and a cephalic phase hypothesis. Psychol. Rev. 84:89.

Ritter, R. C., M. Roelke and M. Neville. 1978. Glucoprivic feeding behavior in absence of other signs of glucoprivation. Amer. J. Physiol. 234:E617.

Ritter, R. C., P. G. Slusser and S. Stone. 1981. Glucoreceptors controlling feeding and blood glucose: location in the hindbrain. Science 213:451.

Riviere, P. and L. Bueno. 1987. Influence of regimen and insulinemia on orexigenic effects of GRF_{1-44} in sheep. Physiol. Behav. 39:347.

Robinson, D. W. 1974. Food intake regulation in pigs. III. Voluntary food selection between protein-free and protein-rich diets. Brit. J. Nutr. 130:522.

Rogers, Q. R. and P. M. B. Leung. 1973. The influence of amino acids on the neuroregulation of food intake. Fed. Proc. 32:1709.

Rogers, Q. R. and P. M. B. Leung. 1977. The control of food intake: when and how are amino acids involved? In: M. R. Kare and O. Miller (Ed.) The Chemical Senses and Nutrition. pp 213–249. Academic Press, New York.

Rothwell, N. J. and M. J. Stock. 1982. Effects of chronic restriction on energy balance, thermogenic capacity and brown adipose tissue activity in the rat. Biosci. Rep. 2:543.

Ruckebusch, Y. and C. H. Malbert. 1986. Stimulation and inhibition of food intake in sheep by centrally-administered hypothalamic releasing factors. Life Sci. 38:929.

Russek, M. 1963. An hypothesis on the participation of hepatic glucoreceptors in the control of food intake. Nature 197:79.

Shimazu, T. 1981. Central nervous system regulation of liver and adipose tissue metabolism. Diabetologia 20:343.

Shimazu, T. 1986. Neuronal control of intermediary metabolism. In: S. Lightman and B. Everitt (Ed.) Neuroendocrinology. p. 304. Blackwell Scientific Publ., Oxford.

Shimazu, T., M. Noma and M. Y. Saito. 1986. Chronic infusion of norepinpherine into the ventromedial hypothalamus induces obesity in rats. Brain Res. 369:215.

Skewes, P. A., D. M. Denbow, P. M. Lacy and H. P. Van Krey. 1986. Alteration of food intake following intracerebroventricular administration of plasma from free-feeding domestic fowl. Physiol. Behav. 36:295.

Stanley, B. G., D. R. Daniel, A. S. Chin and S. F. Leibowitz. 1985. Paraventricular nucleus injections of peptide YY and neuropeptide Y preferentially enhance carbohydrate ingestion. Peptides 6:1205.

Stanley, B. G. and S. F. Leibowitz. 1985. Neuropeptide Y injected in the paraventricular hypothalamus: a powerful stimulant of feeding behavior. Proc. Natl. Acad. Sci. USA 82:3940.

Steffens, A. B., G. J. Mogenson and J. A. F. Stevenson. 1972. Blood glucose, insulin and free fatty acids after stimulation and lesions of the hypothalamus. Amer. J. Physiol. 222:1446.

Steffens, A. B., A. J. Scheurink and P. G. M. Luiten. 1985. Interference of the nutritional condition of the rat with peripheral glucose regulation determined by CNS mechanisms. Physiol. Behav. 35:405.

Stellar. E. 1954. The physiology of motivation. Psychol. Rev. 61:5.

Swanson, L. W. 1986. Organization of mammalian neuroendocrine system. In: F. E. Bloom (Ed.) Handbook of Physiology. Sect. 1, Vol. IV, pp 317–363. American Physiological Society, Bethesda.

Tanikawa, K. 1968. Ultrastructural Aspects of the Liver and Its Disorders. pp. 50–55. Igaku Shoin, Tokyo.

Tatemoto, K. 1982. Isolation and characterization of peptide YY (PYY), a candidate gut hormone that inhibits pancreatic exocrine secretion. Proc. Natl. Acad. Sci. USA 79:2514.

Trayhurn, P., J. B. Douglas and M. M. McGuckin. 1982. Brown adipose tissue is suppressed during lactation in mice. Nature 298:59.

Truett, G. E. 1985. Hypothalamic glucose metabolism in lactating rats. M. S. Thesis. Univ. of Georgia, Athens.

Vaccarino, F. J., F. E. Bloom, J. Rivier, W. Vale and G. F. Koob. 1985. Stimulation of food intake in rats by centrally administered growth hormone-releasing factor. Nature 314:167.

Vandermeerschen-Doize, F. and R. Paquay. 1984. Effects of continuous long-term intravenous infusion of long-chain fatty acids on feeding behavior and blood components of adult sheep. Appetite 5:137.

Wang, G. H. 1925. The changes in the amont of daily food intake of the albino rat during pregnancy and lactation. Amer. J. Physiol. 71:736.

Wangsness, P. J. and G. H. Soroka. 1979. Effect of energy concentration of milk on voluntary food intake of lean and obese piglets. J. Nutr. 108:595.

Weiss, B. and R. P. Maickel. 1968. Sympathetic nervous control of adipose tissue lipolysis. Int. J. Neuropharmacol. 7:395.

Wheeland, R. A., R. J. Martin and A. H. Meier. 1976. The effect of prolactin and CB154 on in vivo lipogenesis in the Japanese quail, and of photostimulation on enzyme patterns in the white throated sparrow. Comp. Biochem. Physiol. 53B:379.

Wilson, P. N. and D. R. Osbourn. 1960. Compensatory growth after undernutrition in animals and birds. Biol. Rev. Cambridge Philos. Soc. 35:324.

Woods, S. C. and D. Porte, Jr. 1974. Neural control of the endocrine pancreas. Physiol. Rev. 54:596.

Woods, S. C., G. J. Taborsky, Jr. and D. Porte, Jr. 1986. Central nervous system control of nutrient homeostasis. In: F. E. Bloom (Ed.) Handbook of Physiology. Sect. 1, Vol. IV, pp 365–411. American Physiological Society, Bethesda.

Wurtman, R. J. and J. P. Fernstrom. 1976. Commentary: control of brain neurotransmitter synthesis by precursor availability and nutritional status. Biochem. Pharmacol. 25:1691.

Yoshimatsu, H., A. Niijima, Y. Oomura, K. Yamabe and T. Katufuchi. 1984. Effects of hypothalamic lesion on pancreatic nerve activity in the rat. Brain Res. 303:147.

Young, J. B., E. Saville, N. Rothwell, J. Stock and M. J. Landsberg. 1982. Effect of diet and cold on norepinephrine turnover in brown adipose tissue of the rat. J. Clin. Invest. 69:1061.

Young, W. S., III and M. J. Kuhar. 1980. Noradrenergic $alpha_1$ and $alpha_2$ receptors: light microscopic autoradiographic localization. Proc. Natl. Acad. Sci. USA 77:1696.

Central Regulation of Growth Hormone Secretion

WILLIAM J. MILLARD

1. Introduction

It is well recognized that body growth in mammals is modulated by a combination of genetic, nutritional, and endocrine factors. Any alteration in any of these parameters will result in varying degrees of abnormal growth. The endocrine system, by virtue of the controlled release of a host of hormonal agents, has a major impact on both growth and development of an organism. Principal among the hormones associated with growth is growth hormone (GH).

GH is a single-chained polypeptide of approximately 21 to 22 kDa. Although the structure of GH is species specific, there is some degree of conservatism within the molecule. GH is released from the anterior pituitary gland from specialized cells called somatotropes. The effects of GH on growth are varied with both direct and indirect effects (via the release of lower-molecular-weight substances termed somatomedins) being observed. It is beyond the scope of this chapter to review the biological effects of GH on growth. Interested readers are referred to Spencer (this volume) and the review by Isaksson *et al.*, 1985.

The central regulation of GH secretion is the major focus of this chapter. The patterns of GH secretion will be discussed initially followed by the hypothalamic and neuropharmocological mechanisms involved in GH control. The chapter will conclude with a review of the feedback processes regulating GH secretion. Although a vast majority of investigations of GH regulation have

WILLIAM J. MILLARD • Department of Pharmacodynamics, College of Pharmacy, University of Florida, Gainesville, Florida 32610.

utilized the rat as the experimental model, a consideraton of the neural regulation of GH secretion in other animals including domestic farm animals will be presented in this chapter whenever possible.

2. Patterns of GH Secretion

Utilizing sequential blood sampling techniques, GH has been found to be released from the anterior pituitary in an intermittent, pulsatile fashion (Martin and Millard, 1986; Millard *et al.*, 1987a, b). Episodic GH secretion has been characterized in humans (Finklestein *et al.*, 1972) and a wide variety of animals including baboons (Steiner *et al.*, 1978), monkeys (Quabbe *et al.*, 1981), sheep (Davis *et al.*, 1977; Klindt *et al.*, 1985), cattle (Anfinson *et al.*, 1975), goats (Hart and Buttle, 1975), pigs (Klindt *et al.*, 1983; Molina *et al.*, 1986), and rats (Jansson *et al.*, 1985; Millard *et al.*, 1987a, b).

Studies in fetal sheep have demonstrated that GH secretion is pulsatile as early as day 110 of gestation, indicating that GH secretion in mammals is pulsatile even in the very early stages of development (Gluckman and Parsons, 1985). In rats, episodic GH secretion is observed prior to puberty, between 25 and 30 days of age, is maximal during early adulthood, and decreases with age (Sonntag *et al.*, 1980; Jansson *et al.*, 1985), similar to reports in humans (Finklestein *et al.*, 1972) and sheep (Klindt *et al.*, 1985).

In adult rats, plasma GH levels fluctuate dramatically throughout the day, with males displaying GH secretory patterns that are distinct from those of females (Jansson *et al.*, 1985; Millard *et al.*, 1987a, b). Male animals exhibit a low-frequency, high-amplitude pattern of GH secretion (bottom panel, Fig. 1). Individual bursts of GH occur every 3–4 hr and are separated by prolonged trough periods during which plasma GH levels remain very low. Female rats, on the other hand, display a high-frequency, low-amplitude GH secretory profile with GH pulses occurring every 1–2 hr (top panel, Fig. 1). Individual GH pulse amplitudes are lower than those of males; the GH trough periods are shortened and contain higher GH levels. The physiological significance of episodic GH secretion has not been clearly defined. However, it is apparent in rodents that differences in GH output between males and females are critical factors in determining the sex differences in adult body size and body length (Isaksson *et al.*, 1985; Jansson *et al.*, 1985). Whether this holds true for other mammalian species has not been clarified.

It is now known that these surges of GH secretion are the result of a complex interaction of neural influences, mediated by specific excitatory and inhibitory peptides released from the hypothalamus. This hypothalamic output is, in turn, regulated by neural connections from extrahypothalamic regions of the brain (see below). Thus, moment-to-moment GH secretion is controlled by a complex interaction of a number of neuropeptides and neurotransmitters.

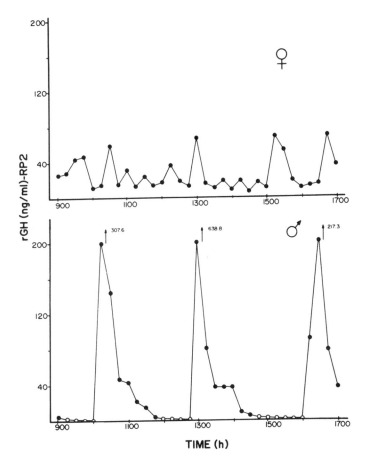

FIGURE 1. Sexually dimorphic GH secretory profiles in adult male and female rats. Animals were studied at 90 days of age with blood samples being taken every 15 min for 8 hr. Note the high-frequency, low-amplitude profile in females and the low-frequency, high-amplitude pattern in males.

3. Hypothalamic Peptides Involved in GH Regulation

Neural regulation of GH secretion is accomplished principally by two hypothalamic peptides: somatostatin, a peptide that inhibits GH release, and GH-releasing factor (GRF), a peptide that stimulates GH secretion. Each peptide is derived from specific neural centers within the hypothalamus.

3.1. Somatostatin

Somatostatin, first isolated from sheep hypothalami (Brazeau *et al.*, 1973), has been found to inhibit GH secretion in all species tested (Reichlin, 1983;

Martin and Millard, 1986). The inhibitory action of somatostatin is not confined to GH, since the peptide also inhibits basal thyrotropin (TSH) secretion from the pituitary and both glucagon and insulin release from the pancreas (Reichlin, 1983; Patel, 1987).

Initial studies indicated that somatostatin was comprised of 14 amino acids and that its complete structure was conserved across all mammalian species. Today it is known that somatostatin is, in fact, one member of a multigene family of peptides. The gene for somatostatin is localized on chromosome 3 in humans (Naylor *et al.*, 1983). Somatostatin, like the vast majority of all other biologically active neuropeptides, is the result of sequential, programmed processing from high-molecular-weight precursor peptides called prohormones.

To date there are six known peptides derived from a 92-amino-acid prosomatostatin molecule (Goodman *et al.*, 1983; Patel, 1987). Two peptides with demonstrated biological activity are the 14-amino-acid form —S-14 and a 28-amino-acid form (S-28). These peptides are located at the COOH terminus of the prosomatostatin molecule. S-28 is an NH_2-terminal extension of S-14 and appears more potent than S-14 in its ability to inhibit GH secretion (Reichlin, 1983; Patel, 1987). Further, the biological activity of somatostatin resides in the ring structure since breakage of the disulfide bond results in complete loss of activity. The other four peptides of prosomatostatin (9K, 8K, 5K, and S-$28_{[1-12]}$) have unknown biological functions (Fig. 2).

3.2. GH-Releasing Factor (GRF)

GRF is the first example of the structure of a hypophysiotropic hormone being isolated from tissue other than neural tissue. The original clue to the structure of GRF came from two human pancreatic tumors that gave rise to acromegaly. Two peptides, one containing 44 amino acids and the other 40 amino acids, with GH-releasing activity were subsequently isolated and se-

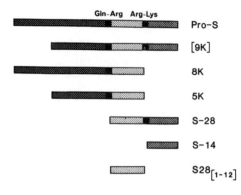

FIGURE 2. Molecular heterogeneity of the prosomatostatin molecule. Six known peptides have been derived from prosomatostatin (pro-S). Somatostatin-28 (S-28) and somatostatin-14 (S-14) occupy the COOH terminus. The other four peptides—9K, 8K, 5K, and S28$_{[1-12]}$—have no known biological activity. From Patel (1987), with permission.

quenced from these pancreatic tumors (Guillemin *et al.*, 1982; River *et al.*, 1983). The two peptides have identical sequences except at the COOH terminus where the 44-amino-acid peptide has four more amino acid residues and possesses an amidated COOH terminus [GRF(1–44)-NH$_2$]. The COOH terminus is nonamidated in the 40-amino-acid peptide [GRF(1–40)-OH]. Subsequently, both of these forms of pancreatic GRF have been isolated from human hypothalamic issue (Frohman and Jansson, 1986).

In addition to human GRF, the peptide has been isolated and sequenced from the hypothalami of a number of other mammalian species (Fig. 3). GRF peptides from domestic farm animals exhibit nearly complete homology with human GRF (Frohman and Jansson, 1986). Rat GRF, a 43-amino-acid peptide, shows 67% homology with the corresponding NH$_2$-terminal 43 residues in human GRF.

Like somatostatin, GRF is a product of programmed processing from a higher-molecular-weight precursor peptide (Gubler *et al.*, 1983; Frohman and Jansson, 1986). Two precursor GRF polypeptides (107 and 108 amino acid residues) have been isolated by molecular cloning techniques. GRF is inserted in the middle of the precursor peptide with its sequence extending from amino acid residue 32 to residue 75. The flanking signal peptides in the GRF precursor moiety have no known biological function. Moreover, restriction analysis of the genomic DNA suggests that there is a single GRF gene (located on human chromosome 20) that encodes for both the pancreatic tumor and hypothalamic GRF peptides in humans (Frohman and Jansson, 1986).

GRF increases GH secretion in a dose-dependent manner in humans and a wide variety of animal species (Frohman and Jansson, 1986). Intravenous administration of both forms of human GRF [GRF(1–44)-NH$_2$ and GRF(1–40)-OH] and biologically active fragments of GRF has been found to stimulate GH secretion in cattle (Moseley *et al.*, 1984; Enright *et al.*, 1987). Further, the administration of GRF to goats (Hart *et al.*, 1984), sheep (Hart *et al.*, 1985a,b), and poultry (Scanes and Harvey, 1984) has stimulated GH secretion. In rats, both human GRF forms and rat GRF have been found to cause a dose-dependent rise in GH *in vitro* and *in vivo* (Badger *et al.*, 1984; McCormick *et al.*, 1985; Frohman and Jansson, 1986). The effects of GRF at the pituitary are inhibited by somatostatin and modulated to some extent by gonadal steroids (see below). However, GRF does not interfere with the binding of somatostatin to receptors (Frohman and Jansson, 1986; Wehrenberg *et al.*, 1986), suggesting that GRF and somatostatin interact on different receptors at the anterior pituitary.

3.3. Other Peptides That Increase GH Secretion

Comparison of the structure of GRF with other biologically active peptides reveals its close homology with the glucagon–secretion family (Frohman and

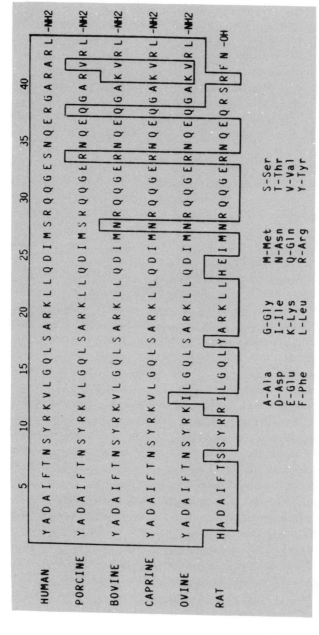

FIGURE 3. Comparison of the amino acid makeup of GRF peptides from various mammalian species. Except for the rat, note the high degree of homology in each GRF molecule compared to human GRF. Single-letter abbreviations for the amino acids are indicated at the bottom of the figure. From Frohman and Jansson (1986), with permission.

Table I
Peptides with GH-Releasing
Activity

TRH	Vasopressin
LHRH	α-MSH
Substance P	Opioid peptides
Neurotensin	β-Endorphin
Galanin	Enkephalins

Jansson, 1986; Martin and Millard, 1986). This family of peptides include glucagon, secretin, vasoactive intestinal polypeptide, and gastric inhibitory peptide. Each of these peptides structurally related to GRF possesses some intrinsic GRF activity but lacks the potency of GRF.

A number of other biologically active peptides have been found to stimulate GH secretion (Table I). Although each peptide does have some GRF activity in one or another physiological test system, their potencies do not rival that of GRF and may only be of significance in pathological conditions (Martin and Millard, 1986).

Noteworthy among the peptides listed in Table I are the opioid peptides and galanin. The opioid peptides (β-endorphin and enkephalins) and the opiate derivative morphine have been found to be potent stimulators of GH secretion in humans and rats (Martin and Millard, 1986). However, the action of these compounds does not occur directly on the somatotrope but primarily through GRF. This is evidenced by the blockade of opiate-induced GH secretion in animals passively immunized with antiserum to GRF (Wehrenberg et al., 1986).

Gluckman et al. (1980) found that intravenous administration of β-endorphin to fetal sheep increased plasma GH levels and that the increase was more pronounced in younger fetuses. Thus, there appears to be an opiate-mediated facilitatory mechanism for GH in ruminants, but it is not known whether the response occurs through the release of GRF.

Galanin, because it is localized in the hypothalamic arcuate nucleus in close proximity to GRF neurons (Everitt et al., 1986), has stimulated interest as a possible modulatory factor in the regulation of GH secretion. The central administration of galanin to conscious adult rats has induced GH secretion (Ottlecz et al., 1986; Murakami et al., 1987a). However, the effects of this peptide appear to be mediated via GRF since galanin has been found ineffective in eliciting GH secretion in cultured pituitary cells and its action is blocked in vivo by passive immunization with antiserum to GRF.

Recently, several synthetic peptides with structural similarity to the enkephalins have been found to possess GH-releasing activity in vivo and in vitro in rats (Badger et al., 1984; McCormick et al., 1985). These peptides are ineffective in stimulating GH secretion in goats *in vivo* but do show some GH-releasing activity in cultured goat pituitary cells (Hart *et al.*, 1984). The physiological significance of these synthetic peptides on GH regulation is unclear.

4. Hypothalamic Regions Involved in GH Regulation

Studies, using destructive focal lesions and immunocytochemical techniques, have identified two functional hypothalamic regions involved in modulating rhythmic GH secretion (Fig. 4). The first region, the anterior periventricular/medial preoptic area, contains somatostatin neuronal perikarya whose axons initially course laterally and then turn caudally and medially to terminate

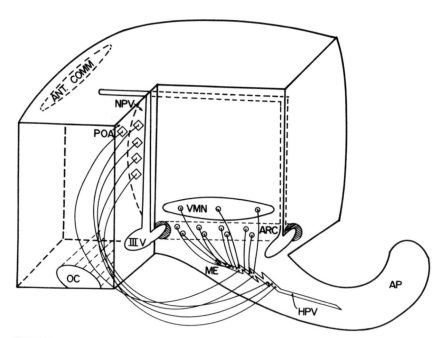

FIGURE 4. Schematic of the hypothalamic–pituitary axis illustrating the centers of cell bodies for somatostatin and GRF and their fiber tracts to the median eminence. Diamond-shaped neurons represent the somatostatinergic cell bodies located in the periventricular nucleus (NPV) and medial preoptic area (POA); circles denote GRF cell bodies located in the arcuate (ARC) and ventromedial nuclei (VMN). Note the course of somatostatin fibers, which initially course laterally, then turn caudally and medially to terminate in the external zone of the median eminence (ME). ANT. COMM, anterior commissure; AP, anterior pituitary; HPV, hypophyseal-portal vessels; OC, optic chiasm; III V, third ventricle.

in the external layer of the median eminence (Makara *et al.*, 1983; Reichlin, 1983; Martin and Millard, 1986). Somatostatin is released from these nerve terminals directly into hypophyseal portal vessels where the peptide is carried to the pituitary to affect GH secretion.

Discrete lesions of this hypothalamic somatostatin-containing region result in an increase in plasma GH levels in rats and a greater than 80% reduction of median eminence somatostatin levels (Epelbaum *et al.*, 1977; Martin and Millard, 1986). In rats with complete hypothalamic deafferentation (a technique that transects the somatostatin fibers), periodic rises of plasma GH persist but basal GH levels are elevated and the GH pulse frequency is increased (Willoughby *et al.*, 1977a). However, the effect of hypothalamic deafferentation on GH secretion varies with the animal tested. Molina *et al.* (1986) recently reported that pulsatile GH secretion is reduced in prepubertal pigs after hypothalamic deafferentation.

The arcuate and ventromedial nuclei (ARC–VMN) comprise the second hypothalamic region involved in GH control (Fig. 4). A number of groups have utilized immunohistochemical techniques to localize GRF-immunoreactive neurons and fiber tracts in the mammalian brain. Bloch *et al.*, (1983) demonstrated specific staining of GRF cell bodies in the ARC region of the hypothalamus in both the human and monkey; with GRF fibers projecting to the median eminence. As in the primate, clusters of GRF cell bodies have been observed in the ARC nucleus of the rat with a rich innervation of GRF fibers in the external zone of the median eminence (Jacobowitz *et al.*, 1983).

Electrical stimulation of these same nuclei elicits GH secretion, whereas both electrolytic and monosodium glutamate-induced lesions of the ARC–VMN region suppress spontaneous GH secretion in the rat (Tannenbaum and Ling, 1984; Martin and Millard, 1986; Tannenbaum, 1987). GH secretion is also suppressed by lesioning of the ARC–VMN in both fetal (Gluckman and Parsons, 1985) and adult (Wroblewska and Domanski, 1981) sheep.

Further evidence for a role of the hypothalamus in regulating GH secretion in domestic farm animals is the observation that transections of the median eminence in pigs (Klindt *et al.*, 1983) and lesions of the median eminence in fetal sheep (Gluckman and Parsons, 1985) reduce episodic GH secretion.

5. Interaction of Somatostatin and GRF in Episodic GH Secretion

The availability of specific antisera against both somatostatin and GRF has provided a method to study the function of GRF and somatostatin in the regulation of GH secretion. Passive immunization with antisomatostatin serum in both the baboon and rat causes an elevation in basal (trough) plasma GH levels without blunting GH pulse amplitudes (Steiner *et al.*, 1978; Terry and Martin,

1981), whereas administration of GRF antiserum abolishes spontaneous GH pulses (Wehrenberg *et al.*, 1986; Frohman and Jansson, 1986).

More direct evidence of the phasic interaction of GRF and somatostatin to produce the pulsatile pattern of GH secretion has been found by Plotsky and Vale (1985). These investigators demonstrated, through direct measurements of the peptides in portal blood, that GRF concentrations were maximal during GH secretory episodes. Portal levels of somatostatin showed moderate reductions during peak GRF surges. Thus, it appears that each GH secretory episode is initiated by the concurrent release of GRF and inhibition of somatostatin release from the hypothalamus.

Further, work from Tannenbaum's laboratory (Tannenbaum and Ling, 1984; Tannenbaum, 1987) has demonstrated that the intravenous administration of GRF during expected GH peaks (when portal somatostatin levels are lowest) causes a more marked response than during GH trough periods (when somatostatin levels are highest). This work has led to the proposal that GRF and somatostatin are tonically released into the portal vasculature, but that superimposed upon this tonic release is a rhythmic surge of each peptide in a "predator–prey" relationship (see Fig. 5). Taken together, these data suggest that somatostatin is responsible for the low trough levels and GRF is responsible for the temporal fluctuations in spontaneous GH secretion.

6. Neuropharmacological Regulation of GH Secretion

For full integration of episodic GH secretion, the timed release of somatostatin and GRF is modulated by putative neurotransmitters whose neuronal cell bodies are generally located outside the hypothalamus. There is evidence that converging neuronal afferents may act to either stimulate or inhibit these peptidergic neurosecretory cells. The neuropharmacological regulation of GH secretion has been extensively studied and a number of comprehensive reviews have been written (Martin *et al.*, 1978a; Martin and Millard, 1986).

The following section will only summarize the current views on neurotransmitter regulation of GH secretion with a focus on the major and most extensively studied putative neurotransmitters: norepinephrine, dopamine, and serotonin (5-hydroxytryptamine). Readers wishing a more detailed review of this field are referred to the comprehensive reviews listed above.

6.1. Norepinephrine (NE)

Of the neurotransmitter systems involved in GH regulation, there is little doubt that the NE systems are the most important. Neuronal perikarya that give rise to hypothalamic NE are located in the brain stem and traverse the brain to

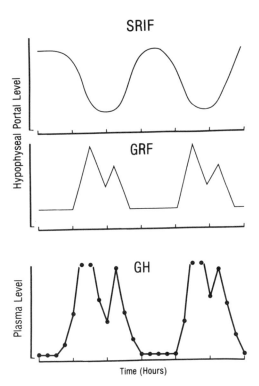

FIGURE 5. Schematic interrelationship of phasic somatostatin (SRIF) and GRF release in determining episodic GH secretion into the blood. Note that each peptide is 180° out of phase with each other in a predator–prey relationship. From Tannenbaum and Ling (1984), with permission.

enter the hypothalamus primarily through the ventral noradrenergic bundle (Moore and Bloom, 1979).

Through the use of specific NE precursors, NE depleting agents and synthesis inhibitors, and α-adrenoreceptor agonists and antagonists, a number of investigators have demonstrated that NE is involved largely in stimulating GH secretion in a wide variety of mammalian species including humans, monkeys, dogs, rabbits, and rats (Martin and Millard 1986). NE precursors and α-adrenoreceptor agonists increase GH secretion, whereas reduction in NE activity by NE depletion and synthesis blockade or α-adrenoreceptor antagonists cause a reduction in GH secretion (Martin et al., 1978a; Krulich et al., 1982).

The majority of experimental evidence indicates that the stimulatory effect of NE on GH secretion is the result of stimulation of GRF release from the hypothalamus. This is evidenced by the fact that clonidine-induced GH secretion is abolished in rats and rabbits passively immunized with GRF antiserum (Wehrenberg et al., 1986 Frohman and Jansson, 1986).

NE regulation of GH secretion in agricultural animals has not been thoroughly investigated. The infusion of L-dopa (a precursor to NE and dopamine) did not influence GH secretion in mature cycling ewes, but did block arginine-induced GH secretion; whereas the adrenoreceptor antagonist phenoxybenzamine resulted in an increase in circulating levels of GH (Davis and Borger, 1973). Further, the administration of reserpine, a drug that depletes central catecholamine and indoleamine stores, did not alter GH secretion in cattle (Bauman *et al.*, 1977). Taken together, these data indicate that adrenergic mechanisms lower or have no effect on GH secretion in ruminants; an effect opposite to that found in all other mammalian species tested.

The role of β-adrenergic mechanisms in regulating GH secretion is unclear and is probably of only minor importance. The infusion of the β-adrenoreceptor antagonist propranolol increases GH release in humans and baboons (Martin and Millard, 1986) but does not effect pulsatile GH secretion in rats (Martin *et al.*, 1978b). These data indicate that β-adrenergic control is either inhibitory or has no role in GH regulation.

6.2. Dopamine (DA)

The major input of DA within the hypothalamus arises from cell bodies located in the ARC and NPV, termed the tuberoinfundibular DA pathway (Bjorklund *et al.*, 1973). Cell bodies from these nuclei send axons to terminate in the external zone of the median eminence.

In humans the administration of the DA receptor agonists apomorphine and nomifensine has been associated with a rise in GH secretion (Martin and Millard, 1986). This rise was blocked by prior administration of DA receptor antagonists, indicating the specificity of the response.

The role DA plays in modulating GH secretion in the rat is unclear. In the unanesthetized rat, administration of the potent DA receptor blocking agent butaclamol did not alter spontaneous GH secretion (Willoughby *et al.*, 1977b). Further, the inhibition of episodic GH secretion by prior catecholamine synthesis blockade was not restored by apomorphine administration (Martin *et al.*, 1978a). Thus, it appears that there is no tonic regulation of GH secretion by DA in the rodent. The role DA plays in modulating GH secretion in agricultural animals has not been investigated.

6.3. Serotonin (5-HT)

The major serotonergic input to the hypothalamus arises from cell bodies located in the brain stem raphe nuclei (Van De Kar and Lorens, 1979). Considerable evidence has accumulated to support a facilitatory role of 5-HT on GH secretion in most mammalian species. The administration of the 5-HT precursors tryptophan or 5-hydroxytryptophan (5-HTP) has been reported to increase

GH secretion in humans and rats (Martin, *et al.*, 1978a,b). Marti-Henneberg *et al.* (1980) demonstrated further that the administration of the 5-HT reuptake blocker fluoxetine induced a significant rise in GH secretion in fetal sheep. However, there was a decreased responsiveness to the agent with advancing fetal age. The authors proposed that this decreased responsiveness to serotonergic agents coincides with the maturation of the inhibitory regulation of the GH secretory axis.

The use of 5-HT receptor antagonists cyproheptadine or methysergide has demonstrated a reduction in GH secretion in humans and rats (Arnold and Fernstrom, 1978; Martin and Millard, 1986). There is evidence that this reduction in GH secretion induced by serotonergic antagonists may be the result of a stimulation of somatostatin release from the hypothalamus. Arnold and Fernstrom (1980) found that the reduction in GH secretion observed after administration of the 5-HT receptor antagonist metergoline was reversed by prior administration of antiserum to somatostatin, indicating that 5-HT mechanisms normally act to inhibit somatostatin release.

However, there is also evidence that serotonergic mechanisms act through a stimulation of GRF. Murakami *et al.* (1986) recently demonstrated that injection of antiserum to GRF blocks the GH rise induced by either 5-HT or 5-HTP.

6.4. Other Neurotransmitters

A number of studies have indicated that other putative neurotransmitters, such as acetylcholine and GABA, may play a role in the modulation of GH secretion. However, because of the severe toxic side effects of drugs that alter the release of acetylcholine and GABA, it is hard to determine whether the known effects are due to a direct effect on GH secretion or indirectly to stress responses.

Direct administration of cholinergic agents into the brain induces GH secretion, implying that acetylcholine acts within the hypothalamus to stimulate GH secretion (Vijayan and McCann, 1981). However, these effects appear to act via a dopaminergic mechanism, as prior administration of a DA antagonist blocks the effect of cholinergic agents.

The bulk of the evidence in humans and rats indicates that GABA stimulates GH secretion (Murakami *et al.*, 1985; Willoughby *et al.*, 1986; Martin and Millard, 1986). Evidence suggests that GABA acts via the stimulation of GRF, since GABA-induced GH secretion is abolished in animals passively immunized with antiserum to GRF (Murakami *et al.*, 1985).

There is also evidence of a GABA involvement in GH control in fetal sheep. Gluckman (1984) found that the GABA antagonist picrotoxin markedly stimulated GH secretion in the ovine fetus. These data suggest that GABA tonically inhibits GH secretion in fetal sheep.

7. Gonadal Steroid Modulation of GH Secretion

In addition to the putative neurotransmitters, there is increasing evidence that gonadal steroids play a role in the regulation of GH secretion and may, in fact, be the basis for the sexually dimorphic patterns of GH secretion seen in the rat (Jansson *et al.*, 1985; Millard *et al.*, 1987b).

Recent studies from our laboratory indicate that steroids modulate basal GH levels and individual GH pulse amplitudes as well as alter the frequency of GH secretory spikes (Millard *et al.*, 1987b). There is also increasing evidence that gonadal steroids can also affect GH secretion by direct actions on the pituitary (Evans *et al.*, 1985; Ho *et al.*, 1986; Wehrenberg *et al.*, 1986).

Gonadal steroids also play a role in the growth and development of agricultural animals, perhaps through modulation of GH secretion. This subject has been reviewed in detail *by Ford and Klindt* (this volume).

8. Feedback Regulation of GH Secretion

Although integration of GH secretion appears to be controlled to some extent by modulatory mechanisms of neurotransmitters and gonadal steroids on both somatostatin and GRF release, another important area of GH regulation has been demonstrated: that of feedback control of GH secretion. This feedback control of GH can occur at three different levels: (1) by actions of the end organ substances somatomedins (long-loop feedback), (2) through direct actions of GH on its own release (short-loop feedback), and (3) through the actions of somatostatin and GRF on each other or on their own release (ultrashort-loop feedback).

Long-loop feedback control of GH secretion has been observed in rats by the demonstration that intracerebroventricular administration of somatomedins results in a marked decrease in GH secretion (Abe *et al.*, 1983; Tannenbaum, 1987). This decrease in GH secretion appears to be mediated by both a stimulation of somatostatin and an inhibition of release of GRF from the hypothalamus (Berelowitz *et al.*, 1981; Shibasaki *et al.*, 1986). In addition, there is evidence that the somatomedins may also act by directly inhibiting the release of GH from the pituitary (Berelowitz *et al.*, 1981; Abe *et al.*, 1983). Taken together, these data indicate that the somatomedins regulate GH secretion by acting both within the CNS and directly at the level of the pituitary.

The bulk of the evidence of short-loop feedback control of GH indicates that GH inhibits its own secretion by actions within the CNS (Patel, 1979; Chihara *et al.*, 1981; Tannenbaum, 1987). Further, this inhibitory action of GH on its own release appears to be caused by a direct stimulation of somatostatin release from the hypothalamus (Patel, 1979; Chihara *et al.*, 1981). One report indicates that centrally administered GH may also blunt its own secretion via a reduction in GRF release from the hypothalamus (Conway *et al.*, 1985). Thus GH may regulate its own secretion via actions on both peptidergic systems.

Initial evidence for a ultrashort-loop feedback control of GH secretion came from the work of Lumpkin *et al.* (1981) who demonstrated that intraventricular somatostatin causes a stimulation of GH secretion. Murakami *et al.* (1987b) subsequently found that prior treatment with antiserum to GRF blunts the GH increase induced by intraventricular somatostatin, indicating that somatostatin may act to stimulate the release of GRF.

Further, there is evidence that centrally administered GRF, in low doses, suppresses GH (Lumpkin *et al.*, 1985) and that the effect of GRF is mediated via a stimulation of the release of somatostatin (Aguila and McCann, 1985; Katakami *et al.*, 1986). Thus, it appears that GH secretion may also be modulated by a direct interaction of the two neuropeptidergic systems upon each other.

The physiological significance of why the GH secretory axis has at least three demonstrated levels of feedback control is not clear. However, due to the complexity by which GH regulates bodily processes (via indirect and direct mechanisms), we can only surmise that a complex feedback control system is needed for ultimate integration of GH secretion. Studies are needed to clarify the importance of these feedback systems in GH regulation.

ACKNOWLEDGMENTS. Supported by NIH Grant HD 22199. I thank Mary Healy and Victoria Patterson for secretarial assistance in preparing the manuscript.

References

Abe, H., M. E. Molitch, J. J. Van Wyk and L. E. Underwood. 1983. Human growth hormone and somatomedin C suppresses spontaneous release of growth hormone in unanesthetized rats. Endocrinology 113:1319.

Aguila, M. C. and S. M. McCann. 1985. Stimulation of somatostatin release in vitro by synthetic human growth hormone-releasing factor by a nondopaminergic mechanism. Endocrinology 117:762.

Anfinson, M. S., S. L. Davis, E. Christian and D. O. Everson. 1975. Episodic secretion of growth hormone in steers and bulls: an analysis of frequency and magnitude of secretory spikes occurring in a 24 hour period. Proc. West. Sect. Amer. Soc. Anim. Sci. 26:175.

Arnold, M. A. and J. D. Fernstrom. 1978. Serotonin receptor antagonists block a natural, short-term surge in serum growth hormone levels. Endocrinology 103:1159.

Arnold, M. A. and J. D. Fernstrom. 1980. Administration of antisomatostatin serum to rats reverses the inhibition of pulsatile growth hormone secretion produced by injection of metergoline but not yohimbine. Neuroendocrinology 31:194.

Badger, T. M., W. J. Millard, G. F. McCormick, C. Y. Bowers and J. B. Martin. 1984. The effects of growth hormone-releasing peptides on GH secretion in perfused pituitary cells of adult male rats. Endocrinology 115:1432.

Bauman, D. E., R. J. Collier and H. A Tucker. 1977. Effect of reserpine on serum prolactin, growth hormone, and glucocorticoids in dairy cows. Proc. Soc. Exp. Biol. Med. 155:189.

Berelowitz, M., M. Szabo, L. A. Frohman, S. L. Firestone, L. Chu and R. L. Hintz. 1981. Somatomedin-C mediates growth hormone negative feedback by effects on both the hypothalamus and the pituitary. Science 212:1279.

Bjorklund, A., R. Y. Moore, A. Nobin and U. Steineri. 1973. The organization of tuberohypophyseal and reticulo-infundibular catecholamine neuron systems in the rat brain. Brain Res. 51:171.

Bloch, B., P. Brazeau, N. Ling, P. Bohlen, F. Esch, W. B. Wehrenberg, R. Benoit, F. Bloom and R. Guillemin. 1983. Immunohistochemical detection of growth hormone releasing factor in brain. Nature 301:607.

Brazeau, P., W. Vale, R. Burgus, N. Ling, M. Butcher, J. Rivier and R. Guillemin. 1973. Hypothalamic polypeptide that inhibits the secretion of immunoreactive pituitary growth hormone. Science 179:77.

Chihara, K., N. Minamilani, H. Kaji, A. Arimura and T. Fujata. 1981. Intraventricularly injected growth hormone stimulates somatostatin release into rat hypophysial portal blood. Endocrinology 109:2279.

Conway, S., S. M. McCann and L. Krulich. 1985. On the mechanism of growth hormone autofeedback regulation: possible role of somatostatin and growth hormone-releasing factor. Endocrinology 117:2284.

Davis, S. L. and M. L. Borger. 1973. Hypothalamic catecholamine effects on plasma levels of prolactin and growth hormone in sheep. Endocrinology 92:303.

Davis, S. L., D. L. Ohlson, J. Klindt and M. S. Anfinson. 1977. Episodic growth hormone secretory patterns in sheep: relationship to gonadal steroids. Amer. J. Physiol. 233:E519.

Enright, W. J., S. A. Zinn, L. T. Chapin and H. A. Tucker. 1987. Growth hormone response of bull calves to growth hormone-releasing factor. Proc. Soc. Exp. Biol. Med. 184:483.

Epelbaum, J., J. O. Willoughby, P. Brazeau and J. B. Martin. 1977. Effects of brain lesions and hypothalamic deafferentation on somatostatin distribution in the rat brain. Endocrinology 101:1495.

Evans, W. S., R. J. Kreig, E. R. Limber, D. L. Kaiser and M. O. Thorner. 1985. Effects of in vivo gonadal hormone environment on in vitro hGRF-40-stimulated GH release. Amer. J. Physiol. 249:E276.

Everitt, B. J., B. Meister, T. Hokfelt, T. Melander, L. Terenius, A. Rokaeus, E. Thoedorsson-Norheim, G. Dockray, J. Edwardson, C. Cuello, R. Elde, M. Goldstein, H. Hemmings, C. Ouimet, I. Walaas, P. Greengard, W. Vale, E. Weber, J.-Y. Wu and K.-J. Chang. 1986. The hypothalamic arcuate nucleus–median eminence complex: immunohistochemistry of transmitters, peptides and DARPP-32 with special reference to coexistence in dopamine neurons. Brain Res. Rev. 11:97.

Finklestein, J. W., H. P. Roffwarg, R. M. Boyar, J. Kream and L. Hellman. 1972. Age-related change in the twenty-four hour spontaneous secretion of growth hormone. J. Clin. Endocrinol. Metab. 35:665.

Frohman, L. A. and J. O. Jansson. 1986. Growth hormone-releasing hormone. Endocrinol. Rev. 7:223.

Gluckman, P. D. 1984. Functional maturation of the neuroendocrine system in the perinatal period: studies of the somatotropic axis in the ovine fetus. J. Dev. Physiol. 6:301.

Gluckman, P. D., C. Marti-Henneberg, S. L. Kaplan, C. H. Li and M. M. Grumbach. 1980. Hormone ontogeny in the ovine fetus. X. The effects of β-endorphine and naloxone on circulating growth hormone, prolactin, and chorionic somatomammotropin. Endocrinology 107:76.

Gluckman, P. D. and Y. Parsons. 1985. Growth hormone secretion in the fetal sheep following stereotaxic lesioning of the fetal hypothalamus. J. Dev. Physiol. 7:25.

Goodman, R. H., D. C. Aron and B. A. Roos. 1983. Rat pre-prosomatostatin. Structure and processing by microsomal membranes. J. Biol. Chem. 258:5570.

Gubler, U., J. J. Monahan, P. T. Lomedico, R. S. Bhatt, K. J. Collier, B. J. Hoffman, P. Bohlen, F. Esch, N. Ling, F. Zeytin, P. Brazeau, M. S. Poonian and L. P. Gage. 1983. Cloning and sequence analysis of cDNA for the precursor of human growth hormone-releasing factor, somatocrinin. Proc. Natl. Acad. Sci. USA 80:4311.

Guillemin, R., P. Brazeau, P. Bohlen, F. Esch, N. Ling and W. B. Wehrenberg. 1982. Growth hormone-releasing factor from a human pancreatic tumor that caused acromegaly. Science 218:585.

Hart, I. C. and H. L. Buttle. 1975. Growth hormone levels in the plasma of male goats throughout the year. J. Endocrinol. 67:137.

Hart, I. C., P. M. E. Chadwick, A. Coert, S. James and A. D. Simmonds. 1985a. Effect of different growth hormone-releasing factors on the concentrations of growth hormone, insulin and metabolites in the plasma of sheep maintained in positive and negative energy balance. J. Endocrinol. 105:113.

Hart, I. C., P. M. E. Chadwick, A. Coert, S. James and A. D. Simmonds. 1985b. Effect of intravenous bovine growth hormone or human pancreatic growth hormone-releasing factor on milk production and plasma hormones and metabolites in sheep. J. Endocrinol. 105:189.

Hart, I. C., S. James, B. N. Perry and A. D. Simmonds. 1984. Effect of intravenous administration of growth hormone-releasing factor (hpGRF-44) and Tyr-D-Ala-Trp-D-Phe-NH₂ on plasma hormones and metabolites in goats. J. Endocrinol. 103:173.

Ho, K. Y., D. A. Leong, Y. N. Sinha, M. L. Johnson, W. S. Evans and M. O. Thorner. 1986. Sex-related differences in GH secretion in rat using reverse hemolytic plaque assay. Amer. J. Physiol. 250:E650.

Isaksson, O. G. P., S. Eden and J.-O. Jansson. 1985. Mode of action of pituitary growth hormone on target cells. Annu. Rev. Physiol. 47:483.

Jacobowitz, D. M., H. Schulte, G. P. Chrousos and D. L. Loriaux. 1983. Localization of GRF-like immunoreactive neurons in the rat brain. Peptides 4:521.

Jansson, J. O., S. Eden and O. Isaksson. 1985. Sexual dimorphism in the control of growth hormone secretion. Endocrinol. Rev. 6:128.

Katakami, H., A. Arimura and L. A. Frohman. 1986. Growth hormone (GH)-releasing factor stimulates hypothalamic somatostatin release: an inhibitory feedback effect on GH secretion. Endocrinology 118:1872.

Klindt, J., J. J. Ford, J. G. Berardinelli and L. L. Anderson. 1983. Growth hormone secretion after hypophysial stalk transection in pigs. Proc. Soc. Exp. Biol. Med. 172:508.

Klindt, J., D. L. Ohlson, S. L. Davis and B. D. Schanbacher. 1985. Ontogeny of growth hormone, prolactin, luteinizing hormone and testosterone secretory patterns in the ram. Biol. Reprod. 33:436.

Krulich, L., M. A. Mayfield, M. K. Steele, B. A. McMillen, S. M. McCann and J. I. Koenig. 1982. Differential effects of pharmacological manipulations of central α_1- and α_2-adrenergic receptors on the secretion of thyrotropin and growth hormone in male rats. Endocrinology 110:796.

Lumpkin, M. D., N. Negro-Vilar and S. M. McCann. 1981. Paradoxical elevation of growth hormone by intraventricular somatostatin: possible ultrashort-loop feedback. Science 211:1072.

Lumpkin, M. D., W. K. Samson and S. M. McCann. 1985. Effects of intraventricular growth hormone-releasing factor on growth hormone release: further evidence for ultrashort loop feedback. Endocrinology 116:2070.

McCormick, G. F., W. J. Millard, T. M. Badger, C. Y. Bowers and J. B. Martin. 1985. Dose–response characteristics of various peptides with growth hormone-releasing activity in the unanesthetized rat. Endocrinology 117:97.

Makara, G. B., M. Palkovits, F. Antoni and J. Z. Kiss. 1983. Topography of somatostatin-immunoreactive fibers to the stalk–median eminence of the rat. Neuroendocrinology 37:1.

Marti-Henneberg, C., P. D. Gluckman, S. L. Kaplan and M. M. Grumbach. 1980. Hormone ontogeny in the ovine fetus. XI. The serotonergic regulation of growth hormone and prolactin secretion. Endocrinology 107:1273.

Martin, J. B., P. Brazeau, G. S. Tannenbaum, J. O. Willoughby, J. Epelbaum, L. C. Terry and D. Durand. 1978a. Neuroendocrine organization of growth hormone regulation. In: S. Reichlin, R. J. Baldessarini and J. B. Martin (Ed.) The Hypothalamus. pp 329–357. Raven Press, New York.

Martin, J. B., D. Durand, W. Gurd, G. Faille, J. Audet and P. Brazeau. 1978b. Neuropharmacological regulation of episodic growth hormone and prolactin secretion in the rat. Endocrinology 102:106.

Martin, J. B. and W. J. Millard. 1986. Brain regulation of growth hormone secretion. J. Anim. Sci. 63 (Suppl. 2):11.

Millard, W. J., T. O. Fox, T. M. Badger and J. B. Martin. 1987b. Gonadal steroid modulation of growth hormone secretory patterns in the rat. In: R. F. Robbins and S. Melmed (Ed.) Acromegaly. Plenum Press, New York, pp. 139–150.

Millard, W. J., D. M. O'Sullivan, T. O. Fox and J. B. Martin. 1987a. Sexually dimorphic patterns of growth hormone secretion. In: W. F. Crowley, Jr., and J. Hofler (Ed.) Episodic Secretion of Hormones. Wiley and Sons, New York, pp. 287–304.

Molina, J. R., J. Klindt, J. J. Ford and L. L. Anderson. 1986. Growth hormone and prolactin secretion after hypothalamic deafferentation in pigs. Proc Soc. Exp. Biol. Med. 183:163.

Moore, R. Y. and F. E. Bloom. 1979. Central catecholamine neuron systems: anatomy and physiology of the norepinephrine and epinephrine systems. Ann U. Rev. Nuerosci. 2:113.

Moseley, W. M., L. F. Krabill, A. R. Friedman and R. F. Olsen. 1984. Growth hormone response of steers injected with synthetic human pancreatic growth hormone-releasing factors. J. Anim. Sci. 58:430.

Murakami, Y., Y. Kato, Y. Kabayama, T. Inoue, H. Koshiyama and H. Imura. 1987b. Involvement of hypothalamic growth hormone (GH)-releasing factor in GH secretion induced by intracerebroventricular injection of somatostatin in rats. Endocrinology 120:311.

Murakami, Y., Y. Kato, Y. Kabayama, K. Tojo, T. Inoue and H. Imura. 1985. Involvement of growth hormone-releasing factor in growth hormone somatostatin induced by gamma-aminobutyric acid in conscious rats. Endocrinology 117:787.

Murakami, Y., Y. Kato, Y. Kabayama, K. Tojo, T. Inoue and H. Imura. 1986. Involvement of growth hormone (GH)-releasing factor in GH secretion induced by serotonergic mechanisms in conscious rats. Endocrinology 119:1089.

Murakami, Y., Y. Kato, H. Koshiyama, T. Inoue, N. Yanaihara and H. Imura. 1987a. Galanin stimulates growth hormone (GH) secretion via GH-releasing factor (GRF) in conscious rats. Eur. J. Pharmacol. 136:415.

Naylor, S. L., A. Y. Sakaguchi, L. Shen, G. L. Bell, W. J. Rutter and T. B. Shows. 1983. Polymorphic human somatostatin gene is located on chromosome 3. Proc. Natl. Acad. Sci. USA 80:2686.

Ottlecz, A., W. K. Samson and S. M. McCann. 1986. Galanin: evidence for a hypothalamic site of action to release growth hormone. Peptides 7:51.

Patel, Y. C. 1979. Growth hormone stimulates hypothalamic somatostatin. Life Sci. 24:1589.

Patel, Y. C. 1987. Somatostatin. In: D. K. Ludecke and G. Tolis (Ed.) Growth Hormone, Growth Factors, and Acromegaly. pp 21–36. Raven Press, New York.

Plotsky, P. M. and W. Vale. 1985. Patterns of growth hormone-releasing factor and somatostatin secretion into the hypophysial–portal circulation of the rat. Science 230:461.

Quabbe, H. J., M. Gregor, C. Bunge-Vogt, A. Eckof and J. Witt. 1981. Twenty-four-hour pattern of growth hormone secretion in the rhesus monkey: studies including alterations in the sleep wake/cycle and sleep stage cycle. Endocrinology 109:513.

Reichlin, S. 1983. Somatostatin. N. Engl. J. Med. 309:1495, 1556.

Rivier, J., J. Spiess, M. Thorner and W. Vale. 1983. Characterization of a growth hormone-releasing factor from a human pancreatic islet tumor. Nature 400:276.

Scanes, C. G. and S. Harvey. 1984. Stimulation of growth hormone secretion by human pancreatic growth hormone-releasing factor and thyrotropin-releasing hormone in anesthetized chickens. Gen. Comp. Endocrinol. 56:198.

Shibasaki, T., N. Yamauchi, M. Hotta, A. Masuda, T. Imaki, H. Demura, N. Ling and K. Shizume. 1986. In vitro release of growth hormone-releasing factor from rat hypothalamus: effect of insulin-line growth factor-1. Regul. Pept. 15:47.

Sonntag, W. E., R. W. Steger, L. J. Forman and J. Meites. 1980. Decreased pulsatile release of growth hormone in old male rats. Endocrinology 107:1875.

Steiner, R. A., J. K. Stewart, J. Barber, D. Koerker, C. J. Goodner, A. Brown, P. Illner and C. C. Gale. 1978. Somatostatin: a physiological role in the regulation of growth hormone secretion in the adolescent male baboon. Endocrinology 102:1587.

Tannenbaum, G. S. 1987. Interactions of growth hormone-releasing factor and somatostatin in the regulation of rhythmic secretion of growth hormone. In: D. K. Ludecke and G. Tolis (Ed.) Growth Hormone, Growth Factors, and Acromegaly. pp 37–53. Raven Press, New York.

Tannenbaum, G. S. and N. Ling. 1984. The interrelationship of growth hormone (GH)-releasing factor and somatostatin in generation of the ultradian rhythm of GH secretion. Endocrinology 115:1952.

Terry, L. C. and J. B. Martin. 1981. The effect of lateral hypothalamic–medial forebrain stimulation and somatostatin antiserum on pulsatile growth hormone secretion in freely behaving rats. Endocrinology 109:622.

Van De Kar, L. D. and S. A. Lorens. 1979. Differential serotonergic innervation of individual hypothalamic nuclei and other forebrain regions by the dorsal and medial midbrain raphe nuclei. Brain Res. 162:45.

Vijayan, E. and S. M. McCann. 1981. Acetylcholine (Ach) induced alterations of plasma growth hormone (GH) in normal and pimozide-treated ovariectomized rats. Brain Res. Bull. 7:11.

Wehrenberg, W. B., F. Esch, A. Baird, S.-Y. Ying, P. Bohlen and N. Ling. 1986. Growth hormone-releasing factor: a new chapter in neuroendocrinology. Horm. Res. 24:82.

Willoughby, J. O., P. Brazeau and J. B. Martin. 1977b. Pulsatile growth hormone and prolactin: effects of (+) butaclamol, a dopamine receptor blocking agent. Endocrinology 101:1298.

Willoughby, J. O., L. C. Terry, P. Brazeau and J. B. Martin. 1977a. Pulsatile growth hormone, prolactin and thyrotropin secretion in rats with hypothalamic deafferentation. Brain Res. 127:137.

Willoughby, J. O., P. M. Jervois, M. F. Menadue, W. W. Blessing. 1986. Activation of GABA receptors in the hypothalamus stimulates secretion of growth hormone and prolactin, Brain Res. 374:119.

Wroblewska, B. and E. Domanski. 1981. Role of the medial-basal hypothalamus in the secretion of growth hormone during pregnancy and lactation in ewes. J. Endocrinol. 89:349.

Mechanisms of Action for Somatotropin in Growth

R. DEAN BOYD AND DALE E. BAUMAN

1. Introduction

Physiological processes such as growth are carefully orchestrated and controlled by complex interactions among a multiplicity of hormones (Bell *et al.*, 1987). Somatotropin (ST) is a key somatotropic hormone and homeorhetic control. It exhibits regulatory effects on metabolism and consequently affords control over how absorbed nutrients are partitioned for growth and lactation. ST plays a pivotal role in metabolism not only during nutrient adequacy, but also during periods where nutrients are severely deficient or poorly utilized (e.g., starvation, diabetes). The metabolic and endocrine "axis" for determining how ST effects are manifested, in the two extremes cited, is the subject of an excellent review by Phillips (1986). This chapter assumes relative nutrient adequacy, commensurate with the needs for normal growth in farm animals.

The somatotropic effects of ST have been known for over 50 years since the classical work of Evans and Simpson (1931) who demonstrated that chronic administration of a crude bovine ST extract increased weight gain in rats. Lee and Schaffer (1934) found that ST led to an increase in protein accretion concurrent with a reduction in fat deposition. Until recently, little progress was made in characterizing production responses and identifying mechanisms by which ST acts, particularly in farm animals, because the supply of ST was limited to that which could be extracted with varying purity from the pituitary gland of slaughtered animals. This situation was resolved by advances in extraction and purification procedures and by recent developments in recombi-

R. DEAN BOYD AND DALE E. BAUMAN ● Department of Animal Science, Cornell University, Ithaca, New York 14853-4801.

nant-DNA methods of synthesis. Accordingly, significant new information is rapidly becoming available with regard to effects on animal performance and metabolism. Results are impressive and convince us that ST represents a remarkable opportunity to alter the rate, efficiency, and composition of growth in farm animals.

ST exerts its effects on a wide range of biological processes (see recent summaries by Hart and Johnsson, 1986; Phillips, 1986). It directly or indirectly stimulates proliferation of several cell types and metabolic processes associated with skeletal growth and protein accretion. It exhibits profound effects on lipid, carbohydrate, and mineral metabolism. Indeed, the physiological effects of ST are so varied that it is difficult to express a unifying hypothesis concerning its mechanisms of action (Paladini *et al.*, 1983).

In this chapter, we will emphasize the effect of ST on the nature of growth and mechanisms by which the effects of ST are mediated. It will include a discussion on the chemistry of ST so that one may appreciate, given current information, the complexity in describing both what ST is and the mechanisms by which this family of proteins accomplish the numerous biological effects ascribed to it. This is an important prelude that demonstrates the merit of both *in vitro* and *in vivo* approaches in delineating metabolic effects. We will discuss selected aspects of the performance response to exogenous ST in order to establish a working model exemplary of the effects on nutrient partitioning. Particular emphasis will be given to the mechanisms of ST action, including situations where inconsistencies have been observed among researchers and between *in vitro* and *in vivo* results. Since ST affects processes in a highly coordinated manner, understanding its effects offers a unique opportunity to gain insight into the mechanisms of growth and homeorhetic control of metabolism.

2. Chemical Nature of Somatotropin

The chemical nature and molecular biology of ST have been discussed in a number of reviews (Paladini *et al.*, 1979, 1983; Lewis *et al.*, 1980, 1987; Chawla *et al.*, 1983; Lewis, 1984). The best available information is on human ST (hST) with relatively little pertaining to farm animal species. Highly purified pituitary ST preparations are still heterogeneous, mainly due to the multiplicity of components rather than to the presence of impurities (Lewis *et al.*, 1980; Paladini *et al.*, 1983; Lewis, 1984). hST appears to be a family or complex of peptides consisting of the major 22-kDa form with one or more structural variants (e.g., 20 kDa) and posttranslational modifications such as monomeric aggregates (two to five), disulfide dimers, and proteolytically derived "fragments" (Lewis, 1984, 1987). Lewis (1984) hypothesized that the variant forms of ST may have individual properties that, when combined, account for the multiple biological effects ascribed to ST. Further, they may provide a diversity of structures that serve as precursor molecules for the production of a variety of smaller peptides. Each may have a defined physiological effect or somatotropic property.

Although insufficient evidence exists at the present time to warrant speculation concerning the presence and physiological importance of genetic variants and peptide "fragments" in domestic livestock, an understanding of the chemistry is crucial to engender appropriate research models and appreciation for the complexity in defining mechanisms.

2.1. Variants of Somatotropin

The major physiological form of ST (22 kDa) is a single-chain polypeptide of 191 amino acid residues with two loops formed by disulfide bonds (see Chawla *et al.*, 1983, and Nicoll *et al.*, 1986, for primary structure). The naturally occurring structural variant (20 kDa) of hST was first identified in human pituitary extracts (Singh *et al.*, 1974a) by means of electrophoresis. It comprises approximately 15% of the total pituitary ST content, and is the second most abundant hormone in the pituitary (Lewis *et al.*, 1987). It differs from the major form by deletion of 15 amino acids (residues 32–46), but the primary structure is otherwise identical to the 22-kDa form. Although variants have not been extensively examined in species other than the human, a 20-kDa variant of ST has been detected in the mouse pituitary (Sinha and Gilligan, 1984).

Purified 20- and 22-kDa hST exhibit similar potency in promoting weight gain, and growth of tibial cartilage in hypophysectomized rats (Lewis *et al.*, 1978) and in stimulating increases in somatomedin production (Spencer *et al.*, 1981; Mosier and Lewis, 1982). However, the 20-kDa variant differs in its capacity to bind both antisera and receptors specific for 22-kDa ST from liver, mammary, and adipose tissue preparations (Sigel *et al.*, 1981; Wohnlich and Moore, 1982; Hizuka *et al.*, 1982; Closset *et al.*, 1983; Smal *et al.*, 1986). This discrepancy between biopotency and receptor binding is difficult to reconcile unless there are separate receptors or differences between forms in clearance rate. To date, no receptor specific for 20-kDa ST has been unequivocally demonstrated but this variant has been shown to have a slower clearance rate (two- to threefold; Baumann *et al.*, 1986), which may contribute to a higher than expected relative biological activity.

A mechanism for the production of ST variants was discussed by Chawla *et al.* (1983) and Lewis *et al.* (1987). It involves alternative splicing in the processing of precursor mRNA for hST. For the 20-kDa form, 45 nucleotides normally associated with the exon, are removed along with those of the B intron. When the mRNA is translated, residues 32–46 are missing (Lewis *et al.*, 1987). Although a 20-kDa variant has not been identified for livestock species, Hampson and Rottman (1986) isolated an mRNA for bovine ST (bST) from the pituitary gland that, if translated, would result in a 27-kDa variant. The D intron had not been removed during splicing and only the first 50 nucleotides of exon 5 were present. If translated, the amino acid sequence would differ from that of 22-kDa bST from residue 125 to the COOH terminus and the chain would be 42 amino acids longer. Thus, alternative splicing of precursor mRNA may prove to be a general means for production of ST variants (Lewis *et al.*, 1987).

Although evidence exists showing that alternative processing of mRNA could result in a variant form, the physiological relevance is not evident. Further, "minor" constituents in a pituitary isolate may not be of minor physiological significance. Potential differences in clearance rate, receptor affinity, or receptor number in addition to variability in the proportion of a form secreted are possible equilibrating considerations. For example, considerable variability exists in the 22-kDa/20-kDa ratio in plasma during both provocative and spontaneous secretory episodes. The 20-kDa form may account for 5–30% of the immunoreactivity. Variability is perhaps even higher when ST is of basal concentrations since the 20-kDa form has been shown to range from being virtually absent to being the major form (Baumann *et al.*, 1983, 1985).

2.2. Fragments of Somatotropin

Early attempts to identify the "active center(s)" of ST, through controlled enzymatic cleavage of "native" ST (human, bovine, ovine), have given rise to the concept that different regions of the molecule may also account for the multiple biological actions ascribed to the hormone (Kostyo, 1974; Paladini *et al.*, 1979, 1983; Lewis *et al.*, 1980; Lewis, 1984). The relevance of this work is that ST may not only undergo enzymatic modification, but may be an entity in which several independently active molecules are present and available through at least partial cleavage. The isolation, structure, and biological activities of enzymatically modified ST forms have been extensively reviewed (Lewis *et al.*, 1980; Paladini *et al.*, 1983; Lewis, 1984; Baumann *et al.*, 1985).

The region of the hST molecule between residues 134 and 150 is very susceptible to enzymatic cleavage. The net result of proteolysis in this region is to remove a small peptide sequence, resulting in a two-chain structure linked by a disulfide bridge. Lewis's group (Singh *et al.*, 1974b) isolated three enzymatically modified forms from pituitaries of cadavers and observed enhanced biological activities for growth and lactogenic assays. Each had undergone cleavage in this region.

Baumann *et al.*, (1986) demonstrated that immunoreactive substances distinct from previously identified hST forms (22 and 20 kDa; aggregates of two to five monomers) were present in the circulation and concluded, on the basis of molecular size and immunochemical behavior, that they were hST fragments. This finding was significant since it showed that immunoreactive fragments exist *in vivo*. It is conceivable that a number of biologically active peptides exist, perhaps each with affinities for specific receptors.

Somatomedinlike activity has also been observed when enzymatic cleavage of ST was generated *in vitro* (Liberti and Miller, 1978). Somatomedins (Sm) are low-molecular-weight peptides, which are believed to mediate many of the metabolic actions of ST (see Steele and Elsasser, this volume). It is widely held that they are structurally distinct from ST. Although some reservation exists concerning generation of Sm-like activity from the ST molecule, results of studies by Maciag *et al.* (1980), and Liberti and Durham (1983) give considerable weight to this thesis.

In concert with the suggestion of potentially active fragments of ST is the report of a cleaved form of prolactin. There is extensive structural homology and overlap in biological effects between ST and prolactin. This fragment is reported to be secreted from the pituitary gland of rats in culture (Mittra, 1980a, b, 1984), is a 16-kDa "fragment," and exhibits mitogenic properties in mammary epithelium.

The physiological significance of enzymatic processing of ST *in vivo* remains unclear, but it adds to the possibilities arising from ST metabolism. Possible sites of processing include the pituitary gland, blood, liver, and other target organs or tissues. At least some of the biological effects observed for particular fragments are not observed when the intact and highly purified 22 kDa form is administered to animals (e.g., hyperglycemia, discussed in Section 4.2). However, Baumann and associates (1986) argue that the well-recognized disparity between bioassayable and immunoassayable ST in blood (Ellis *et al.*, 1978; Vodian and Nicoll, 1977, 1979) could be at least partially accounted for not only by ST dependent growth factors (somatomedins), but also by ST fragments. Biological activity in excess of radioimmunoassayable ST is generated within minutes after ST secretion (Vodian and Nicoll, 1977, 1979).

If enzymatic cleavage occurs and has physiological relevance, intriguing questions arise concerning the regulatory mechanisms of this process, whether forms can be produced with enhanced activity for specific processes and to what extent ST expresses its specific growth-promoting actions through generation of small peptides collectively known as the somatomedins as compared to "fragments" exhibiting somatomedinlike activity. Clearly, additional studies are needed to determine what role(s), if any, these modified forms play in mediation of ST's effects.

2.3. Receptor Mediation of Somatotropin Effects

Receptors are recognition molecules, specific for a particular ligand. It is generally accepted that recognition by a target tissue requires binding of the hormone to its respective receptor in the plasma membrane (Flint, 1982). This is a necessary prelude to the biological response for tissue(s) unless the effects of the hormone are mediated by another substance. Response may vary according to age or physiological state by varying receptor number or responsiveness to the hormone without altering its concentration. By far the greatest detail with respect to peptide hormone–receptor function and biochemical response has been generated for insulin. Excellent reviews specific for livestock have been written by Flint (1982), Etherton (1982), and Weekes (1986). Using insulin, Flint has provided an example of the differential changes in tissue responsiveness to accommodate changes in the vastly different needs of physiological states (e.g., pregnancy, lactation). ST research is in a more infantile state relative to insulin in the receptor area, but is also providing marvelous examples of dynamic physiological control.

Response to ST is a function of age, a fact established both *in vitro* and *in vivo* for a number of species (Nutting, 1976; Goodman and Coiro, 1981; Gluckman *et al.*, 1983; Maes *et al.*, 1983). Growth during the fetal and early postnatal period appears to be largely independent of pituitary ST, whereas a progressive increase in response begins to occur during the neonatal period. The mechanism for this "evolution" is unclear, but may be the result of an appearance of ST receptors or binding sites. For example, the number of binding sites for ST is low during the fetal and early neonatal periods, but begins to increase during the latter so that relatively high binding capacity is observed in older lambs (Gluckman *et al.*, 1983; Freemark *et al.*, 1986). Similar development of binding capacity has been reported for the rat (Maes *et al.*, 1983).

Hughes and Friesen (1985) aptly characterized the frustrations of many who are involved in research pertaining to mechanisms of ST action when they stated that, "although more than a decade has passed since development of a radioreceptor assay for growth hormone (Lesniak *et al.*, 1973), characterization of the ST receptor has only recently progressed beyond rather superficial binding studies." This relatively slow progress in characterization of the hormone–receptor scenario can be attributed to several problems, including the difficulty of defining an appropriate biological response. The latter is an essential criterion that provides a means of establishing whether binding sites are, indeed, receptors for ST. This work is fraught with paradoxes between effects manifested *in vitro* and those observed *in vivo*, thus making a thorough review and critique of the literature timely. It is further plagued by unresolved questions in chemistry such as whether multiple forms and potentially biologically active "fragments" exist or have any physiological relevance. Obviously, advances in the chemistry of the ST family and a correct definition of biological responses are an essential prelude to ST receptor approach.

The generalized relation between the presence of tissue receptors and biological response requires clarification in that the effects of ST are, at least partially, mediated through the Sm. According to the Sm hypothesis (Daughaday, 1981), ST effects are mediated by production and (or) release of Sm [e.g., insulinlike growth factor I and II (IGF-I, IGF-II)] from liver and perhaps other tissues which result in cell proliferation or changes in metabolism. Although this may be an oversimplification of the mechanism by which effects of ST are mediated, indirect mediators and the possibility of different genomic (forms) or biologically active "fragments" justify consideration of a heterogeneous receptor population (Barnard et al., 1984; Isaksson et al., 1985). For example, despite similar growth-promoting activity, 20-kDa hST is relatively ineffective in inhibiting 22 kDa[^{125}I]hST binding in liver membranes of rats and rabbits (Sigel et al., 1981; Wohnlich and Moore, 1982). This apparent paradox was addressed by the authors cited, who interpreted this to indicate the "presence" of novel receptors for 20-kDa ST.

Specific binding sites for ST exist on plasma membranes of many organs and tissues in addition to the liver (see review by Kostyo, 1986). Methods of identification encompass in vivo distribution of and in vitro binding to isolated cells or membranes of the organs and tissues of various laboratory animals (e.g., mice, rats) using radioiodinated hST and bST. The heterologous binding assay is potentially problematic because of the concern for possible species differences in ST receptor structure and binding affinity (Nicoll et al., 1986). Nevertheless, tissues identified as having ST receptors include adipose tissue, adipocytes and preadipose cell lines, liver membranes and hepatocytes, chondrocytes, various skeletal muscles, and cultured lymphocytes. Unfortunately, little published information exists for livestock in general. To date, binding of ST to bovine mammary tissue has not been demonstrated (Baumrucker, 1986); however, this must be viewed with caution given the potential for ST receptor presence in undifferentiated mammary cells. On the other hand, an indirect effect is possible. Receptors for both Sm types are present in bovine mammary tissue and IGF-I has been shown to have a direct effect on mammary epithelial cell growth (Dehoff et al., 1988). Receptors for IGF-I are also present on membranes of mammary cells of swine (Gregor and Burleigh, 1985).

The central involvement of insulin in stimulating tissue anabolism and inhibiting "nutrient" mobilization makes regulation of insulin secretion or action an ideal locus for homeorhetic regulation (Weekes, 1986). Modification of insulin action through receptor function (binding) or postreceptor mechanisms may be one locus for homeorhetic controls such as ST (Weekes, 1986). A change in binding or cellular recognition could result in a host of metabolic changes contrary to the insulin stimulus, but we are not aware of any reports demonstrating that chronic administration of ST adversely affects receptor number or affinity to insulin.

3. Biological Response to Somatotropin

The growth-promoting effects of exogenous ST in various livestock species have been reviewed (Hart and Johnsson, 1986; see Beermann, this volume). A number of approaches may be taken to increase endogenous ST synthesis and release. These are beyond the scope of this review; however, rational targets for exploitation have been discussed by Convey (1987). Another possible and novel approach involves the use of monoclonal antibodies to ST to enhance the effectiveness of ST (Aston *et al.*, 1986). Although the binding of antibodies to hormones is often associated with inhibition of hormonal activity, Aston and co-workers reported that binding by a monoclonal form to hST results in enhancement of its somatogenic activity *in vivo*.

A dose–response relationship has been observed for exogenous ST treatment in swine (Boyd *et al.*, 1986; Etherton *et al.*, 1986; Campbell *et al.*, 1987), dairy heifers (Moseley *et al.*, 1987; Crooker and Bauman, unpublished data), and lactating cows (Bauman and McCutcheon, 1986) using various response criteria. Selected aspects of these *in vivo* studies are used below to establish a working model, exemplary of the effects of ST on nutrient partitioning as a prelude to the discussion on mechanisms.

3.1. Nutrient Partitioning

The effects of exogenous ST on energy and protein metabolism appear to be primarily associated with postabsorptive use of nutrients. Administration of ST to growing cattle (Car *et al.*, 1967; Eisemann *et al.*, 1986c) or lactating cows (Peel *et al.*, 1981; Tyrrell *et al.*, 1982) had no effect on apparent digestibilities of energy or nitrogen with the exception of one report in which a small effect was noted (Moseley *et al.*, 1982). Although these results do not rule out the possibility that ST treatment exerts subtle effects on digestion, it appears unlikely that it acts to any large extent through changes in the digestive process.

Effects of ST have also been examined using calorimetry techniques. Eisemann *et al.* (1986c) determined that energy retained as protein was greater in bST-treated beef cattle, but total heat production was not altered. Thus, there was no change in the efficiency of metabolizable energy use with bST treatment. The same conclusion was reached following calorimetry studies with lactating cows that had received bST (Tyrrell *et al.*, 1982). Therefore, ST alters nutrient partitioning but the partial efficiency with which metabolizable energy is used for specific tissue processes is not affected.

The hypothesis that was widely held during the last decade was that ST effects were predominately associated with enhanced lipid mobilization (Machlin, 1975). Thus, response to exogenous ST would be expected to be greatest

Table I

Relative Response of Swine to Porcine Somatotropin

Variable	25–55 kg BW[a,b]		Percent difference	45–100 kg BW[a,c]		Percent difference
	Control	pST		Control	pST	
Average daily gain (g/day)	905	1052	+16	950	1100	+16
Feed:gain ratio	2.57	1.96	+23	3.02	2.18	+28
Protein gain (g/day)[d]	110	151	+37	96	148	+54
Lipid gain (g/day)	283	193	−32	297	93	−69
Ash gain (g/day)	20.4	36.8	+79	16.1	26.2	+63

[a] Wt. constant initiation and termination.
[b] Campbell et al. (1988); 100 μg pST/kg BW injected i.m.
[c] Boyd et al. (1986, unpublished data); 120 μg pST/kg BW injected s.c.
[d] Accretion rate for eviscerated carcasses determined by comparative slaughter.

in an animal during the later phase of growth in which substantial lipid reserves exist. However, recent studies have demonstrated that this hypothesis is incorrect and that the mechanism is considerably more complex. For example, substantial increases in growth rate, efficiency of gain, and shifts in carcass composition of swine occur with ST administration over a major portion of the growth curve (25–100 kg) with the magnitude of the response being similar (Table 1), despite markedly different propensities for fat accretion. Studies with cattle have also shown a response to exogenous ST in the early phases of postnatal growth (Sandles and Peel, 1987). These data suggest that ST affects tissues and nutrient partitioning more broadly and that it is a major limit to protein accretion and growth even during the early stages of postnatal growth.

3.2. Pattern of Administration

Most studies with growing animals have employed daily subcutaneous or intramuscular injections of exogenous ST. This results in a blood profile in which ST reaches peak concentration 2 to 6 hr postinjection, followed by a progressive decline to baseline during the remaining 24-hr period (Bauman and McCutcheon, 1986). This represents a marked departure from the episodic pattern normally exhibited in both laboratory and farm animal species (Martin, 1976). Unfortunately, few studies have addressed the significance of administrative pattern relative to responsiveness. Moseley et al. (1982) studied the effect of pattern of ST administration on cattle and observed that continuous infusion (i.v.), pulsatile injections (i.v., every 4 hr), or a combination of the two produced identical increases in nitrogen retention. Similarly, variable patterns of ST administration had no differential effects on increases in milk yield of dairy cows (Fronk et al., 1983; McCutcheon and Bauman, 1986a) or increases in epiphyseal plate width in hypophysectomized rats (Schlechter et al., 1986a). In contrast, other studies with hypophysectomized rats found that an episodic pattern of ST administration gave greater rates of growth than did continuous infusion (Cotes et al., 1980; Jansson et al., 1982a,b; Clark et al., 1985; Isaksson et al., 1985).

3.3. Dose–Response Relationships: Growth Model

The most extensive and definitive work to date on dose–response relationships has been with growing swine. The extent to which ST alters nutrient partitioning and the diversity of processes affected in growing animals are illustrated in Fig. 1. This array of variables describes the nature of changes in performance and suggests that the "optimal" dose differs according to the criterion. Increases in body weight are largely a function of increases in fat, protein, and water with mineral deposition (i.e., ash) contributing a minor part by

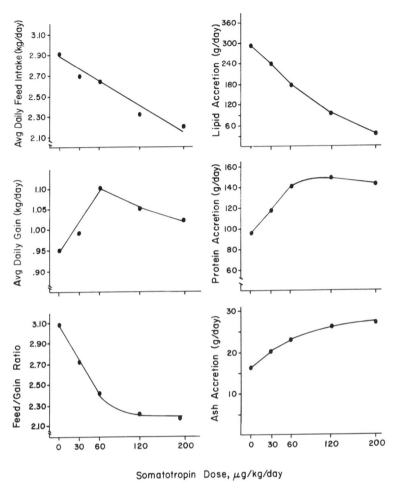

FIGURE 1. Dose–response curves for various growth and carcass criteria in growing swine treated chronically with excipient or pituitary-derived somatotropin (Boyd *et al.*, 1986, and unpublished data; 45–100 kg body weight). Accretion rates for lipid, protein, and ash determined for the eviscerated carcass by comparative slaughter.

comparison. Lipid accretion declined from approximately 300 to 35 g/day at the highest ST dose, representing a reduction in daily accretion rate of approximately 88%. Equally impressive is the fact that protein (and associated water) was increased sufficiently to not only replace the reduction in fat, but further to result in a substantial increase in growth rate (+19% at 60 μg/kg body weight). This alone makes the earlier postulate concerning lipid mobilization untenable, as marked decreases in the rate of lipid accretion must be offset by reciprocal increases in "lean" to prevent a decrease in the rate of gain. The quadratic nature of the gain response occurred because protein accretion reached a plateau (55% increase) even though lipid accretion progressively declined with increasing dose. Failure to achieve a larger increment in the growth response may not be due to ST optimization for protein accretion since theoretical calculations suggest that protein intake may have become limiting for the two highest pST doses. In this and another long-term study with swine, there was no evidence of "down-regulation" or decreased responsiveness with time (Boyd, unpublished data).

Altering the growth pattern toward a greater proportion of lean tissue and less fat dramatically increases the efficiency of gain (Fig. 1). This marked advantage in energetic efficiency is not due to a concomitant difference in efficiency for protein accretion, as compared to fat, since the cost for accretion and maintenance of protein is relatively similar to that of fat. Although estimates vary, approximately 11.5 kcal of metabolizable energy (ME) is required per g of protein accrued compared to 12.8 kcal ME/g of fat. Using values of 5.7 kcal/g of protein and 9.5 kcal/g of fat, the energetic efficiency for fat accretion is 74% (i.e., 9.5/12.8) versus 51% for protein (Thorbek, 1977; Webster et al., 1979). Consequently, a smaller proportion of the energy utilized for fat accretion is lost as heat. The added energy expenditure, required for protein accrued, reflects differences in costs associated with synthesis and maintenance over and above that required for lipid. Thus, the advantage in efficiency for growth is primarily due to the water associated with gain of lean tissue since it is approximately 75% water. Efficiency of gain failed to increase beyond which an increase in protein and associated water was observed. The linear decline in feed intake is consistent with changes in carcass constituents. In fact, the response for diet intake $(r = 0.99)$ parallels and is likely a function of the decline in nutrient use for lipid accretion.

The net effect of ST was to elevate protein accretion (and decrease fat) beyond an intrinsic "limit" and to an unprecedented level for castrate and female pigs. The level achieved is in close approximation to that normally achieved by the intact male (see Campbell, 1987). The biological potential for ST-enhanced protein accretion and therefore growth remains unclear. However, using the ST-treated boar (60 kg) as the best approximation of potential for protein accretion, Campbell (1987) estimated that the potential was in excess of 235 g/day.

These results suggest some additional considerations. First, the fact that the dose–response curves (e.g., lipid, protein) differ in shape is evidence that ST has specific effects on and coordinates metabolism among tissues. For example, although the increased propensity for muscle accretion potentially diverts glucose from adipose deposition, the apparent difference in curve shape for the two tissue masses suggests that ST cannot be merely affecting muscle while "passively" affecting adipose (i.e., "push–pull" concept). This implied coordination among tissue types suggests that differential growth is accommodated by the diversion of nutrients which may include establishment of a new "set point" for energy intake. This is an illustration of the reciprocal type of control exhibited in homeorhesis and for metabolic regulation in general.

Second, the dramatic change in carcass composition and energy intake requires that nutrient fortification of the diet be carefully considered. This has not received sufficient consideration in the ST literature. The marked shift in nitrogen, lipid, and ash accretion, coincident with improved efficiency of gain, mandates an increase in the protein–calorie relationship, mineral concentration, and perhaps other nutrients. One must appreciate that performance and carcass responses are a function of diet adequacy and that diminished or even adverse effects might be expected if dietary input fails to match specific nutrient requirements for tissue accretion.

4. Mechanisms of Action

4.1. General Considerations

Prerequisite to a consideration of the mechanism of action of ST is a discussion of some of the difficulties in evaluating studies. First, most researchers have used exogenous ST, often of uncertain chemical purity. Purification techniques have varied, resulting in preparations that may include other hormones, biologically active fragments, chemical and conformational artifacts of ST, and endotoxins. Several instances will be discussed where this has led to misleading conclusions. As previously discussed, even highly purified preparations can be heterogeneous. In addition, variants exist but the implications have not been thoroughly explored across species and biological systems. Other factors that may complicate evaluation of the mechanism of action include the use of heterologous systems (most frequently bST or hST with rat tissues) and pharmacological levels of hormone. Finally, most of the laboratory animal studies have used hypophysectomized animals and investigated the mechanism of action for ST in a situation where one is attempting to restore to normal growth an animal that has a number of endocrine abnormalities. In contrast, large-animal studies have predominately used intact animals and investigated the mechanisms of ST action under conditions where one is attempting to enhance normal growth.

These complications have led to some paradoxes that make it difficult to develop a conceptual basis for the action of ST. Nevertheless, the data demonstrate general aspects that have been consistently observed and are central to the mechanism of action. First, effects of ST are associated with the regulation of nutrient partitioning. The lack of ST effects on bioenergetic efficiency of specific processes and the digestibility of dietary dry matter, energy, and nitrogen were discussed earlier.

Second, the effects of ST can be broadly categorized into those associated with cell proliferation and those that are "metabolic." Effects on cell proliferation have been documented in bone (see Chapter 2), adipose tissue (see Chapter 3), and muscle (see Chapter 4). These could be direct and/or may be indirectly mediated by the production of ST-dependent IGFs.

Metabolic effects of ST have been demonstrated for the metabolism of all major nutrient groups (mineral, carbohydrate, protein, lipid), but have been most extensively characterized for adipose tissue. Regulation of nutrient use involves both homeostasis and homeorhesis (Bauman and Currie, 1980; Bauman, 1984). Homeostatic controls operate minute to minute to accommodate the need for steady-state conditions. Homeorhesis differs from the latter by orchestrating changes in metabolism to support changing priorities for nutrients among tissues for each physiological state. Current data demonstrate that ST is a homeorhetic control and therefore a chronic regulator of metabolism. One potential mechanism by which a homeorhetic control orchestrates the change in nutrient partitioning involves changes in response of tissues to homeostatic signals (Bauman et al., 1982; Bauman, 1984). Results are limited but the mechanisms identified thus far are consistent with this postulate.

Third, the concept that metabolic effects of ST are chronic is a relatively recent idea and has been emphasized in several reviews (Bauman and McCutcheon, 1986; Etherton and Walton, 1986). Since ST is involved in homeorhetic (long-term) control of metabolism rather than acute maintenance of homeostasis, metabolic effects must involve cellular alterations that have relatively long half lives (hours to days as opposed to seconds or minutes). In concert with this concept, nitrogen retention in growing cattle gradually increases the first few days of ST treatment, becoming statistically significant by days 7 to 10 (Moseley et al., 1982). A similar pattern is observed for milk yield in dairy cows treated with ST (Bauman and McCutcheon, 1986). Examples can likewise be given for mineral, carbohydrate, and lipid metabolism which support the concept of chronic regulation by ST. Attention must also be given to the relevance of acute metabolic effects with regard to mechanism of action. These effects have been reported, especially in older work, and emphasized as key components of the mechanism in some recent reviews (Davidson, 1987; Gluckman et al., 1987). Attempts to ascribe some acute responses from ST treatment to the mechanism of action have given rise to some paradoxes between production and metabolic responses. We attempt to reconcile this on the basis of complications cited above and through alternative hypotheses.

4.2. Carbohydrate Metabolism

The dramatic shifts that occur in body composition with ST treatment are a clear indication that the utilization of carbohydrates by specific organs must be altered. This has been investigated in whole-animal studies by using glucose and insulin tolerance tests. These studies have demonstrated dramatic reductions in exogenous glucose or insulin challenges in a wide variety of animals treated with ST. Results from studies with growing pigs treated with pST are shown in Fig. 2 (Wray-Cahen *et al.*, 1987b). Reduced rates of glucose clearance in response to insulin or glucose challenges have also been observed for adult human males treated with hST for 4 days (Rosenfeld *et al.*, 1982) and sheep treated with bST for 2 days (Hart *et al.*, 1984). However, the development of this shift in carbohydrate metabolism occurs only after several days of treatment. Although there were no alterations in obese mice 24 hr after hST (22 kDa) treatment, a reduction in the clearance of a glucose challenge occurred after 3 days of treatment with either the 22- or 20-kDa variant of hST (Kostyo *et al.*, 1984, 1985). Lewis *et al.*, (1977) found no effect on glucose tolerance in dogs 10 hr after administration of physiological doses of highly purified hST. In contrast, others using less purified hST have demonstrated a change in glucose tolerance in dogs after several days of ST administration (Altszuler *et al.*, 1968). Although the mechanism by which ST alters tissue response to insulin is not known, Rosenfeld *et al.* (1982) concluded that the reduction in insulin response was at a postreceptor site based on observations

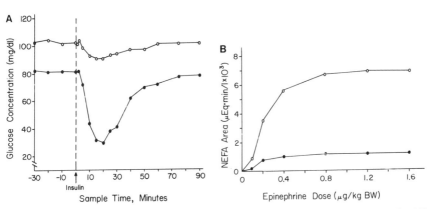

FIGURE 2. Effect of chronic somatotropin treatment on response to homeostatic signals. (A) depicts blood glucose response to an insulin challenge (1.0 μg/kg body weight) in growing pigs treated with excipient or pST (Wray-Cahen *et al.*, 1987b). (B) depicts the response in blood concentration of nonesterified fatty acids to an epinephrine challenge (varying doses) in lactating cows treated with excipient or bST (Sechen, Dunshea, and Bauman, unpublished data).

that insulin receptor number and affinity in myocytes were relatively unaffected in ST-treated humans.

The ability of ST to alter glucose clearance and peripheral tissue response to insulin requires consideration with regard to two aspects. First, many studies have used doses of exogenous ST that are unphysiological, being several fold greater than required to obtain growth effects (see review by Lewis *et al.*, 1987). Wray-Cahen *et al.* (1987a) addressed this in pigs using a daily pST dose (120 μg/kg body weight) consistent with *in vivo* growth responses (see Fig. 1). They also used the euglycemic clamp technique, a more powerful method for examining effects on glucose homeostasis than the acute tolerance tests (DeFronzo *et al.*, 1979). Similar to earlier studies, Wray-Cahen *et al.* (1987a) reported a dramatic reduction in tissue uptake of glucose in response to insulin infusion with values in pigs treated with pST for 4 weeks averaging only 28% of the control group.

A second critical consideration relates to the purity of the ST preparations. A number of studies have observed that ST administration causes acute (i.e., within minutes or a few hours) hyperglycemic and hyperinsulinemic responses or gives immediate (within a few hours) changes in insulin and glucose tolerance in laboratory and farm animals (e.g., see Cotes *et al.*, 1949; Campbell, 1955; Wallace and Bassett, 1966; Davis *et al.*, 1969; Bourne *et al.*, 1977). Louis and Conn (1972) were among the first to suggest that these acute effects were due to a contaminant. Subsequently, Lewis *et al.* (1977, 1980, 1981) demonstrated that these acute effects were reduced by further purification of hST (20 or 22 kDa) or enhanced by proteolysis of highly purified hST. These results were extended by Larson *et al.* (1978, 1980) who showed that whereas a contaminant of hST exhibited acute hyperglycemic and hyperinsulinemic effects *in vivo* and stimulated release of insulin from pancreatic cells *in vitro*, highly purified 22 kDa hST had no acute effect *in vivo* or *in vitro*. Other recent studies have failed to demonstrate any acute effects of bST on circulating concentrations of glucose or insulin in growing cattle (Eisemann *et al.*, 1986a; Peters, 1986) and sheep (Johnsson and Hart, 1985; Laarveld *et al.*, 1986) or lactating cows (see review by Bauman and McCutcheon, 1986). The only recent exception is a study in which glucose increases were observed in sheep immediately following an injection of bST (Hart *et al.*, 1984). However, it seems likely that the pH extremes (pH 11 and 3.6) used to solubilize the ST in that study would lead to chemical and conformational artifacts. Overall, the results are reasonably consistent and suggest that the acute hyperglycemic and hyperinsulinemic effects of ST observed in earlier studies were due to contaminants including fragments and chemical artifacts of ST as discussed in reviews by Lewis (1984) and Lewis *et al.* (1980, 1987).

Large animals, because of their size, have allowed for extensive sampling and a complete characterization of blood levels of insulin and glucose with chronic ST treatment. In growing cattle receiving daily treatment with bST, circulating levels of insulin are increased (Eisemann *et al.*, 1986a; Peters, 1986). Similar effects have been observed in growing pigs treated with pST (Wray-Cahen *et al.*, 1987a,b) and lambs treated with bST (Johnsson and Hart, 1985). In all these studies, ST treatment resulted in enhanced performance, and the elevation in blood insulin concentration was a chronic rather than an acute transitory response that followed changes in circulating ST as a result of the daily administration of exogenous ST. Studies by Eisemann *et al.* (1986a) demonstrated that although blood concentrations of insulin were maintained at a higher level during the period of ST treatment, insulin concentrations still oscillated with food consumption in a normal manner.

Blood concentrations of glucose were unaltered in growing cattle receiving chronic ST treatment (Eisemann *et al.*, 1986a; Sandles and Peel, 1987), but both basal and fasting levels were significantly greater in ST-treated pigs (Wray-Cahen *et al.*, 1987b). We are unaware of any studies that have examined the effect of chronic ST treatment on glucose kinetics in growing pigs, sheep, or cattle. However, this has been examined in dairy cows. ST treatment of dairy cows has no effect on circulating concentrations of glucose or insulin (see review by Bauman and McCutcheon, 1986), but does result in a modest increase in glucose irreversible loss rate (due to increased gluconeogenesis) and a reduction in glucose oxidation to CO_2 (Bauman *et al.*, 1988). These metabolic adaptations quantitatively accounted for the extra glucose needed as a result of the increase in milk yield with bST treatment (Bauman *et al.*, 1988).

The chronic shift in insulin and glucose tolerance is frequently referred to as a "diabetogenic" effect of somatotropin and is discussed as an abnormality in metabolism. We have chosen not to use the term "diabetogenic" for two reasons. First, this term has also been used to describe the supposed acute hyperglycemic and hyperinsulinemic effects of ST and this has led to confusion. Second, we do not feel that this represents an abnormality and the overall effects of ST are not consistent with diabetes. Glucose and insulin tolerance tests or glucose clamp techniques are merely measuring the ability of the body to remove an excessive carbohydrate load via insulin-sensitive tissues. Adipose tissue is insulin sensitive and the majority of excess carbohydrate is stored as fat. Somatotropin alters nutrient partitioning such that less nutrient is deposited as fat. A component of the mechanism involves an alteration in adipose tissue response to insulin (see Section 4.4). The changes in tissue response to this and other homeostatic signals are merely the physiological mechanism to allow a redirection of nutrient partitioning.

4.3. Bone and Mineral Metabolism

Mineral metabolism must also be regulated by somatotropin as illustrated by the enhanced rate of ash accretion in pigs receiving ST (Fig. 1). Although not a major component of the increased rate of gain in ST-treated animals, bone growth is of obvious importance to structural integrity. Current data demonstrate that ST facilitates bone growth by stimulating chondrocyte proliferation in the epiphyseal plate and by enhancing bone mineralization.

Early studies which examined the mechanism of action of ST on bone growth observed no direct effects on *in vitro* cartilage metabolism, whereas responses were observed with serum obtained from ST-treated animals (Daughaday, 1981). Thus, it was postulated that the effects of ST on growth, particularly skeletal growth, were mediated by ST-dependent plasma factors (somatomedins) produced in the liver (Daughaday, 1981). Consistent with this hypothesis, recent studies have demonstrated that infusion of IGF-I to hypophysectomized rats increases tibial epiphyseal growth (Schoenle *et al.*, 1982b; Russell and Spencer, 1985; Isgaard *et al.*, 1986).

Isaksson and co-workers (1982) were the first to challenge the completeness of this concept. They infused ST to the growth plate of the proximal tibia of hypophysectomized rats and observed accelerated longitudinal bone growth in that leg as compared to the contralateral leg, which received saline treatment. Subsequent work, utilizing a similar *in vivo* technique, has verified these results and clearly demonstrated that ST has a direct effect on bone growth (Russell and Spencer, 1985; Isgaard *et al.*, 1986; Schlechter *et al.*, 1986a; Nilsson *et al.*, 1987). Somatotropin receptors have been demonstrated in chondrocytes, which were isolated from the epiphyseal plate (Eden *et al.*, 1983) and stimulation of DNA synthesis has been observed in chondrocytes cultured with ST (Madsen *et al.*, 1983).

Tissue proliferation is dependent on two different cellular processes, differentiation and multiplication (Isaksson *et al.*, 1986). Green and co-workers (1985) proposed a dual effector theory for ST action on cell proliferation. Applying this theory to bone growth, ST would stimulate the differentiation of prechondrocytes to chondrocytes in the growth plate and then cells would undergo clonal expansion, which is dependent upon somatomedins (Green *et al.*, 1985; Isaksson *et al.*, 1986). Thus, ST could have both direct and indirect (via local production of somatomedins) effects on bone growth as portrayed in Fig. 3. This has not been extensively examined, but recent studies support this concept. Only cells at the proximal part of the growth plate bind ST (Isaksson *et al.*, 1985). This region corresponds to the germinal cell layer and is consistent with the theory that ST exerts a direct effect on longitudinal bone growth through action on chondrocyte progenitor cells in the growth plate. Pretreatment with ST for 24 hr *(in vivo)* markedly potentiates the stimulatory effect of IGF-I on *in vitro* colony formation of epiphyseal chondrocytes (Lindahl *et al.*, 1987a,b). Insulinlike growth factor I is present in proliferative chondrocytes (Stracke *et al.*, 1984) with the number of cells in the growth plate that contain IGF-I being directly regulated by ST (Nilsson *et al.*, 1986). A recent study has demonstrated that local administration of antibodies to IGF-I neutralized the stimula-

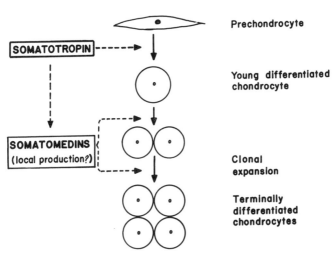

FIGURE 3. Dual effector model for somatotropin action on bone growth. Adapted from Green *et al.* (1985) and Isaksson *et al.* (1986).

tory effect of local ST administration on bone growth (Schlechter *et al.*, 1986b). This suggests that IGF-I, produced locally in response to ST, acts as an autocrine/paracrine growth factor, thereby promoting the proliferation of chondrocytes.

Somatotropin also exhibits metabolic effects on mineral metabolism. Obviously, absorption rates of several minerals must be altered with ST treatment to accommodate the increase in bone growth (Fig. 1). Braithwaite (1975) demonstrated that ST treatment of nearly mature wether sheep for 14 days resulted in increases in skeletal retention of calcium and phosphorus. He characterized changes in calcium kinetics and observed that ST treatment increased gut absorption, accretion into bone, and resorption from bone. Others have demonstrated that at least a portion of the effects of ST on calcium homeostasis are mediated by effects on the vitamin D cascade. For instance, administration of ST to hypophysectomized rats or ST-deficient children increased circulating concentrations of 1,25-dihydroxyvitamin D_3 (Spencer and Tobiassen, 1981; Burstein *et al.*, 1983; Gray *et al.*, 1983), levels of intestinal calcium-binding protein (Bruns *et al.*, 1983), and calcium absorption (Henneman *et al.*, 1960; Finkelstein and Schachter, 1962; Mainoya, 1975).

Wongsurawat *et al.* (1984) proposed that the mechanism by which ST affected calcium homeostasis was by direct or indirect modulation of renal metabolism of 25-hydroxyvitamin D_3. In hypophysectomized rats, treatment with ST increased the activity of 1α-hydroxylase (Spencer and Tobiassen, 1981;

Wongsurawat *et al.*, 1984; Gray and Garthwaite, 1985). This is the renal enzyme that converts 25-hydroxyvitamin D_3 to the active form, 1,25-dihydroxyvitamin D_3. Also, ST must be present for the normally inherent increase in 1,25-dihydroxyvitamin D_3 synthesis that occurs in response to dietary phosphate deprivation (Gray *et al.*, 1983; Gray and Garthwaite, 1985). From work with ST-deficient children receiving chronic ST therapy, Burstein *et al.* (1983) concluded that ST sensitizes the kidney to a cAMP-mediated effect of parathormone. Spanos *et al.* (1981) demonstrated that the addition of ST to primary kidney cell cultures from chickens increased the activity of 1α-hydroxylase. However, we are not aware of any similar investigations with mammalian kidney cells. Alternatively, effects could be indirect via IGF-I as postulated by Gray (1987) after observing that circulating concentrations of 1,25-dihydroxyvitamin D_3 and IGF-I were highly correlated during various dietary and endocrine manipulations. However, a cause-and-effect relationship has not been established.

4.4. Adipose Tissue Lipid Metabolism

In relation to overall shifts in nutrient partitioning, ST effects on lipid metabolism are a major component. Body lipid accretion represents the balance between synthesis and mobilization. Treatment with ST affects both processes in adipose tissue and these chronic alterations in lipid metabolism relate to the "metabolic effects" of ST and its role as a homeorhetic control of nutrient partitioning.

The reduction in lipid synthesis that occurs with ST treatment appears to be a function of biochemical changes in adipose tissue. Goodman (1963) first demonstrated that the lipogenic capacity of adipose tissue, assessed by *in vitro* incubations, was severalfold lower in hypophysectomized rats that received chronic treatment with ST. More recent studies with normal growing pigs have shown that the activities of several enzymes associated with lipogenesis were substantially reduced in adipose tissue from the ST-treated group (Magri *et al.*, 1987). Schoenle *et al.* (1983) reported that the glucose transport system in adipocytes was also affected, because the basal rate of glucose transport decreased in a dose-dependent manner when hypophysectomized rats were infused with varying doses of ST for 6 days. The effect on glucose transport was maximal after 3 days of *in vivo* treatment and Schoenle *et al.* (1982a) proposed that the mechanism involved an ST-dependent "limiting factor" for glucose transport. Studies by this group have also provided some additional insight into the mechanism of action of ST with the observation that 6-day infusions of IGF-I or IGF-II had no effect on glucose transport in adipocytes (Schoenle *et al.*, 1983). Thus, the effects of ST on adipose tissue may be direct and appear not to involve Sm. This is in obvious contrast to the involvement of Sm in the ST response in bone and muscle as discussed in other sections.

The reduction in lipid synthesis is related, at least in part, to changes in adipose tissue response to insulin. This has been clearly demonstrated in a series of culture studies with adipose tissue from growing cattle (Etherton *et al.*, 1987) and pigs (Walton *et al.*, 1986). No effects on insulin-stimulated lipogenesis were observed with ST addition in short-term cultures (pST with porcine tissue and bST with bovine tissue). However, after 48–50 hr of culture, ST antagonized the ability of insulin to maintain lipogenesis. This antagonism was potentiated when hydrocortisone was included in the culture medium. A similar antagonism to insulin has been observed with ovine adipose tissue (obtained from growing and lactating animals) chronically incubated with oST (Vernon, 1982; Vernon and Finley, 1986). These *in vitro* results are consistent with *in vivo* studies (euglycemic clamp and insulin/glucose tolerance) that demonstrated a marked decrease in glucose uptake in response to insulin in ST-treated animals.

Insulin is a homeostatic signal involved in short-term regulation of energy balance. Reduced insulin-stimulated lipogenesis in ST-treated animals would decrease the proportion of nutrients deposited as fat in adipose tissue. The biological changes that cause the reduced insulin response have not been well established. One possibility is that ST changes the binding affinity and/or reduces insulin receptor number. This has not been investigated for adipose tissue but seems unlikely to be a major component of the mechanism because chronic ST treatment also increases the antilipolytic response to insulin (see the subsequent discussion). Rather, the effects of ST appear to center on intracellular biochemical processes. In addition to the previously discussed changes in enzyme activities and glucose transport, several studies have suggested that ST alters the activation of key enzymes. Bornstein *et al.* (1983) observed that ST addition to adipocytes or hepatocytes from normal rats decreased acetyl CoA carboxylase activity by 40–50%, this being related to an increase in the inactive, phosphorylated form of the enzyme. However, these results are not entirely consistent with the aforementioned effects of ST in chronic cultures of adipose tissue. Bornstein *et al.* (1983) observed changes a few minutes after adding ST, with maximum inhibition of acetyl CoA carboxylase activity occurring within 30 min.

Recent studies have demonstrated a temporal pattern of enzyme response that is more consistent with the chronic effects of ST. Schaffer (1985) examined the effect of ST additions on the activity of lipogenic enzymes (ATP-citrate lyase, NADP-malate dehydrogenase, and fatty acid synthetase) in rat hepatocyte cultures. (Note: liver is an important site of lipogenesis in the rat.) Compared to cells incubated in basal medium (which included insulin, dexamethasone, and T_3), there was no effect after 1 day of culture, but thereafter decreases (of 20 to 52%) in these lipogenic enzymes were observed when ST was included. Similar studies involving chronic cultures (48 hr) with sheep adipocytes demonstrated that ST did not block the increase in fatty acid synthetase activity and only partially prevented the increase in total acetyl CoA carboxylase observed with insulin and dexamethasone (Vernon *et al.*, 1988). However, ST or actinomycin D prevented the conversion of acetyl CoA carboxylase to the active state (Vernon *et al.*, 1988). Collectively, these results suggest that the mechanism by which ST decreases rates of lipid synthesis involves both a decrease in amount of several key enzymes involved in lipogenesis and an impairment of the transduction system whereby insulin (and possibly other homeostatic signals) activates enzymes such as acetyl CoA carboxylase.

An interesting paradox is that ST addition to adipocyte incubations also acutely stimulates glucose uptake, increases rates of glucose oxidation to CO_2 and incorporation into fatty acids, as well as inhibiting lipolysis. These acute insulinlike effects have been examined in laboratory animals and responses appear related to the ST status of the animal. If the adipose tissue comes from an animal that is ST deficient, the *in vitro* stimulation of glucose uptake and use by addition of ST is immediately followed by a period of refractoriness. If the tissue is from an animal with normal ST status, tissue is not responsive and several hours of culture in the absence of ST is required before this refractoriness is lost. Goodman and Coiro (1981) have reviewed this area and discussed these effects in relation to whole-animal physiology. They concluded that in an animal that has circulating ST, adipose tissue would always be refractory to these specific acute effects of ST. The preceding discussion clearly demonstrates that such a conclusion is warranted because the chronic effects of ST lead to a reduction rather than an increase in glucose uptake and lipid accretion. However, the physiological significance of adipose tissue having acute lipogenic and antilipolytic responses to ST with refractoriness of the responses being induced by ST is not obvious.

The shift in lipid mobilization that occurs with ST treatment also involves an alteration of adipose tissue response to homeostatic signals that stimulate lipolysis. For example, ST-treated cattle (both restricted and ad libitum fed) exhibit a marked increase in response to an acute epinephrine challenge as determined by the increase in circulating levels of nonesterified fatty acids (NEFA) (Peters, 1986). In growing pigs, an epinephrine challenge also gives a greater rise in circulating NEFA in ST-treated animals than observed for controls (Wray-Cahen *et al.*, 1987b). This effect was first demonstrated in dairy cows regardless of whether the milk production increase with ST treatment caused cows to be in a positive or negative energy balance (Sechen *et al.*, 1985; McCutcheon and Bauman, 1986b). A subsequent study quantified the epinephrine dose–response profile. A severalfold increase in circulating concentrations of NEFA and glycerol was observed across a wide range of epinephrine doses in ST-treated dairy cows (Fig. 2; Sechen, Dunshea, and Bauman, unpublished results). Vernon *et al.* (1987) have investigated the fall in norepinephrine-stimulated lipolysis that occurs in lactating rats when the litter is removed. They observed that this decrease in adipose tissue responsiveness with weaning could be prevented by injections of ST and that the decline could be mimicked in lactating rats by administration of antiserum to growth hormone. Other studies have examined the antilipolytic effects of insulin in animals receiving ST treatment. After several days of treatment with ST, an acute insulin challenge results in a more dramatic reduction in circulating concentrations of glycerol and/ or NEFA in both growing pigs (Wray-Cahen *et al.*, 1987b) and lactating cows (Sechen *et al.*, 1985; Sechen, Dunshea, and Bauman, unpublished results). Thus far, the cellular mechanism by which ST alters the lipolytic response to homeostatic signals has not been examined.

Overall, results demonstrate that ST treatment results in a reduced response of adipose tissue to homeostatic signals that stimulate lipogenesis and an enhanced responsiveness to homeostatic signals that affect lipolysis. Thus, animals treated with ST maintain the ability to respond to homeostatic controls but the magnitude of response to specific signals is altered. The net result is that less nutrients are partitioned to adipose tissue, thereby reducing body fat accretion. The extent to which reduced lipid accretion is due to reduced synthesis versus enhanced lipid mobilization is undoubtedly related to energy balance (Bauman and McCutcheon, 1986). In a growing animal when energy intake is high (positive energy balance), the major effect of ST on adipose tissue is probably to reduce rates of lipid synthesis. Circulating concentrations of NEFA are unaffected when bST-treated cattle are in a positive energy balance (Peters, 1986), but there is some indication that NEFA levels may be elevated under the same conditions in pST-treated pigs (Chung *et al.*, 1985; Wray-Cahen *et al.*, 1987b). On the other hand, when energy intake is restricted, lipogenic rate is already low. In this instance, the major effect of ST is on lipid mobilization (Machlin, 1972). Under conditions of restricted intake, ST-treated cattle have chronically elevated NEFA levels (Eisemann *et al.*, 1986a; Peters, 1986) and the turnover rates and oxidation rates of NEFA are increased (Eisemann *et al.*, 1986a). This relationship has been more extensively studied in dairy cattle where ST treatment causes chronic elevations of NEFA and changes in milk fat composition only when animals are in a negative energy balance (see review by Bauman and McCutcheon, 1986). In this situation, the NEFA turnover rate and rate of oxidation to CO_2 are increased in dairy cows and highly correlated with the degree of negative energy balance (Bauman *et al.*, 1988).

It is frequently stated that ST has acute lipolytic effects based on the fact that a number of early studies observed increases in NEFA or glycerol with ST administrations. This lipolytic response occurred within minutes or a few hours after *in vivo* ST administration or ST addition to *in vitro* adipose incubations in a range of species. However, this acute effect of ST was controversial and not observed in many studies. Goodman and Schwartz (1974) reviewed the literature and pointed out the inconsistencies in the lipolytic responses obtained in various laboratories, the extremely high doses frequently used, the variations of lipolytic potency of different preparations, and the discrepancies between *in vivo* and *in vitro* responses. They concluded at that time that ST acts as a modulator or amplifier of physiological stimuli for lipid mobilization rather than as an acute stimulator of lipolysis. With only a few exceptions, most recent studies have not supported the concept that ST has an acute lipolytic effect.

Undoubtedly, a major portion of the controversy relates to the purity of the ST preparations used. Lee *et al.* (1974) demonstrated that the acute lipolytic stimulation elicited by bST in rat adipose tissue resulted from TSH contamination. In contrast to earlier reports, more recent studies have also observed that more highly purified preparations of bST, pST, and hST (20 and 22 kDa) have no acute lipolytic effects *in vivo* or *in vitro* with rat or rabbit adipose preparations (Tutwiler and Kirsch, 1976; Frigeri *et al.*, 1979; Frigeri, 1980; Bowden *et al.*, 1985). Similarly, hST is without effect on lipolysis in short-term incubations of human adipose tissue (Nyberg and Smith, 1977). In the case of farm animals, *in vitro* studies using homologous systems have failed to observe any acute lipolytic effects in sheep or pig adipose tissue (Duquette *et al.*, 1984; Walton *et al.*, 1986). Similarly, *in vivo* studies, where the typical ST effects on performance occurred, have found no acute effects on NEFA or glycerol in growing cattle (Eisemann *et al.*, 1986a; Peters, 1986) or lactating cows (see review by Bauman and McCutcheon, 1986). However, some groups have continued to observe an acute lipolytic effect in incubations of rat adipose tissue with hST (Goodman and Grichting, 1983; Goodman, 1984) and sheep injected with bST (Hart *et al.*, 1984). Machlin (1972) compared several different pST preparations and observed a wide variation in lipolytic activity as assessed by release of glycerol with short-term incubations of rat and rabbit adipose tissue. He further purified one of these preparations and found that the *in vitro* lipolytic activity was abolished but the pST preparation still gave a marked increase in growth rate and reduced carcass fat accretion in pigs treated for 21 days. Thus, several conclusions are apparent. First, the preponderance of recent data demonstrates that ST has no acute lipolytic effects. Previously reported effects were partly due to hormone contaminants and also due to fragments and/or chemical and conformational artifacts introduced in the purification process. Second, the growth and lactational responses with exogenous ST treatment occur regardless of whether the ST preparation possesses any acute lipolytic effects, so that the mechanisms involved in the performance responses are obviously not dependent on such acute effects.

Growth hormone also affects adipocyte proliferation *in vitro* (see Chapter 3). Specifically, when 3T3 preadipose cells are cultured with ST, cell differentiation occurs followed by the appearance of lipogenic enzymes and the accumulation of triglyceride (Morikawa *et al.*, 1982a,b; Nixon and Green, 1984a). These *in vitro* results represent a paradox with *in vivo* data. An increase in the number of adipocytes would logically lead to an increase in lipid accretion. Indeed, adipose hyperplasia can occur throughout the postnatal growth period and studies with rats have demonstrated that nutritional manipulations which cause fat cell hyperplasia are associated with an increase in fat accretion (Bertrand *et al.*, 1984). Yet *in vivo* studies involving a wide range of species have consistently demonstrated that growth hormone treatment of postnatal animals results in a decrease in body fat accumulation. Similar results have been observed in the prenatal period. Hypophysectomy or decapitation of fetal pigs leads to an increase in fat accretion that is characterized by increased adipocyte size and enhanced activity of lipogenic enzymes (Hausman and Thomas, 1984; Hausman *et al.*, 1987). Hypophysectomy of fetal lambs leads to a substantial increase in subcutaneous fat and a modest increase in internal fat, which does not occur if the hypophysectomized lambs are given ST therapy (Stevens and Alexander, 1986). Adipose cell numbers have not been systematically characterized in any of the *in vivo* studies with ST treatment but clearly the reduction in fat accretion appears at odds with the proliferation of adipocytes observed *in vitro*.

4.5. Muscle and Protein Metabolism

Studies have consistently demonstrated that treatment of hypophysectomized laboratory animals or intact farm animals with exogenous ST causes a remarkable increase in lean tissue growth. Muscle growth is accomplished by two fundamental processes, cell proliferation and protein accretion (Allen *et al.*, 1979; see Chapter 4). Available evidence suggests that ST affects both components.

Cell proliferation can result either from the generation of new muscle fibers (hyperplasia) or as a consequence of the fusion of new cells with existing muscle fibers (hypertrophy). Hypophysectomy or decapitation of pig fetuses at about mid pregnancy results in a decrease in the number of muscle fibers (Hausman *et al.*, 1982, 1987). Thus, ST may be involved in muscle hyperplasia in prenatal animals, although ST replacement therapy has not been employed to demonstrate the specificity of this effect. In contrast, the limited available data indicate that ST effects in the postnatal animal are related to muscle hypertrophy. In hypophysectomized rats, ST treatment increased the total DNA content of muscle (Beach and Kostyo, 1968; Goldspink and Goldberg, 1975), but not the ratio of DNA content to cytoplasmic mass (Beach and Kostyo, 1968). Similar observations were made in rats with ST-secreting tumors (Prysor-Jones and Jenkins, 1980; McCusker and Campion, 1986). These results are consistent with the recruitment of new nuclei into muscle fibers.

There is general agreement that the source of the new nuclei in growing muscle of the postnatal animal is the satellite cells (Allen *et al.*, 1979; Campion, 1984). Thus, ST treatment of postnatal animals appears to result in satellite cell proliferation and incorporation of daughter nuclei into the adjacent myofibers.

The specific mechanism(s) by which ST leads to proliferation of muscle has been investigated. Florini (1985) reviewed the studies by his group and others involving myoblasts from various species. The results have uniformly demonstrated a lack of any direct effect of ST with cultured myoblast cells, as assessed by various indices of differentiation and proliferation. Nixon and Green (1984b) also observed no direct effect of ST on some of the same cell lines. However, they did observe a direct effect on one myoblast cell line ($10T_{1/2}$) in which ST promoted myogenesis and the formation of multinucleated muscle cells. They suggested that some cell lines, such as the commonly used L_6 line, may have undergone selection pressure for growth hormone independence during their evolution in culture or that different myoblast subtypes may exist, which differ in their dependence of ST (Nixon and Green, 1984b). Another consideration is that studies have concentrated almost exclusively on myogenic cells of fetal or neonatal origin. As previously discussed, the dramatic effects of ST on muscle growth *in vivo* occur during the postnatal period and would therefore, logically involve the proliferation of satellite cells. However, ST also has no direct effect on satellite cell proliferation or actin synthesis in satellite cell-derived myotubes (Allen *et al.*, 1983, 1986).

In contrast to the results with ST, studies with myoblast cell lines and satellite cells clearly demonstrate a mitogenic response to several somatomedins including IGF-I (Allen *et al.*, 1984; Florini, 1985, 1987; Dodson *et al.*, 1985). The dual effector theory of ST action in cell proliferation is based on the direct effect of ST on the target cell, in this case myoblasts or satellite cells, followed by an indirect effect mediated via somatomedins. The bulk of the evidence, as discussed above, would not support this theory for skeletal muscle. Instead, results are consistent with the original postulate of Daughaday (1981) that ST has no direct effects on proliferation of muscle cells but rather effects are indirectly mediated by ST-dependent production of somatomedins in other tissues, probably liver, which are transported to muscle via the circulatory system. However, there are some inconsistencies for which we do not have any adequate explanations. First, skeletal muscle does possess specific binding sites for ST (Kostyo, 1986), although the cell type involved has not been characterized. Second, muscle, or at least regenerating skeletal muscle cells, produces IFG-I. Jennische and co-workers (1987) demonstrated that following irreversible damage to skeletal muscle (extensor digitorum longus muscle) of adult rats, IGF-I was expressed in satellite cells within 24 hours after injury and subsequently expressed in the progeny of satellite cells, myoblasts, and immature muscle cells.

Accretion of tissue protein represents the net effect of two processes: synthesis and protein degradation. Protein accretion may be affected by altering the rate of either or both of these processes (Lindsay, 1983; Reeds and Palmer, 1986). The extent to which the increase in protein accretion with ST treatment is related to alterations in protein synthesis and/or degradation rates has been examined in only a few studies. Studies with hypophysectomized rats have shown that treatment with bST increases rates of protein synthesis in the soleus muscle but does not alter rates of protein degradation (Goldberg, 1969; Goldberg et al., 1980). The increase in protein synthesis was observed for both soluble and myofibrillar proteins (Goldberg, 1969). Two studies have examined this in intact beef cattle where bST administration resulted in increased nitrogen retention. Working with beef heifers fed adequate protein but near-maintenance levels of intake, Eisemann et al. (1986a) observed that whole-body rates of protein synthesis were higher and leucine oxidation rates were lower with bST treatment. By using protein deposition rates, they calculated that whole-body rates of protein degradation were not different. In a second experiment with beef steers, animals received radioactive leucine and the fractional rates of protein synthesis tended to be higher in all tissues examined with bST treatment (Eisemann et al., 1986b). Pell and Bates (1987) also measured turnover of collagen and noncollagen protein in skeletal muscle of bST-treated lambs and observed that the increase in net muscle growth could be accounted for by increased rates of muscle protein synthesis.

Several groups have also reported that ST causes an acute stimulation of amino acid transport and protein synthesis in muscle. This has been most extensively examined in rat diaphragm muscle. Effects are similar to the acute insulinlike responses in adipose tissue (increased lipogenesis and decreased lipolysis) in that they are transitory and a refractoriness immediately develops if ST is present. (Albertsson-Wikland and Isaksson, 1978; Albertsson-Wikland et al., 1980). Thus, the physiological significance is not obvious and these acute effects appear to be unrelated to the increased protein accretion with chronic ST treatment.

5. Summary and Perspectives

During the last decade, our knowledge of the role of ST in metabolism and the regulation of growth has increased tremendously. Although much remains to be determined about the mechanisms of ST action, much is known about how tissue and biochemical processes are altered. It is clear that we have entered into an era where scientific discovery will be immense.

Since the physiological effects of ST are so varied, it is difficult to develop a working hypothesis for its mechanisms of action. Delineation of the mechanisms will, of necessity, coincide with advances in chemical definition of the family. Research is needed to determine to what extent "members" evolve because of genomic considerations as compared to metabolism of the parent or precursor molecule. Most investigators have focused on the 22-kDa form of ST, but it is possible that both variants and biologically active peptides (arising from enzymatic modification subsequent to translation) exist in farm animals. Although the physiological relevance of multiple forms is still a matter of conjecture, it is conceivable that different forms and alterations in structure may be required to accomplish the many known mitogenic and metabolic actions ascribed to this family. Further, the possibility for a heterogeneous and even dynamic population of receptors, differing with physiological state and perhaps even among organs, must also be considered (Nicoll et. al., 1986). Our knowledge in these areas is meager and deserving of considerable attention.

Changes in animal production traits resulting from ST treatment cannot be ascribed to acute effects on metabolism but rather involve processes requiring chronic exposure. The effects of ST can be broadly categorized as to those associated with cell proliferation and those that are "metabolic" in nature. The mechanisms appear to involve both direct action of ST and indirect action mediated by Sm. The presence of specific receptors on chondrocytes isolated from the epiphyseal plate (Eden et al., 1983), adipocytes (Fagin et al., 1980), and muscle (Kostyo, 1986) suggests that ST acts at the local level. With respect to cell proliferation per se, ST may stimulate local production of IGF-I, which in turn initiates the mitogenic response. In the case of muscle, the presence of ST receptors has not been reported on myoblasts, satellite cells, or myofibers. Limited data also indicate that effects of ST on protein accretion in muscle involve an increase in the rate of protein synthesis, whereas the dramatic effects on lipid accretion involve alterations in both lipid synthesis and mobilization. The relative importance or magnitude of response for each of the lipid processes is related to energy balance.

In postulating that nutrient partitioning is regulated by both chronic (homeorhetic) and acute (homeostatic) signals, we proposed that homeorhetic controls coordinate overall direction by altering the responsiveness of tissues to a variety of homeostatic signals (Bauman and Currie, 1980; Bauman, 1984). This thesis appears consistent for adipose tissue, in that ST treatment results in decreased tissue response to those homeostatic signals that stimulate lipogenesis but increases tissue response to those that affect rates of lipolysis. This is an example of a pivotal role ST plays in dynamic physiological control. Study of other tissues in relation to this thesis and identification of specific cellular processes affected are directions for future research.

References

Albertsson-Wikland, K., S. Eden and O. Isaksson. 1980. Analysis of early responses to growth hormone on amino acid transport and protein synthesis in diaphragms of young normal rats. Endocrinology 106:291.

Albertsson-Wikland, K. and O. Isaksson. 1978. Time course of the effect of growth hormone in vitro on amino acid and monosaccharide transport and on protein synthesis in diaphragm of young normal rats. Endocrinology 102:1445.

Allen, R. E., M. V. Dodson, L. K. Boxhorn, S. L. Davis and K. L. Hossner. 1986. Satellite cell proliferation in response to pituitary hormones. J. Anim. Sci. 62:1596.

Allen, R. E., M. V. Dodson and L. S. Luiten. 1984. Regulation of skeletal muscle satellite cell proliferation by bovine pituitary fibroblast growth factor. Exp. Cell Res. 152:154.

Allen, R. E., K. C. Masak, P. K. McAllister and R. A. Merkel. 1983. Effect of growth hormone, testosterone and serum concentration on actin synthesis in cultured satellite cells. J. Anim. Sci. 56:833.

Allen, R. E., R. A. Merkel and R. B. Young. 1979. Cellular aspects of muscle growth: myogenic cell proliferation. J. Anim. Sci. 49:115.

Altszuler, N., I. Rathgeb, B. Winkler and R. C. de Bodo. 1968. The effects of growth hormone on carbohydrate and lipid metabolism in the dog. Ann. N.Y. Acad. Sci. 148:441.

Aston, R., A. T. Holder, M. A. Preece and J. Ivanyi. 1986. Potentiation of the somatogenic and lactogenic activity of human growth hormone with monoclonal antibodies. J. Endocrinol. 110:381.

Barnard, R., K. Munro, R. Duplock, P. Bundesen, D. Rylatt and M. Waters. 1984. Evidence for structural heterogeneity of the growth hormone receptor using monoclonal antibody probes. Proc. Endocrinol. Soc. (Aust.) 27:115.

Bauman, D. E. 1984. Regulation of nutrient partitioning. In: F. M. C. Gilchrist and R. I. Machie (Ed.) Herbivore Nutrition in the Subtropics and Tropics. pp 505–524. The Science Press, Craighall, South Africa.

Bauman, D. E. and W. B. Currie. 1980. Partitioning of nutrients during pregnancy and lactation: a review of mechanisms involving homeostasis and homeorhesis. J. Dairy Sci. 63:1514.

Bauman, D. E., J. H. Eisemann and W. B. Currie. 1982. Hormonal effects on partitioning of nutrients for tissue growth: role of growth hormone and prolactin. Fed. Proc. 41:2538.

Bauman, D. E. and S. N. McCutcheon. 1986. The effects of growth hormone and prolactin on metabolism. In: L. P. Milligan, W. L. Grovum and A. Dobson (Ed.) Proceedings, VI International Symposium on Ruminant Physiology: Control of Digestion and Metabolism in Ruminants. pp 436–455. Prentice–Hall, Englewood Cliffs, NJ.

Bauman, D. E., C. J. Peel, W. D. Steinhour, P. J. Reynolds, H. F. Tyrrell, A. C. G. Brown and G. L. Haaland. 1988. Effect of bovine somatotropin on metabolism of lactating dairy cows: influence on rates of irreversible loss and oxidation of glucose and nonesterified fatty acids. J. Nutr. 118:1031.

Baumann, G., J. G. MacCart and K. Amburn. 1983. The molecular nature of circulating growth hormone in normal and acromegalic man: evidence for a principal and minor monomeric form. J. Clin. Endocrinol. Metab. 56:946.

Baumann, G., M. W. Stolar and K. Amburn. 1985. Molecular forms of circulating growth hormone during spontaneous episodes and in the basal state. J. Clin. Endocrinol. Metab. 60:1216.

Baumann, G., M. W. Stolar and T. A. Buchanan. 1986. The metabolic clearance, distribution, and degradation of dimeric and monomeric growth hormone (GH): implications for the pattern of circulating GH forms. Endocrinology 119:1497.

Baumrucker, C. R. 1986. Insulin-like growth factor I (IGF-I) and insulin stimulates DNA synthesis in bovine mammary tissue explants from pregnant cows. J. Dairy Sci. 69 (Supl. 1):120 (Abstr.).

Beach, R. K. and J. L. Kostyo. 1968. Effect of growth hormone on the DNA content of muscles of young hypophysectomized rats. Endo. 82:882.

Bell, A. W., D. E. Bauman and W. B. Currie. 1987. Regulation of nutrient partitioning and metabolism during pre- and post-natal growth. J. Anim. Sci. 65 (Suppl. 2):186.

Bertrand, H. A., C. Stacy, E. J. Masoro, B. P. Yu, I. Murata and H. Maeda. 1984. Plasticity of fat cell number. J. Nutr. 114:127.

Bornstein, J., F. M. Ng, D. Heng and K. P. Wong. 1983. Metabolic actions of pituitary growth hormone. I. Inhibition of acetyl CoA carboxylase by human growth hormone and a carboxyl terminal part sequence active through a second messenger. Acta Endocrinol. 103:479.

Bourne, R. A., H. A. Tucker and E. M. Convey, 1977. Serum growth hormone concentrations after growth hormone or thyrotropin releasing hormone in cows. J. Dairy Sci. 60:1629.

Bowden, C. R., K. D. White, U. J. Lewis and G. F. Tutwiler. 1985. Highly purified human growth hormone fails to stimulate lipolysis in rabbit adipocytes in vitro or in rabbits in vivo. Metabolism 4:237.

Boyd, R. D., D. E. Bauman, D. H. Beermann, A. F. DeNeergaard, L. Souza and W. R. Butler. 1986. Titration of the porcine growth hormone dose which maximizes growth performance and lean deposition in swine. J. Anim. Sci. 63 (Suppl. 1):218 (Abstr.).

Braithwaite, G. D. 1975. The effect of growth hormone on calcium metabolism in the sheep. Brit. J. Nutr. 33:309.

Bruns, M. E. H., S. S. Vollmer, D. E. Bruns and J. G. Overpeck. 1983. Human growth hormone increases intestinal vitamin D-dependent calcium binding protein in hypophysectomized rats. Endocrinology 113:1387.

Burstein, S., I. -W. Chen and R. C. Tsang. 1983. Effects of growth hormone replacement therapy on 1,25-dihydroxyvitamin D and calcium metabolism. J. Clin. Endocrinol. Metab. 56:1246.

Campbell, J. 1955. Diabetogenic actions of growth hormone. In: R. W. Smith, O. H. Gaebler and C. N. H. Long (Ed.) The Hypophyseal Growth Hormone, Nature and Actions. pp 270–285. McGraw–Hill Book Co., New York.

Campbell, R. G. 1987. Energy and protein metabolism. Proc. First Aust. Pig Conf. p. 43.

Campbell, R. G., T. J. Caperna, N. C. Steele and A. D. Mitchell. 1988. Interrelationships between energy intake and exogenous porcine growth hormone administration on the performance, body composition, and protein and energy metabolism of growing pigs weighing 25 to 55 kilograms liveweight. J. Anim. Sci. 66:1643.

Car, M., A. Zindar and T. Fillipan. 1967. An effect of the treatment of young steers with STH (growth hormone) upon nitrogen retention in intensive feeding. Vet. Arh. 37:173.

Campion, D. R. 1984. The muscle satellite cell: a review. Int. Rev. Cytology 87:225.

Chawla, R. K., J. S. Parks and D. Rudman. 1983. Structural variants of human growth hormone: biochemical, genetic and clinical aspects. Annu. Rev. Med. 34:519.

Chung, C. S., T. D. Etherton and J. P. Wiggins. 1985. Stimulation of swine growth by porcine growth hormone. J. Anim. Sci. 60:118.

Clark, R. G., J.-O. Jansson, O. Isaksson and I. C. A. F. Robinson, 1985. Intravenous growth hormone: growth responses to patterned infusions in hypophysectomized rats. J. Endo. 104:53.

Closset, J., J. Smal, F. Gomez and G. Hennen. 1983. Purification of the 22,000- and 20,000-mol. wt. forms of human somatotropin and characterization of their binding to liver and mammary binding sites. Biochem. J. 214:885.

Convey, E. M. 1987. Advances in animal science: potential for improving meat animal production. Proc. Cornell Nutr. Conf. p. 1.

Cotes, P. M., W. A. Bartlett, R. E. Gaines Das, P. Flecknell and R. Termeer. 1980. Dose regimens of human growth hormone: effects of continuous infusion and of a gelatin vehicle on growth in rats and rate of absorption in rabbits. J. Endocrinol. 87:303.

Cotes, P. M., E. Reid and F. G. Young. 1949. Diabetogenic action of pure anterior pituitary growth hormone. Nature 164:209.

Daughaday, W. H. 1981. Growth hormone and the somatomedins. In: W. H. Daughaday (Ed.) Endocrine Control of Growth. pp 1–24. Elsevier, Amsterdam.

Davidson, M. B. 1987. Effect of growth hormone on carbohydrate and lipid metabolism. Endocrinol. Rev. 8:115.

Davis, S. L., U. S. Garrigus and F. C. Hinds. 1969. Metabolic effects of growth hormone and diethylstilbestrol in lambs. II. Effects of daily ovine growth hormone injections on plasma metabolites and nitrogen-retention in fed lambs. J. Anim. Sci. 30:236.

DeFronzo, R. A., J. D. Tobin and R. Andreas. 1979. Glucose clamp technique: a method for quantifying insulin secretion and resistance. Amer. J. Physiol. 237:E214.

Dehoff, M. H., R. G. Elgin, R. J. Collier and D. R. Clemmons. 1987. Both type I and II insulin-like growth factor receptor binding increase during lactogenesis in bovine mammary tissue. Endocrinology 122:2412.

Dodson, M. V., R. E. Allen and K. L. Hossner. 1985. Ovine somatomedin, multiplication-stimulating activity, and insulin promote skeletal muscle satellite cell proliferation in vitro. Endocrinology 117:2357.

Duquette, P. F., C. G. Scanes and L. A. Muir. 1984. Effects of ovine growth hormone and other anterior pituitary hormones on lipolysis of rat and ovine adipose tissue in vitro. J. Anim. Sci. 58:1191.

Eden, S., O. G. P. Isaksson, K. Madsen and U. Friberg. 1983. Specific binding of growth hormone to isolated chondrocytes from rabbit ear and epiphyseal plate. Endocrinology 112:1127.

Eisemann, J. H., A. C. Hammond, D. E. Bauman, P. J. Reynolds, S. N. McCutcheon, H. F. Tyrrell and G. L. Haaland. 1986a. Effect of bovine growth hormone administration on metabolism of growing Hereford heifers: protein and lipid metabolism and plasma concentrations of metabolites and hormones. J. Nutr. 116:2504.

Eisemann, J. H., A. C. Hammond, T. S. Rumsey and D. E. Bauman. 1986b. Tissue protein synthesis rates in beef steers injected with placebo or bovine growth hormone. J. Anim. Sci. 63 (Suppl. 1):217 (Abstr.).

Eisemann, J. H., H. F. Tyrrell, A. C. Hammond, P. J. Reynolds, D. E. Bauman, G. L. Haaland, J. P. McMurtry and G. A. Varga. 1986c. Effect of bovine growth hormone administration on metabolism of growing Hereford heifers: Dietary digestibility, energy and nitrogen balance. J. Nutr. 116:157.

Ellis, S., M. A. Vodian and R. E. Grindeland. 1978. Studies on the bioassayable growth hormone-like activity of plasma. Recent Prog. Horm. Res. 34:213.

Etherton, T. D. 1982. The role of insulin–receptor interactions in regulation of nutrient utilization by skeletal muscle and adipose tissue: a review. J. Anim. Sci. 54:58.

Etherton, T. D., C. M. Evock, C. S. Chung, P. E. Walton, M. N. Sillence, K. A. Magri and R. E. Ivy. 1986. Stimulation of pig growth performance by long-term treatment with pituitary porcine growth hormone (pGH) and a recombinant pGH. J. Anim. Sci. 63 (Suppl. 1):219.

Etherton, T. D., C. M. Evock and R. S. Kensinger. 1987. Native and recombinant bovine growth hormone antagonize insulin action in cultured bovine adipose tissue. Endocrinology 121:699.

Etherton, T. D. and P. E. Walton. 1986. Hormonal and metabolic regulation of lipid metabolism in domestic livestock. J. Anim. Sci. 63 (Suppl. 2):76.

Evans, H. M. and M. E. Simpson. 1931. Hormones of the anterior hypophysis. Amer. J. Physiol. 98:511.

Fagin, K. D., S. L. Lackey, C. R. Reagan and M. DiGirolamo. 1980. Specific binding of growth hormone by rat adipocytes. Endocrinology 107:608.

Finkelstein, J. D. and D. Schachter. 1962. Active transport of calcium by intestine: effects of hypophysectomy and growth hormone. Amer. J. Physiol. 203:873.

Flint, D. J. 1982. The role of insulin receptors in insulin action. Rep. Hannah Res. Inst. p 111.

Florini, J. R. 1985. Hormonal control of muscle growth. J. Anim. Sci. 61 (Suppl. 2):21.

Florini, J. R. 1987. Hormonal control of muscle growth. Muscle and Nerve 10:577.

Freemark, M., M. Comer and S. Handwerger. 1986. Placental lactogen and GH receptors in sheep liver: striking differences in ontogeny and function. Amer. J. Physiol. 251:E238.

Frigeri, L. G. 1980. Absence of in vitro dexamethasone-dependent lipolytic activity from highly purified growth hormone. Endocrinology 107:738.

Frigeri, L. G., S. M. Peterson and U. J. Lewis. 1979. The 20,000-dalton structural variant of human growth hormone: lack of some early insulin-like effects. Biochem. Biophys. Res. Commun. 91:778.

Fronk, T. J., C. J. Peel, D. E. Bauman and R. C. Gorewit. 1983. Comparison of different patterns of exogenous growth hormone administration on milk production in Holstein cows. J. Anim. Sci. 57:699.

Gluckman, P. D., B. H. Breier and S. R. Davis. 1987. Physiology of the somatotropic axis with particular reference to the ruminant. J. Dairy Sci. 70:442.

Gluckman, P. D., J. H. Butler and T. B. Elliott. 1983. The ontogeny of somatotropic binding sites in ovine hepatic membranes. Endocrinology 112:1607.

Goldberg, A. L. 1969. Protein turnover in skeletal muscle. I. Protein catabolism during work-induced hypertrophy and growth induced with growth hormone. J. Biol. Chem. 244:3217.

Goldberg, A. L., M. Tischler, G. DeMartino and G. Griffin. 1980. Hormonal regulation of protein degradation and synthesis in skeletal muscle. Fed. Proc. 39:31.

Goldspink, D. F. and A. L. Goldberg. 1975. Influence of pituitary growth hormone on DNA synthesis in rat tissues. Am. J. Physiol. 228:302.

Goodman, H. M. 1963. Effects of chronic growth hormone treatment on lipogenesis by rat adipose tissue. Endocrinology 72:95.

Goodman, H. M. 1984. Biological activity of bacterial derived human growth hormone in adipose tissue of hypophysectomized rats. Endocrinology 114:131.

Goodman, H. M. and V. Coiro. 1981. Induction of sensitivity to the insulin-like action of growth hormone in normal rat adipose tissue. Endocrinology 108:113.

Goodman, H. M. and G. Grichting. 1983. Growth hormone and lipolysis: a reevaluation. Endocrinology 113:1697.

Goodman, H. M. and J. Schwartz. 1974. Growth hormone and lipid metabolism. In: R. O. Greep and E. B. Astwood (Ed.) Handbook of Physiology. Sect. 7: Endocrinology, IV, The Pituitary Gland and Its Neuroendocrine Control, Part 2. pp 211–231. American Physiological Society, Washington, DC.

Gray, R. W. 1987. Evidence that somatomedins mediate the effect of hypophosphatemia to increase serum 1,25-dihydroxyvitamin D_3 levels in rats. Endocrinology 121:504.

Gray, R. W. and T. L. Garthwaite. 1985. Activation of renal 1,25-dihydroxyvitamin D_3 synthesis by phosphate deprivation: evidence for a role for growth hormone. Endocrinology 116:189.

Gray, R. W., T. L. Garthwaite and L. S. Phillips. 1983. Growth hormone and triiodothyronine permit an increase in plasma 1,25 $(OH)_2D$ concentrations in response to dietary phosphate deprivation in hypophysectomized rats. Calcif. Tissue Int. 35:100.

Green, H., M. Morikawa and T. Nixon. 1985. A dual effector theory of growth hormone action. Differentiation 29:195.

Gregor, P. and B. D. Burleigh. 1985. Presence of high affinity somatomedin/insulin-like growth factor receptors in porcine mammary gland. Proc. 67th Annu. Meet. Endocrinol. Soc., June 19–21. p 223 (Abstr.).

Hampson, R. K. and F. M. Rottman. 1986. A potential variant of bovine growth hormone resulting from non-splicing of an intron. Fed. Proc. 45:1703.

Hart, I. C., P. M. E. Chadwick, T. C. Boone, K. E. Langley, C. Rudman and L. Souza. 1984. A comparison of the growth-promoting, lipolytic, diabetogenic and immunological properties of pituitary and recombinant-DNA-derived bovine growth hormone (somatotropin). Biochem. J. 224:93.

Hart, I. C. and I. D. Johnsson. 1986. Growth hormone and growth in meat producing animals. In: P. J. Buttery, D. B. Lindsay and N. B. Haynes (Ed.) Control and Manipulation of Animal Growth. pp 135–159. Butterworths, London.

Hausman, G. J., D. R. Campion and G. B. Thomas. 1982. Semitendinosis muscle development in fetally decapitated pigs. J. Anim. Sci. 55:1330.

Hausman, G. J., E. J. Hentges and G. B. Thomas. 1987. Differentiation of adipose tissue and muscle in hypophysectomized pig fetuses. J. Anim. Sci. 64:1255.

Hausman, G. J. and G. B. Thomas. 1984. Histochemical and cellular aspects of adipose tissue development in decapitated pig fetuses: an ontogeny study. J. Anim. Sci. 58:1540.

Henneman, P. H., A. P. Forbes, M. Moldawer, E. F. Dempsey and E. L. Carroll. 1960. Effects of human growth in man. J. Clin. Invest. 39:1223.

Hizuka, N., C. M. Hendricks, G. N. Pavlakis, D. H. Hamer and P. Gorden. 1982. Properties of human growth hormone polypeptides: purified from pituitary extracts and synthesized in monkey kidney cells and bacteria. J. Clin. Endocrinol. Metab. 55:545.

Hughes, J. P. and H. G. Friesen. 1985. The nature and regulation of the receptors for pituitary growth hormone. Annu. Rev. Physiol. 47:469.

Isaksson, O. G. P., S. Eden and J. -O. Jansson. 1985. Mode of action of pituitary growth hormone on target cells. Annu. Rev. Physiol. 47:483.

Isaksson, O. G. P., S. Eden, J. -O. Jansson, A. Lindahl, J. Isgaard and A. Nilsson. 1986. Sites of action of growth hormone on somatic growth. J. Anim. Sci. 63 (Suppl. 2):48.

Isaksson, O. G. P., J. -O. Jansson and I. A. M. Gause. 1982. Growth hormone stimulates longitudinal bone growth directly. Science 216:1237.

Isgaard, J., A. Nilsson, A. Lindahl, J. -O. Jansson and O. G. P. Isaksson. 1986. Effects of local administration of GH and IGF-I on longitudinal bone growth in the rat. Amer. J. Physiol. 250:E367.

Jansson, J. -O., K. Albertsson-Wikland, S. Eden, K. -G. Thorngren and O. Isaksson. 1982a. Circumstantial evidence for a role of the secretory pattern of growth hormone in control of body growth. Acta Endocrinol. 99:24.

Jansson, J. -O., K. Albertsson-Wikland, S. Eden, K. -G. Thorngren and O. Isaksson. 1982b. Effect of frequency of growth hormone administration on longitudinal bone growth and body weight in hypophysectomized rats. Acta Physiol. Scand. 114:261.

Jennische, E., A. Skottner and H. -A. Hansson. 1987. Satellite cells express the trophic factor IGF-I in regenerating skeletal muscle. Acta Physiol. Scand. 129:9.

Johnsson, I. D. and I. C. Hart. 1985. The effects of exogenous bovine growth hormone and bromocriptine on growth, body development, fleece weight and plasma concentrations of growth hormone, insulin and prolactin in female lambs. Anim. Prod. 41:207.

Kostyo, J. L. 1974. Search for the active core of growth hormone. Metabolism 23:885.

Kostyo, J. L. 1986. Growth hormone: receptors and mode of action. In L. A. Schuler and N. L. First (Ed.) Regulation of Growth and Lactation in Animals. Univ. of Wisconsin Biotech. Ser. (No. 1). p 35.

Kostyo, J. L., C. M. Cameron, K. C. Olson, A. J. S. Jones and R. -C. Pai. 1985. Biosynthetic 20-kilodalton methionyl-human growth hormone has diabetogenic and insulin-like activities. Proc. Natl. Acad. Sci. USA 82:4250.

Kostyo. J. L., S. E. Gennick and S. E. Sauder. 1984. Diabetogenic activity of native and biosynthetic human growth hormone in obese (ob/ob) mouse. Amer. J. Physiol. 246:E356.

Laarveld, B., D. E. Kerr and R. P. Brockman. 1986. Effects of growth hormone on glucose and acetate metabolism in sheep. Comp. Biochem. Physiol. 83A:499.

Larson, B. A., T. L. Williams, U. J. Lewis and W. P. VanderLaan. 1978. Insulin secretion from pancreatic islets: effect of growth hormone and related proteins. Diabetologia 15:129.

Larson, B. A., T. L. Williams, M. O. Showers and U. J. Lewis. 1980. Interaction between glucose and naturally occurring diabetogenic substance in the induction of insulin release from islets of hypophysectomized rats. J. Endocrinol. 84:281.

Lee, V., J. Ramachandran and C. H. Li. 1974. Does bovine growth hormone possess rapid lipolytic activity? Arch. Biochem. Biophys. 161:222.

Lee, M. O. and N. K. Schaffer. 1934. Anterior pituitary growth hormone and the composition of growth. J. Nutr. 7:337.

Lesniak, M. A., J. Roth, P. Gorden and J. R. Gavin III. 1973. Human growth hormone radiore-ceptor assay using cultured human lymphocytes. Nature New Biol. 241:20.

Lewis, U. J. 1984. Variants of growth hormone and prolactin and their posttranslational modifi-cations. Annu. Rev. Physiol. 46:33.

Lewis, U. J. 1987. Variant forms of growth hormone. In: Beltsville Symposia in Agricultural Research. XII. Biomechanisms Regulating Growth and Development: Keys to Progress. Ag-ricultural Research Service, USDA, Beltsville, MD (Abstr. 3).

Lewis, U. J., J. T. Dunn, L. F. Bonewald, B. K. Seavey and W. P. VanderLaan. 1978. A naturally occurring structural variant of human growth hormone. J. Biol. Chem. 253:2679.

Lewis, U. J., E. Markoff, F. L. Culler, A. Hayek and W. P. VanderLaan. 1987. Biologic prop-erties of the 20K-dalton variant of human growth hormone: a review. Endocrinol. Jpn. 34 (Suppl. 1):73.

Lewis, U. J., R. N. P. Singh and G. F. Tutwiler. 1981. Hyperglycemic activity of the 20,000-dalton variant of human growth hormone. Endocrinol. Res. Commun. 8:155.

Lewis, U. J., R. N. P. Singh, G. F. Tutwiler, M. B. Sigel, E. F. VanderLaan and W. P. VanderLaan. 1980. Human growth hormone: a complex of proteins. Recent Prog. Horm. Res. 36:477.

Lewis, U. J., R. N. P. Singh, W. P. VanderLaan and G. F. Tutwiler. 1977. Enhancement of the hyperglycemic activity of human growth hormone by enzymic modification. Endocrinology 101:1587.

Liberti, J. P. 1981. Isolation and somatomedin activity of bovine growth hormone fragment 87–124. Biochim. Biophys. Acta 657:239.

Liberti, J. P. and L. A. Durham III. 1983. Bovine growth hormone fragment (1–133) has in-vitro somatomedin-like activity. J. Endocrinol. 96:195.

Liberti, J. P. and M. S. Miller. 1978. Somatomedin-like effects of biologically active bovine growth hormone fragments. Endocrinology 103:29.

Lindahl, A., J. Isgaard, L. Carlsson and O. G. P. Isaksson. 1987a. Diffferential effects of growth hormone and insulin-like growth factor I on colony formation of epiphyseal chondrocytes in suspension culture in rats of different ages. Endocrinology 121:1061.

Lindahl, A., J. Isgaard and O. G. P. Isaksson. 1987b. Growth hormone in vivo potentiates the stimulatory effect of insulin-like growth factor-1 in vitro on colony formation of epiphyseal chondrocytes isolated from hypophysectomized rats. Endocrinology 121:1070.

Lindsay, D. B. 1983. Growth and fattening. In: J. A. F. Rook and P. C. Thomas (Ed.) Nutritional Physiology of Farm Animals. pp 261–313. Longman Inc., New York.

Louis, L. H. and J. W. Conn. 1972. Diabetogenic polypeptide from human pituitaries similar to that excreted by proteinuric diabetic patients. Metabolism 21:1.

Machlin, L. J. 1972. Effect of porcine growth hormone on growth and carcass composition of the pig. J. Anim. Sci. 35:794.

Machlin, L. J. 1975. Role of growth hormone in improving animal production. In: Anabolic agents in animal production, FAO/WHO Symp. pp 43–52. Stuttgart, Germany.

Maciag, T., R. Forand, S. Ilsley, J. Cerundolo, R. Greenlee, P. R. Kelley and E. Canalis. 1980. The generation of sulfation factor activity by proteolytic modification of growth hormone. J. Biol. Chem. 255:6064.

Madsen, K., U. Friberg, P. Roos, S. Eden and O. Isaksson. 1983. Growth hormone stimulates the proliferation of cultured chondrocytes from rabbit ear and rat rib growth cartilage. Nature (Lond.) 304:545.

Maes, M., R. deHertogh, P. Watrin-Granger and J. M. Ketelslegers. 1983. Ontogeny of liver somatotropic and lactogenic binding sites in male and female rats. Endocrinology 113:1325.

Magri, K. A., R. Gopinath and T. D. Etherton. 1987. Inhibition of lipogenic enzyme activities by porcine growth hormone (pGH). J. Anim. Sci. 65 (Suppl. 1):258 (Abstr.).

Mainoya, J. R. 1975. Effects of bovine growth hormones, human placental lactogen, and ovine prolactin on intestinal fluid and ion transport in the rat. Endocinology 96:1165.

Martin, J. B. 1976. Brain regulation of growth hormone secretion. In: L. Martini and W. F. Ganong (Ed.) Frontiers in Neuroendocrinology. Vol. 4, p 129. Raven Press, New York.

McCusker, R. H. and D. R. Campion. 1986. Effect of growth hormone secreting tumours on skeletal muscle cellularity in the rat. J. Endocrinol. 111:279.

McCutcheon, S. N. and D. E. Bauman. 1986a. Effect of pattern of administration of bovine growth hormone on lactational performance of dairy cows. J. Dairy Sci. 69:38.

McCutcheon, S. N. and D. E. Bauman, 1986b. Effect of chronic growth hormone treatment on responses to epinephrine and thyrotropin-releasing hormone in lactating cows. J. Dairy Sci. 69:44.

Mittra, I. 1980a. A novel "cleaved prolactin" in the rat pituitary: Part I. Biosynthesis, characterization and regulatory control. Biochem. Biophys. Res. Commun. 95:1750.

Mittra, I. 1980b. A novel "cleaved prolactin" in the rat pituitary: Part II. In vivo mammary mitogenic activity of its N-terminal 16K moiety. Biochem. Biophys. Res. Commun. 95:1760.

Mittra, I. 1984. Somatomedins and proteolytic bioactivation of prolactin and growth hormone. Cell 38:347.

Morikawa, M., T. Nixon and H. Green. 1982a. Growth hormone and the adipose conversion of 3T3 cells. Cell 29:783.

Morikawa, M., H. Green and U. J. Lewis. 1982b. Activity of human growth hormone and related polypeptides on the adipose conversion of 3T3 cells. Mol. Cell. Biol. 4:228.

Moseley, W. M., J. Huisman and E. J. VanWeerden. 1987. Serum growth hormone and nitrogen metabolism responses in young bull calves infused with growth hormone-releasing factor for 20 days. Domest. Anim. Endocrinol. 4:51.

Moseley, W. M., L. F. Krabill and R. F. Olsen. 1982. Effect of bovine growth hormone administered in various patterns on nitrogen metabolism in the Holstein steer. J. Anim. Sci. 55:1062.

Mosier, H. D. and U. J. Lewis. 1982. The 20,000-dalton variant of human growth hormone. Effect on bioassayable somatomedin activity in serum. Horm. Metab. Res. 14:440.

Nicoll, C. S., J. F. Tarpey, G. L. Mayer and S. M. Russell. 1986. Similarities and differences among prolactins and growth hormones and their receptors. Amer. Zool. 26:965.

Nilsson, A., J. Isgaard, A. Lindahl, A. Dahlstom and O. G. P. Isaksson. 1986. Regulation by growth hormone of a number of chondrocytes containing IGF-I in rat growth plate. Sci. 233:571.

Nilsson, A., J. Isgaard, A. Lindahl, L. Peterson and O. Isaksson. 1987. Effects of unilateral arterial infusion of GH and IGF-I on tibial longitudinal bone growth in hypophysectomized rats. Calcif. Tissue Int. 40:91.

Nixon, B. T. and H. Green. 1984a. Contribution of growth hormone to the adipogenic activity of serum. Endocrinology 114:527.

Nixon, B. T. and H. Green. 1984b. Growth hormone promotes the differentiation of myoblasts and preadipocytes generated by azacytidine treatment of 10T1/2 cells. Proc. Natl. Acad. Sci. USA 81:3429.

Nutting, D. F. 1976. Ontogeny of sensitivity to growth hormone in rat diaphragm muscle. Endocrinology 98:1273.

Nyberg, G. and U. Smith. 1977. Human adipose in culture. VII. The long-term effect of growth hormone. Horm. Metab. Res. 9:22.

Paladini, A. C., C. Pena and E. Poskus. 1983. Molecular biology of growth hormone. CRC Crit. Rev. Biochem. 15:25.

Paladini, A. C., C. Pena and L. A. Retegui. 1979. The intriguing nature of the multiple actions of growth hormone. Trends Biochem. Sci. 4:256.

Peel, C. J., D. E. Bauman, R. C. Gorewit and C. J. Sniffen. 1981. Effect of exogenous growth hormone on lactational performance in high yielding dairy cows. J. Nutr. 111:1662.

Pell, J. M. and P. C. Bates. 1987. Collagen and non-collagen protein turnover in skeletal muscle of growth hormone-treated lambs. J. Endocrinal. 115:R1.

Peters, J. P. 1986. Consequences of accelerated gain and growth hormone administration for lipid metabolism in growing beef steers. J. Nutr. 116:2490.

Phillips, L. S. 1986. Nutrition, somatomedins, and the brain. Metabolism 35:78.

Prysor-Jones, R. A. and J. S. Jenkins. 1980. Effect of excessive secretion of growth hormone on tissues of the rat, with particular reference to the heart and skeletal muscle. J. Endocrinol. 85:72.

Reeds, P. J. and R. M. Palmer. 1986. The role of prostaglandins in the control of muscle protein turnover. In: P. J. Buttery, D. B. Lindsay and N. B. Haynes (Ed.) Control and Manipulation of Animal Growth. pp 161–185. Butterworths, London.

Rosenfeld, R. G., D. M. Wilson, L. A. Dollar, A. Bennett and R. L. Hintz. 1982. Both human pituitary growth hormone and recombinant DNA-derived human growth hormone cause insulin resistance at a postreceptor site. J. Clin. Endocrinol. Metab. 54:1033.

Russell, S. M. and E. M. Spencer. 1985. Local injections of human or rat growth hormone or of purified human somatomedin-C stimulate unilateral tibial epiphyseal growth in hypophysectomized rats. Endocrinology 116:2563.

Sandles, L. D. and C. J. Peel. 1987. Growth and carcass composition of pre-pubertal dairy heifers treated with bovine growth hormone. Anim. Prod. 44:21.

Schaffer, W. T. 1985. Effects of growth hormone on lipogenic enzyme activities in cultured hepatocytes. Amer. J. Physiol. 248:E719.

Schlechter, N. L., S. M. Russell, S. Greenberg, E. M. Spencer and C. S. Nicoll. 1986a. A direct growth effect of growth hormone in rat hindlimb shown by arterial infusion. Amer. J. Physiol. 250:E231.

Schlechter, N. L., S. M. Russell, E. M. Spencer and C. S. Nicoll. 1986b. Evidence suggesting that the direct growth-promoting effect of growth hormone on cartilage in vivo is mediated by local production of somatomedin. Proc. Natl. Acad. Sci. USA 83:7932.

Schoenle, E., J. Zapf and E. R. Froesch. 1982a. Glucose transport in adipocytes and its control by growth hormone in vivo. Amer. J. Physiol. 242:E368.

Schoenle, E., J. Zapf, R. E. Humbel and E. R. Froesch. 1982b. Insulin-like growth factor I stimulates growth in hypophysectomized rats. Nature (Lond.) 296:252.

Schoenle, E., J. Zapf and E. R. Foresch. 1983. Regulation of rat adipocyte glucose transport by growth hormone: no mediation by insulin-like growth factors. Endocrinology 112:384.

Sechen, S. J., S. N. McCutcheon and D. E. Bauman. 1985. Response to metabolic challenges in lactating dairy cows during short-term bovine growth hormone treatment. J. Dairy Sci. 68 (Suppl. 1):170 (Abstr).

Sigel, M. B., N. A. Thorpe, M. S. Kobrin, U. J. Lewis and W. P. VanderLaan. 1981. Binding characteristics of a biologically active variant of human growth hormone (20K) to growth hormone and lactogen receptors. Endocrinology 108:1600.

Singh, R. N. P., B. K. Seavey and U. J. Lewis. 1974a. Heterogeneity of human growth hormone. Endocrinol. Res. Commun. 1:449.

Singh, R. N. P., B. K. Seavey, V. P. Rice, T. T. Lindsey and U. J. Lewis. 1974b. Modified forms of human growth hormone with increased biological activities. Endocrinology 94:883.

Sinha, Y. N. and T. A. Gilligan. 1984. A "20K" form of growth hormone in the murine pituitary gland. Proc. Soc. Exp. Biol. Med. 177:465.

Smal, J., J. Closset, G. Hennen and P. DevMeyts. 1986. The receptor binding properties of the 20K variant of human growth hormone explain its discrepant insulin-like and growth promoting activities. Biochem. Biophys. Res. Commun. 134:159.

Spanos, E., D. J. Brown, J. C. Stevenson and I. Macintyre. 1981. Stimulation of 1,25-dihydroxycholecalciferol production by prolactin and related peptides in intact renal cell preparations in vitro. Biochim. Biophys. Acta 672:7.

Spencer, E. M., L. J. Lewis and U. J. Lewis. 1981. Somatomedin generating activity of the 20,000-dalton variant of human growth hormone. Endocrinology 109:1301.

Spencer, E. M. and O. Tobiassen. 1981. The mechanism of the action of growth hormone on vitamin D metabolism in the rat. Endocrinology 108:1064.

Stevens, D. and G. Alexander. 1986. Lipid deposition after hypophysectomy and growth hormone treatment in the sheep fetus. J. Developmental Physiol. 8:139.

Stracke, H., A. Schulz, D. Moeller, S. Rossol and H. Schatz. 1984. Effect of growth hormone on osteoblasts and demonstration of somatotomedin-C/IGF-I in bone organ culture. Acta Endocrinol. 107:16.

Thorbek, G. 1977. The energetics of protein deposition during growth. Nutr. Metab. 21:105.

Tutwiler, G. F. and T. J. Kirsch. 1976. Noncyclic AMP-mediated lipolytic effect of bovine and porcine diabetogenic proteins. Biochem. Med. 15:149.

Tyrrell, H. F., A. C. G. Brown, P. J. Reynolds, G. L. Haaland, C. J. Peel, D. E. Bauman and W. D. Steinhour. 1982. Effect of growth hormone on utilization of energy by lactating Holstein cows. In: A. Ekern and F. Sundstol (Ed.) Energy Metabolism of Farm Animals. pp 46–49. Publ. No. 29, Eur. Assoc. Anim. Prod.

Vernon, R. G. 1982. Effects of growth hormone on fatty acid synthesis in sheep adipose tissue. Int. J. Biochem. 14:255.

Vernon, R. G., M. Barber, E. Finley and M. R. Grigor. 1988. Endocrine control of lipogenic enzyme activity in adipose tissue from lactating ewes. Proc. Nutr. Soc. 47:100A (Abstr.).

Vernon, R. G. and E. Finley. 1986. Endocrine control of lipogenesis in adipose tissue from lactating sheep. Biochem. Soc. Trans. 14:635.

Vernon, R. G., E. Finley and D. J. Flint. 1987. Role of growth hormone in the adaptations of lipolysis in rat adipocytes during recovery from lactation. Biochem. J. 242:931.

Vodian, M. A. and C. S. Nicoll. 1977. Growth hormone releasing factor and the bioassay–radioimmunoassay paradox revisited. Acta Endocrinol. 86:71.

Vodian, M. A. and C. S. Nicoll. 1979. Evidence to suggest that rat growth hormone is modified when secreted by the pituitary gland. J. Endocrinol. 80:69.

Wallace, A. L. C. and J. M. Bassett. 1966. Effect of sheep growth hormone on plasma insulin concentration in sheep. Metabolism 15:95.

Walton, P. E., T. D. Etherton and C. M. Evock. 1986. Antagonism of insulin action in cultured pig adipose tissue by pituitary and recombinant porcine growth hormone: potentiation by hydrocortisone. Endocrinology 118:2577.

Webster, A. J. F., G. Lobley, P. J. Reeds and J. D. Pullar. 1979. Protein mass, protein synthesis and heat loss in the Zucker rat. Proc. Nutr. Soc. 37:21A (Abstr.).

Weekes, T. E. C. 1986. Insulin and growth. In: P. J. Buttery, D. B. Lindsay and N. B. Haynes (Ed.) Control and Manipulation of Animal Growth. pp 187–206. Butterworths, London.

Wohnlich, L. and W. V. Moore. 1982. Binding of a variant of human growth hormone to liver plasma membranes. Horm. Metab. Res. 14:138.

Wongsurawat, N., H. J. Armbrecht, T. V. Zenser, L. R. Forte and B. B. Davis. 1984. Effects of hypophysectomy and growth hormone treatment on renal hydroxylation of 25-hydroxycholecalciferol in rats. J. Endocrinol. 101:333.

Wray-Cahen, C. D., R. D. Boyd, D. E. Bauman, D. R. Ross and K. D. Fagin. 1987a. Somatotropin's effect on metabolic response to the euglycemic clamp in growing swine. In: Beltsville Symposia in Agricultural Research. XII. Biomechanisms Regulating Growth and Development: Keys to Progress. p 2. Agricultural Research Service, USDA, Beltsville, MD (Abstr.).

Wray-Cahen, D., R. D. Boyd, D. E. Bauman, D. A. Ross and K. Fagin. 1987b. Metabolic effects of porcine somatotropin (pST) in growing swine. J. Anim. Sci. 65 (Suppl. 1):261 (Abstr.).

Regulation of Somatomedin Production, Release, and Mechanism of Action

NORMAN C. STEELE AND THEODORE H. ELSASSER

1. Origin of the Somatomedin Hypothesis

Early studies revealed that a pituitary factor, growth hormone (GH), was involved in long bone growth and nitrogen retention and, without conclusive evidence, accepted theories suggested that GH interacted directly with target tissues to induce such adaptations in metabolism. The elegant studies by Salmon and Daughaday (1957) described a factor in normal serum that, when added to cartilage explants, stimulated the incorporation of $^{35}SO_4$ into proteoglycans. Serum from hypophysectomized rats failed to stimulate sulfate incorporation; however, when such rats were treated with GH, serum stimulation of sulfate uptake was observed within 24 hr. In contrast, direct addition of GH to the explant media, either in the presence or absence of hypophysectomized rat serum, failed to stimulate sulfate incorporation. Based on the bioassay response, the serum component was named *sulfation factor*.

With the advent of specific antisera for insulin, Foesch *et al.* (1963) reported that serum added to adipose tissue explant cultures stimulated glucose uptake. Preincubation of serum with insulin antiserum, a treatment that neutralizes specific insulin effects, failed to significantly depress glucose uptake. Again based on bioassay endpoint, the serum factor was named *nonsuppressible insulinlike activity* (NSILA). Acid ethanol extraction of serum resulted in a

NORMAN C. STEELE AND THEODORE H. ELSASSER • United States Department of Agriculture, Agricultural Research Service, Beltsville Agricultural Research Center, Livestock and Poultry Science Institute, Beltsville, Maryland 20705

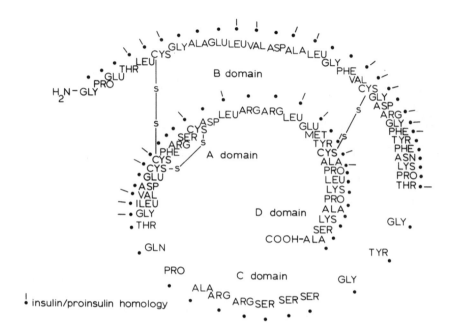

FIGURE 1. Primary structure of human IGF-I (Rinderknecht and Humbel, 1978). Points with proinsulin homology are shown with ticks. Destruction of disulfide bridges abolishes biological activity.

small portion of NSILA present in the aqueous phase, NSILA-s (soluble), with an estimated M_r of 7500 (Fig. 1). The majority of NSILA precipitated in acid ethanol, retained bioassay activity, and had an estimated M_r of 200,000. This material was named NSILA-p (particulate). This high-molecular-weight form is a complex material possessing NSILA activity that by chemical characterization is only partially due to NSILA-s.

Cell biology research (Dulak and Temin, 1973) gave rise to a family of macromolecules isolated from serum-free media conditioned by the BRL-3A rat liver cell line. These proteins, *multiplication-stimulating activity* (MSA), replaced insulin as a growth factor for chick embryo fibroblasts and stimulated thymidine incorporation. Although this material possessed sulfation activity in the hypophysectomized rat cartilage bioassay, subsequent amino acid sequence analysis (Rinderknecht and Humbel, 1978) revealed a distinct growth factor with an M_r of 7484.

For further background information, the reader should consult Foesch *et al.* (1986).

Spencer *et al.* (1983) reconciled the confusing somatomedin nomenclature (i.e., somatomedin versus NSILA-s versus MSA) and suggested that somatomedin C and IGF-I are the same peptide with a pI of 8.4 (Fig. 1), and IGF-II and somatomedin A (Hall *et al.*, 1979) are the same peptide with a neutral pI. Several other peptides with somatomedin bioactivity have been reported, so this nomenclature, IGF-I and IGF-II, is overly simplistic. However, for the purpose of this review, only these two major forms are considered.

2. Methods of Analyses

Bioassays, radiometric (radioreceptor and radioimmunological procedures) and physicochemical methods can be used to quantify somatomedins. No single method is ideal for all circumstances.

A bioassay measured activity in terms of biological action versus some standardized preparation of defined activity. The original technique for quantifying IGFs involved protein synthesis by cartilage tissue *in vitro* using $^{35}SO_4$ incorporation. Chondrocytes or intracostal cartilage plates of chicks, rats, or pigs will incorporate the tracer in a dose-responsive manner when incubated with serum or serum fractions. The sulfation assay lacks specificity because both IGF-I and IGF-II activities are measured. The assay does not rely on the release of IGFs from carrier proteins prior to assay and can be stimulated or inhibited by several intrinsic substances present in serum [amino acids, thyroid hormones, and IGF inhibitors (Phillips *et al.*, 1979)]. Due to the insulin activity of IGFs, fat tissue incubated in the presense of serum or serum fractions shows stimulation of glucose uptake, oxidation, and lipid synthesis (Froesch *et al.*, 1963). As in the sulfation assay, this assay lacks specificity, will respond in the presence of binding proteins, and is subject to intrinsic interferences. Mitogenic activity, as measured by thymidine incorporation in fibroblasts, is similar to other bioassays, but because these cells are readily carried as continuous cultures, some utility of a standard cell line can be gained (Pierson and Temin, 1972).

Bioassay characterization of growth peptides forms an essential component in characterization (Zapf, 1983). Only through such methods can one ensure that immunologically or receptor-recognized materials are appropriately active.

Early investigations that attempted to characterize the physical properties of IGF-I were filled with discrepancies concerning the size of the molecule. Reported M_r values ranged from 7000 to 300,000. Daughaday (1877), Hintz and Liu (1977), Zapf *et al.* (1978), and Copeland *et al.* (1980) then demonstrated that IGF-I circulates tightly bound to several presumed transport binding proteins (Fig. 2). The carrier protein exists as several molecular species and only the high-molecular-weight, acid-labile, form (150,000 M_r) appears under

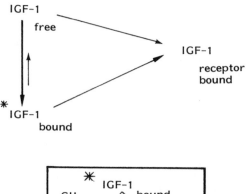

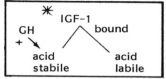

FIGURE 2. Interrelationships between free plasma protein-bound and receptor-bound IGF-I. Heavier arrows signify higher probability for affinity-dependent kinetic transfer of IGF-I between molecular species.

GH regulation (Furlanetto, 1980). The low-molecular-weight protein (35,000–55,000 M_r) does not appear to be a degradation product of the larger protein.

Estimates of the binding protein size have varied greatly by analysis method (Wilkins and D'Ercole, 1985; Scamurra *et al.*, 1986; McCusker *et al.*, 1986). Both of these carriers exist in serum in varying degrees of saturation with IGF-I across several species. Due to their high affinity for IGF, binding protein(s) confound analyses by competing with assay ligand for all forms of IGF-I, including iodinated tracer (Hintz and Liu, 1977), causing an artificially high measure of IGF-I (Scamurra *et al.*, 1986). This effectively prevents a tracer from interacting with ligand, resulting in the precipitation of fewer counts and giving the appearance of greater IGF-I mass. Therefore, in order to measure IGF-I, it is necessary to liberate the hormone from the binding proteins by some physical means. In addition, this affects the kinetics of binding associations so that it is impossible to effect quantitative recovery of added nonlabeled IGF-I from plasma and establish parallel dose displacement between standard concentrations of hormone and increasing volumes of plasma.

Several approaches have been taken to remedy the binding protein problem. One approach is to dissociate IGF-I from the binding protein. This can be accomplished by acidification of plasma or serum to a pH less than 3.5 with HC1, acetic acid, formic acid, or glycylglycine-HC1 buffers (Underwood *et al.*, 1982). Another approach is to add heparin to the plasma (Clemmons *et al.*, 1983), but the results have been variable among different species. Beyond this stage, elimination of the binding protein effects can be achieved by physically separating the molecular species via acidified gel permeation chromatography (Sephadex G-50; Furlanetto et al., 1977), HPLC (Svoboda and Van Wyk, 1984), or by denaturing the binding proteins with acidified ethanol (Daughaday et al., 1980) and removing the interfering proteins by centrifugation, or with rapid dilution from an acid environment with assay buffer and direct addition to assay tubes. Effectively achieving at least a 1 : 320 dilution, the reassociation kinetics of the IGF-I with the residual binding protein are significantly diminished with the third technique so that more than 90% of the available IGF-I can be measured. Acidification liberates IGF-I by altering noncovalent bonds between the molecules and also apparently degrades the higher-molecular-weight binding protein (Copeland et al., 1980).

Radioligand assay techniques utilize specific receptor fractions isolated from target tissue, immunological assays direct antibody formation against the specific IGF ligand, and competitive binding assay techniques are used that take advantage of the binding protein. Ranging from crude to semipurified, plasma membrane subfractions have been utilized to quantify IGFs. IGF-I concentrations have been determined using human placental membrane preparations (Marshall et al., 1974) and the assay method distinguished hypo- from hyperpituitary patient serum. Rat liver and placental membranes are the tissues of choice for development of IGF-II radioreceptor assays (Rechler et al., 1980). The advantage of such assays is that common interfacing sources of variation (other growth factors and thyroid hormones) do not cross-react in the assay. One major limitation, depending on the species, is the relative affinity of the binding protein for the ligand compared to the receptor affinity. A common recommendation in the application of radioreceptor assay techniques is the prior extraction of a sample to remove the binding protein prior to analysis (Daughaday et al., 1980, 1982).

Similar to analysis of adrenal steroids and thyroid hormones, the binding protein can be utilized as an analysis technique for IGFs. Since serum is a source of binding protein, some extraction is necessary prior to analysis. Human serum binding protein has a greater affinity for IGF-II than for IGF-I, whereas rat serum binding protein has a greater affinity for IGF-I than for IGF-II. Therefore, using IGF-I or IGF-II as tracer, a relatively specific competitive binding assay can be developed (Zapf et al., 1975; Schalch et al., 1978).

To achieve specificity in analysis, only radioimmunoassay procedures distinguish IGF-I and IGF-II. Furlanetto et al. (1977) described a nonequilibrium assay using a polyclonal antiserum developed in the rabbit against partially purified human IGF-I. The antiserum, identical or similar to the antiserum distributed by the National Pituitary Agency, has been used by many investigators. Depending on the species, the binding protein can confound even the immunoassay because unsaturated binding sites can cause spuriously high values or, depending on the binding protein, competition may be unfavorable for antibody access. This implies that some method of sample pretreatment is still preferred to avoid problems. Likewise, monoclonal antisera directed against human IGF-I require perturbation of the IGF–binding protein complex (Baxter *et al.*, 1982). Hintz *et al.* (1983) compared antisera generated against IGF-I C and D domains (Fig. 1) and found essentially no differences in mass compared to antisera generated to the intact peptide. Antisera generated to the 1–7 amino acid region of IGF-II do not cross-react with IGF-I. The concept of region-specific antisera for the IGFs is the method of choice in the development of radioimmunoassay procedures.

Recently, Baxter and co-workers (Scott *et al.*, 1985) reported that high-pressure chromatographic techniques could resolve both IGF and carrier protein in acidified hepatocyte media. Both IGF and the carrier protein reportedly measured 6.8 pI, and the binding protein produced by hepatocytes measured 50,000 M_r. The chromatographic technique offers several advantages over indirect analyses. Total recovery was obtained with the simultaneous assay of both proteins. Application of this technique increases quantitative analysis of IGF-related variables and experimental variables, and provides physicochemical analysis of IGF and carrier proteins from species other than the rodent.

Binding protein interferences were introduced in the previous discussion of analysis methods. Physiologically, binding protein increases serum IGF half-life markedly, suggesting that binding protein is a mechanism for delivery to target tissues and/or for protection against the insulin effects of IGF. Binding protein may represent a disposal mechanism by which all the consequences of IGF action on target tissues are autocrine in nature and the carrier serves as a scavenger prior to degradation (D. Clemmons, personal communication). Suffice it to say, binding proteins can be a formidable interference in the analysis of circulating IGF and appropriate methods must be validated to avoid such interference.

3. Gene Expression and the Insulin Peptide Family

Sequence homology between insulin, relaxin, IGF-I, and IGF-II suggests a high degree of conservation in genes encoding these peptides. The latter exhibit insulinlike activities in bioassays, yet insulin and relaxin are double-chain molecules whereas the IGFs are single-chain molecules. Both insulin and relaxin arise from cleavage of a single-stranded prohormone (i.e., proinsulin and prorelaxin) and both possess a connecting peptide unnecessary for biological activity. Therefore, IGFs can be regarded as prohormonelike molecules in which the connecting peptide remains integrated, but not essential for biological action. Yang *et al.* (1985a) composed a diagram describing the primary structural homology of these peptides with both IGF-I and IGF-II possessing propeptide segments.

The IGF-I gene spans more than 45 kb on chromosome 12 (Rotwein *et al.*, 1986) with exons separated by intervening sequences of 1.5 to >21 kb. This large gene size is a hindrance to mRNA synthesis and may account for the low level of gene expression. Additionally, two cDNA probes that encode a 195- and 153-amino-acid IGF-I precursor have been identified. Therefore, two IGF mRNAs are generated through alternative RNA processing of the original gene transcript. The role of this alternative RNA processing in relation to tissue-specific IGF-I production or age-dependent IGF-I response to GH is unknown.

The gene encoding IGF-II resides on chromosome 11 and, similar to that for IGF-I, produces an mRNA that generates a preprohormone, then a prohormone, followed by the M_r 7484 peptide (Yang *et al.*, 1985a,b). Immunological probes have identified a 22-kDa prepro-IGF-II that is processed intracellularly to the 20-kDa proIGF-II. Upon secretion, the proIGF-II is processed further to the mature IGF-II species.

A comparison of the gene structures for the insulin peptide family (Fig. 3) supports the theory that all four peptides arose from a common ancestral gene. Indeed, comparing IGF-I and IGF-II, the DNA sequence homology of 61 and 49% respectively within exons 2 and 3, the presence of two introns within these genes, and the similarity of their amino acid sequences provide strong evidence (Rotwein *et al.*, 1986) that the IGFs are closely related. However, the lack of similarity between insulin and relaxin gene structures and IGF gene structures is not consistent with a common gene ancestor, unless both insulin and relaxin have further evolved from the parent gene.

Research in the area of molecular biology for IGFs is rapidly expanding. Needed data concern the regulation of gene expression and posttranlational processing of peptides. Now, only correlative physiological data can be cited as evidence that gene expression is tightly controlled and functions in response to animal age and/or growth velocity.

4. Site of Synthesis

Using a perfused, recirculating liver preparation, Schwander *et al.* (1983) demonstrated that [^{35}S]cysteine was rapidly incorporated into a M_r 7500 peptide that immunoprecipitated with rabbit IGF-I antisera. Production increased linearly following a 30 min lag period and IGF-I could not be extracted from liver. Similar studies performed on hypophysectomized rat liver revealed a 90% decrease in IGF-I production; treatment of the hypophysectomized rat with GH for 8 days restored IGF-I production to normal. Vassilopoulou-Sellin and Phil-

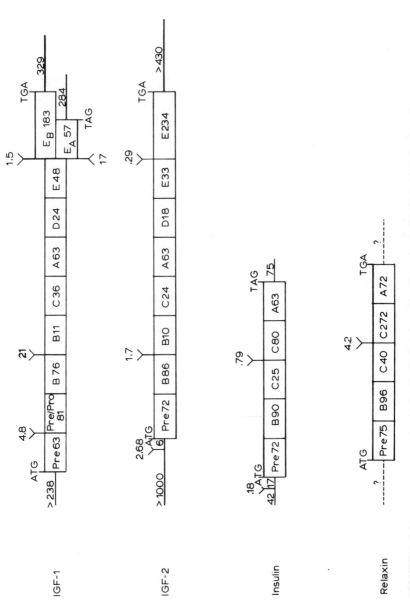

FIGURE 3. Primary gene structure for the insulin family of peptides (Rotwein *et al.*, 1986). Note the large intron regions of IGF-I and IGF-II compared to that of insulin.

lips (1982) identified a 30-kDa IGF-I-like peptide in acid extracts of rat liver by cartilage bioassay. Similar processing of spleen, muscle, and brain tissues did not reveal IGF-I-like material, suggesting specificity as to the site of gene expression and production.

Fetal liver explant cultures also secrete an IGF-II peptide as determined by specific radioimmunoassay (Rechler *et al.*, 1979). The material was secreted in considerable quantities by the explant culture but secretion was inhibited by cycloheximide, suggesting that synthesis is dependent on new protein synthesis. Recently, Graham *et al.* (1986) demonstrated with a cDNA probe specific for rat IGF-II that mRNA for IGF-II was present only in liver of neonatal rats and was absent from older rat liver. These data are consistent with the view that IGF-II is the fetal/neonatal somatomedin (Sm).

Spencer (1979) examined Sm production by regenerating liver cells in serum-free media. Using bioassay and radioreceptor techniques, monolayer hepatocyte cultures secreted a high-molecular-weight (40,000), acid-labile Sm into the media. This secretion was stimulated by the addition of GH to the media. The specificity, IGF-I versus IGF-II, of peptide synthesis could not be concluded from this study.

Based on these references selected from a volume of evidence, liver tissue is considered the major, but not the sole Sm production site on the basis of synthetic capacity and organ mass. Explant cultures of brain and pituitary tissues have demonstrated that IGF-like material is secreted by both lobes of the pituitary, the hypothalamus, the cerebral cortex, and the cerebellum and that synthesis is, in part, sensitive to cycloheximide inhibition (Binoux *et al.*, 1982). Production rate per unit tissue weight was similar for liver and brain tissues; therefore, sheer organ mass would differentiate total synthetic capacity. Using both radioreceptor and immunoassay techniques, D'Ercole *et al.* (1980) reported that media conditioned by explants of neonatal mouse lung, brain, kidney, intestine, and heart contained similar amounts of Sm per unit of tissue explant. Pancreatic B-cell cultures (Rabinovitch *et al.*, 1982), Sertoli cell cultures (Ritzen *et al.*, 1982), porcine granulosa cell cultures (Hammond *et al.*, 1984), bovine cartilage tissue (Kato *et al.*, 1981), and numerous continuous cell lines synthesize Sm-like peptides. As literature accumulates, one could conclude that all tissue has the potential for Sm production and fundamental differences must reside in how individual tissues respond to external stimuli and the fate of Sm produced from such numerous sites.

Both Sara *et al.* (1980) and Daughaday *et al.* (1982) reported that circulating IGF-II levels are 100-fold greater in fetal than in maternal circulation. In rats, IGF-I levels are low at birth and increase to adult levels by 4 weeks of age. Such developmental regulation can also be observed in livestock species using appropriate sampling protocols.

In addition to developmental regulation, several endocrine factors have been reported to have regulatory roles. GH has a stimulatory effect on IGF production both *in vivo* (e.g., Salmon and Daughaday, 1957; Chung *et al.*, 1985) and *in vitro* (e.g., Spencer, 1979). GH status of the animal and circulating IGF concentration are poorly related to developmental regulation. For example, neonates have relatively great circulating quantities of GH not accompanied by increased IGF-I concentration. A classic example compared GH and IGF-I concentrations in standard, miniature, and toy poodles (Eigenmann *et al.*, 1984). Despite vast differences in mature body size, circulating GH and sensitivity to GH secretagogues were quite similar in all three breeds. IGF-I concentrations, however, are correlated with mature size. Genetic constraints on IGF-I gene expression may impair the translation of the GH signal to the endpoint of long bone growth as mediated by IGF-I. A similar phenomenon observed in human pygmies (Merimee *et al.*, 1981) also represents a normal pattern of GH secretion not accompanied by the production of IGF-I. These examples emphasize how imprecise our current knowledge of "GH secretion–gene activation–IGF-I response" is and underscore that critical knowledge gaps exist for research to unravel, but in no sense is this to imply that GH is a correlative response or casual precursor to IGF-I production. Schoenle *et al.* (1982) reported a dose response in body weight gain by hypophysectomized rats treated with IGF-I comparable to responses elicited by GH. In this study, IGF-II was devoid of growth-promoting activity.

Placental lactogen (PL), a hormone chemically related to both GH and prolactin, has been proposed as a stimulus important during gestation for the production of IGF-II. Hurley *et al.* (1977) reported that PL was as effective as GH in the stimulation of Sm concentration in the hypophysectomized rat model. Adams *et al.* (1983) found that PL stimulates IGF-II production by fetal rat fibroblasts, but that PL stimulates IGF-I, not IGF-II, production by adult rat fibroblasts. The model proposed by Hall and Sara (1983) still serves as a guide for the role of PL and GH because these factors influence circulating Sm concentration (Fig. 4).

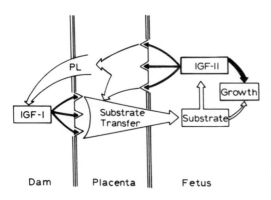

FIGURE 4. Diagram of Hall and Sara (1983) showing that maternal IGF-I acting on placental receptors facilitates substrate transfer to the fetus. In turn, IGF-II signals through placental lactogen (PL) to amplify the IGF-I stimulation of substrate delivery. This coordination model predicts that fetal growth controls fetal energy delivery.

Prolactin's influence on IGF production is unclear. Bala *et al.* (1978) reported that prolactin increased sulfation activity by hypophysectomized rat cartilage explants. Clemmons *et al.* (1981) compared human patients with distinctly different types of pituitary tumors. In patients with deficient GH secretion and normal prolactin secretion, circulating IGF-I concentration was reduced, whereas in patients with deficient GH secretion and elevated prolactin secretion, IGF-I concentration was approximately normal. Results from both studies are subject to the criticism that prolactin may mimic GH by cross-reacting with hepatic GH receptors.

5. Somatomedin Research in Domestic Animals

As discussed, Sm's are a class of insulinlike peptides currently under intensive investigation in animal science because of their enormous influence on bone, muscle, fat, and cartilage growth and their inherent propensity for altering the composition of gain. Of particular importance is Sm-C, also referred to as IGF-I, because of the interrelationship between IGF-I and GH. GH is a major factor in the regulatory scheme of IGF-I expression. Several of the actions of GH on bone growth and tissue deposition are assumed to occur through subsequent IGF-I-dependent mechanisms. Interpreting the nature of the IGF-I and GH signals and their impact on growth confronts investigators in the area of growth physiology.

Several of the metabolic and physiological effects of IGF-I on cell function are listed in Table I. To date, all isolated metabolic actions that have been studied in association with Sm's are directly or indirectly anabolic in nature. Numerous investigations in *in vitro* culture systems have demonstrated increases in nutrient transport into cells to support increased tissue-product syn-

Table I
Cell-Associated Activities Affected by IGF-I

Activity	IGF-I effect	Cells studied
Aromatase	Increase	Rat granulosa cell
Protein and mRNA turnover	Increase	Swine granulosa cell
[^{3}H]thymidine incorporation	Increase	Rat myoblast and fibroblast
Glucose metabolism	Increase	Rat, swine, cattle adipocyte
Lipid synthesis	Increase	Rat, swine adipocyte
Lipolysis	Decrease	Rat, swine adipocyte
Sulfate incorporation	Increase	Rat, chick chondrocytes
mRNA, DNA	Increase	Rat, chick chondrocytes
Ribosomal RNA/polyribosomes	Increase	Several growing cell lines

thesis (e.g., increased lipid synthesis, increased protein synthesis, increased RNA production of tissue-specific gene product), as well as increased ribosomal protein synthesis and formation of polyribosomes, decreases in lipolysis, and increases in sulfate incorporation into cartilage to support subsequent bone growth. In contrast, GH action is both anabolic, demonstrated by increased protein synthesis and nitrogen retention, and catabolic, demonstrated clearly in its lipolytic actions and ability to block the lipogenic actions of other hormones such as insulin (Underwood *et al.*, 1972; Etherton and Walton, 1986).

Research on Sm's in domestic animals has lagged behind that in human subjects or rodent models. The lag developed because of several factors, the most difficult of which to overcome, the lack of an available source of purified hormone for investigation, has been partially relieved with the availability of genetically engineered IGF-I and several antibody preparations for immunoassay methods. Because of the limited quantity of IGF-I and the expense associated with obtaining this hormone, much of the work done in domestic animals has been of a characterizational nature revolving around measurement of plasma concentrations of IGF-I/Sm-C, determination of factors that regulate plasma concentrations, and developmental studies aimed at the ontogeny of Sms. Adding to the lag in domestic animal research is the expense associated with the amounts of hormone needed to conduct a trial and the practical concern that much of the administered hormone would be bound by the circulating Sm binding proteins (Fig. 2) and made unavailable for bioactivity.

Recent work by Honegger and Humbel (1986) has demonstrated that the amino acid sequences of bovine and human IGF-I are identical. Knowledge that the amino acid sequence and tertiary structure of Sm's are highly conserved across species has allowed the validation of analytical procedures to measure plasma and tissue concentrations of IGF-I using antisera directed at the human IGF-I peptide. Because almost all of the available antisera for radioimmunoassay determination of Sm-C is directed against the human antigen, the bovine assay is essentially a homologous assay. In addition, comparable potencies and specificities of natural, solid-phase synthetic and recombinant forms of IGF-I (Schalch *et al.*, 1984) have allowed for better validation and thus lent more uniformity to the measurement of IGF-I. Several laboratories are now successfully measuring Sm's in blood fluids and tissue extracts in several domestic species including cattle, sheep, swine, poultry, and fish. Another significant aspect of the IGF-I assay that has allowed for successful validation and increased uniformity of assay values obtained by different investigators was the recognition and elimination of the effects of the IGF-I plasma binding protein on the assay.

Some of the first work associated with IGF-I and domestic animals was aimed at determining whether a significant correlation existed between growth and the plasma concentrations of circulating IGF-I. Lund-Larsen and Bakke

(1975) investigated the relationship between serum GH, Sm activity, and rate of weight gain and backfat thickness in swine genetically selected for enhanced expression of these growth characteristics. In these studies, serum Sm activity was measured by bioassay where sulfate incorporation into growing cartilage was the quantifiable endpoint. Pigs selected for the higher rates of gain and backfat thickness had more serum Sm activity (indicated by the higher incorporation of radiosulfate into cartilage) than either control of low-line selected pigs. In additional studies, Lund-Larsen *et al.* (1977) explored the theory that serum concentrations of Sm's could be a useful indicator of growth performance in animals. They measured linear growth, weight gain, cross-sectional muscle area, and feed utilization efficiency in bulls and correlated with serum concentrations of testosterone and Sm's. Their hypothesis was based on the premise that faster-growing bulls would display higher serum concentrations of anabolic hormones than slower-growing animals. Their results showed a negative correlation between both testosterone and Sm concentrations in serum and feed utilization efficiency. However, rate of gain and linear growth were positively correlated to Sm levels. They concluded that Sm may be a useful parameter to screen animals and predict potential growth performance.

The idea that Sm could serve as a predictive index of growth capacity and nutrient partitioning became less popular following publication of data concerning correlations between serum GH and Sm concentrations and growth rate of swine and obese Zucker rats. Using a validated bioassay, Gahagan *et al.* (1980) demonstrated that Sm concentrations did not differ between fast- and slow-growing strains of pigs although GH concentrations were significantly lower in the slow-growing Ossabaw pigs. In the same study, plasma concentrations of GH and insulin in lean and genetically obese Zucker rats were measured. The obese animals had significantly reduced serum concentrations of GH and Sm but higher concentrations of insulin compared to their lean counterparts. This finding, in addition to that presented earlier, begins to demonstrate the complexity associated with Sm's and growth, especially in terms of composition of gain in animals. It is not surprising that measurements of blood concentrations of Sm might not be an optimal indicator of potential growth because several other nutritional and hormonal factors figure into the regulatory scheme of Sm and could bias the interpretation of the measurements (Fig. 5).

Hormonal and nutritional interrelationships that regulate plasma concentrations of IGF-I in laboratory and domestic species are being pursued. The protein and energy content of diets significantly influence circulating plasma concentrations of Sm in laboratory animals (Prewitt *et al.*, 1982; Isley *et al.*, 1984) and in domestic species, (Elsasser *et al.*, 1985; Scamurra *et al.*, 1986; Butler and Gluckman, 1986; Rosebrough, unpublished data). Of all the factors known to influence circulating concentrations of Sm's, diet is one of the most effective. A generalized scheme depicting the relationship between Sm-C and nutri-

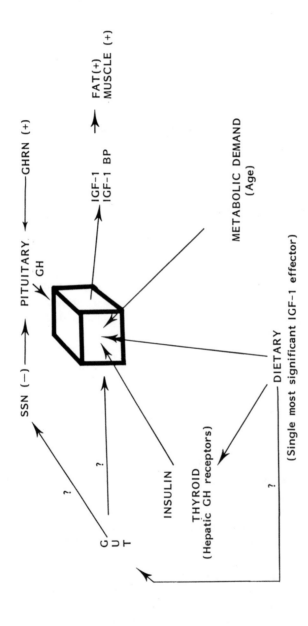

FIGURE 5. Numerous determinants factor into the regulatory scheme for IGF-I. Key factors include regulation by GH and plane of nutrition. Of interest is the fact that nutritional effects can, under particular circumstances, override normal regulation by GH. Dietary effects may manifest themselves through effects on GH secretion while effects on GH receptor and postreceptor function can be mediated by changing insulin and thyroid levels concurrent with plane of nutrition. Relatively nothing is known about direct effects of nutrients on genetic regulation or the effects of potential regulatory signals of peripheral origin. Note that no specific tissue is referenced, although liver is of particular importance, since several tissues in the body are capable of producing IGF-I.

Table II

Relative Changes and Differences in Circulating Plasma and Serum Concentrations in Several Domestic Species in Varying Nutritional and Physiological States Measured under Similar RIA Conditions

Animal model	Size/age	Comments[a]	Plasma IGF-I (ng/ml)
Beef cattle	320–360 kg	15 g/day + N balance	85–100
		28–32 g/day + N balance	120–160
		+N balance/exogenous bGH	150–221
	420–450 kg	10–15 g/day + N balance	60–80
Dairy steers	4–5 months	>40 g/day + N balance	75–100
		10 to 15 g/day + N balance	60–80
		−5 to 5 g/day + N balance	20–50
Lactating Holstein		Variable energy/protein status	8–100
Wethers	1 year	+ energy/+ protein balance	200–300
Chicken	6–8 weeks	+ energy/+ protein balance	13–25
Swine	3 day–100 kg	+ protein/+ age	100–1000

[a] + signifies an increase.

tion is presented in Fig. 5. Suboptimal nutrition significantly lowers plasma Sm-C. In fact, with severe nutrient restriction, animals can become refractory to the somatomedogenic effect of GH so that no significant increase in circulating plasma concentrations of IGF-I can be measured following administration of exogenous GH.

Circulating plasma concentrations of IGF-I can fluctuate widely in association with the relative physiological and pathological state of an animal and correlations between circulating concentrations of GH and IGF-I and size are not always straightforward. For example, concentrations of IGF-I in plasma and sera from several domestic species vary considerably by species, breed, relative rate of maturity, sex, stage of lactation, nutrient balance, and so forth. General trends in livestock species are summarized in Table II. Of particular interest is the extreme variation in plasma concentrations of IGF-I in animals that have a poor energy and nitrogen balance in contrast to positive balance. Additional effects of disease suggest an important role for Sm-C in redirecting utilization of nutrients by tissues and assist in imparting the schedule of prioritization by which tissues most critical for survival are sustained.

Eigenmann et al. (1984) developed a hypothesis using a canine model that relates plasma concentrations of IGF-I to mature body size. They established a high correlation between elevated levels of IGF-I in larger breeds of poodles. Recent data from several species show that this relationship breaks down as it becomes apparent that some of the highest levels of circulating IGF-I can be measured in relatively small animals such as the guinea pig (Froesch et al., 1986). Additional data concerning domestic animals suggest that the relative differences in measurable concentrations of IGF-I are related to rate of maturation, stage of maturation, and metabolic need (Elsasser and Hammond, unpublished observations).

A significant body of data suggests that plasma concentrations of IGF-I vary with age (Steele *et al.*, unpublished data; Gluckman *et al.*, 1979). Typical of several species, IGF-I levels rise, as determined by Gluckman's laboratory using sheep, throughout the growing phase of maturation but then decline with increasing age. There are few data that address the ontogeny of the rise and fall with age. Perhaps it is associated with the metabolic demands of animals as they grow. Purely speculatively, it is an interesting possibility that IGF-I levels are higher in younger animals so as to effect the relatively greater protein deposition activities that accompany muscle accretion and taper off as fat accretion becomes the dominant component of gain. No data are available that compare the relative sensitivity of protein and lipid tissues to the effects of IGF-I. Perhaps muscle is more sensitive and, with a higher plasma level of the hormone, muscle accretion is favorably directed. Then fat cells develop additional sensitivity as the need for protein accretion declines. Much more research needs to be performed relating the age-dependent tissue responses to IGF-I and the age-related changes in plasma IGF-I.

Data concerning age-related influences on plasm IGF-I suggest that even with optimal nutrition, very young (neonatal) animals are refractory to GH stimulating effects on Sm production. Information on the nature of the age- and diet-associated changes in circulating Sm-C suggests that the state of hormone receptors on Sm-C-producing tissues is partially responsible for the changes in responsiveness to GH. Evidence favoring this receptor hypothesis includes: (1) the time that Sm-C begins to increase in the circulation in sheep coincides with the age-related changes in the liver during which hepatic membrane receptors for GH are first measurable (Gluckman *et al.*, 1979); (2) nutritional modulation of GH receptors in laboratory species in association with diabetic and thyroid-deficient states is well correlated with changes in plasma Sm-C (Maes *et al.*, 1984); and (3) changes in plasma Sm-C were evident in cattle fed diets differing in protein and energy content but plasma concentrations of GH and insulin did not differ although thyroxine levels varied with the protein and energy content of the diets (Elsasser *et al.*, 1986; Breier *et al.*, 1986). Similarly, Elsasser *et al.* (1986) demonstrated that the incremental change in Sm response to intramuscular injection of 0.1 mg bovine GH/kg body weight was not affected by dietary protein or energy within a range of dietary conditions that produced 0.7 to 1.1 kg of gain per day. However, with more extreme planes of nutrition, significant effects on the IGF-I response to GH were observed. The cumulative data suggest that GH is a necessary mediator of Sm expression in cattle but that dietary effects can override the regulatory actions of GH on Sm, even to the extreme situation where, in the face of undernutrition, the

Table III
Relative Directional Changes in Circulating IGF-I Associated with Changes in Dietary Protein, Energy Status, and Plane of Nutrition[a]

	IGF-I		
	Cattle	Swine	Chicken
Protein ↑	↑	↑	↑
Energy ↑	—	↑	—
Total intake ↑	↑	↑	↑

[a] Arrow indicates change in dietary intake and corresponding directional change of circulating IGF-I.

response to GH is abolished. Additional data demonstrated that finishing steers failed to respond to exogenous GH when fed to effect a state of negative nitrogen balance. The basal response of Sm-C to undernutrition appears to be in part dependent on the age or perhaps the body composition of the animal. Younger animals responded to starvation with a dramatic fall in circulating concentrations of Sm-C (Elsasser et al., 1985). Finishing steers had less dramatic reductions in Sm with feed restriction but lost the ability to respond to GH.

In general, several domestic animals including cattle, swine, and poultry respond to increases in dietary protein with elevations in radioimmunoassayable IGF-I (see Table III). An energy component factors into the relationship because adequate dietary energy must be present to effect efficient use of protein. Data of Campbell et al. (1985) imply that only when energy is adequate, is gain proportional to protein intake. When the energy level is readjusted, animals gain on a different curve but protein level is determinative. Perhaps it is in association with the nature of gain that the dietary composition affects circulating concentrations of IGF-I.

References

Adams, S. O., S. P. Nissley, S. Handwerger and M. M. Rechler. 1983. Developmental patterns of insulin-like growth factor-I and -II syntheses and regulation in rat fibroblasts. Nature 302:150.

Bala, R. M., C. Wright, A. Bardai and G. R. Smith. 1978. Somatomedin bioactivity in serum and amniotic fluid during pregnancy. J. Clin. Endocrinol. Metab. 46:649.

Baxter, R. D., S. Axiak and R. L. Raison. 1982. Monoclonal antibody against somatomedin-C/insulin-like growth factor-1. J. Clin. Endocrinol. Metab. 54:474.

Binoux, M. S., S. Hardouin, C. Lassarre and P. Hossenlopp. 1982. Evidence for production by the liver of two IGF binding proteins with similar molecular weights but different affinities for IGF-I and IGF-II. Their relations with serum and cerebrospinal fluid IGF binding proteins. J. Clin. Endocrinol. Metab. 55:600.

Breier, B. H., J. J. Bass, J. H. Butler and P. D. Gluckman. 1986. The somatotrophic axis in young steers: influence of nutritional status on pulsatile release of growth hormone and circulating concentrations of insulin-like growth factor 1. Journal of Endocrinology 111:209.

Butler, J. H. and P. D. Gluckman. 1986. Circulating insulin-like growth factor-binding proteins in fetal, neonatal and adult sheep. J. Endocrinol. 109:333.

Campbell, R. G., M. R. Taverner and D. M. Curic. 1985. Effects of sex and energy intake between 48 and 90 kg live weight on protein deposition in growing pigs. Anim. Prod. 40:497.

Chung, C. S., T. D. Etherton and J. P. Wiggins. 1985. Stimulation of swine growth by porcine growth hormone. J. Anim. Sci. 60:118.

Clemmons, D. R., L. E. Underwood, P. G. Chatelain and J. J. Van Wyk. 1983. Liberation of immunoreactive somatomedin-C from its binding proteins by proteolytic enzymes and heparin. J. Clin. Endocrinol. Metab. 56:384.

Clemmons, D. R., L. E. Underwood, E. C. Ridgeway, B. Kliman and J. J. Van Wyk. 1981. Hyperprolactinemia is associated with increased immunoreactive somatomedin C in hypopituitarism. J. Clin. Endocrinol. Metab. 52:731.

Copeland, K. C., L. E. Underwood and J. J. Van Wyk. 1980. Induction of immunoreactive somatomedin-C in human serum by growth hormone: dose response relationships and effect on chromatographic profiles. J. Clin. Endocrinol. Metab. 50:690.

Daughaday, W. 1977. Hormonal regulation of growth by somatomedin and other growth factors. Clinics in Endocrinology and Metabolism. 6:117.

Daughaday, W. 1983. The somatomedin hypothesis: origins and recent developments in insulin-like growth factors. In: E. M. Spencer (Ed.) Insulin-Like Growth Factors/Somatomedins. pp 3–11. Walter de Gruyter, Berlin.

Daughaday, W. H., I. K. Mariz and S. L. Blethen. 1980. Inhibition of access of bound somatomedin to membrane receptor and immunobinding sites: a comparison of radioreceptor and radioimmunoassay of somatomedin in native and acid-ethanol-extracted serum. J. Clin. Endocrinol. Metab. 51:781.

Daughaday, W. H., K. A. Parker, S. Borowsky, B. Trivedia and M. Kapedia. 1982. Measurement of somatomedin related peptides in fetal, neonatal and maternal rat serum by insulin-like growth factor (IGF) I radioimmunoassay, IGF-II radioreceptor assay (RRA) and multiplication-stimulating activity RRA after acid-ethanol extraction. Endocrinology 110:575.

D'Ercole, A. J., G. T. Applewhite and L. E. Underwood. 1980. Evidence that somatomedin is synthesized by multiple tissues in the fetus. Dev. Biol. 75:315.

Dulak, N. and H. M. Temin. 1973. A partially purified polypeptide fraction from rat liver cell conditioned medium with multiplication stimulating activity for embryo fibroblasts. J. Cell. Physiol. 81:153.

Eigenmann, J. E., D. F. Patterson and E. R. Froesch. 1984. Body size parallels insulin-like growth factor I levels but not growth hormone secretory capacity. Acta Endocrinol. 106:448.

Elsasser, T. H., A. C. Hammond, T. S. Rumsey and R. Fayer. 1985. Changes in plasma growth hormone, somatostatin-like immunoreactivity and somatomedin-C concentrations in steers during infection with *Sarcocystis cruzi*. Abstr. Proc. Endocrinol. Soc. p 29.

Elsasser, T. H., T. S. Rumsey and A. C. Hammond. 1986. Dietary regulation of somatomedin-C in beef cattle. J. Nutr. 116:XXXVI.

Elsasser, T. H., M. Pello and T. S. Rumsey. 1987. Somatomedin-C (SMC) and binding protein responses to growth hormone in cattle. J. Anim. Sci. 65 (Suppl. 1):256.

Etherton, D. and P. E. Walton. 1986. Hormonal and metabolic regulation of lipid metabolism in domestic livestock. J. Anim. Sci. 63 (Suppl. 2):76.

Froesch, E. R., H. Burgi, E. B. Ramseier, P. Bally and A. Labhart. 1963. Antibody-suppressible and nonsuppressible insulin-like activities in human serum and their physiological significance. An insulin assay with adipose tissue of increased precision and specificity. J. Clin. Invest. 42:1816.

Froesch, E. R., C. Schmid, I. Zangger, E. Schoenle, E. Eigenmann and J. Zapf. 1986. Effects of IGF/somatomedins on growth and differentiation of muscle and bone. J. Anim. Sci. 63 (Suppl. 2):57.

Furlanetto, R. W. 1980. The somatomedin-C binding protein: Evidence for a heterologous subunit structure. Endocrinology 51:12.

Furlanetto, R. W., L. E. Underwood, J. J. Van Wyk and A. J. D'Ercole. 1977. Estimation of somatomedin-C levels in normals and patients with pituitary disease. J. Clin. Invest. 60:648.

Gahagan, J. H., R. J. Martin and R. M. Leach. 1980. Serum somatomedin activity in two animal models as measured using chick epiphyseal plate cartilage bioassay. Proc. Soc. Exp. Biol. Med. 163:455.

Gluckman, P. D., K. Uthne, D. M. Styne, S. L. Kaplan, A. M. Rudolph and M. M. Grumbach. 1979. Hormone ontogeny in the ovine fetus. IV. Serum somatomedin activity in the fetal and neonatal lamb and pregnant ewe: correlation with maternal and fetal growth hormone, prolactin, and chorionic somatomammotropin. Pediatr. Res. 14:194.

Graham, D. E., M. M. Rechler, A. L. Brown, R. Frunzio, J. A. Romanus, C. B. Bruni, H. J. Whitfield, S. P. Nissley, S. Seelig and S. Berry. 1986. Coordinate developmental regulation of high and low molecular weight mRNA's for rat insulin-like growth factor II. Proc. Natl. Acad. Sci. USA 83:4519.

Hall, K., J. Brandt, G. Enberg and L. Fryklund. 1979. Immunoreactive somatomedin A in human serum. J. Clin. Endocrinol. Metab. 48:271.

Hall, K. and V. R. Sara. 1983. Growth and somatomedins. Vitam. Horm. (N.Y.) 40:175.

Hammond, J. M., J. L. F. Baranao, D. A. Skaleris, M. M. Rechler and A. P. Knight. 1984. Somatomedin (Sm) production by cultured porcine granulosa cells (GC). J. Steroid Biochem. 20:1597.

Hintz, R. L. and F. Liu. 1977. Demonstration of specific plasma protein binding sites for somatomedin. J. Clin. Endocrinol. Metab. 45:988.

Hintz, R. L., F. Liu, D. Chang and E. R. Rinderknecht. 1983. The use of synthetic peptides for the development of radioimmunoassays for insulin-like growth factors. In: E. M. Spencer (Ed.) Insulin-Like Growth Factors/Somatomedins. pp 169–176. Walter de Gruyter, Berlin.

Honegger, A. and R. E. Humbel. 1986. Insulin-like growth factors I and II in fetal and adult bovine serum. J. Biol. Chem 251:569.

Hurley, T. W., A. J. D'Ercole, S. Handwerger, L. E. Underwood, R. W. Furlanetto and R. E. Fellows. 1977. Ovine placental lactogen induces somatomedin: a possible role in fetal growth. Endocrinology 101:1635.

Isley, W. L., L. E. Underwood and D. R. Clemmons. 1984. Changes in plasma somatomedin-C in response to ingestion of diets with variable protein and energy content. J. Parent. Ent. Nutr. 8:407.

Kato, Y., Y. Nomura, M. Tsuji, M. Konoshita, H. Ohmae and F. Suzuki. 1981. Somatomedin-like peptide(s) isolated from fetal bovine cartilage (cartilage-derived factor): isolation and some properties. Proc. Natl. Acad. Sci. USA 78:6831.

Lund-Larsen, T. R. and H. Bakke. 1975. Growth hormone and somatomedin activities in lines of pigs selected for rate of gain and thickness of backfat. Acta Agr. Scand. 25:231.

Lund-Larsen, T. R., A. Sundby, V. Kruse and W. Velle. 1977. Relations between growth rate, serum somatomedin and plasma testosterone in young bulls. J. Anim. Sci. 44:189.

Maes, M., L. E. Underwood, G. Gerard and J. M. Ketelslegers. 1984. Relationship between plasma somatomedin-C and liver somatogenic binding sites in neonatal rats during malnutrition and after short and long term refeeding. Endocrinology 115:786.

McCusker, R. H., D. R. Campion and D. R. Clemmons. 1986. The ontogeny and regulation of insulin-like growth factor-somatomedin (IGF-1) binding proteins in fetal and maternal porcine plasma and serum. J. Anim. Sci. 63 (Suppl. 1):242.

Marshall, R. N., L. E. Underwood, S. J. Viona, D. B. Foushee and J. J. Van Wyk. 1974. Characterization of the insulin and somatomedin-C receptors in human placental cell membranes. J. Clin. Endocrinol. Metab. 39:283.

Merimee, T. J., J. Zapf and E. R. Froesch. 1981. Dwarfism in the pygmy: an isolated deficiency of insulin-like growth factor I. N. Engl. J. Med. 305:965.

Phillips, L. S., D. S. Belosky, L. A. Reichard, H. S. Young, S. R. Berkowitz and M. F. Pertcheck. 1979. Somatomedin(s) and somatomedin inhibitor(s) in normal and diabetic serum. In: G. Giordano. J. J. Van Wyk and F. Minuto (Ed.) Somatomedins and Growth. pp 209–212. Academic Press, New York.

Pierson, R. W. and H. M. Temin. 1972. The partial purification from calf serum of a fraction with multiplication-stimulating activity for chicken fibroblasts in culture and with non-suppressible insulin-like activity. J. Cell. Physiol. 79:319.

Prewitt, T. E., A. J. D'Ercole, B. R. Switzer and J. J. Van Wyk. 1982. Relationship of serum immunoreactive somatomedin-C to dietary protein and energy in growing rats. J. Nutr. 112:144.

Rabinovitch, A., C. Quigley, T. Russell, Y. Patel and D. H. Mintz. 1982. Insulin and multiplication stimulating activity (an insulin-like growth factor) stimulate islet beta cell replication in neonatal rat pancreatic monolayer cultures. Diabetes 31:160.

Rechler, M. M., H. J. Eisen, O. Z. Higa, S. P. Nissley, A. C. Moses, E. E. Schilling, I. Fennoy, C. B. Bruni, L. S. Phillips and K. L. Baird. 1979. Characterization of a Somatomedin (Insulin-like Growth Factor) Synthesized by Fetal Rat Liver Organ Cultures. J. Biol. Chem. 254:7942.

Rechler, M. M., J. Zapf, S. P. Nissley, E. R. Froesch, A. C. Moses, J. M. Podskalny, E. E. Schilling and R. E. Humbel. 1980. Interactions of insulin-like growth factors I and II and multiplication-stimulating activity with receptors and serum carrier proteins. Endocrinology 107:1451.

Rinderknecht, E. and R. E. Humbel. 1978. Primary structure of human insulin-like growth factor II. FEBS Lett. 89:283.

Ritzen, E. M., B. Froysa, B. Gustafsson, G. Westerholm and E. Diczfalusy. 1982. Improved in vitro Bioassay of Follitropin. Horm. Rest. 16:42.

Rotwein, P., K. M. Pollock, D. K. Didier and G. G. Krivi. 1986. Organization and sequence of the human insulin-like growth factor I gene. J. Biol. Chem. 261:4828.

Salmon, W. D., Jr. and W. H. Daughaday. 1957. A hormonally controlled serum factor which stimulates sulfate incorporation by cartilage in vivo. J. Lab. Clin. Med. 49:825.

Sara, V. R., K. Hall, P. Lins and L. Fryklund. 1980. Serum levels of immunoreactive somatomedin A in the rat: some developmental aspects. Endocrinology 107:622.

Scamurra, R. S., N. C. Steele, T. J. Caperna and M. L. Failla. 1986. Comparison of swine serum pretreatment methods for quantifying insulin-like growth factor I (IGF-I). J. Anim. Sci. 63 (Suppl. 1):224.

Schalch, D. S., U. E. Heinrich, J. G. Koch, C. J. Johnson and R. J. Schlueter. 1978. Nonsuppressible insulin-like activity (NSILA). I. Development of a new sensitive competitive protein-binding assay for determination of serum levels. J. Clin. Endocrinol. Metab. 46:664.

Schalch, D., D. Reismann, C. Emler, R. Humbel, C. H. Li, M. Peters and E. Lau. 1984. Insulin-like growth factor 1/somatomedin-C (IGF-1/SMC): comparison of natural, solid phase synthetic and recombinant DNA analog peptides in two radioligand assays. Endocrinology 115:2490.

Schoenle, E., J. Zapf, R. E. Humbel and E. R. Froesch. 1982. Insulin-like growth factor stimulates growth in hypophysectomized rats. Nature 296:252.

Schwander. J. C., C. Hauri, J. Zapf and E. R. Froesch. 1983. Synthesis and secretion of insulin-like growth factor and its binding protein by the perfused rat liver: dependence on growth hormone status. Endocrinology 113:297.

Scott, C. D., J. L. Martin and R. C. Baxter. 1985. Production of insulin-like growth factor I and its binding protein by adult rat hepatocytes in primary culture. Endocrinology 116:1094.

Spencer, E. M. 1979. Synthesis by cultured hepatocytes of somatomedin and its binding protein. FEBS Lett. 99:157.

Spencer, E. M., M. Ross and B. Smith. 1983. The identity of human insulin-like growth factors I and II with somatomedins C and A and homology with IGF I and II. In: E. M. Spencer (Ed.) Insulin-Like Growth Factors/Somatomedins. pp 81–96. Walter de Gruyter, Berlin.

Svoboda, M. E. and J. J. Van Wyk. 1984. Somatomedins. In: W. S. Hancock (Ed.) CRC Handbook of HPLC for the Separation of Amino Acids, Peptides and Proteins. pp 439–444. CRC Press, Boca Raton, FL.

Underwood, L. E., A. J. D'Ercole, K. C. Copeland, J. J. Van Wyk, T. Hurley and S. Handwerger. 1982. Development of a heterologous radioimmunoassay for somatomedin-C in sheep blood. J. Endocrinol. 93:31.

Underwood, L. E., R. L. Hintz, S. J. Voina and J. J. Van Wyk. 1972. Human somatomedin, the growth hormone dependent sulfation factor, is antilipolytic. J. Clin. Endocrinol. Metab. 35:194.

Vassilopoulou-Sellin, R. and L. S. Phillips. 1982. Extraction of somatomedin activity from rat liver. Endocrinology 110:582.

Wilkins, J. R. and A. J. D'Ercole. 1985. Affinity labeled plasma somatomedin-C/insulin-like growth factor I binding proteins. J. Clin. Invest. 75:1350.

Yang, Y. W. -H., M. M. Rechler, S. P. Nissley and J. E. Coligan. 1985b. Biosynthesis of rat insulin-like growth factor II. J. Biol. Chem. 260:2578.

Yang, Y. W. -H., J. A. Romanus, T. -Y. Liu, S. P. Nissley and M. M. Rechler. 1985a. Biosynthesis of rat insulin-like growth factor II. J. Biol. Chem. 260:2570.

Zapf, J. 1983. Determination of insulin-like growth factors: a survey of methods. In: E. M. Spencer (Ed.) Insulin-Like Growth Factors/Somatomedins. pp 155–168. Walter de Gruyter, Berlin.

Zapf, J., G. Jagars and E. R. Froesch. 1978. Evidence for the existence in human serum of large molecular weight nonsuppressible insulin-like activity (NSILA) different from small molecular weight forms. FEBS Lett. 90:135.

Zapf, J., M. Waldvogel and E. R. Froesch. 1975. Binding of nonsuppressible insulin-like activity to human serum: evidence for a carrier protein. Arch. Biochem. Biophys. 168:638.

Sexual Differentiation and the Growth Process

J. JOE FORD AND JOHN KLINDT

1. Introduction

As adults, males are larger than females in most species with which we are familiar, but this generalization is not appropriate for all species (Ralls, 1976). In cattle, sheep, and swine, testicular secretions (testosterone and its metabolites) are associated with the greater size of males, but few discussions of growth in domestic farm animals address the total impact of these steroids on developmental processes. The influence of testicular steroids on growth and muscling during pubertal development is well documented (Tucker and Merkel, 1987), but when steers are produced by castration shortly after birth, they are not exposed to testicular secretions during postnatal development. Why then do steers grow faster and larger than heifers? A second point that has perplexed animal scientists is the inconsistency among cattle, sheep, and swine relative to body growth after castration of young males. From *Hammond's Farm Animals* (Hammond *et al.*, 1971) we quote, "While (at equal body weight) the castrated male, in sheep or cattle, has a higher proportion of muscle and less fat than the female, in pigs the position is reversed." Trenkle and Marple (1983) reiterated this point: "The inconsistent ranking of the barrow as compared with the steer and wether is not easily explained." We also are unable to fully explain this issue but speculate that this may relate to the time when sexual differentiation of the growth process occurs.

In mammalian species, males are the heterogametic sex. Development of

J. JOE FORD AND JOHN KLINDT • United States Department of Agriculture, Agricultural Research Service, Roman L. Hruska U.S. Meat Animal Research Center, Clay Center, Nebraska 68933.

the masculine phenotype is dependent upon responses to testicular secretions (Jost and Magre, 1984). Genetic sex is determined at conception, but feminine phenotypic characteristics are expressed unless the embryo is impacted by testicular secretions. Understanding of sexual differentiation originated with studies on morphological development of the urogenital tract but has been extended to sex differences in gonadotropin secretion, sexual behavior, nonreproductive behaviors, hepatic enzyme activity, and so forth. In genetic male embryos, there is a testis determining gene on the Y chromosome that is separate from the coding region for H-Y antigen (Simpson *et al.*, 1986; Page *et al.*, 1987). Newly formed testes secrete anti-Mullerian hormone and testosterone, which directly or through specific metabolites, alter development from feminine to masculine characteristics. The potential for growth processes to be similarly regulated has received less attention by investigators and is, therefore, more difficult to document in farm mammals. Certainly when cattle, sheep, and swine are gonadally intact and genetic potential for growth is not restricted by environmental factors, males are larger than females as adults. Thus, growth in these species is sexually dimorphic.

Bulls and rams have two periods of elevated testicular steroid secretion during development compared to three such periods in boars (D'Occhio and Ford, 1988). The prenatal period of maximal testicular activity is at 1½–3 months of gestation in bulls, 1–2½ months in rams, and 1–2 months in boars. Boars experience increased secretion of testicular steroids during the first month of neonatal development; this period of elevated testicular steroidogenesis is absent in bulls and rams. All three species secrete increasing concentrations of testicular steroids throughout pubertal development: after 5 months of age in bulls and after 3 months of age in rams and boars.

Gonadal steroids have two general actions that affect sexual dimorphism: permanent modification of processes (organizational actions) or activation of existing processes. Steroidal modifications (differentiation) are generally limited to somewhat discrete periods during early development when a specific process is maximally sensitive to alteration by a steroid. After this unique period in development, the steroid is ineffective or less effective for induction of the alteration. An example of the organizational action of androgen is its effect on development of the external genitalia of genetic females. Exposure to unusually high amounts of testosterone during the embryonic stage causes formation of a penis and scrotum in females; however, if exposure does not occur until the midfetal stage or later, these genital organs will not develop. Sexual differentiation generally occurs during perinatal development in species with short gestation periods and during prenatal development in species with long gestation periods (Goy and McEwen, 1980). The second way in which gonadal steroids act to induce dimorphic traits is through activation of specific processes. Expression of the dimorphic character is dependent upon the presence of the gonadal steroid, and after castration such processes become deactivated. Activational effects of gonadal steroids occur throughout an animal's life and are exemplified by the cyclic occurrence of sexual receptivity in female, adult, domestic mammals. After ovariectomy this cyclicity ceases, but it can be restored by appropriate treatment with ovarian steroids. Growth and traits thought to be associated with growth (e.g., hormone secretion, enzyme activities) are influenced by gonadal steroids through both modification and activation.

We will summarize sexual differences in growth by species and then conclude with a discussion of sexual differences in specific endocrine processes that are associated with this dimorphism in growth. The primary interest in livestock production is lean tissue growth but many studies report changes in total body weight. Thus, carcass composition will be emphasized when data are available.

2. Rats

Male rats are 5–8% heavier at birth than females (Slob and van der Werff ten Bosch, 1975), and this difference increases throughout development (Fig. 1A). If male rats are castrated at birth, they do not achieve as great a weight as males castrated at 21 days of age. Thus, testicular secretions between birth and day 21 enhance postnatal growth potential. In contrast to males, female rats that are castrated at birth and those castrated at 21 days of age have similar rates of body growth. Castrated females grow to a larger size than intact females because estrogens have an inhibitory effect on body growth in this species (Beatty, 1979). Since body weight of castrated females approaches that of males castrated at birth, it is concluded that growth potential of these males has not undergone sexual differentiation.

Additional support for the concept that growth in male rats has a sexually differentiated component is apparent from studies with females in which early postnatal treatment with testosterone produces an increase in skeletal growth and body weight (Fig. 1B). Postnatally, day 2 to 3 of age is the period of greatest sensitivity to testosterone. No increase in growth was observed when low dosages of testosterone were administered to female rats on day 4 or 5 (Tarttelin et al., 1975). Both intact and ovariectomized females show accelerated rates of growth after early neonatal testosterone treatment, but weight at maturity is greater in ovariectomized animals (Slob and van der Werff ten Bosch, 1975; Perry et al., 1979). Additionally, the increase in rate of growth of females after neonatal testosterone treatment is not solely the consequence of increased fat deposition because body composition is similar in control and treated females (Perry et al., 1979).

In summary, male rats grow to a larger weight prenatally than females,

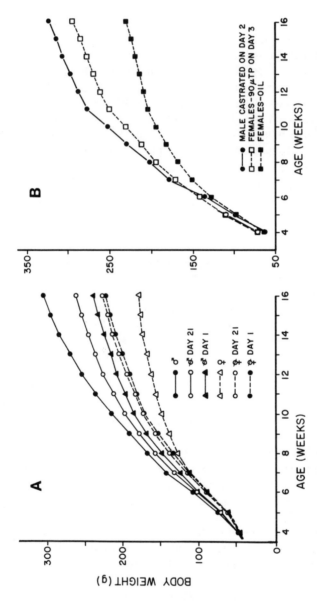

Figure 1. (A) Influence of age at castration on body weights in male and female rats. Adapted from Slob and van der Werff ten Bosch (1975). (B) Influence of neonatal castration of male rats and neonatal testosterone treatment of females on body weights. Adapted from Tarttelin *et al.* (1975).

but their potential for postnatal growth is similar to that of females. In females, the potential for postnatal body growth is sequestered by estrogens throughout adulthood unless their ovaries are removed. In postnatal males, testosterone secretion from the testes during the first few days of life irreversibly elevates rate of growth to a point greater than that of females. Then, testicular steroid secretion associated with puberty and adulthood increases body growth further. This final phase is likely activation of existing growth processes because prolonged treatment of pubertal females with testosterone does little to enhance their rate of body growth (Beatty, 1973). More definitive dose–response studies in females and neonatally castrated males are required before this latter statement can be fully accepted. A primary question that remains is whether ovariectomized females that are treated neonatally with testosterone and that receive long-term testosterone treatment at the appropriate dosage throughout postnatal development will increase in body weight similar to that of intact males.

3. Cattle and Sheep

Similarities are apparent when growth processes in cattle and sheep are compared to those in rats. Birth weight is 6–8% greater in bulls than in heifers (Price and Wiltbank, 1978) and for rams the advantage in birth weight is 5–6% relative to ewes (Sidwell and Miller, 1971). For bulls, some of this difference is due to an increase of 1.8 days in length of gestation. In both cattle and sheep, castrated males are superior to intact females for growth rate and proportion of muscle in the carcass (Berg and Butterfield, 1976; Seideman et al., 1982). Unlike its positive effect on growth rate in female rats, ovariectomy of heifers or ewes has either no effect or a negative impact on body growth and its influence on composition appears negligible (Bradfield, 1968; Ray et al., 1969; Hubard Ocariz et al., 1970; Klindt et al., 1987). Estrogens, in appropriate dosages, are growth stimulants, and the responses they produce are more consistent in castrated males than in intact females (Berende and Ruitenberg, 1983). Differentiation of growth potential of steers and wethers appears similar to that of male rats castrated after 1 week of age. These three groups of males grow more rapidly and to greater size than intact or ovariectomized females of their respective species.

In male rats, testosterone-dependent differentiation of neuroendocrine regulation of gonadotropin secretion, sexual behavior, and growth potential occurs during the first few days after birth (Tarttelin et al., 1975; Feder, 1981a,b). In sheep, testosterone-dependent differentiation of gonadotropin secretion and sexual behavior is associated with days 50 to 80 of gestation (Short, 1974; Clarke, 1982), whereas in cattle, the time for differentiation of these processes has not been clearly delineated but is thought to occur after day 60 of gestation (D'Occhio and Ford, 1988). Consequently, we conclude that in steers and wethers, growth processes undergo sexual differentiation prenatally. Thus, growth potential of these castrated males is expected to be greater than that of females of the respective species.

In support of these conclusions are observations that heifers and ewes exposed to testosterone prenatally grow more rapidly during postnatal development than do control females. Heifers that were exposed to testosterone prenatally from either days 40 to 60 or days 80 to 110 of gestation were 9.5% heavier than control heifers at 1 year of age (Putney et al., 1984; DeHaan et al., 1987a). Postnatal growth in ewes is improved more than 10% by prenatal testosterone treatment (DeHaan et al., 1987b; Jenkins et al., 1987; Klindt et al., 1987). Growth rate of prenatally testosterone-treated ewes is similar to that observed in wethers. Collectively, the hypothesis on growth potential that was developed from studies with rats appears consistent with results from studies with cattle and sheep. In conflict with this conclusion is the observation that ovariectomy reduced growth in prenatally testosterone-treated ewes but not in control ewes (Klindt et al., 1987). These data are indicative of some prenatal modification that results in enhanced growth in response to postnatal ovarian secretions. This is contrary to what is expected by the rat growth model, and further studies are needed to resolve this issue. Prenatal testosterone treatment also reduced fat deposition in carcasses of ewes that were slaughtered at 55 kg live weight (DeHaan et al., 1987b; Jenkins et al., 1987). Sex differences in growth potential of cattle and sheep are likely the consequence of prenatal modifications, but which specific components are altered remain unknown.

4. Swine

Sex differences in growth of pigs are less apparent than those observed in rats, cattle, and sheep. Birth weight of male pigs is only 2.6% greater than that of females (Bereskin et al., 1973). Differences in growth rate of barrows, boars, and gilts are inconsistent among studies, and differences are not apparent until after 45 kg body weight (Kay and Houseman, 1975; Fuller, 1981). Barrows consistently have greater appetites than boars, but boars convert feed to gain more efficiently than barrows. Gilts are intermediate for these two traits. Castration of either sex leads to increased backfat thickness, shorter carcass length, and less muscling (Hammond and Murray, 1937; Wallace, 1944; Hines, 1968). Relative sex differences in carcass composition change with stage of maturity (Hansson et al., 1975; Walstra, 1980). At lighter slaughter weights, carcass composition of intact gilts is more like that of boars, but at heavier slaughter weights, carcasses of gilts increase in percentage fat such that they are more like carcasses of barrows.

That barrows are similar to ovariectomized gilts in length and carcass fatness is indicative of lack of sexual differentiation of appropriate growth processes. Studies on the timing of differentiation of sexual behavior lend support to this idea. A component of the differentiation processes of sexual behavior in boars is associated with the pubertal increase in testicular steroid secretion (Ford, 1982; Ford and Christenson, 1987). This is in contrast to the situation in bulls and rams where these events occur prenatally. Boars castrated at 2 months of age or earlier show female-typical sexual behavior when treated with estrogen as adults, whereas males castrated after 6 months show significantly less female-typical sexual behavior after estrogen treatment. Therefore, if differentiation of sexual behavior is associated with the timing for differentiation of growth processes, steers and wethers would be expected to be superior to heifers and ewes for body growth and carcass muscling whereas barrows would be similar to ovariectomized gilts. Intact, prepubertal gilts apparently respond positively to their low secretion of ovarian estrogen with the result being longer, leaner carcasses. This explanation of sex-related differences in the growth of pigs is speculative, but if correct, we would expect that barrows would have longer, learner carcasses than ovariectomized females if castration of both sexes was delayed until 6 months of age. Unfortunately, this question is academic and its answer would be of little benefit to production practices as pigs reach market weight at or before this age.

5. Chickens

Avian females are the heterogametic sex; therefore, masculine phenotype is the intrinsic form of development while females represent the sexually differentiated phenotype (Adkins-Regan, 1981). For chickens, cockerels are slightly larger ($< 2\%$) than pullets at hatching, and they grow more rapidly to larger body weights (Wilson, 1952). Castration of males produces variable responses in growth rate; capons grow as large as roosters, but their carcasses are fatter (Parkes and Marshall, 1960; Hammond et al., 1971). Bilaterally gonadectomized pullets do not grow as large as capons. The basis for sex differences in growth of chickens has not been investigated in detail.

6. Sexual Differences in Growth-Related Endocrine Processes

Numerous endocrine processes are different for males and females, and many of these are maintained over a broad phylogenetic range. Basal metabolic rate is greater in bulls than in steers (Brody, 1945), and oxygen consumption is increased by estradiol treatment of ovariectomized female rats (Bartness and

Table I

Estimates of Growth Hormone Secretion in Males and Females

	Weight or age	Sex[a]	Mean (ng/ml)	Baseline (ng/ml)	Peaks		Reference
					No./hr	Amplitude (ng/ml)	
Rats	22 days	M	13.7			37[b]	Eden (1979)
		F	11.0			31[b]	
	30 days	M	36.4			217[b]	
		F	59.7			91[b]	
	Adult	M		6.8	0.36	870[c]	Millard et al. (1987)
		M, d1 cast.		14.7	0.58	454[c]	
		F		28.9	0.73	296[c]	
		F, d1 TP		34.4	0.64	190[c]	
Chickens	10 weeks	M	19.8				Hoshino et al. (1982)
		F	33.0				
	30 weeks	M	11.8				
		F	31.0				
	Adults	M	23				Harvey et al. (1979)
		F	18				

Species		Sex					Reference
Swine	50 kg	M	4.9	1.8	0.33	15[b]	Arbona et al. (1986)
		F	3.3	1.4	0.23	10[b]	
	23 weeks	M	2.6	1.9	0.21	6[c]	Klindt et al. (unpublished data)
		M, cast.	2.5	2.5	0.08	3[c]	
Sheep	7½ months	M	2.6	2.4	0.5	7[c]	Davis et al. (1984)
		F	1.2	1.1	0.3	3[c]	
	9–12 months	M	10.2	8.8	0.18	24[b]	Davis et al. (1977)
		M, cast.	3.6	2.7	0.38	7[b]	
Cattle	2 months	M, cast.	23.3	13.5	0.54	26[c]	Klindt and Maurer (1986)
		F	12.7	8.0	0.51	14[c]	
	6 months	M	10.9				Keller et al. (1979)
		F	7.7				

[a] M, male; F, female; d, day; cast, castrated; TP, testosterone propionate.
[b] Absolute magnitude of the secretory peak.
[c] Magnitude of the secretory peak above baseline.

Wade, 1984). Activity of hepatic enzymes that respond to and metabolize steroid hormones differs according to sex. Hepatic concentrations of prolactin receptors and activities of some carbohydrate-metabolizing enzymes in the liver differ between the sexes (Lee *et al.*, 1986). Growth hormone (GH) secretion is generally greater in males than in females (Table I). The difference between the sexes with regard to GH is not just in circulating concentration, as pattern of secretion also differs. Pattern of GH secretion is sexually dimorphic in adult rats but not in early postnatal animals (Edén, 1979). Moreover, individual somatotropes from pituitary glands of mature male rats secrete more GH than those from females (Ho *et al.*, 1986; Hoeffler and Frawley, 1986). Such differences in GH secretion may be the cause of higher concentrations of somatomedins or insulinlike growth factors (IGFs) in males. In the past, research emphasis on sex differences in body growth of farm animals has been limited primarily to postnatal effects of gonadal steroid hormones. The consequences of differentiation have received much less attention than the activational effects of gonadal steroids.

Because neonatal (days 1–3) testosterone treatment of female rats enhances growth through its differentiating action (Slob and van der Werff ten Bosch, 1975; Tarttelin *et al.*, 1975; Perry *et al.*, 1979), studies were extended to identify endocrine changes that occur in concert with this enhanced growth. Jansson *et al.* (1985) and Jansson and Frohman (1987a) observed in adult male rats that neonatal gonadectomy (1–2 days) reduced mean and maximal plasma concentrations of GH compared to those observed in intact males, but nadir concentrations were increased and were similar to those of females. Administration of testosterone at the time of neonatal castration of male rats, increased pulse height and mean plasma GH concentrations to those observed in adult males; however, minimum GH concentrations and longitudinal bone growth remained different from those of untreated adult males. These results are interpreted as evidence for differentiation of the regulation of GH secretion in male rats.

Treatment of female rats with testosterone at the time of neonatal ovariectomy resulted in partially masculinized GH patterns in adulthood, whereas administration of testosterone to intact, neonatal, female rats was without major effect on adult GH secretory patterns (Millard *et al.*, 1987; Jansson and Frohman, 1987b). Implantation of adult female rats with testosterone produced male-typical secretory patterns of GH and after removal of the implants, GH patterns returned to those typical of female rats. Administration of testosterone to adult females and males that were gonadectomized neonatally also partially masculinized GH secretory parameters. Millard *et al.* (1987) have assigned greater significance to the activational actions of gonadal steroids, whereas Jansson and Frohman (1987a,b) have emphasized the ability of testosterone during neonatal development to differentiate patterns of GH secretion. It seems apparent that increased body growth of intact, neonatally androgenized female rats is not associated with major changes in the pattern of GH secretion.

The rationale for assuming that pattern of GH secretion could be related to growth was based on studies with hypophysectomized rats. Administration of a dose of exogenous GH to hypophysectomized rats as four injections per day was more efficacious in stimulation of body growth than was the same daily dose administered as one, two, or eight injections (Jansson et al., 1985). These results suggest that there is an optimal pattern of GH secretion for maximal rate of body growth. Because hypophysectomized rats are deficient in hormones other than GH, there may be limitations of the applicability of these results to normal animals.

While testosterone will induce the masculine pattern of GH secretion, it will also synergize with GH for promotion of body growth. In hypophysectomized male rats, body weight gain was greater with GH plus testosterone than with GH alone (Simpson et al., 1944; Steinetz et al., 1972; Jansson et al., 1983). Testosterone may affect body gain directly because skeletal muscle has testosterone receptors (Florini, 1985; Mooradian et al., 1987). In contrast to its effect in males, testosterone failed to potentiate the growth response to GH in hypophysectomized female rats (Simpson et al., 1944). What remains to be investigated is whether testosterone will synergize with GH to increase body growth in hypophysectomized females that were treated neonatally with testosterone.

Norstedt and Palmiter (1984) administered GH to GH-deficient mice such that pattern could be classified as masculine or feminine. Masculine pattern was achieved through twice daily injections of GH and feminine pattern was achieved through constant infusion of the same daily dose. Hepatic levels of messenger RNA for specific proteins and prolactin binding were quantitated. Twice daily injections of GH produced masculine hepatic responses, whereas feminine hepatic responses were obtained with constant infusion of GH. Others observed that pulsatile administration of GH to GH-deficient mice elevated the activity of the male isozyme of testosterone 16 α-hydroxylase in the liver, a cytochrome P-450 enzyme (Noshiro and Negishi, 1986); whereas constant infusion of GH had no effect on the low levels of activity of this isozyme (typical of females). Gustafsson et al. (1983a,b) and Jansson et al. (1985) concluded through a variety of experimental approaches that hepatic metabolism of gonadal steroids is different between sexes in rats. Constant infusion of GH into hypophysectomized male rats resulted in feminine patterns of hepatic metabolism of gonadal steroids. These observations indicate that hepatic metabolism of gonadal steroids is sexually dimorphic and that regulation of these enzymes is mediated through the sexually dimorphic pattern of GH secretion. Activity of some of these enzymes is influenced by both neonatal and postpubertal testosterone secretion, whereas others are influenced only by postpubertal testosterone secretion.

Hormonal regulation of hepatic function is mediated through receptors specific for each hormone. Steroids act through intracellular receptors whereas protein and peptide hormones act through membrane-associated receptors on the cell surface. Presence of some receptors and postreceptor events is sexually dimorphic. For example, adipose tissue from male and female rats has receptors for estradiol, whereas progesterone receptors have been identified in adipose tissue from females but not from males (Wade, 1985).

Sexual dimorphism of some hepatic functions that are exhibited in postpubertal animals often are not evident in prepubertal animals. Estrogen modulates the production of many hepatic proteins. In rats, hepatic estrogen binding is not detectable until 20 days of age (Lucier et al., 1984) after which there are two classes of estrogen binding: (1) classical high-affinity low-capacity receptors ($Kd \sim 0.5$ nM) and (2) high-capacity low-affinity (HCLA) binding proteins ($Kd \sim 0.3$ μM). The level of HCLA binding is low in all prepubertal rats but higher in adult males than in adult females. Both sexes exhibit binding of estrogen to nuclear receptors, but in males this response is inhibited by the HCLA binding proteins, which sequester estrogens and interfere with their binding to nuclear estrogen receptors. Testosterone differentiates, as well as activates, synthetic processes for these HCLA binding proteins.

Hepatic prolactin receptor numbers or activities also differ between the sexes. The number of prolactin receptors is low in adult male rats, high in females, and these sex differences become apparent at puberty (Gustafsson et al., 1983a). Estrogen increases the number of prolactin receptors, but the effect requires the presence of an intact pituitary. This effect of estrogen is apparently the consequence of its ability to modify GH secretion to a female-typical pattern. GH, when infused continuously, increases prolactin binding in livers of hypophysectomized rats (Gustafsson et al., 1983a) and dwarf mice (Norstedt and Palmiter, 1984). In GH-deficient dwarf mice, administration of GH as twice daily injections has minimal effect on hepatic prolactin binding sites. When the same dose of GH is infused continuously, prolactin binding sites increase over 13-fold. These results indicate that sexual dimorphic pattern of GH secretion programs the number of hepatic prolactin binding sites.

Factors associated with prolactin secretion and activity are of interest because prolactin is potentially a growth-regulating hormone. In sheep, increased length of photoperiod increases prolactin concentrations and rate of growth (Forbes et al., 1979; Schanbacher et al., 1985; Tucker and Merkel, 1987), and characteristics of prolactin secretion are correlated with measures of body growth (Klindt et al., 1985). It is unknown if the control of prolactin receptors in sheep is the same as that in rats and mice. A primary difference between these species is that chronic estrogen treatment suppresses growth in rats but stimulates growth in sheep.

IGFs or somatomedins are mediators of at least a portion of the ability of GH to promote growth (Froesch et al., 1986). In chickens and sheep, IGF concentrations are greater in males than in females (Hoshino et al., 1982; Van Vliet et al., 1983). In chickens, GH concentrations are generally greater in males than in females (Harvey et al., 1979); however, Hoshino et al. (1982) reported the reverse. Body growth rate of chickens, in general, correlates well with plasma IGF concentrations but is not always correlated with plasma GH concentrations (Huybrechts et al., 1985; Buyse et al., 1987). In some strains of chickens, plasma GH concentration is normal but the liver has reduced GH receptor numbers and the plasma IGF concentration is low (Hoshino et al., 1982; Leung et al., 1987). Regulation of synthesis and secretion of IGFs by the liver or other tissues is potentially a basis for sex differences in growth rate.

Regulation of GH secretion in meat animals by gonadal steroids is both similar and dissimilar to that described for the rat (Table I). Pattern of GH secretion in rams and bulls is characterized by discrete episodes of release, and after castration the frequency of the episodes increases while their amplitude decreases (Anfinson et al., 1975; Davis et al., 1977). In prepubertal heifers the pattern of GH secretion is qualitatively similar to that seen in bulls, but post-pubertally GH secretion in heifers is no longer characterized by discrete episodes of release (Anfinson et al., 1975; Zinn et al., 1986). The episodic nature of GH secretion is preserved in postpubertal males (Anfinson et al., 1975). Berg and Butterfield (1976) evaluated accretion of carcass constituents in bulls, steers, and heifers and observed that between approximately 300 and 360 kg body weight, rate of fat accretion in heifers increased dramatically. No such change in fat accretion was observed in bulls or steers. Thus, this increase in fat deposition in heifers may be related to the abolition of GH secretory peaks that occurs with the onset of puberty. Increased GH secretion is associated with decreased fat deposition in lambs (Muir et al., 1983; Klindt et al., 1985).

In heifers, GH secretory patterns appear to be altered by exposure to androgen during fetal development (Putney et al., 1984). In androgenized heifers, basal GH concentration and total 12-hr secretion of GH were lower than in control heifers. The design of the experiment did not allow for determination of whether these changes resulted from modification of sexually differentiated processes or from alterations in gonadal steroid secretion. Of note is the relationship between GH profiles and body growth: the faster-growing androgenized heifers had lower plasma concentrations of GH.

Because ewe lambs that are exposed to testosterone during early fetal development grow more rapidly than untreated ewe lambs, we hypothesized that growth was stimulated by a more optimal pattern of GH and/or prolactin secre-

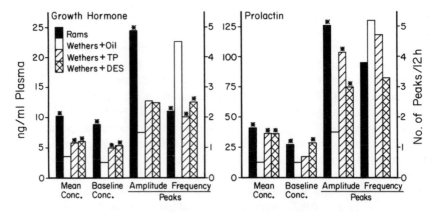

Figure 2. Effect of castration and androgenic [testosterone propionate (TP), 100 mg/day] or estrogenic [diethylstilbestrol (DES), 2 mg/day] hormones on secretory parameters of GH and PRL in male sheep. Amplitude of secretory peaks is mean maximal concentration above zero. $*p < 0.05$ as compared to wethers + oil. Adapted from Davis *et al.* (1977, 1978).

tion (Klindt *et al.*, 1987). Secretory patterns of GH and prolactin were determined at 85 and 135 days of age in intact or ovariectomized and normal or androgenized prepubertal ewe lambs. Neither prenatal androgen treatment nor ovariectomy had any effect on parameters of GH or prolactin secretion. These results suggest that in prepubertal ewe lambs, GH and prolactin secretory patterns are likely not the consequence of differentiating or activating actions of gonadal steroids.

A clear activational effect of gonadal steroids on GH secretion has been observed in wethers. Estrogens, particularly diethylstilbestrol, have long been hypothesized to have anabolic actions through stimulation of GH secretion (Preston, 1975). Davis *et al.* (1977, 1978) observed major changes in the secretory patterns of GH and prolactin in wethers due to the activational activities of estrogen or androgen (Fig. 2). Whereas the effects of estrogen and androgen were similar for mean concentrations of prolactin, the patterns of secretion were different. Muir *et al.* (1983) observed increased growth of wethers treated with diethylstilbestrol before an increase in circulating concentrations of GH was detected. Consequently, diethylstilbestrol may increase growth independent of effects on GH secretion, through subtle effects on GH secretion, or through an interaction of GH and/or prolactin secretion with both having potential impact on hepatic function.

Presently, we are unable to fully explain how sexual differences in a variety of endocrine processes interact to influence rate of growth. Currently, IGFs are postulated as a terminal step in endocrine regulation of growth. Thus, their production by the liver or other tissues and regulation of this production by patterns of GH secretion and hepatocyte receptors for hormones become primary candidates for subsequent research. Because weight of the liver at constant body weight is highly related to an animal's fasting heat production and growth rate (Koong *et al.*, 1985), determination of whether changes in hepatic function are a cause or a consequence of changes in body growth rate is needed. Full understanding of how processes of sexual differentiation can be manipulated at specified periods of development will allow for greater control over the rate and composition of growth in farm animals. In our efforts to improve efficiency of muscle growth, we must remember that body weight is the total of all tissues and an individual tissue may respond differently to an animal's endocrine status.

7. Conclusions

In cattle, sheep, and swine, mature body weight is greater in males than in females. This sex difference is the consequence of testicular steroid secretions. After neonatal castration of bulls and rams, growth rate is reduced relative to intact males but continues at a faster rate than that of females of the respective species. Furthermore, carcasses of steers and wethers contain more muscle and less fat than carcasses of heifers and ewes when compared at similar weights. Testicular steroids and their metabolites cause these sex differences by modification of growth processes in males during prenatal development (steers and wethers > heifers and ewes) and by activation of existing processes (bulls and rams > steers and wethers). For swine the situation is different in that carcasses of gilts are leaner than those of barrows. We speculate that growth processes in swine do not undergo sexual differentiation until pubertal development; thus, those sex differences that are apparent during the finishing phase of normal swine production reflect primarily activational effects of gonadal steroids.

Presently, our understanding of growth-related endocrine processes that are modified during sexual differentiation is incomplete; thus, we are unable to develop an explanation for all aspects of sex differences in body growth. It is obvious that liver function and GH secretion can be influenced greatly by activational effects of gonadal steroids at specific stages of development and in specific species. But the association of differentiated components of liver function and GH secretion with body growth rate is less obvious. Moreover, we have little appreciation for the interaction between gonadal steroids, liver function, GH secretion, and ultimately IGF production. Greater understanding of these interactions will likely improve the response of animals treated with exogenous GH. Without this understanding, broad extrapolations across species and sexes within a species may lead to less than the desired results.

References

Adkins-Regan, E. 1981. Early organizational effects of hormones: an evolutionary perspective. In: N. T. Adler (Ed.) Neuroendocrinology of Reproduction: Physiology and Behavior. pp 159–228. Plenum Press, New York.

Anfinson, M. S., S. L. Davis, E. Christian and D. O. Everson. 1975. Episodic secretion of growth hormone in steers and bulls: an analysis of frequency and magnitude of secretory spikes occurring in a 24 hour period. J. Anim. Sci., Proc. West. Sect. 26:175.

Arbona, J. R., D. N. Marple, D. R. Mulvaney, J. L. Sartin, O. H. Rahe, T. J. Prince, D. L. Kuhlers and S. B. Jungst. 1986. Secretory patterns of growth hormone in swine selected for growth. J. Anim. Sci. 63 (Suppl. 1):228.

Bartness, T. J. and G. N. Wade. 1984. Effects of interscapular brown adipose tissue denervation on body weight and energy metabolism in ovariectomized and estradiol-treated rats. Behav. Neurosci. 98:674.

Beatty, W. W. 1973. Postneonatal testosterone treatment fails to alter hormonal regulation of body weight and food intake of female rats. Physiol. Behav. 10:627.

Beatty, W. W. 1979. Gonadal hormones and sex differences in nonreproductive behaviors in rodents: organizational and activational influences. Horm. Behav. 12:112.

Berende, P. L. M. and E. J. Ruitenberg. 1983. Modifying growth: an example of possibilities and limitations. In: L. Peel and D. E. Tribe (Ed.) Domestication, Conservation and Use of Animal Resources. pp 191–233. Elsevier, Amsterdam.

Bereskin, B., C. E. Shelby and D. F. Cox. 1973. Some factors affecting pig survival. J. Anim. Sci. 36:821.

Berg, R. T. and R. M. Butterfield. 1976. New Concepts of Cattle Growth. Sydney Univ. Press, Sydney.

Bradfield, P. G. E. 1968. Sex differences in the growth of sheep. In: G. A. Lodge and G. E. Lamming (Ed.) Growth and Development of Mammals. pp 92–108. Plenum Press, New York.

Brody, S. 1945. Bioenergetics and Growth with Special Reference to the Efficiency Complex in Domestic Animals. Reinhold Publ. Corp., New York.

Buyse, J., E. Decuypere, P. J. Sharp, L. M. Huybrechts, E. R. Kühn and C. Whitehead. 1987. Effect of corticosterone on circulating concentrations of corticosterone, prolactin, thyroid hormones and somatomedin C and on fattening in broilers selected for high or low fat content. J. Endocrinol. 112:229.

Clarke, I. J. 1982. Prenatal sexual development. Oxford Rev. Reprod. Biol. 4:101.

Davis, S. L., K. L. Hossner and D. L. Ohlson. 1984. Endocrine regulation of growth in ruminants. In: J. F. Roche and D. O'Callaghan (Ed.) Manipulation of Growth in Farm Animals. pp 151–178. Martinus Nijhoff Publ. The Hague.

Davis, S. L., D. L. Ohlson, J. Klindt and M. S. Anfinson. 1977. Episodic growth hormone secretory patterns in sheep: relationship to gonadal steroid hormones. Amer. J. Physiol. 233:E519.

Davis, S. L., D. L. Ohlson, J. Klindt and M. S. Anfinson. 1978. Episodic patterns of prolactin and thyrotropin secretion in rams and wethers: influence of testosterone and diethylstilbestrol. J. Anim. Sci. 46:1724.

DeHaan, K. C., L. L. Berger, D. J. Kesler, F. K. McKeith, D. L. Thomas and T. G. Nash. 1987b. Effect of prenatal androgenization on lamb performance and carcass composition. J. Anim. Sci. 65 (Suppl. 1):85.

DeHaan, K. C., L. L. Berger, D. J. Kesler, F. K. McKeith, D. B. Faulkner and G. F. Cmarik. 1987a. Effect of prenatal androgenization on performance of steers and heifers. J. Anim. Sci. 65 (Suppl. 1):242.

D'Occhio, M. J. and J. J. Ford. 1988. Sexual differentiation and adult sexual behavior in cattle, sheep and swine: the role of gonadal hormones. In: J. M. A. Sitsen (Ed.) Handbook of Sexology. Vol. 6. pp. 209–230. Elsevier, Amsterdam.

Edén, S. 1979. Age- and sex-related differences in episodic growth hormone secretion in the rat. Endocrinology 105:555.

Feder, H. H. 1981a. Hormonal actions on the sexual differentiation of the genitalia and the gonadotropin-regulation systems. In: N. T. Adler (Ed.) Neuroendocrinology of Reproduction: Physiology and Behavior. pp 89–126. Plenum Press, New York.

Feder, H. H. 1981b. Perinatal hormones and their role in the development of sexually dimorphic behaviors. In: N. T. Adler (Ed.) Neuroendocrinology of Reproduction: Physiology and Behavior. pp 127–157. Plenum Press, New York.

Florini, J. R. 1985. Hormonal control of muscle cell growth. J. Anim. Sci. 61 (Suppl. 2):21.

Forbes, J. M., P. M. Driver, W. B. Brown and C. G. Scanes. 1979. The effect of daylength on the growth of lambs: 2. Blood concentrations of growth hormone, prolactin, insulin and thyroxine, and the effect of feeding. Anim. Prod. 29:43.

Ford, J. J. 1982. Testicular control of defeminization in male pigs. Biol. Reprod. 27:425.

Ford, J. J. and R. K. Christenson. 1987. Influences of pre- and postnatal testosterone treatment on defeminization of sexual receptivity in pigs. Biol. Reprod. 36:581.

Froesch, E. R., C. Schmid, I. Zangger, E. Schoenle, E. Eigenmann and J. Zapf. 1986. Effects of IGF/somatomedins on growth and differentiation of muscle and bone. J. Anim. Sci. 63 (Suppl. 2):57.

Fuller, M. F. 1981. Sex differences in the nutrition and growth of pigs. In: W. Haresign (Ed.) Recent Advances in Animal Nutrition—1980. pp 157–169. Butterworths, London.

Goy, R. W. and B. S. McEwen. 1980. Sexual Differentiation of the Brain. MIT Press, Cambridge, MA.

Gustafsson, J.-A., A. Mode, G. Norstedt, P. Eneroth and T. Hokfelt. 1983a. Central control of prolactin and estrogen receptors in rat liver—expression of a novel endocrine system, the hypothalamo–pituitary–liver axis. Annu. Rev. Pharmacol. Toxicol. 23:259.

Gustafsson, J.-A., A. Mode, G. Norstedt and P. Skett. 1983b. Sex steroid induced changes in hepatic enzymes. Annu. Rev. Physiol. 45:51.

Hammond, J., Jr., I. L. Mason and T. J. Robinson. 1971. Hammond's Farm Animals (4th Ed.). Edward Arnold, London.

Hammond, J. and G. N. Murray. 1937. The body proportions of different breeds of bacon pigs. J. Agrc. Sci. 27:394.

Hansson, I., K. Lundström and B. Malmros. 1975. Effect of sex and weight on growth, feed efficiency and carcass characteristics of pigs. 2. Carcass characteristics of boars, barrows and gilts, slaughtered at four different weights. Swed. J. Agrc. Res. 5:69.

Harvey, S., C. G. Scanes, A. Chadwick and N. J. Bolton. 1979. Growth hormone and prolactin secretion in growing domestic fowl: influence of sex and breed. Brit. Poult. Sci. 20:9.

Hines, R. H. 1968. The interaction of restricted feed intake and sex on swine performance and carcass quality. Anim. Breed. Abstr. 36:267.

Ho, K. Y., D. A. Leong, Y. N. Sinka, M. L. Johnson, W. S. Evans and M. O. Thorner. 1986. Sex-related differences in GH secretion in rat using reverse hemolytic plaque assay. Amer. J. Physiol. 250:E650.

Hoeffler, J. P. and L. S. Frawley. 1986. Capacity of individual somatotropes to release growth hormone varies according to sex: analysis by reverse hemolytic plaque assay. Endocrinology 119:1037.

Hoshino, S., M. Wakita, M. Suzuki and K. Yamamoto. 1982. Changes in a somatomedin-like factor and immunoassayable growth hormone during growth of normal and dwarf pullets and cockerels. Poult. Sci. 61:777.

Hubard Ocariz, J. L., A. Littlejohn and I. S. Robertson. 1970. A comparison of entire and ovariectomized beef heifers treated with ethylestrenol. J. Agrc. Sci. 74:349.

Huybrechts, L. M., D. B. King, T. J. Lauterio, J. Marsh and C. G. Scanes. 1985. Plasma concentrations of somatomedin-C in hypophysectomized, dwarf and intact growing domestic fowl as determined by heterologous radioimmunoassay. J. Endocrinol. 104:233.

Jansson, J.-O., S. Edén and O. Isaksson. 1983. Sites of action of testosterone and estradiol on longitudinal bone growth. Amer. J. Physiol. 244:E135.

Jansson, J.-O., S. Edén and O. Isaksson. 1985. Sexual dimorphism in the control of growth hormone secretion. Endocrine Rev. 6:128.

Jansson, J.-O. and L. A. Frohman. 1987a. Differential effects of neonatal and adult androgen exposure on the growth secretory pattern in male rats. Endocrinology 120:1551.

Jansson, J.-O. and L. A. Frohman. 1987b. Inhibitory effect of the ovaries on neonatal androgen imprinting of growth hormone secretion in female rats. Endocrinology 121:1417.

Jenkins, T. G., J. Klindt and J. J. Ford. 1987. Effect of alteration of sexual differentiation upon growth, feed efficiency and empty body composition. J. Anim. Sci. 65 (Suppl. 1):248.

Jost, A. and S. Magre. 1984. Testicular development phases and dual hormonal control of sexual organogenesis. In: M. Serio, M. Mottas, M. Zanisi and L. Martini (Ed.) Sexual Differentiation: Basic and Clinical Aspects. pp 1–15. Raven Press, New York.

Kay, M. and R. Houseman. 1975. The influence of sex on meat production. In: D. J. A. Cole and R. A. Lawrie (Ed.) Meat. pp 85–108. Butterworths, London.

Keller, D. G., V. G. Smith, G. H. Coulter and G. J. King. 1979. Serum growth hormone concentrations in Hereford and Angus calves: effects of breed, sire, sex, age, age of dam, and diet. Can. J. Anim. Sci. 59:367.

Klindt, J., T. G. Jenkins and J. J. Ford. 1987. Prenatal androgen exposure and growth and secretion of growth hormone and prolactin in ewes postweaning. Proc. Soc. Exp. Biol. Med. 185:201.

Klindt, J., T. G. Jenkins and K. A. Leymaster. 1985. Relationships between some estimates of growth hormone and prolactin secretion and rates of accretion of constituents of body gain in rams. Anim. Prod. 41:103.

Klindt, J. and R. R. Maurer. 1986. Reciprocal cross effects on growth hormone and prolactin secretion in cattle: influence of genotype and maternal environment. J. Anim. Sci. 62:1660.

Koong, L. J., C. L. Ferrell and J. A. Nienaber. 1985. Assessment of interrelationships among levels of intake and production, organ size and fasting heat production in growing animals. J. Nutr. 115:1383.

Lee, V. M., B. Szepesi and R. J. Hansen. 1986. Gender-linked differences in dietary induction of hepatic glucose-6-phosphate dehydrogenase, 6-phosphogluconate dehydrogenase and malic enzyme in the rat. J. Nutr. 116:1547.

Leung, F. C., W. J. Styles, C. I. Rosenblum, M. S. Lilburn and J. A. Marsh. 1987. Diminished hepatic growth hormone receptor binding in sex-linked dwarf broiler and leghorn chickens. Proc. Soc. Exp. Biol. Med. 184:234.

Lucier, G. W., C. L. Thompson, T. C. Sloop and R. C. Rumbaugh. 1984. Sexual differentiation of rat hepatic estrogen and androgen action. In: M. Serio, M. Motta, M. Zanisi and L. Martini (Ed.) Sexual Differentiation: Basic and Clinical Aspects. pp 209–222. Raven Press, New York.

Millard, J. W., T. O. Fox, T. M. Badger and J. B. Martin. 1987. Gonadal steroid modulation of growth hormone secretory patterns in the rat. In: R. J. Robbins and S. Melmed (Ed.) Acromegaly: A Century of Scientific and Clinical Progress. pp 139–150. Plenum Press, New York.

Mooradian, A. D., J. E. Morley and S. G. Korenman. 1987. Biological actions of androgens. Endocrine Rev. 8:1.

Muir, L. A., S. Wien, P. F. Duquette, E. L. Riches and E. H. Cordes. 1983. Effects of exogenous growth hormone and diethylstilbestrol on growth and carcass composition of growing lambs. J. Anim. Sci. 56:1315.

Norstedt, G. and R. Palmiter. 1984. Secretory rhythm of growth hormone regulates sexual differentiation of mouse liver. Cell 36:805.

Noshiro, M. and M. Negishi. 1986. Pretranslational regulation of sex-dependent testosterone hydroxylases by growth hormone in mouse liver. J. Biol. Chem. 261:15923.

Page, D. C., R. Mosher, E. M. Simpson, E. M. C. Fisher, G. Mardon, J. Pollack, B. McGillivray, A. de la Chapelle and L. G. Brown. 1987. The sex-determining region of the human Y chromosome encodes a finger protein. Cell 51:1091.

Parkes, A. S. and A. J. Marshall. 1960. The reproductive hormones in birds. In: A. S. Parkes (Ed.) Marshall's Physiology of Reproduction. Vol. I, Part 2, pp 583–706. Longmans, Green and Co., New York.

Perry, B. N., A. McCracken, B. J. A. Furr and H. J. H. MacFie. 1979. Separate roles of androgen and oestrogen in the manipulation of growth and efficiency of food utilization in female rats. J. Endocrinol. 81:35.

Preston, R. L. 1975. Biological responses to estrogen additives in meat producing cattle and lambs. J. Anim. Sci. 41:1414.

Price, T. D. and J. N. Wiltbank. 1978. Dystocia in cattle: a review and implications. Theriogenology 9:195.

Putney, D. J., W. E. Beal and G. A. Good. 1984. The effect of prenatal androgen exposure on sexual differentiation and growth hormone secretion in female calves. 10th Int. Congr. Anim. Reprod. Artif. Insem. Vol. II:5.

Ralls, K. 1976. Mammals in which females are larger than males. Rev. Biol. 51:245.

Ray, D. E., W. H. Hale and J. A. Marchello. 1969. Influence of season, sex and hormonal growth stimulants on feedlot performance of beef cattle. J. Anim. Sci. 29:490.

Schanbacher, B. D., W. Wu, J. A. Nienaber and G. L. Hahn. 1985. Twenty-four-hour profiles of prolactin and testosterone in ram lambs exposed to skeleton photoperiods consisting of various light pulses. J. Reprod. Fertil. 73:37.

Seideman, S. C., H. R. Cross, R. R. Oltjen and B. D. Schanbacher. 1982. Utilization of the intact male for red meat production: a review. J. Anim. Sci. 55:826.

Short, R. V. 1974. Sexual differentiation of the brain of the sheep. In: M. G. Forest and J. Bertrand (ed.) The Sexual Endocrinology of the Perinatal Period. pp 121–142. INSERM, Paris.

Sidwell, G. M. and L. R. Miller. 1971. Production in some pure breeds of sheep and their crosses. II. Birth weights and weaning weights of lambs. J. Anim. Sci. 32:1090.

Simpson, M., W. Marx, H. Becks and H. M. Evans. 1944. Effect of testosterone propionate on the body weight and skeletal system of hypophysectomized rats: synergism with pituitary growth hormone. Endocrinology 35:309.

Simpson, E., P. Chandler, R. Hunt, H. Hogg, K. Tomonari and A. McLaren. 1986. H-Y status of X/X Sxr[i] male mice: in vivo tests. Immunology 57:345.

Slob, A. K. and J. J. van der Werff ten Bosch. 1975. Sex differences in body growth in the rat. Physiol. Behav. 14:353.

Steinetz, B. G., T. Giannina, M. Butler and F. Popick. 1972. The role of growth hormone in the anabolic action of methandrostenolone. Endocrinology 90:1396.

Tarttelin, M. F., J. E. Shryne and R. A. Gorski. 1975. Patterns of body weight change in rats following neonatal hormone manipulation: a "critical period" for androgen-induced growth increases. Acta Endocrinol. 79:177.

Trenkle, A. and D. N. Marple. 1983. Growth and development of meat animals. J. Anim. Sci. 57 (Suppl. 2):273.

Tucker, H. A. and R. A. Merkel. 1987. Applications of hormones in the metabolic regulation of growth and lactation in ruminants. Fed. Proc. 46:300.

Van Vliet, G., D. M. Styne, S. L. Kaplan and M. M. Grumbach. 1983. Hormone ontogeny in the ovine fetus. XVI. Plasma immunoreactive somatomedin C/insulin-like growth factor I in the fetal and neonatal lamb and in the pregnant ewe. Endocrinology 113:1716.

Wade, G. N. 1985. Actions of gonadal steroids on adipose tissue in rodents. In: J. Vague, P. Björntorp, B. Guy-Grand, M. Rebuffé-Scrive and P. Vague (Ed.) Metabolic Complications of Human Obesities. pp 105–114. Excerpta Medica, Amsterdam.

Wallace, L. R. 1944. The influence of sex upon carcass quality and efficiency of food utilisation. Proc N. Z. Soc. Anim. Prod. 4:64.

Walstra, P. 1980. Growth and Carcass Composition from Birth to Maturity in Relation to Feeding Level and Sex in Dutch Landrace Pigs. H. Veenman and B. V. Zonen, Wageningen.

Wilson, P. N. 1952. Growth analysis of the domestic fowl. I. Effect of plane of nutrition and sex on live-weights and external measurements. J. Agr. Sci. 42:369.

Zinn, S. A., R. W. Purchas, L. T. Chapin, D. Petitclerc, R. A. Merkel, W. G. Bergen and H. A. Tucker. 1986. Effects of photoperiod on growth, carcass composition, prolactin, growth hormone and cortisol in prepubertal and postpubertal Holstein heifers. J. Anim. Sci. 63:1804.

CHAPTER 15

Potential Mechanisms for Repartitioning of Growth by β-Adrenergic Agonists

H. J. MERSMANN

1. Introduction

A major goal of meat animal production is to generate efficiently an acceptable meat product as a source of protein for humans. In the last several years, there has been considerable interest in using β-adrenergic agonists to repartition growth in animals raised for meat production (Ricks *et al.*, 1984a); oral administration of the β-adrenergic agonists clenbuterol or cimaterol to growing cattle (Ricks *et al.*, 1984b; Hanrahan *et al.*, 1986; Williams *et al.*, 1986; Miller *et al.*, 1988), chickens (Dalrymple *et al.*, 1984a; Hanrahan *et al.*, 1986), pigs (Dalrymple *et al.*, 1984b; Ricks *et al.*, 1984a; Jones *et al.*, 1985; Hanrahan *et al.*, 1986; Mosher *et al.*, 1986), and sheep (Baker *et al.*, 1984; Beermann *et al.*, 1985, 1986b; Thornton *et al.*, 1985; Hamby *et al.*, 1986; Hanrahan *et al.*, 1986; Bohorov *et al.*, 1987) increases carcass muscle mass and decreases carcass fat mass. Nitrogen retention increases in calves fed clenbuterol (Williams *et al.*, 1986). The shift toward more muscle and less fat deposition by clenbuterol led to the description of this phenomenon as repartitioning of growth and the designation of such compounds as repartitioning agents. At this time there is one experiment in very young pigs in which an adrenergic agonist did not change

H. J. MERSMANN • United States Department of Agriculture, Agricultural Research Service, Roman L. Hruska U.S. Meat Animal Research Center, Clay Center, Nebraska 68933. Mention of a trade name, proprietary product, or specific equipment does not constitute a guarantee or warranty of the product by the U.S. Department of Agriculture and does not imply its approval to the exclusion of other products that may also be suitable.

body composition; the reason for the negative result is unknown (Mersmann *et al.*, 1987). The effects of β-adrenergic agonists on rate of gain and feed consumption are less pronounced but, in some studies, in some species, there is a modest increase in rate of gain and/or a decrease in feed consumption to yield improved gain/feed. Regardless, if the efficiency were to be calculated on the basis of product produced, i.e., lean meat or carcass muscle mass, the efficiency of production is markedly improved even when there is no change in gain or feed consumption.

2. Biology of Adrenergic Hormones and Neurotransmitters

Much of this chapter will be devoted to a discussion of the biology of adrenergic hormones and neurotransmitters because the as yet unknown mechanism(s) of the β-adrenergic agonists to repartition animal growth will most likely be mediated through adrenergic receptors to influence various aspects of physiological function and metabolism. The reader is referred to several more thorough discussions of the biochemistry, physiology, and pharmacology of adrenergic agents (Gilman *et al.*, 1980; Martin, 1985; Cooper *et al.*, 1986).

2.1. Endogenous Adrenergic Agents

In animals, there are several structurally related substances, the catecholamines, that function as hormones or as nervous system chemical transmitters. Portions of the brain and the sympathetic nervous system have as their major chemical mediator or neurotransmitter, norepinephrine or noradrenaline. This rather simple catecholamine (Fig. 1) consists of a ring with OH groups at positions 3 and 4, the catechol nucleus. At position 1, a two-carbon chain is attached with a nitrogen at the terminus. The carbon adjacent to the ring (β-carbon) has a OH and the amine is a primary amine. The adrenal medulla, a tissue embryonically derived from the nervous system, produces the hormone epinephrine or adrenaline, which has the same structure (Fig. 2) as norepinephrine except that the amine has a CH_3 in place of one of the H groups. Dopamine, a compound similar in structure to norepinephrine but with no OH on the side chain β-carbon (Fig. 2), acts as a neurotransmitter in the brain and is released from many peripheral nerves (Lackovic and Relja, 1983). Although

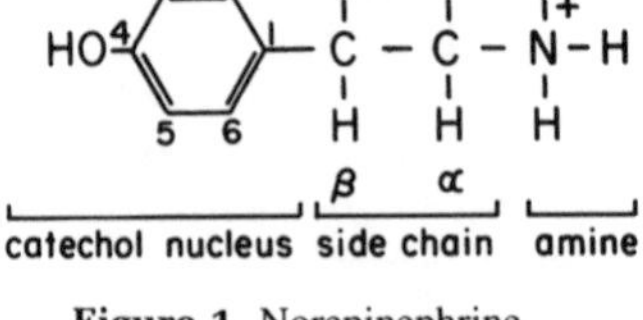

Figure 1. Norepinephrine.

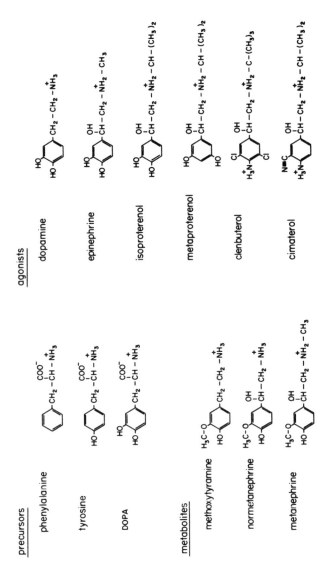

Figure 2. Structure of precursors, metabolites, and selected β-adrenergic agonists.

epinephrine is considered a hormone, the neurotransmitters norepinephrine and dopamine are present in the blood plasma, suggesting that these substances also may act as hormones at peripheral target organs. In most mammalian species, the plasma concentration ranges from 5×10^{-10} to 5×10^{-9} M but can increase markedly. The plasma concentration of norepinephrine is two to nine times greater than epinephrine, and the dopamine concentration is variable but less than norepinephrine (Buhler *et al.*, 1978). For example, the plasma concentrations (pg/ml) of norepinephrine, epinephrine, and dopamine in resting animals are, respectively, 152, 56, and 91 in cattle; 509, 175, and 84 in rats; 203, 64, and 98 in humans; and 609, 73, and 276 in cats (Buhler *et al.*, 1978). Catecholamines are usually estimated by high-pressure liquid chromatography coupled with fluorometric or electrochemical detection or by more sensitive radioenzymatic methods (Holly and Makin, 1983).

2.2. Synthesis and Removal

The three major catecholamines are sequentially synthesized (structures in Fig. 2) in the same pathway (Axelrod, 1971; Ungar and Phillips, 1983). The precursor amino acid, phenylalanine, is hydroxylated to produce tyrosine, which is then hydroxylated (at the 3-position) to produce dopa (dihydroxyphenylalanine), a compound with the catechol nucleus. In many situations, tyrosine hydroxylase appears to be the rate-limiting enzyme for catecholamine biosynthesis. Decarboxylation of dopa produces dopamine, hydroxylation of dopamine at the β-carbon of the side chain produces norepinephrine, and methylation of the amine of the latter yields epinephrine. The latter methylation by the enzyme phenylethanolamine N-methyltransferase, only occurs in selected cells, usually the chromaffin cells, and consequently most of this hormone is produced in the adrenal medulla (Ungar and Phillips, 1983). Because of the sequential nature of synthesis of the three major catecholamines, it is not surprising that a nerve that uses norepinephrine as a neurotransmitter also may release some dopamine or that the adrenal medulla may release some norepinephrine.

As with all neurotransmitters and hormones, the released material must be rapidly dissipated to terminate the action. Released neurotransmitters may reenter the nerve ending via specific uptake or transport mechanisms (Axelrod, 1971; Iversen, 1974; Trendelenburg, 1979) or be diluted by diffusion away from the effector site. The latter is probably the source of much of the circulating noradrenaline and dopamine. The catecholamine may also be metabolized and ultimately excreted (Axelrod, 1971). The enzyme COMT (catechol-O-methyltransferase) methylates the 3—OH on the catechol ring to produce the inactive metabolites methoxytyramine, normetanephrine, and metanephrine (Fig. 2) of dopamine, norepinephrine, and epinephrine, respectively (Guldberg and Marsden, 1975). The use of this enzyme coupled with a radioactive methyl group donor provides the basis for many of the radioenzymatic assay methods for the catecholamine hormones or neurotransmitters (Holly and Makin, 1983). The enzyme MAO (monoamine oxidase) provides the other degradation route for catecholamines by removal of the amine group from the side chain. After further metabolism, the products are excreted. Little is known about the biosynthesis, distribution, degradation, or uptake of catecholamines in animals raised for meat production. The most information available is for pigs (Stanton, 1986).

2.3. Adrenergic Receptors

The pharmacology of catecholamines has been investigated for many years, mostly using various organ or tissue preparations *in vitro*. Many analogues of norepinephrine have been synthesized by changing substituents and positions on the ring, by changing the ring structure, by lengthening and substituting on the side chain, and by substituting on the amine. Some of the analogues are more effective or potent than norepinephrine or epinephrine, some are selective for certain functions or tissues, and some are inhibitory (antagonists or blockers) to norepinephrine or epinephrine function.

The divergence of physiological and metabolic functions modulated by norepinephrine and its many agonistic and antagonistic analogues has led to a complex set of observations that continue to grow (a comprehensive treatise on adrenergic function is available—Szekeres, 1980). In order to make these observations more systematic and interpretable, the concept of different types of receptors on the effector cells was proposed. Ahlquist (1948) classified adrenergic receptors into α and β subtypes: increase in heart rate and contractility, bronchodilation, and stimulation of fat cell lipolysis are considered β-adrenergic receptor-controlled functions whereas contraction of gut sphincters and cerebral, skin, and salivary gland arterioles are classified as α-adrenergic receptor-controlled functions. As more observations and analogues accumulated, the need for more complexity became obvious and the concept of a β_1 and β_2 subtype of the β-adrenergic receptor was proposed (Lands *et al.*, 1967). The idea that heart (β_1) and lungs (β_2) have different receptors provided the impetus to synthesize compounds that would preferentially cause bronchodilation with little stimulation of the heart. Thus, compounds were produced, e.g., terbutaline, to treat asthmatics without the cardiac side effects of the more traditional compounds, e.g., epinephrine or isoproterenol. Many compounds have been classified as either β_1- or β_2-adrenergic agonists or antagonists based on their differential potency and efficacy to affect heart contractility compared to bronchodilation. More recently, the α-adrenergic receptor has been classified into α_1 and α_2 subtypes (Johansson, 1984).

In the middle 1970s, techniques were developed to assess β-adrenergic receptors in membranes of various cell types by specific binding of tritiated or iodinated β-adrenergic antagonists with a high degree of specific radioactivity. Competitive binding techniques using agonists and antagonists with demonstrated specificity for receptor subtypes led to another approach for receptor classification. α-Adrenergic receptors and receptor subtypes also can be assessed and classified by binding techniques.

Classification of receptors into subtypes and classification of adrenergic agonists and antagonists as specific for a receptor subtype must be interpreted cautiously. First, much of the classification of norepinephrine analogues as specific for β_1-adrenergic receptors has used rat or guinea pig heart, tissues that have exclusively or at least predominantly β_1-adrenergic receptors; classification of analogues as specific for β_2-adrenergic receptors has used lung tissue from several species, tissues that may have predominantly β_2 receptors (Minneman et al., 1979; O'Donnell and Wanstall, 1981; Kenakin, 1984). In other species or other tissues within a species, the receptor distribution appears to be less clear so that various proportions of both β_1 and β_2-adrenergic receptors are present in a given tissue. Thus, the affinity for binding, the ability to control cAMP production, or the ability to control tissue-specific physiological function by purported β_1- or β_2-specific adrenergic agents may not be clearly delineated. Furthermore, the receptor in a given tissue from a particular species may have structural specificity for agonists or antagonists different from that observed in tissues or species used to classify the norepinephrine analogues as purportedly specific for β_1- or β_2-adrenergic receptors. It should be noted that there is currently no adrenergic agonist available that is specific for the β_1-adrenergic receptor, e.g., dobutamine is sometimes considered β_1-adrenergic specific but in reality is not (Malta et al., 1985). The classification of receptors into subtypes or of adrenergic agonists or antagonists as specific for a receptor subtype can easily become confused and probably is less useful as more tissues or cell types in different species are examined.

Second, binding phenomena must meet several criteria before the apparent binding is valid and can be designated as specific to the β-adrenergic receptor (Williams and Lefkowitz, 1978). The binding must be rapid and reversible. Nonspecific binding should not comprise a major portion of the total binding. Specific binding must be saturable. The binding constant obtained from kinetic observations should approximate the binding constant from saturation phenomena. The potency of the naturally occurring L forms of norepinephrine or epinephrine should be greater than for the D forms. The order and relationship of potency for competitive binding by a series of norepinephrine analogues should be the same as for mediation by the receptor of the biological phenomena in that tissue or cell. Many reports of adrenergic receptor binding do not meet these criteria. When ligand binding is used to measure receptor number or type in intact cells, the particular ligand used may profoundly affect the observations depending on the hydrophilic or hydrophobic nature of the ligand. For example, in fat cells valid binding data are obtained with hydrophilic ligands but not with hydrophobic ligands, such as the commonly used [^{3}H]dihydroalprenolol (Lacasa et al., 1986).

The β-adrenergic receptors from several sources have been purified (Lef-kowitz *et al.*, 1983) and the amino acid sequences of the human and hamster β₂-adrenergic receptors have been deduced from cDNA sequences (Kobilka *et al.*, 1987). The more recent work indicates that the mammalian β_1- and β_2-adrenergic receptors are similar in structure with a peptide of about 64 kDa that binds the adrenergic agents. After proteolysis the peptide maps for β_1- and β_2-adrenergic receptors are distinct (Stiles *et al.*, 1983). Antibody for one receptor subtype cross-reacts with the other receptor subtype (Moxham et al., 1986). Consequently, the receptor subtypes can be distinguished readily by pharma-cological means but structural differences appear to be subtle.

2.4. Coupling of Adrenergic Receptors to Intracellular Function

The β-adrenergic receptor functions almost universally by coupling to the enzyme adenylate cyclase to produce intracellular cAMP (Lefkowitz *et al.*, 1982, 1983; Birnbaumer *et al.*, 1985; Northrup, 1985; Levitzki, 1986). Attach-ment of cAMP to the regulatory subunit of a protein kinase enzyme causes cleavage of this inhibitory subunit from the catalytic subunit, which is then active to phosphorylate various cellular proteins (Fig. 3). Phosphorylation of some enzymes causes activation, e.g., glycogen phosphorylase or hormone-sensitive lipase, whereas phosphorylation of other enzymes causes inactivation, e.g., glycogen synthase.

The β-adrenergic receptor is coupled to adenylate cyclase with a regula-tory protein (g or N protein) that binds GTP and functions to activate the ade-nylate cyclase enzyme (Robdell, 1980; Birnbaumer *et al.*, 1985; Northrup, 1985). Hydrolysis of the bound GTP is a function of the g protein to produce g pro-

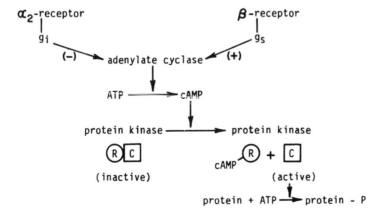

Figure 3. Adrenergic regulation by control of cAMP.

tein–GDP, an inactive form. There is a specific g protein [designated g_s (or N_s)] to stimulate adenylate cyclase and this protein seems to couple not only β-adrenergic receptors but also other membrane receptors that stimulate adenylate cyclase and cAMP production. Both β_1- and β_2-adrenergic receptors are coupled to adenylate cyclase in the same manner.

α_2-Adrenergic receptors are coupled to adenylate cyclase through a g protein (g_i or N_i) that inhibits rather than stimulates adenylate cyclase (Birnbaumer et al., 1985). Thus, there is a molecular basis for the opposite effects produced in a cell upon stimulation of β-adrenergic and α_2-adrenergic receptors by control of cAMP production via g_s or g_i proteins. For example, adrenergic stimulation of human adipose tissue lipolysis is by β-adrenergic receptors through interaction with g_s and subsequent production of cAMP whereas stimulated α_2-adrenergic receptors interact with g_i to lower the cAMP concentration and lipolytic rate (Fain and Garcia-Sainz, 1983; Exton, 1985).

The α_1-adrenergic receptor appears to operate through coupling to phosphatidylinositol metabolism and stimulates production of particular inositol phosphate molecules (e.g., 1,4,5-inositol triphosphate) and diglyceride (Fain and Garcia-Sainz, 1983; Downes and Michell, 1985; Exton, 1985). Both of these products are active intracellular effectors and at least part of the α_1-adrenergic effects are mediated by changes in intracellular Ca^{2+} concentration.

Regulation of receptor number or density on a cell provides another form of modulation of function. The ontogenetic development of receptors on a particular cell type plus the intracellular mechanisms to execute an effect are properties of that cell type in a particular species. Most receptors, including adrenergic receptors, develop later than most metabolic processes of the cell, e.g., undifferentiated muscle cells do not exhibit functional β-adrenergic receptors until after fusion to form myotubes (Parent et al., 1980). Likewise, neonatal mammals, including pigs, have poor adrenergic control of cardiac function, possibly the result of low receptor density; adrenergic control develops after birth as receptor density increases whereas receptor affinity does not increase during this period (Schumacher et al., 1982; Stanton, 1986). Many aspects of adrenergic control are modified during fetal and neonatal development; some of this change is understood in pigs (Stanton, 1986) but not in cattle or sheep. Receptor type can change with development, e.g., the rat hepatic adrenergic receptor initially is predominantly a β-adrenergic receptor but with development, the α-adrenergic receptor becomes predominant (Morgan et al., 1983).

Another, more acute mechanism to control receptor density is the desensitization process, sometimes called down-regulation. Exposure of many cell types to an adrenergic agonist will result in removal of receptors from the cell surface, thus making the cell less sensitive to stimulation. Desensitization is evident in vitro and in vivo after chronic exposure to an adrenergic agonist. In some cells, desensitization is rapid and obvious within a few minutes of exposure to the adrenergic agonist (Lefkowitz *et al.*, 1983; Sibley and Lefkowitz, 1985; Levitzski, 1986).

2.5. Physiological Effects of Adrenergic Agonists

There is considerable opportunity for adrenergic control of physiological and metabolic function because the sympathetic nervous system has input to most organs, not only epinephrine but also norepinephrine and dopamine are present in the blood plasma, and adrenergic receptors are almost universally distributed on cells (Arnold, 1980; Gilman *et al.*, 1980; Martin, 1985). Although both norepinephrine and epinephrine stimulate α- and β-adrenergic receptors, there is some distinction between the two agonists. Epinephrine generally stimulates α-adrenergic receptors with greater potency than does norepinephrine. Furthermore, the latter seems to have slightly greater potency for β_1- than $\beta2$-adrenergic receptors, whereas epinephrine has a potency of β_2 slightly $> \beta_1$. Thus, although these two agonists may interact with a physiological system in a similar manner, sometimes one will be effective and the other not, sometimes one will be more potent than the other, and occasionally they will have opposite effects (see Table 8-1 in Martin, 1985). The question of physiological control is further complicated because most cell types do not have exclusively β_1- or β_2-adrenergic receptors and also have α-adrenergic receptors. The antagonism between α- and β-adrenergic function can be particularly baffling. For example, infusion of epinephrine into pigs increases blood pressure for the first few minutes (presumably an α-adrenergic function) but then the pressure is decreased (presumably a β-adrenergic function); norepinephrine has similar effects (Mersmann, 1987a).

It is beyond the scope of this discussion to delve into any of the wide variety of physiological and metabolic functions controlled by adrenergic receptors, but suffice it to say that not only are heart rate and contractile force changed by adrenergic stimulation but also blood pressure, blood vessel constriction or dilation, blood flow, airway constriction or dilation, motility of the gut and uterus, gut and urinary bladder sphincter tone, skeletal muscle contraction, sweating, piloerection, and many other physiological functions (Szekeres, 1980; Gilman *et al.*, 1980; Martin, 1985; Cooper *et al.*, 1986). Adrenergic control provides major input into regulation of glycogen metabolism, glycolysis, gluconeogenesis, lipolysis, and lipogenesis (Ellis, 1980).

The use of synthetic norepinephrine analogues to study adrenergic control of physiological or metabolic function is extremely complex. The effects are difficult to predict because of different organ and species distribution of receptor subtypes, density of receptors, and divergent structural requirements for effector activity of analogues between tissues within a species or within a tissue between species. Furthermore, different pharmacodynamics of the exogenous agent between species including absorption, distribution, metabolism, and excretion, can cause major interspecific differences in function *in vivo*.

3. Administration of β-Adrenergic Agonists to Animals

3.1. Effects

The change observed in muscle or fat accretion when clenbuterol or cimaterol (structures in Fig. 2) is administered varies with the species. With these two compounds, chickens respond with the smallest and cattle or sheep with the largest percentage change. The response may be governed partly by the potential for increase in muscle mass in the particular species and partly by the diverse physiology and pharmacology in individual species. Thus, as more norepinephrine analogues are tested, a specific analogue may be active in only one species or selectively repartition growth in one compared to another species.

Clenbuterol is usually designated a β_2-adrenergic agonist (O'Donnell, 1976); it stimulates bronchodilation in horses and inhibits uterine contraction in pregnant cattle. The repartitioning of growth will almost certainly not be limited to this specific receptor subtype because of species and tissue receptor specificity, as discussed previously. Because there are so many existing and potential analogues of norepinephrine, it is anticipated that there will be a number of adrenergic agents that repartition animal growth. The initial discovery of the repartitioning effect of clenbuterol was by American Cyanamid Co. after many years of effort to alter lipid deposition and animal growth (references by Baker, Dalrymple, and Ricks).

The finding, in meat-producing animals, of increased muscle mass and decreased fat mass may be the result of direct adrenergic interaction with these tissues but also could easily result from a variety of indirect effects on many other tissues or a combination of both direct and indirect effects. Change in some physiological, endocrine, or metabolic function after administration of an exogenous adrenergic agent does not distinguish between primary and secondary effects. Furthermore, acute effects must be separated from chronic effects. Extensive live animal studies can generate a picture of some of the fluctuations in physiological or metabolic function but species variation and compound variation will make it difficult to provide a thorough description of mechanism for many years.

3.2. Adipose Tissue

Adipose tissue (Lafontan and Berlan, 1985) including porcine (Bocklen *et al.*, 1986) and bovine (Jaster and Wegner, 1981) has β-adrenergic receptors and stimulation of these receptors causes activation of hormone-sensitive li-pase, which cleaves triacylglycerol to glycerol and free fatty acids (Fain and Garcia-Sainz, 1983; Vernon and Clegg, 1985). Stimulation of adipose tissue lipolysis would provide a mechanism to explain the reduced adipose tissue mass in animals administered analogues of norepinephrine. When studied *in vitro*, many norepinephrine analogues stimulate lipolysis in adipose tissue from many species but there is species specificity so that stimulation by an analogue in tissue from one species does not predict efficacy or potency in another species. For example, isoproterenol and norepinephrine have equal efficacy for stimu-lation of lipolysis in rat fat cells whereas in guinea pig, dog, and human fat cells, norepinephrine is less efficacious than isoproterenol (Milavec-Krizman and Wagner, 1978). In porcine adipose tissue, only a few select and structur-ally related analogues (the catecholamines and compounds related to metapro-terenol—Fig. 2) are active (Mersmann, 1984b); to the extent studied, this struc-tural specificity is verified in the production of cAMP (Hu *et al.*, 1987) so that it seems to reside in the receptor coupled to adenylate cyclase. The specificity is operative *in vivo*, as well (Mersmann, 1987a). Stimulation of adipose tissue α_2-adrenergic receptors causes inhibition of lipolysis in most mammalian spe-cies (Fain and Garcia-Sainz, 1983; Lafontan and Berlin, 1985), but not in the rat or pig (Mersmann, 1984a).

Of the meat-producing animals, much is known about adrenergic control of lipolysis in adipose tissue from the pig (Mersmann, 1986). Lipolysis is ad-renergically regulated in bovine and ovine adipose tissue but less is known about details of this regulation than in pigs (Vernon, 1981; Jaster and Wegner, 1981; Blum *et al.*, 1982). Effects of norepinephrine analogues on adipose tis-sue *in vitro* or *in vivo* could result from stimulation of α- as well as β-adren-ergic receptors. Consequently, a multitude of effects could occur in adipose tissue alone, depending on the specificity of the norepinephrine analogue for the α- or β-adrenergic receptor or for the receptor subtypes in the particular species under consideration. Clenbuterol stimulates lipolysis in chick (Camp-bell and Scanes, 1985) and ovine (Thornton *et al.*, 1985) adipose tissue *in vitro*. Thus, the decrease in adipose tissue mass observed in chickens and sheep after oral administration of clenbuterol could be explained, at least partly, by direct stimulation of adipose tissue lipolysis. This mechanism probably is not operative in pigs because clenbuterol does not stimulate lipolysis or increase cAMP in adipose tissue from pigs (Hu *et al.*, 1987; Mersmann, 1987a).

Less is known regarding inhibition of anabolism by adrenergic hormones in adipose tissue. Norepinephrine analogues depress *de novo* synthesis of fatty acid and also depress triacylglycerol biosynthesis *in vitro* (Saggerson, 1985). If operative *in vivo*, these effects would complement the adrenergic stimulation of lipolysis to provide not only increased degradation but also decreased syn-thesis of adipose tissue lipid. Norepinephrine analogues inhibit incorporation of glucose into rat (Saggerson, 1985) and acetate into sheep (Thornton *et al.*, 1985) adipose tissue lipids as well as acetate incorporation into lipid in chick hepatocytes (Campbell and Scanes, 1985), the site of *de novo* fatty acid biosynthesis in this species. However, in porcine adipose tissue there is no evidence for such inhibition (Rule *et al.*, 1987). Clenbuterol inhibits lipogenesis in sheep adipose tissue and chick liver but not in pig adipose tissue *in vitro*. Norepinephrine analogues inhibit triacylglycerol biosynthesis in rat (Saggerson, 1985) and pig adipose tissue (Rule *et al.*, 1987). If operative *in vivo,* inhibition of this pathway could contribute to the decreased adiposity in various species. Clenbuterol does not affect porcine adipose tissue triacylglycerol biosynthesis *in vitro* (Rule *et al.*, 1987).

It is difficult to verify the inhibition of adipose tissue anabolic function by β-adrenergic agonists *in vivo*. In sheep (Coleman *et al.*, 1988) and cattle (Miller *et al.*, 1987), chronic feeding of a norepinephrine analogue elicits the expected carcass changes but inhibition of lipogenesis *in vitro* in adipose tissue obtained from treated animals is observed only in cattle. In my opinion, it is probably the exceptional case to demonstrate such an effect. If operative, the degree of inhibition of anabolic effects or stimulation of catabolic effects must be very small *in vivo* because over a long feeding period (4 to 12 weeks), the reduction in carcass fat is only 10 to 15%. The small change *in vivo* coupled with the insensitivity of the techniques *in vitro* will preclude detection in most cases.

Evidence for adrenergic stimulation of lipolytic function *in vivo* is abundant in many species; most is from acute injection of infusion or norepinephrine analogues and is implied from an increase in plasma concentration of free fatty acids or glycerol (Baetz *et al.*, 1973; Gregory *et al.*, 1980; Blum *et al.*, 1982; Thompson, 1984). In sheep fed cimaterol (Beermann *et al.*, 1985, 1987) and cattle fed clenbuterol (Eisemann *et al.*, 1988), plasma free fatty acid concentration is elevated. In pigs, acute infusion of clenbuterol increases plasma free fatty acid concentration even though clenbuterol has no effect on lipolytic function in porcine adipose tissue *in vitro,* implying an indirect effect that may result from changes in plasma metabolite or metabolic hormone concentration, from transformation of clenbuterol to an active metabolite by a tissue other than adipose tissue, or from changes in blood flow (Mersmann, 1987a). The change in plasma free fatty acid concentration does not appear to result from release of endogenous adrenergic hormones because it occurs in reserpinized pigs, expected to be depleted of these hormones (C. Y. Hu and H. J. Mersmann, unpublished observations).

3.3. Muscle

Although traditional thought about the catabolic function of adrenergic agonists would not lead to a direct role in stimulation of muscle accretion, retrospective examination of information about adrenergic effects on muscle published in the last decade or so indicates anabolic effects of norepinephrine analogues. An early clue is the β-adrenergic agonist-induced increase in heart size possibly attributable to a work-induced increase in mass (Stanton *et al.*, 1969). An anabolic effect is observed in salivary glands and this cannot be attributed to work-induced growth (Brenner and Stanton, 1970). Adrenergic receptors are present on membranes from muscle cell lines (Parent *et al.*, 1980; Pittman and Molinoff, 1983) and skeletal muscle (Williams *et al.*, 1984) including pig muscle (Bocklen *et al.*, 1986).

Incubation of skeletal muscle with norepinephrine analogues increases not only catabolic processes such as glycogenolysis but also anabolic processes such as amino acid transport and incorporation into protein (Deshaies *et al.*, 1981; Nutting, 1982). Protein degradation is decreased in muscle incubated with adrenergic agonists (Garber *et al.*, 1976; Li and Jefferson, 1977). All of these effects appear to be mediated by β-adrenergic mechanisms. Clenbuterol causes hypertrophy of rat skeletal muscle (Emery *et al.*, 1984; Reeds *et al.*, 1986) including denervated muscle (Zeman *et al.*, 1987). The increase in muscle size appears to result from an increase in cell size with no change in cell number; there are changes in fiber characteristics but these early studies indicate the changes or timing may be specific to certain muscles (Maltin *et al.*, 1986) as also observed in sheep (Bohorov *et al.*, 1987). In sheep (Beermann *et al.*, 1985; Hamby *et al.*, 1986) or cattle (Miller *et al.*, 1988) fed clenbuterol or cimaterol, muscle fiber size is increased; the studies do not agree as to whether hypertrophy is restricted to a particular fiber type. In rats given clenbuterol, Emery *et al.* (1984) demonstrated increased protein synthesis; Reeds *et al.* (1986) in rats and Bohorov *et al.* (1987) in sheep could not confirm this effect. Reeds *et al.* (1986) and Bohorov *et al.* (1987) suggested (not estimated) that protein degradation is inhibited in the clenbuterol-fed rats and sheep, respectively.

The rudimentary studies regarding adrenergic effects on muscle suggest that selected norepinephrine analogues might act directly on skeletal muscle cells but it is not yet clear whether such effects are operative to produce the increased muscle accretion in meat-producing animals fed β-adrenergic agonists.

3.4. Other Mechanisms

One of the major mechanisms to be considered that does not involve direct action on the adipose or muscle cell is stimulation of blood flow. β-Adrenergic agonists stimulate both chronotropic and inotropic mechanisms in the heart so that given no change in vascular resistance, cardiac output would be increased. β-Adrenergic agonists also cause vasodilation so that the cardiac and vascular effects can be complementary. Blood flow may be regulated on a tissue-specific basis as evidenced in rats wherein infusion of isoproterenol decreases blood flow to liver, kidney, and white fat whereas blood flow increases to heart, skeletal muscle, and brown fat (Wickler *et al.*, 1984). Other studies indicate that sympathetic nervous system stimulation yields decreased blood flow in white adipose tissue (Fredholm, 1985). Differential changes in blood flow to various organs including skeletal muscle (increased) and adipose tissue (decreased) could account for some or all of the repartitioning effect of β-adrenergic agonists. Heart rate is increased acutely in calves fed clenbuterol (Williams *et al.*, 1986) and heart rate and hind limb blood flow are increased about 70 and 85% upon initial exposure of steers to clenbuterol whereas, after 9 days of feeding, heart rate and blood flow are still elevated about 25 and 40%, respectively (Eisemann *et al.*, 1988). Similar studies in sheep, with cimaterol as the adrenergic agent, indicate that hind limb blood flow is elevated 44 and 14% after 2 and 4 weeks of chronic feeding (Beermann *et al.*, 1986a). Thus, although there is some desensitization to chronic administration of clenbuterol or cimaterol, the increased heart rate and blood flow probably are maintained for a considerable period of time. In cattle, hind limb blood flow is preferentially increased relative to blood flow to the portal-drained viscera (Eisemann *et al.*, 1988), contributing to the preferential growth of skeletal muscle. There could be considerable species variation in the changes observed in blood flow because of species-specific receptor characteristics, the agonist used, and the adaptation or desensitization to chronic administration of the agonists.

   Another possible mechanism to influence the repartitioning effects is the adrenergic influence on release of a variety of metabolic hormones. Adrenergic hormones inhibit insulin release in mammals, including the pig (Hertelendy *et al.*, 1966), so that a potential mechanism for β-adrenergic agonists to decrease the accretion of fat may be to inhibit insulin release and thus decrease lipogenesis and increase lipolysis (Vernon, 1981; Mersmann, 1986; Mersmann and Hu, 1987). In sheep chronically fed cimaterol, plasma insulin and somatomedin concentrations are lower than in control sheep (Beermann *et al.*, 1985, 1987). Plasma somatotropin and thyroxine concentrations are elevated in sheep chronically fed cimaterol; prolactin, triiodothyronine, and cortisol concentrations are not changed (Beermann *et al.*, 1985, 1987). Secretion of many hormones may be influenced positively or negatively by adrenergic agonists (Huang and McCann, 1983) so that multiple effects may occur depending on the agonist and the species.

   β-Adrenergic agonists may stimulate cell replication through production of cAMP (Whitfield *et al.*, 1979); however, DNA concentration is not increased in muscle from sheep fed cimaterol (Beermann *et al.*, 1987). Adrenergic agonists may stimulate genes to increase hyperplasia as evidenced by isoproterenol stimulation of the c-fos gene (an oncogene homologue) in rodent salivary glands; the gene mRNA concentration is not correlated with the hyperplastic or hypertrophic response induced by isoproterenol (Barka *et al.*, 1986). Studies such as this indicate the possibility of direct adrenergic regulation of gene function but it is not clear whether this is a primary response or a secondary response of cells to increased or decreased metabolic demand with consequent change in enzyme or structural protein content. Also, the gene product of RAS protooncogenes may be directly involved in adenylate cyclase regulation (Garte, 1985). Because β- and α_2-adrenergic agonists operate through a regulatory cascade, an expansive level of function is available without alteration of gene function or enzyme concentration. However, interaction at the gene level may be important to maintain altered function during chronic administration of adrenergic agonists.

There are a multitude of physiological and metabolic effects that could occur in animals fed adrenergic agonists. Many of these could contribute directly or indirectly to the repartitioning effect. Although a small amount of knowledge is beginning to accrue, not enough is known to postulate a mechanism of action for this effect. Some norepinephrine analogues could act on the central nervous system to reduce feed intake (Garattini and Samanin, 1984) but there is little evidence for this type of effect by clenbuterol or cimaterol in the meat-producing animals; with other norepinephrine analogues, the anorectic function could be profound. Nothing is known about the digestive functions or maintenance requirements in animals treated with adrenergic agonists. Finally, because of the diversity of adrenergic function, the mechanism(s) for the repartitioning effect probably will be complex and somewhat species specific.

Note added in proof: There have been advances in many topics discussed in this chapter since the completion of the literature review. Recent reviews on the adrenergic receptor (Lefkowitz and Caron, 1988) and on g-proteins (Casey and Gilman, 1988) will give the reader access to recent literature on these topics. There are several review articles on effects of β-adrenergic agonists on growing meat-producing animals including discussions of and references to two new norepinephrine analogs, ractopamine and L-644-969 (Beerman, 1987; Convey et al., 1987; Mersmann, 1987b; Smith, 1987). Additional papers have been published regarding cimaterol effects on skeletal muscle in sheep (Kim et al., 1987), growth and body composition in mice selected for rapid gain (Eisen et al., 1988) and thermogenesis and body composition in lean and obese rats (Rothwell and Stock, 1987). Finally, the proceedings of a European workshop on the effects of β-adrenergic agonists on animal growth and body composition has been published (Hanrahan, 1987).

ACKNOWLEDGMENTS. This chapter is dedicated to Dr. H. C. Stanton who introduced me to adrenergic pharmacology and who provided encouragement and an atmosphere conducive to my growth as an experimental biologist.

References

Ahlquist, R. P. 1948. A study of the adrenotropic receptors. Amer. J. Physiol. 153:586.

Arnold, A. 1980. Sympathomimetic amine-induced responses of effector organs subserved by alpha-, beta$_1$-, and beta$_2$-adrenoceptors. In: L. Szekeres (Ed.) Adrenergic Activators and Inhibitors, Part I, Handbook of Experimental Pharmacology. Vol. 54/I, pp 63–88. Springer-Verlag, Berlin.

Axelrod, J. 1971. Noradrenaline: fate and control of its biosynthesis. Science 173:598.

Baetz, A. L., D. A. Witzel and C. K. Graham. 1973. Certain metabolic responses of swine to epinephrine and norepinephrine infusion. Amer. J. Vet. Res. 34:497.

Baker, P. K., R. H. Dalrymple, D. L. Ingle and C. A. Ricks. 1984. Use of a β-adrenergic agonist to alter muscle and fat deposition in lambs. J. Anim. Sci. 59:1256.

Barka, T., R. M. Gubits and H. M. van der Noen. 1986. β-Adrenergic stimulation of c-fos gene expression in the mouse submandibular gland. Mol. Cell. Biol. 6:2984.

Beermann, D. H., W. R. Butler, V. K. Fishell, E. N. Bergman and J. P. McCann. 1986a. Preliminary observations on the effects of cimaterol on haert rate, blood flow, plasma insulin concentration and net glucose uptake in the hindquarters of growing lambs. J. Anim. Sci. 63 (Suppl. 1):225 (Abstr.).

Beermann, D. H. 1987. Effects of beta adrenergic agonists on endocrine influence and cellular aspects of muscle growth. Proc. Recip. Meat Conf. 40:57.

Beermann, D. H., W. R. Butler, D. E. Hogue, V. K. Fishell, R. H. Dalrymple, C. A. Ricks and C. G. Scanes. 1987. Cimaterol-induced muscle hypertrophy and altered endocrine status in lambs. J. Anim. Sci. 65:1514.

Beermann, D. H., D. R. Campion and R. H. Dalrymple. 1985. Mechanisms responsible for partitioning tissue growth in meat animals. Proc. Recip. Meat Conf. 38:105.

Beermann, D. H., D. E Hogue, V. K. Fishell, R. H. Dalrymple and C. A Ricks. 1986b. Effects of cimaterol and fishmeal on performance, carcass characteristics and skeletal muscle growth in lambs. J. Anim. Sci. 62:370.

Birnbaumer, L., J. Codina, R. Mattera, R. A. Cerione, J. D. Hildebrandt, T. Sunyer, F. J. Rojas, M. G. Caron, R. J. Lefkowitz and R. Iyengar. 1985. Structural basis of adenylate cyclase stimulation and inhibition by distinct guanine nucleotide regulatory proteins. In: P. Cohen and M. D. Houslay (Ed.) Molecular Mechanisms of Transmembrane Signalling. pp 131–182. Elsevier Science Publ., New York.

Blum, J. W., D. Froehli and P. Kunz. 1982. Effects of catecholamines on plasma free fatty acids in fed and fasted cattle. Endocrinology 110:452.

Bocklen, E., S. Flad, E. Muller and H. von Faber. 1986. Comparative determination of beta-adrenergic receptors in muscle, heart and backfat of Pietrain and Large White pigs. Anim. Prod. 43:335.

Bohorov, O., P. J. Buttery, J. H. R. D. Correia and J. B. Soar. 1987. The effect of the β-2-adrenergic agonist clenbuterol or implantation with oestradiol plus trenbolone acetate on protein metabolism in wether lambs. Brit. J. Nutr. 57:99.

Brenner, G. M. and H. C. Stanton. 1970. Adrenergic mechanisms responsible for submandibular salivery glandular hypertrophy in the rat. J. Pharmacol. Exp. Ther. 173:166.

Buhler, H. U., M. Da Prada, W. Haefely and G. B. Picotti. 1978. Plasma adrenaline, noradrenaline and dopamine in man and different animal species. J. Physiol. (London) 276:311.

Campbell, R. M. and C. G. Scanes. 1985. Adrenergic control of lipogenesis and lipolysis in the chicken in vitro. Comp. Biochem. Physiol. 82C137.

Casey, P. J. and A. G. Gilman. 1988. G protein involvement in receptor-effector copuling. J. Biol. Chem. 263:2577.

Coleman, M. E., P. A. Ekeren and S. D. Smith. 1988. Lipid synthesis, fatty acid-binding protein activity and adipose cell growth in adipose tissue from sheep chronically fed a beta-adrenergic agonist. J. Anim. Sci. 66:372.

Convey, E. M., E. Rickes, Y. T. Yang, M. A. McElligott and G. Olson. 1987. Effects of the beta-adrenergic agonist L-644, 969 on growth performance, carcass merit and meat quality. Proc. Recip. Meat Conf. 40:47.

Cooper, J. R., F. E. Bloom and R. H. Roth. 1986. The Biochemical Basis of Neuropharmacology (5th Ed.). Oxford Univ. Press, London. Chapters 9 and 10, pp 203–314.

Dalrymple, R. H., P. K. Baker, P. E. Gingher, D. L. Ingle, J. M. Pensack and C. A. Ricks. 1984a. A repartitioning agent to improve performance and carcass composition of broilers. Poult. Sci. 63:2376.

Dalrymple, R. H., P. K. Baker and C. A. Ricks. 1984b. Repartitioning agents to improve performance and body composition. Proc. Georgia Nutr. Conf. pp 111–118.

Deshaies, Y., J. Willemot and J. Leblanc. 1981. Protein synthesis, amino acid uptake, and pools during isoproterenol-induced hypertrophy of the rat heart and tibialis muscle. Can. J. Physiol. Pharmacol. 59:113.

Downes, C. P. and R. H. Michell. 1985. Inositol phospholipid breakdown as a receptor-controlled generator of second messengers. In: P. Cohen and M. D. Houslay (Ed.) Molecular Mechanisms of Transmembrane Signalling. pp 3–56. Elsevier Science Publ. New York.

Eisemann, J. H., G. B. Huntington and C. L. Ferrell. 1988. Effects of dietary clenbuterol on metabolism of the hindquarters in steers. J. Anim. Sci. 66:342.

Eisen, E. J., W. J. Croom, Jr. and S. W. Helton. 1988. Differential response to the β-adrenergic agonist cimaterol in mice selected for rapid gain and unselected controls. J. Anim. Sci. 66:361.

Ellis, S. 1980. Effects on the metabolism. In: L. Szekeres (Ed.) Adrenergic Activators and Inhibitors, Handbook of Experimental Pharmacology. Vol. 54/I, pp 319–352. Springer-Verlag, Berlin.

Emery, P. W., N. J. Rothwell, M. J. Stock and P. D. Winter. 1984. Chronic effects of β_2-adrenergic agonists on body composition and protein synthesis in the rat. Biosci. Rep. 4:83.

Exton, J. H. 1985. Mechanisms involved in α-adrenergic phenomena. Amer. J. Physiol. 248:E633.

Fain, J. N. and J. A. Garcia-Sainz. 1983. Adrenergic regulation of adipocyte metabolism. J. Lipid Res. 24:945.

Fredholm, B. B. 1985. Nervous control of circulation and metabolism in white adipose tissue. In: A. Cryer and R. L. R. Van (Eds.) New Perspectives in Adipose Tissue: Structure, Function and Development. pp 45–64. Butterworths, London.

Garattini, S. and R. Samanin. 1984. d-Fenfluramine and salbutamol: two drugs causing anorexia through different neurochemical mechanisms. Int. J. Obesity 8 (Suppl. 1):151.

Garber, A. J., I. E. Karl and D. M. Kipnis. 1976. Alanine and glutamine synthesis and release from skeletal muscle. IV. β-Adrenergic inhibition of amino acid release. J. Biol. Chem. 251:851.

Garte, S. J. 1985. Differential effects of phorbol ester on the β-adrenergic response of normal and ras-transformed NIH3T3 cells. Biochem. Biophys. Res. Commun. 133:702.

Gilman, A. G., L. S. Goodman and A. Gilman (Ed.). 1980. In: Goodman and Gilman's The Pharmacological Basis of Therapeutics. Macmillan Publ. Co., Inc., New York. Chapters 4, 8, 9 and 12, pp 56–90, 138–210, 235–257.

Gregory, N. G., J. D. Wood, M. Enser, W. C. Smith and M. Ellis. 1980. Fat mobilisation in large white pigs selected for low backfat thickness. J. Sci. Food Agrc. 31:567.

Guldberg, H. C and C. A. Marsden. 1975. Catechol-o-methyl transferase: pharmacological aspects and physiological role. Pharmacol. Rev. 27:137.

Hamby, P. L., J. R. Stouffer and S. B. Smith. 1986. Muscle metabolism and real-time ultrasound measurement of muscle and subcutaneous adipose tissue growth in lambs fed diets containing a beta-agonist. J. Anim. Sci. 63:1410.

Hanrahan, J. P. (Ed.) 1987. Beta-Agonists and Their Effects on Animal Growth and Carcass Quality. Elsevier Applied Science, New York.

Hanrahan, J. P., J. F. Quirke, W. Bomann, P. Allen, J. C. McEwan, J. M. Fitzsimons, J. Kotzian and J. F. Roche. 1986. β-Agonists and their effects on growth and carcass quality. In: W. Haresign and D. J. A. Cole (Ed.) Recent Advances in Animal Nutrition. pp 125–138. Butterworths, London.

Hertelendy, F., L. J. Machlin, R. S. Gordon, M. Horino and D. M. Kipnis. 1966. Lipolytic activity and inhibition of insulin release by epinephrine in the pig. Proc. Soc. Exp. Biol. Med. 121:675.

Holly, J. M. P. and H. L. J. Makin. 1983. The estimation of catecholamines in human plasma. Anal. Biochem. 128:257.

Hu, C. Y., J. E. Novakofski and H. J. Mersmann. 1987. Hormonal control of porcine adipose tissue fatty acid release and cyclic AMP concentration. J. Anim. Sci. 64:1031.

Huang, X. Y. and S. M. McCann. 1983. Effect of β-adrenergic drugs on LH, FSH, and growth hormone (GH) secretion in conscious, ovariectomized rats. Proc. Soc. Exp. Biol. Med. 174:244.

Iversen, L. L. 1974. Uptake mechanisms for neurotransmitter amines. Biochem. Pharmacol. 23:1927.

Jaster, E. H and T. N. Wegner. 1981. Beta-adrenergic receptor involvement in lipolysis of dairy cattle subcutaneous adipose tissue during dry and lactating state. J. Dairy Sci. 64:1655.

Johansson, P. 1984. α-Adrenoceptors: recent development and some comparative aspects. Comp. Biochem. Physiol. 78C:253.

Jones, R. W., R. A. Easter, F. K. McKeith, R. H. Dalrymple, H. M. Maddock and P. J. Bechtel. 1985. Effects of the β-adrenergic agonist cimaterol (CL 263,780) on the growth and carcass characteristics of finishing swine. J. Anim. Sci. 61:905.

Kenakin, T. P. 1984. The classification of drugs and drug receptors in isolated tissues. Pharmacol. Rev. 36:165.

Kim, Y. S., Y. B. Lee and R. H. Dalrymple. 1987. Effect of the repartitioning agent cimaterol on growth, carcass and skeletal muscle characteristics in lambs. J. Anim. Sci. 65:1392.

Kobilka, B. K., R. A. F. Dixon, T. Frielle, H. G. Dohlman, M. A. Bolanowski, I. S. Sigal, T. L. Yang-Feng, U. Francke, M. G. Caron and R. J. Lefkowitz. 1987. cDNA for the human β_2-adrenergic receptor: a protein with multiple membrane-spanning domains and encoded by a gene whose chromosomal location is shared with that of the receptor for platelet-derived growth factor. Proc. Natl. Acad. Sci. USA 84:46.

Lacasa, D., P. Mauriege, M. Lafontan, M. Berlan and Y. Giudiceli. 1986. A reliable assay for beta-adrenoceptors in intact isolated human fat cells with a hydrophilic radioligand [^{3}H]CGP-12177. J. Lipid Res. 27:368.

Lackovic, Z. and M. Relja. 1983. Evidence for a widely distributed peripheral dopaminergic system. Fed. Proc. 42:3000.

Lafontan, M. and M. Berlan. 1985. Plasma membrane properties and receptors in white adipose tissue. In: A. Cryer and R. L. R. Van (ed.) New Perspectives in Adipose Tissue: Structure, Function and Development. pp 145–182. Butterworths, London.

Lands, A. M., A. Arnold, J. P. McAnliff, F. P. Ludvena and T. G. Brown. 1967. Differentiation of receptor systems activated by sympathomimetic amines. Nature 214:597.

Lefkowitz, R. J., M. G. Caron, T. Michel and J. M. Stadel. 1982. Mechanisms of hormone receptor–effector coupling: the β-adrenergic receptor and adenylate cyclase. Fed. Proc. 41:2664.

Lefkowitz, R. J., J. M. Stadel and M. G. Caron. 1983. Adenylate cyclase-coupled beta-adrenergic receptors: structure and mechanisms of activation and desensitization. Annu. Rev. Biochem. 52:159.

Lefkowitz, R. J. and M. G. Caron. 1988. Adrenergic receptors. J. Biol. Chem. 263:4993.

Levitzki, A. 1986. β-Adrenergic receptors and their mode of coupling to adenylate cyclase. Physiol. Rev. 66:819.

Li, J. B. and L. S. Jefferson. 1977. Effect of isoproterenol on amino acid levels and protein turnover in skeletal muscle. Amer. J. Physiol. 232:E243.

Malta, E., G. A. McPherson and C. Raper. 1985. Selective β₁-adrenoceptor agonists—fact or fiction? Trends Pharmacol. Sci. 6:400.

Maltin, C. A., M. I. Delday and P. J. Reeds. 1986. The effect of a growth promoting drug, clenbuterol, on fibre frequency and area in hind limb muscles from young male rats. Biosci. Rep. 6:293.

Martin, C. R. 1985. Endocrine Physiology. Oxford Univ. Press, London. Chapter 8, pp 270–318.

Mersmann, H. J. 1984a. Absence of α-adrenergic inhibition of lipolysis in swine adipose tissue. Comp. Biochem. Physiol. 79C:165.

Mersmann, H. J. 1984b. Adrenergic control of lipolysis in swine adipose tissue. Comp. Biochem. Physiol. 77C:43.

Mersmann, H. J. 1986. Lipid metabolism in swine. In: H. C. Stanton and H. J. Mersmann (Ed.) Swine in Cardiovascular Research. Vol. 1, pp 75–104. CRC Press, Inc., Boca Raton, FL.

Mersmann, H. J. 1987a. Acute metabolic effects of adrenergic agents in swine. Amer. J. Physiol. 252:E85.

Mersmann, H. J. 1987b. Primer on beta adrenergic agonists and their effect on the biology of swine. In: The Repartitioning Revolution: Impact of Somatotropin and Beta Adrenergic Agonists on Future Pork Production. pp 19–45. Proc. Univ. Illinois Pork Industry Conference.

Mersmann, H. J. and C. Y. Hu. 1987. Factors affecting measurements of glucose metabolism and lipolytic rates in porcine adipose tissue slices in vitro. J. Anim. Sci. 64:148.

Mersmann, H. J., C. Y. Hu, W. G. Pond, D. C. Rule, J. E. Novakofski and S. B. Smith. 1987. Growth and adipose tissue metabolism in young pigs fed cimaterol with adequate or low dietary protein. J. Anim. Sci. 64:1384.

Milavec-Krizman, M. and H. Wagner. 1978. Effects of the adrenergic agonists isoprenaline and noradrenaline and the α-blocking agents dihydroergotamine and phentolamine on the lipolysis in isolated fat cells of the rat, guinea pig, dog and man. Biochem. Pharmacol. 27:2305.

Miller, M. F., D. K. Garcia, M. E. Coleman, P. A. Ekeren, F. Y. Wu, D. K. Lunt, K. A. Wagner, M. Prockner, T. H. Welsh, Jr. and S. B. Smith. 1988. Adipose tissue, longissimus muscle and anterior pituitary growth and function in clenbuterol-fed heifers. J. Anim. Sci. 66:12.

Minneman, K. P., A. Hedberg and P. B. Molinoff. 1979. Comparison of beta adrenergic receptor subtypes in mammalian tissues. J. Pharmacol. Exp. Ther. 211:502.

Morgan, N. G., P. F. Blackmore and J. H. Exton. 1983. Age-related changes in the control of hepatic cyclic AMP levels by α₁- and α₂-adrenergic receptors in male rats. J. Biol. Chem. 258:5103.

Moser, R. L., R. H. Dalrymple, S. G. Cornelius, J. E. Pettigrew and C. E. Allen. 1986. Effect of cimaterol (CL 263,780) as a repartitioning agent in the diet for finishing pigs. J. Anim. Sci. 62:21.

Moxham, C. P., S. T. George, M. P. Graziano, H. J. Brandwein and C. C. Malbon. 1986. Mammalian β₁- and β₂-adrenergic receptors. J. Biol. Chem. 261:14562.

Northrup, J. K. 1985. Overview of the guanine nucleotide regulatory protein systems, N_s and N_i, which regulate adenylate cyclase activity in plasma membranes. In: P. Cohen and M. D. Houslay (Ed.) Molecular Mechanisms of Transmembrane Signalling. pp 91–116. Elsevier Science Publ., New York.

Nutting, D. F. 1982. Anabolic effects of catecholamines in diaphragm muscle from hypophysectomized rats. Endocrinology 110:307.

O'Donnell, S. R. 1976. Selectivity of clenbuterol (NAB 365) in guinea-pig isolated tissues containing β-adrenoceptors. Arch. Int. Pharmacodyn. 224:190.

O'Donnell, S. R and J. C. Wanstall. 1981. Pharmacological approaches to the characterization of β-adrenoreceptor populations in tissues. J. Auton. Pharmac. 1:305.

Parent, J. B., J. F. Tallman, R. C. Henneberry and P. H. Fishman. 1980. Appearance of β-adrenergic receptors and catecholamine-responsive adenylate cyclase activity during fusion of avian embryonic muscle cells. J. Biol. Chem. 255:7782.

Pittman, R. N. and P. B. Molinoff. 1983. Interactions of full and partial agonists with beta-adrenergic receptors on intact L6 muscle cells. Mol. Pharmacol. 24:398.

Reeds, P. J., S. M. Hay, P. M. Dorwood and R. M. Palmer. 1986. Stimulation of muscle growth by clenbuterol: lack of effect on muscle protein biosynthesis. Brit. J. Nutr. 56:249.

Ricks, C. A., P. K. Baker and R. H. Dalrymple. 1984a. Use of repartitioning agents to improve performance and body composition of meat animals. Proc. Recip. Meat Conf. 37:5.

Ricks, C. A., R. H. Dalrymple, P. K. Baker and D. L. Ingle. 1984b. Use of a β-agonist to alter fat and muscle deposition in steers. J. Anim. Sci. 59:1247.

Rodbell, M. 1980. The role of hormone receptors and GTP-regulatory proteins in membrane transduction. Nature 284:17.

Rothwell, N. J. and M. J. Stock. 1987. Influence of clenbuterol on energy balance, thermogenesis and body composition in lean and genetically obese Zucker rats. Intl. J. Obesity 11:641.

Rule, D. C., S. B. Smith and H. J. Mersmann. 1987. Effects of adrenergic agonists and insulin on porcine adipose tissue metabolism in vitro. J. Anim. Sci. 65:136.

Saggerson, E. D. 1985. Hormonal regulation of biosynthetic activities in white adipose tissue. In: A. Cryer and R. L. R. Van (Ed.) New Perspectives in Adipose Tissue: Structure, Function and Development. pp 87–120. Butterworths, London.

Schumacher, W. A., J. R. Sheppard and B. L. Mirkin. 1982. Biological maturation and beta-adrenergic effectors: pre- and postnatal development of the adenylate cyclase system in the rabbit heart. J. Pharmacol. Exp. Ther. 223:587.

Sibley, D. R. and R. J. Lefkowitz. 1985. Molecular mechanisms of receptor desensitization using the β-adrenergic receptor-coupled adenylate cyclase system as a model. Nature 317:124.

Smith, S. B. 1987. Effects of β-adrenergic agonists on cellular metabolism. Proc. Recip. Meat Conf. 40:65.

Stanton, H. C. 1986. Development of cardiovascular control and function. In: H. C. Stanton and H. J. Mersmann (Ed.) Swine in Cardiovascular Research. Vol. 2, pp 49–72. CRC Press, Inc., Boca Raton, FL.

Stanton, H. C., G. Brenner and E. D. Mayfield, Jr. 1969. Studies on isoproterenol-induced cardiomegaly in rats. Amer. Heart J. 77:72.

Stiles, G. L., R. H. Strasser, M. G. Caron and R. J. Lefkowitz. 1983. Mammalian β-adrenergic receptors. J. Biol. Chem. 258:10689.

Szekeres, L. (Ed.). 1980. Adrenergic Activators and Inhibitors. Handbook of Experimental Pharmacology Vol. 54/I. Springer-Verlag, Berlin.

Thompson, G. E. 1984. Dopamine and lipolysis in adipose tissue of the sheep. Q. J. Exp. Physiol. 69:155.

Thornton, R. F., R. K. Tume, G. Payne, T. W. Larsen, G. W. Johnson and M. A. Hohenhaus. 1985. The influence of the β₂-adrenergic agonist, clenbuterol, on lipid metabolism and carcass composition of sheep. Proc. N.Z. Soc. Anim. Prod. 45:97.

Trendelenburg, U. 1979. The extraneuronal uptake of catecholamines: is it an experimental oddity or a physiological mechanism? Trends Pharmacol. Sci. 1:4.

Ungar, A. and J. H. Phillips. 1983. Regulation of the adrenal medulla. Physiol. Rev. 63:787.

Vernon, R. G. 1981. Lipid metabolism in the adipose tissue of ruminant animals. In: W. W.

Christie (Ed.) Lipid Metabolism in Ruminant Animals. pp 279–362. Pergamon Press, Elmsford, N.Y.

Vernon, R. G. and R. A. Clegg. 1985. The metabolism of white adipose tissue in vivo and in vitro. In: A. Cryer and R. L. R. Van (Ed.) New Perspectives in Adipose Tissue: Structure, Function and Development. pp 65–86. Butterworths, London.

Walton, P. E. and T. D. Etherton. 1986. Stimulation of lipogenesis by insulin in swine adipose tissue: antagonism by porcine growth hormone. J. Anim. Sci. 62:1584.

Whitfield, J. F., A. L. Boynton, J. P. Macmanus, M. Sikorska and B. K. Tsang. 1979. The regulation of cell proliferation by calcium and cyclic AMP. Mol. Cell. Biochem. 27:155.

Wickler, S. J., B. A. Horwitz, S. F. Flaim and K. F. LaNoue. 1984. Isoproterenol-induced blood flow in rats acclimated to room temperature and cold. Amer. J. Physiol. 246:R747.

Williams, L. T. and R. J. Lefkowitz. 1978. Receptor Binding Studies in Adrenergic Pharmacology. Raven Press, New York.

Williams, P. E. V., L. Pagliani and G. M. Innes. 1986. The effect of a β-agonist (clenbuterol) on the heart rate, nitrogen balance and some carcass characteristics of veal calves. Livestock Prod. Sci. 15:289.

Williams, R. S., M. G. Caron and K. Daniel. 1984. Skeletal muscle β-adrenergic receptors: variations due to fiber type and training. Amer. J. Physiol. 246:E160.

Zeman, R. J., R. Ludemann and J. D. Etlinger. 1987. Clenbuterol, a β_2-agonist, retards atrophy in denervated muscles. Amer. J. Physiol. 252:E152.

Gene Transfer for Enhanced Growth of Livestock

VERNON G. PURSEL, CAIRD E. REXROAD, JR., AND DOUGLAS J. BOLT

1. Introduction

During the past decade, a series of scientific advances have provided the opportunity for the modification of domestic animals in a way that has never before been possible—that is, the ability to insert individual genes or groups of genes into living cells and have them function in a predictable manner.

A vast number of scientific achievements have served as the brick and mortar to build the foundation of knowledge and technology that supported the first successful insertion of genes into animals. The final breakthrough occurred in 1980 when six groups of researchers introduced cloned genes into mice (see review by Brinster and Palmiter, 1986). The integration of cloned genes into the genome provides the potential for transmitting the foreign gene through the germ line. Such animals are called transgenic (Gordon and Ruddle, 1981).

Even though the genes integrated into random sites within the genome, early studies with transgenic mice established that expression of the gene was frequently obtained in the appropriate tissue (Brinster et al., 1981) and the gene promotor was still responsive to induction (Palmiter et al., 1982b). Subsequent studies established that the gene promoter provided the tissue specificity for expression (Brinster et al., 1983; McKnight et al., 1983; Swift et al., 1984). In addition, genes that contained enhancer sequences could be stimulated to further increase gene expression (Brinster et al., 1983). Most importantly, these

VERNON G. PURSEL, CAIRD E. REXROAD, JR., AND DOUGLAS J. BOLT ● United States Department of Agriculture, Agricultural Research Service, Beltsville Agricultural Research Center, Reproduction Laboratory, Beltsville, Maryland 20705.

early studies demonstrated that a high proportion of the transgenic mice could transmit the newly acquired DNA as a Mendelian trait to offspring.

Insertion of genes into mice has become an extraordinarily powerful tool for the study of gene function, developmental biology, and animal physiology. Some of the foreign genes that have been inserted into mice during the past few years include the globins, α-fetoprotein, elastase, transferrin, myosin, oncogenes, and several of the major histocompatibility complex. However, the potential that gene transfer offers was most dramatically demonstrated when transgenic mice harboring the rat growth hormone (rGH) gene grew much faster and to almost twice their normal size (Palmiter et al., 1982a). Similar gene transfer experiments were subsequently initiated with domestic animals in the hope that they too would be amenable to genetic manipulation and that enhanced growth performance would result (Hammer et al., 1985b).

Alternatively, an animal's characteristics may be modified by transplantation of genetically modified somatic cells. Implants composed of cells whose genomes have been modified by insertion of cloned genes are called transkaryotic (Selden et al., 1987). Animals with a transkaryotic implant would not transmit the genome to their progeny.

Although development of transgenic animals has received the most emphasis to date, the potential for improved livestock production by transkaryotic implantation in animals destined for market deserves serious consideration. Ethical considerations involved in modifying the human genome will influence efforts to use transkaryotic implantation to correct the 3000 genetic diseases afflicting humans. Therefore, one may anticipate that the most successful transkaryotic implantation techniques will be useful in livestock and that some domestic species will be used as models for humans.

The successful use of genetic engineering to enhance growth and efficiency of meat production in livestock depends on many factors. These include identification, isolation, and modification of useful genes or groups of genes that influence growth, successful insertion of these genes into the genome, and control of the expression of the inserted genes in transgenic or transkaryotic animals.

2. Identification of Genes for Transfer

A large amount of variation in growth rate exists between and within breeds of livestock. This variation is quantitative and of moderate heritability. Animal breeders have successfully utilized genetic variation to improve livestock for many years. Unfortunately, the quantitative genetic approach has yielded few clues regarding the fundamental genetic changes that accompanied the selection. As noted in preceding chapters, growth of an animal is an exceedingly complex process; thus, one might expect that numerous genes orchestrate animal growth.

Few single genes have been identified that have major effects on growth. In cattle, the double-muscle gene is responsible for muscle hypertrophy and enhanced yield of lean meat (Hanset, 1982). In pigs, the halothane sensitivity gene *(Hal)* is associated with increased lean meat in the carcass. Pigs homozygous for *Hal* are also susceptible to stress and have a high incidence of pale-soft-exudative (PSE) meat (Smith and Bampton, 1977). In mice containing the autosomal recessive high growth *(hg)* gene, postweaning gains may be up to 50% greater than for contemporary controls (Bradford and Fanula, 1984). Each of these genes offers considerable potential for investigation of growth in animals. However, the specific product of each gene remains to be identified.

A group of peptide hormones—growth hormone-releasing factor (GRF), somatostatin, growth hormone (GH), insulinlike growth factor (IGF-I), insulin, and thyrotropic hormone—work in concert to regulate and coordinate the metabolic pathways responsible for tissue growth. Even though correlations between growth and circulating levels of some of these peptide hormones have often produced conflicting results, the preponderance of data indicates that genetic capacity for growth is related to increased circulating levels of GH and IGF-I (see review by Sejrsen, 1986). Furthermore, injection of pigs, sheep, and cattle with exogenous GH and stimulation of GH production in lambs by immunizing them against somatostatin have generally produced enhanced feed efficiency, slightly increased growth rate, and reduced subcutaneous fat. These findings all suggest the possible usefulness of genes for the peptide hormones for modifying the growth of livestock.

The endocrine system is affected by genetic variation at all levels. Thus, as our understanding of the endocrine regulation of growth increases, we can assume that a number of other genes might be candidates for transfer, such as those encoding receptors of each of the growth-related peptide hormones, or those for the IGF-I carrier proteins.

Control of Gene Expression

Selection of the regulatory DNA sequences for ligation to the structural gene is extremely important because the regulatory sequences largely determine the site, level, and timing of gene expression. One might expect that transfer of the normal 5' and 3' flanking sequences along with the structural gene might produce the most desirable gene expression characteristics for the growth-related peptide hormones. However, results have thus far failed to support this expectation. Hammer *et al.* (1984) produced 32 transgenic mice with integrations of 1 to 250 copies of human (h) GH or rGH genes that contained their own 5' and 3' sequences. Even though pituitaries of some transgenic mice contained mRNA for hGH, serum hGH was not detected, and none of the mice grew larger than controls.

Although use of the normal regulatory sequences warrants additional investigation, expression of the gene in the usual tissue site may not be a prerequisite in the case of peptide hormones because the target tissue is usually independent of the source of hormone production. Palmiter *et al.* (1982a, 1983) recognized this fact and pioneered the use of a fusion gene, which was composed of promoter sequences of one gene fused to a different structural gene. Thus, when the promoter from the mouse metallothionein-1 (MT-1) gene was fused to the rGH structural gene and transferred into mice, rGH was produced in the same tissues as MT-1 instead of the pituitary. Unfortunately, our understanding of regulatory sequences is greatly lacking, and as a consequence, our choice for construction of fusion genes is limited at present to those that are best understood. In addition, until more information is available, it will be difficult to predict the action of regulatory sequences when transferred into a foreign species.

3. Methods of Producing Transgenic Animals

Several methods of inserting cloned genes into the germ line have potential for use in livestock.

3.1. Microinjection into Fertilized Ova

The only technique that has thus far resulted in production of transgenic farm animals is the microinjection of the gene into ova by modifications of the procedure used originally by Gordon *et al.* (1980) in the mouse. A diagram of the technique is shown in Fig. 1.

Fertilized ova that are between the pronuclear stage of development and the second cleavage division are flushed from the oviduct and placed in suitable culture media. Estrous synchronization, superovulation, and induction of ovulation are frequently used to increase the number of ova recovered and to provide the desired stage of ovum development.

The pronuclei are visible in about 80% of sheep ova when examined with interference-contrast microscopy (Hammer *et al.*, 1985b). However, the dense lipid granules present in both pig and cow ova prevent visualization of pronuclei or nuclei by microscopy. Centrifugation of pig and cow ova for about 3 min at 15,000g stratifies the cytoplasm and leaves the pronuclei or nuclei near the center of the ovum below the lipid granules. The nuclei and pronuclei are then visible with interference-contrast microscopy (Wall *et al.*, 1985).

The ova are placed in a microdrop of media, overlaid with silicone oil to

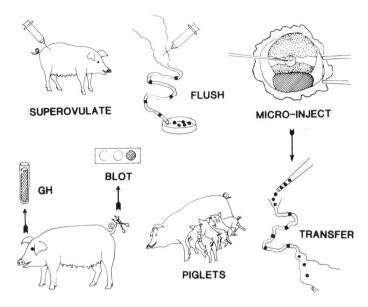

Figure 1. Procedures used to produce transgenic pigs by microinjection. Fertilized ova from su-perovulated gilts are recovered and centrifuged to permit visualization of pronuclei. Linearized gene copies are microinjected and ova are transferred into a surrogate. At birth, tail tissue and blood plasma are assayed for presence of the gene and its product.

prevent evaporation, on a depression slide. Either inverted microscopes or up-right microscopes fitted with long-working-distance objectives are suitable if they are fitted with high-quality interference-contrast optics with 200 to 300 × magnification. The microscope must be equipped with two micromanipulators, one for an egg-holding pipette and the other for an injection pipette. The hold-ing pipette and injection pipette are each fitted with a tube leading to a syringe that permits either gentle suction or carefully controlled fluid injection.

As an ovum is held with light suction by the holding pipette, the tip of the injection pipette is inserted through the zona pellucida and cytoplasm and into the most visible pronucleus. Several hundred copies of the gene are ex-pelled into the pronucleus. The person performing the injection carefully ob-serves the pronucleus and withdraws the pipette when the pronuclear structure has visibly enlarged. Only one pronucleus is injected in pronuclear ova; both nuclei are injected in two-cell ova.

Ova of most species of domestic animals can be surgically transferred into oviducts of surrogate mothers to develop until term. In cattle, transfer of early stage ova has an extremely low rate of success. Therefore, *in vivo* culture in a ligated ewe or doe oviduct appears to be feasible until the ova develop into morulae or blastocysts. At that stage, the embryos can be transferred into a surrogate cow.

The mechanism by which injected DNA integrates into a chromosome is unknown. Usually, the DNA integrates in a single site on one chromosome, but double integrations can occur. Brinster *et al.* (1985) suggested that microinjection may cause a break in a chromosome, permitting subsequent integration.

3.1.1. Factors Affecting Efficiency of Gene Integration

The efficiency of gene insertion by microinjection is affected by many factors. Brinster *et al.* (1985) concluded that the concentration and form of DNA were the most critical factors affecting gene integration in mouse ova. Integration efficiency improved as DNA concentration in injection fluid increased from 0.01 and 0.1 ng/μl to 1–2ng/μl, but poor fetal survival was observed when 10 ng/μl or higher DNA concentrations were injected. Linear molecules had a fivefold higher rate of integration than circular molecules. Gene integration frequency was extremely low after injections into the cytoplasm or injections into both nuclei of two-cell ova rather than injections into one pronucleus of one-cell ova. Integration frequency was slightly higher after injection of male pronuclei (17%) than female pronuclei (13%).

The efficiency of production of transgenic animals by microinjection is much lower for farm animals than for mice. Part of this difference may be due to extensive experience in several laboratories in manipulating mouse ova. More likely, ova of one species or stage of development may not survive culture or micromanipulation as readily as ova of another species or stage of development. For example, two-cell pig ova appear to survive microinjection more frequently than do one-cell pig ova (Pursel *et al.*, 1987a), which is directly contradictory to results in mice (Brinster *et al.*, 1985).

3.1.2. Advantages and Disadvantages of Microinjection

Certainly the low rate of integration is a key disadvantage. Another concern is that multiple copies of genes usually integrate at a single site on a chromosome. For example, the number of copies that have integrated into the genome of pigs varied from 1 to 490 per cell (Hammer *et al.*, 1985b).

One advantage of microinjection is that rate of mosaicism is lower than for other procedures. Progeny of mosaic animals have from 0 to 50% chance of receiving the gene. Wilkie *et al.* (1986) reported that about 30% of transgenic mice are mosaic in the germline, which is a higher incidence than had been reported previously. This finding may mean integration is occurring after first cleavage.

3.2. Retroviral Insertion of Genes

Viral infection is a natural method of introducing DNA into the genome of animals. Retroviruses (RNA viruses) penetrate a host cell, and their RNA genome is copied into linear DNA that is then integrated into the host genome. The host cell may transcribe the integrated viral DNA, resulting in the production of the multiple viral proteins. The three viral coding regions are: gag region, coding for core proteins of the virion; pol region, coding for reverse transcriptase; and env region, coding for envelop glycoproteins. If these coding regions are lacking, infectious viral particles are not produced by the host.

The rationale for construction of an effective viral vector is to replace a coding region in the RNA genome with the gene to be transferred. The replication-defective retrovirus, which contains the new gene, is inserted into the host genome.

3.2.1. Use of Helper Cells

One approach under study is use of a helper cell that can produce the replication-defective viral stock that will enter the target cells. Watanabe and Temin (1983) successfully manipulated a canine cell line (D17) to produce viral proteins but not infectious virus because the encapsidation site responsible for RNA packaging was deleted from the genome. These cells are then transfected with another recombinant viral DNA that contains the encapsidation site along with the desired gene in place of the protein coding regions. Thus, the helper cells are able to produce the desired virions that are capable of only one infectious cycle, i.e., the virions will enter the target cell but cannot replicate after insertion. Thus, helper cells can then be inserted into the blastocyst where virions infect the inner cell mass (Mettus *et al.*, 1986).

Another approach used in mice was culture of zona-free ova on fibroblasts that were producing defective retroviral particles (Rubenstein *et al.*, 1986). After the ova developed into compact morulae, they were transferred into surrogate females.

3.2.2. Advantages and Disadvantages of Retroviral Insertion

The use of defective viruses to introduce desired genes will hopefully result in a high efficiency of gene transfer and the introduction of single copies of the desired gene. Most researchers have found a high incidence of mosaicism in the transgenic animals, which results in a rather low percentage of transgenics in G1 progeny.

One of the main problems to date is the low incidence of gene expression. Another potential problem is that the length of DNA sequence transferred may be limited to under 10 kb.

3.3. Insertion via Pluripotent Cells

Another gene transfer technique under investigation involves transfecting desired genes into stem cells derived from embryos and the subsequent insertion of the stem cells into a blastocyst. Since the stem cells are pluripotent, the embryos that incorporate the stem cells would be chimeric. A portion of these chimeric animals would have the desired gene in the germline and would transmit the gene to progeny (Robertson *et al.,* 1986).

One advantage this method offers is the potential to select for transformed cells that express the desired gene; therefore, only stem cells capable of transferring functional genes would be inserted into blastocysts. Application of this method to domestic animals depends on the ability to modify and maintain pluripotent lines of cells *in vitro.* Thus far, stem cell lines have not been developed for domestic animals.

3.4. Insertion by Transposons

A number of species contain transposable elements (transposons) that might act as vectors for gene transfer similar to retroviruses. Several researchers have utilized transposable P elements to insert genes into *Drosophila,* and the inserted genes were expressed at the appropriate site and developmental stage (see review by O'Hare, 1985).

For P elements to be useful, they would require modification to prevent them from jumping from site to site after their initial insertion. This could be accomplished by making the P element defective in the transposase coding region.

If transposable elements can be devised to function in mammals, one of the main advantages would be insertion of single gene copies into the genome. Another advantage would be that genes inserted by P element vectors have had relatively uniform expression (Spradling and Rubin, 1983), which is in marked contrast to the extreme variability in expression observed in transgenic animals. Whether this is the result of the vector or the species remains to be determined. Transposonlike DNA has been identified in mammals but functionality of these elements has not been demonstrated (Paulson *et al.,* 1985). Khillan *et al.* (1985) successfully incorporated a *Drosophila* P element into mice but it apparently failed to function.

4. Production of Transkaryotic Animals

Although little information is available regarding production of transkaryotic animals, Selden *et al.* (1987) proposed that the ideal gene delivery system would utilize primary cells taken from the host species. Primary fibroblasts, which are readily available and culturable, might be the cell type of choice. Application of the transkaryotic approach would require the rapid transfection of primary cells with the desired gene. Use of electroporation to enhance transfection of cultured cells may be a useful approach. Another worthwhile approach might be use of a replication-defective retrovirus such as discussed above.

The function of the implanted transkaryotic cells will undoubtedly depend on the location and size of the implant and the histocompatibility of the transkaryotic cells with the host animal. Overcoming the problems of implant rejection may be a formidable challenge in farm animals due to the heterologous nature of the population.

5. Transfer of Growth-Related Genes into Livestock

The discovery that transgenic mice harboring MtrGH or MThGH fusion genes produced high serum levels of foreign GH, had an increased rate of growth, and grew to almost twice normal size (Palmiter et al., 1982a, 1983) has provided the impetus for insertion of similar genes into livestock.

5.1. Integration of Growth-Related Genes

Several fusion genes that contain structural sequences for peptide hormones associated with growth have recently been microinjected into fertilized ova of farm animals. The efficiency of producing transgenic animals by this technique has been quite low. In most experiments, less than 1% of microinjected ova developed into transgenic animals (Table I). Notable exceptions were results for rabbit ova (Brem et al., 1985; Hammer et al., 1985b). Nuclear structures are much more visible with interference-contrast microscopy in rabbit ova than in ova of other species of farm animals, an attribute that makes microinjection less difficult and possibly contributes to the higher rate of embryo survival. Additionally, rabbit ova may survive culture and microinjection better than ova of other farm animals.

The number of MThGH copies integrating into pigs varied from 1 to 490 per cell, with most integrations probably at a single site (Hammer et al., 1985b). Southern blot analysis showed that transgenic pigs with MThGH contained intact copies of the fusion gene with many oriented in tandem head-to-tail arrays and some in a head-to-head configuration. These observations agreed with those in mice (Palmiter et al., 1982a).

5.2. Expression of Integrated Genes

The insertion of fusion genes has resulted in immunodetectable levels of hGH and bovine (b) GH in transgenic pigs, bGH in lambs, and hGH in rabbits

Table I
Efficiency of Transferring Growth-Related Genes into Animals

Fusion gene	Species	Embryos injected (no.)	Offspring No.	Offspring %	Transgenic No.	Transgenic %	Expressing No.	Expressing %
MThGH	Mouse[a]		101		33		23	70
	Pig[b]	268	15	5.6	1	0.37		
	Pig[c]	2035	192	9.4	20	0.98	11/18	61
	Rabbit[b]	385	37	9.6	5	1.30		
	Rabbit[c]	1097	218	19.9	28	2.55	4/16	25
	Sheep[c]	1032	73	7.1	1	0.10		
MTbGH	Mouse[d]		65		10		7	70
	Pig[e]	2198	149	6.8	11	0.50	8	73
	Sheep[e]	711	38	5.3	2	0.28	2	100
MToGH	Goat[f]	153	9	5.9	0	0		
	Sheep[g]	436	27	6.2	1	0.23	0	0

[a]Palmiter *et al.* (1983).
[b]Brem *et al.* (1985).
[c]Hammer *et al.* (1985b).
[d]Hammer *et al.* (1985a).
[e]Pursel *et al.* (1987).
[f]Fabricant *et al.* (1987).
[g]Nancarrow *et al.* (1987).

(Table I). The percentage of transgenic pigs and mice that expressed the hGH and bGH genes was remarkably similar: 61% for hGH and 73% for bGH in pigs, and 70% for both genes in mice. In contrast, only 25% of transgenic rabbits with the MThGH fusion gene expressed (Hammer *et al.*, 1985b). No explanation for this difference has been offered.

Expression of the transgene does not appear to be related to the number of copies that integrate into the genome. The proportion of transgenic pigs producing hGH was similar over a wide range of gene copies per cell (Table II). These findings agree with previous results in mice (Palmiter *et al.*, 1983; Hammer *et al.*, 1985a). Also, the plasma concentrations of GH in transgenic

Table II
Relationship of Gene Copies Integrated to Gene
Expression in Transgenic Pigs with MThGH
Fusion Gene

Gene copies per cell	Number of pigs — Transgenic	Number of pigs — Expressing
1	4	2
2–10	7	5
11–100	4	3
>100	3	2

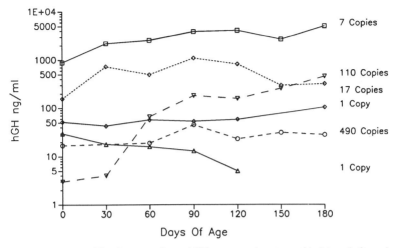

Figure 2. Each line identifies the mean plasma hGH concentration at monthly intervals for a single transgenic pig. The number of gene copies per cell for each pig is indicated at the right.

pigs (Fig. 2) did not appear to be related to the number of gene copies per cell. The results with pigs agreed with previous observations in transgenic mice that expressed hGH, bGH, or hGRF fusion genes (Palmiter *et al.*, 1983; Hammer *et al.*, 1985a; Morrello *et al.*, 1986).

When blood samples were collected at weekly intervals from birth to 6 months of age, plasma hGH concentrations increased in some transgenic pigs, decreased in other pigs, and remained relatively constant in others (Fig. 2). The site of gene integration may play a major role in determining both the level of expression and pattern of gene expression throughout the animal's life.

In the absence of heavy metal induction of the MT promoter, levels of foreign GH and GH mRNA were considerably higher in transgenic mice than in transgenic pigs when both species harbored the same fusion gene. In mice, GH and GH mRNA levels were frequently elevated more than tenfold after zinc stimulation (Palmiter *et al.*, 1983; Hammer *et al.*, 1985a). In contrast, stimulation of transgenic pigs with zinc has not induced more than a doubling of GH (K. F. Miller and V. G. Pursel, unpublished data). Additional research with other genes and regulatory sequences will be required to determine whether these differences are peculiar to these fusion genes or to species differences in general.

5.3. Growth Performance of Transgenic Livestock

The increased growth rate and body size of transgenic mice that produced rGH or hGH (Palmiter *et al.*, 1982a, 1983) raised the expectation that insertion

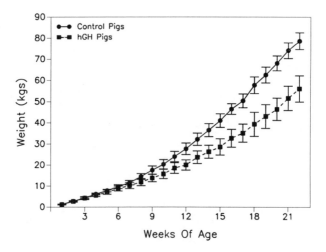

Figure 3. Mean cumulative body weights ± S.E. from birth to 22 weeks of age for six transgenic pigs (■–■) and littermate control pigs (●–●).

of similar fusion genes into livestock might produce comparable results. This expectation has not been realized. When compared on a total body weight basis, transgenic pigs that produced hGH (Fig. 3) or bGH (data not shown) grew slower than their littermates (Pursel *et al.,* 1987b).

Transgenic pigs with elevated plasma hGH or bGH were considerably leaner than their littermates. The lack of fat deposition became visibly evident at about 2 months of age. When transgenic pigs that expressed the hGH gene were compared with littermate control pigs at 90 kg body weight, the average back-fat thickness at the tenth rib was significantly different (7.0 ± 2.1 versus 18.5 ± 1.8 mm; Pursel *et al.,* 1987b). Therefore, if transgenic pigs and their littermates had been compared on a lean-tissue basis, growth would have been similar.

Why elevation of plasma hGH or bGH did not stimulate more rapid growth in pigs remains unanswered. Several pieces of evidence demonstrated that hGH and bGH were physiologically active in the transgenic pigs. These were: (1) transgenic pigs that produced hGH rarely had detectable levels of porcine (p)GH in their plasma after 1 week of age, which indicates that the feedback mechanism was functioning in a normal manner (Bolt *et al.,* 1986); (2) plasma IGF-I concentrations in transgenic pigs were elevated about two- to threefold above those of littermate control pigs (K. F. Miller and V. G. Pursel, unpublished data), which indicates foreign GH was able to bind to GH receptors in the liver to stimulate IGF-I synthesis; (3) insulin and glucose levels were elevated comparable to those found in pigs injected daily with exogenous pGH (K. F. Miller and V. G. Pursel, unpublished data); (4) antilipogenic effects of GH were evident as mentioned above; (5) the lactogenic properties of hGH stimulated visible mammary development that was more noticeable in boars than in gilts (Pursel *et al.*, 1987b); (6) as they approached maturity, transgenic pigs exhibited multiple body features common to acromegaly in humans.

A number of factors had impact on the growth performance of the transgenic pigs. First, expressing transgenics weighted 20% less at birth than their littermates (V. G. Pursel, unpublished data). This factor contributed to high neonatal mortality, and the smaller size left the transgenic pigs at a disadvantage when competing with littermates during nursing. Second, expressing transgenic pigs exhibited a tendency for lethargy and reduced appetite. Pigs producing bGH consumed 20% less feed daily from 30 to 60 kg body weight than their littermates (A. D. Mitchell and V. G. Pursel, unpublished data). Third, several transgenic pigs had persistent diarrhea that lasted 30 to 40 days, which resulted in death of one pig and retarded growth in others (Pursel *et al.*, 1987b).

Other possible physiological factors that prevented transgenic pigs from exhibiting enhanced growth performance were the continuous production of foreign GH in contrast to the episodic released of GH that occur normally (K. F. Miller and V. G. Pursel, unpublished data), and the altered site of GH production: hGH and bGH were produced in the liver, intestine, kidney, testis, and other organs under the influence of the MT promoter whereas pGH is normally produced in the anterior pituitary. Thus, all cells that were actively synthesizing GH would be exposed to vastly higher concentrations of GH than plasma levels would indicate. Whereas organs and tissues of transgenic mice and pigs with the same construct would presumably be exposed to similar excesses in GH, metabolic effects of excess GH might differ among species. For example, excess GH could inhibit transgenic pigs from utilizing one or more essential amino acid, fatty acid, mineral, or vitamin, thereby preventing the genetic potential for growth from being realized and contributing to the multiple health problems occurring in these animals as their growth rate plateaued (Pursel *et al.*, 1987b).

The average birth weight and growth rate of nonexpressing transgenic pigs did not differ from those of littermate control pigs. Also, nonexpressing transgenic pigs did not have any of the health problems observed in pigs that expressed the foreign GH genes (Pursel *et al.*, 1987b). Thus, the adverse effects were caused by the expression of the gene and not merely by the presence of the gene.

Whether integration of ovine (o) GH or bGH fusion genes into sheep or cattle, or integration of a pGH fusion gene into pigs would result in different patterns of growth remains open to conjecture. In any case, one might anticipate that animals exposed to continuously elevated levels of GH would encounter problems similar to those found in acromegaly as they approached maturity.

This situation might be overcome if molecular biologists discover alternative regulatory sequences that can be tightly controlled by external stimuli.

5.4. Germline Transmission of Fusion Genes

For the transgenic approach to be ultimately effective, it is essential that the integrated gene be transmitted to the next generation. Five of six founder transgenic pigs transmitted the MThGH fusion gene to one or more progeny (Pursel *et al.*, 1986). Germline transmission was obtained from both expressing and nonexpressing pigs. Transgenic progeny of an expressing boar also expressed the gene. Transgenic progeny of nonexpressing parents also failed to express the gene. In most cases, the gene was transmitted to about 50% of the progeny when a transgenic was mated with a nontransgenic pig.

Some transgenic pigs were apparently mosaic for the fusion gene. One sow failed to transmit the gene to her progeny. A boar transmitted the gene to only 1 of 33 progeny. Analysis of DNA indicated that the single transgenic offspring had a higher average number of gene copies per cell than the sire, which provided additional evidence of germline mosaicism. Wilkie *et al.* (1986) reported that between 25 and 36% of transgenic mice, which were produced by microinjection, were mosaic for the transferred gene. Thus, presence of the gene in a founder does not ensure that the fusion gene will be transmitted to progeny or that the number of gene copies per cell will be constant from founder to progeny. The high rate of mosaicism in mice that had been injected at the pronucleate stage of development suggests that integration frequently occurs after the first round of chromosomal DNA replication (Wilkie *et al.*, 1986).

5.5. Expression in Animals with Transkaryotic Implants

Production of hGH in mice that had previously received transkaryotic implants of mouse fibroblasts has been reported by Selden *et al.* (1987). The cultured fibroblasts had been stably transfected with an MThGH fusion gene, and the fibroblasts were implanted intraperitoneally, subcutaneously, under the renal capsule, intrahepatically, intraportally, or into the inferior vena cava. hGH was detected in blood serum of transkaryotic mice for 7 to 14 days before the host's immune system destroyed the implanted fibroblasts. When the immune system of mice harboring transkaryotic implants was suppressed with dexamethasone and rabbit antiserum to mouse thymocytes, the implants survived for more than 3 months and hGH was continuously produced. The serum hGH concentrations varied considerably among mice, with the level exceeding 500 ng/ml in one mouse. The effect of elevated GH on growth was not reported. However, GH levels were comparable to those in transgenic mice that exhibited enhanced growth (Palmiter *et al.*, 1983).

6. Conclusions

Recent research has clearly demonstrated that foreign genes can be integrated into the genome of farm animals. The integrated genes were expressed, the gene products were biologically active, and the genes could be transmitted to subsequent generations. Although transgenic pigs with elevated plasma hGH or bGH did not grow faster than littermates, they were dramatically leaner and significantly more efficient in utilization of feed. More research will be required to establish why the growth rate was not enhanced in transgenic pigs. Expansion of research with sheep, goats, and cattle is required to determine whether growth in these species can be enhanced by gene transfer. In addition, use of transkaryotic implants to deliver growth-related peptide hormones into farm animals deserves further investigation.

The paucity of identified genes that affect growth performance and the lack of knowledge regarding the regulation of fusion genes that might impact growth are the main factors that must be overcome before gene transfer can contribute to the genetic improvement of livestock. However, at the present time, the low rate of efficiency of the microinjection method of transferring genes greatly inhibits the transfer of growth-related genes into farm animals. Gradual improvement of the microinjection technique can be expected, but alternative methods of gene transfer, such as use of retroviral vectors, may offer the greatest hope for efficient gene insertion in the future.

ACKNOWLEDGMENTS. We are extremely grateful to our collaborators for their many contributions, to Harold Hawk and Robert Wall for their helpful review of the manuscript, and to Linda Neuenhahn for excellent manuscript preparation.

References

Bolt, D. J., V. G. Pursel, R. E. Hammer, R. J. Wall, R. D. Palmiter and R. L. Brinster. 1986. Plasma concentrations of human growth hormone and porcine growth hormine in transgenic pigs. J. Anim. Sci. 63 (Suppl. 1):220 (Abstr.).

Bradford, G. E. and T. R. Fanula. 1984. Evidence for a major gene for rapid postweaning growth in mice. Genet. Res. (Camb.) 44:293.

Brem, G., B. Brenig, H. M. Goodman, R. C. Selden, F. Graf, B. Kruff, K. Springman, J. Hondele, J. Meyer, E. -L. Winnaker and H. Krausslich. 1985. Production of transgenic mice, rabbits and pigs by microinjection into pronuclei. Z. Zuchthygiene 20:251.

Brinster, R. L., H. Y. Chen, M. E. Trumbauer, M. K. Yagle and R. D. Palmiter. 1985. Factors affecting the efficiency of introducing foreign DNA into mice by microinjecting eggs. Proc. Natl. Acad. Sci. USA 82:4438.

Brinster, R. L., H. Y. Chen, M. E. Trumbauer, M. K. Yagle, A. W. Senear, R. Warren and R. D. Palmiter. 1981. Somatic expression of herpes thymidine kinase in mice following injection of a fusion gene into eggs. Cell 27:223.

Brinster, R. L. and R. D. Palmiter. 1986. Introduction of genes into the germ line of animals. Harvey Lect. 80:1.

Brinster, R. L., K. A. Ritchie, R. E. Hammer, R. L. O'Brien, B. Arp and U. Storb. 1983. Expression of a microinjected immunoglobulin gene in the spleen of transgenic mice. Nature 306:332.

Fabricant, J. D., L. C. Nuti, B. S. Minhas, W. C. Baker, J. S. Capehart, P. Marrack, J. H. Chalmers, M. W. Bradbury and J. E. Womack. 1987. Gene Transfer in goats. Theriogenology 27:229 (Abstr.).

Gordon, J. W. and F. H. Ruddle. 1981. Integration and stable germ line transmission of genes injected into mouse pronuclei. Science 214:1244.

Gordon, J. W., G. A. Scangos, D. J. Plotkin, J. A. Barbosa and F. H. Ruddle. 1980. Genetic transformation of mouse embryos by microinjection of purified DNA. Proc. Natl. Acad. Sci. USA 77:7380.

Hammer, R. E., R. L. Brinster and R. D. Palmiter. 1985a. Use of gene transfer to increase animal growth. Cold Spring Harbor Symp. Quant. Biol. 50:379.

Hammer, R. E., R. D. Palmiter and R. L. Brinster. 1984. Partial correction of murine hereditary growth disorder by germ-line incorporation of a new gene. Nature 311:65.

Hammer, R. E., V. G. Pursel, C. E. Rexroad, Jr., R. J. Wall, D. J. Bolt, K. M. Ebert, R. D. Palmiter and R. L. Brinster. 1985b. Production of transgenic rabbits, sheep and pigs by microinjection. Nature 315:680.

Hanset, R. 1982. Major genes in animal production, examples and perspectives: cattle and pigs. Proc. 2nd World Congr. Genet. Appl. Livestock Prod. VI:439.

Khillan, J. S., P. A. Overbeek and H. Westphal. 1985. Drosophila P element integration in mouse. Dev. Biol. 109:247.

McKnight, G., R. E. Hammer, E. A. Kuenzal and R. L. Brinster. 1983. Expression of the chicken transferrin gene in transgenic mice. Cell 34:335.

Mettus, R. V., R. M. Petters, B. H. Johnson and R. M. Shuman. 1986. In vitro development of porcine blastocysts injected with a retrovirus helper cell line. J. Anim. Sci. 63 (Suppl. 1):236 (Abstr.).

Morrello, D., G. Moore, A. M. Salmon, M. Yaniv and C. Babinet. 1986. Studies on the expression of an H-2K/human growth hormone fusion gene in giant transgenic mice. EMBO J. 5:1877.

Nancarrow, C., J. Marshall, J. Murray, I. Hazelton and K. Ward. 1987. Production of a sheep transgenic with the ovine growth hormone gene. Theriogenology 27:263 (Abstr.).

O'Hare, K. 1985. The mechanism and control of P element transposition in Drosophila melanogaster. Trends Genet. 9:250.

Palmiter, R. D., R. L. Brinster, R. E. Hammer, M. E. Trumbauer, M. G. Rosenfeld, N. C. Birnberg and R. M. Evans. 1982a. Dramatic growth of mice that develop from eggs microinjected with metallothionein–growth hormone fusion genes. Nature 300:611.

Palmiter, R. D., H. Y. Chen and R. L. Brinster. 1982b. Differential regulation of metallothionein–thymidine kinase fusion genes in transgenic mice and their offspring. Cell 29:710.

Palmiter, R. D., G. Norstedt, R. E. Gelinas, R. E. Hammer and R. L. Brinster. 1983. Metallothionein–human growth hormone fusion genes stimulate growth of mice. Science 222:809.

Paulson, K. E., N. Deka, C. W. Schmid, R. Misra, C. W. Schindler, M. G. Rush, L. Kadyk and L. Leinwand. 1985. A transposon-like element in human DNA. Nature 316:359.

Pursel, V. G., R. E. Hammer, D. J. Bolt, R. J. Wall, R. D. Palmiter and R. L. Brinster. 1986. Transgenic swine transmit foreign gene to progeny. J. Anim. Sci. 63 (Suppl. 1):203 (Abstr.).

Pursel, V. G., K. F. Miller, C. A. Pinkert, R. D. Palmiter and R. L. Brinster. 1987a. Development of 1-cell and 2-cell pig ova after microinjection of genes. J. Anim. Sci. 65 (Suppl. 1): 402 (Abstr.).

Pursel, V. G., C. E. Rexroad, Jr., D. J. Bolt, K. F. Miller, R. J. Wall, R. E. Hammer, C. A. Pinkert, R. D. Palmiter and R. L. Brinster. 1987b. Progress on gene transfer in farm animals. Vet. Immunol. Immunopathol. 17:303.

Robertson, E. J., A. Bradley, M. Kuehn and M. Evans. 1986. Germ line transmission of genes introduced into cultured pluripotential cells by retroviral vector. Nature 323:445.

Rubenstein, J. L. R., J. F. Nicolas and F. Jacob. 1986. Introduction of genes into preimplantation mouse embryos by use of a defective recombinant retrovirus. Proc. Natl. Acad. Sci. USA 83:366.

Sejrsen, K. 1986. Endocrine mechanisms underlying genetic variation in growth in ruminants. Proc. 3rd World Congr. Genet. Appl. Livestock Prod. XI:261.

Selden, R. F., M. J. Skoskiewicz, K. B. Howie, P. S. Russell and H. M. Goodman. 1987. Implantation of genetically engineered fibroblasts into mice: implications for gene therapy. Science 236:714.

Smith, C. and P. R. Bampton. 1977. Inheritance of reaction to halothane anesthesia in pigs. Genet. Res. (Camb.) 29:287.

Spradling, A. C. and G. M. Rubin. 1983. The effect of chromosomal position on the expression of the Drosophila xanthine dehydrogenase gene. Cell. 34:47.

Swift, G. H., R. E. Hammer, R. J. MacDonald and R. L. Brinster. 1984. Tissue-specific expression of the rat pancreatic elastase 1 gene in transgenic mice. Cell 38:639.

Wall, R. J., V. G. Pursel, R. E. Hammer and R. L. Brinster. 1985. Development of porcine ova that were centrifuged to permit visualization of pronuclei and nuclei. Biol. Reprod. 32:645.

Watanabe, S. and H. Temin. 1983. Construction of a helper cell line for avian reticuloendotheliosis virus cloning vectors. Mol. Cell. Biol. 3:2241.

Wilkie, T. M., R. L. Brinster and R. D. Palmiter. 1986. Germline and somatic mosaicism in transgenic mice. Dev. Biol. 118:9.

Status of Current Strategies for Growth Regulation

D. H. BEERMANN

1. Introduction

Strategies for the regulation of animal growth have evolved over the last several decades to serve two primary purposes: (1) to increase the efficiency and total amount of meat and milk production, and (2) to improve the nutrient composition of meat animals. The first objective has been pursued with success for over 50 years, providing us with such historical developments as genetic selection for increased rate of gain and improved feed conversion, the use of a large-mature-size breed of sire, the use of nonprotein nitrogen in ruminant diets, the discovery of the growth-promoting effects of antibiotics, antibacterial, and antiparasitic drugs, and more recently the use of the ionophores or rumen modifiers, anabolic and xenobiotic agents. Genetics will probably always be an important aspect of meat animal growth performance, recognizing the value of a diverse gene pool to provide the combination of traits best suited to specific environmental circumstances or specific production requirements. Genetic manipulation now appears to provide great promise for many of the animal science disciplines involved with improving animal performance.

The second objective, to improve body composition or nutrient composition of the meat derived from animals, has been achieved in a minor way with genetic selection for specific single or multiple traits that can be easily and accurately measured in the live animal or carcass, the use of a large-mature-size male breed in a terminal-crossbreeding system, the use of the approved

D. H. BEERMANN • Department of Animal Science, Cornell University, Ithaca, New York 14853-4801.

endogenous hormone and xenobiotic agents in ruminants and, where economical, with restricted feeding in swine. However, the improvement is minor when compared to the repartitioning effects of somatotropin and the β-adrenergic agonists. Immunization of animals against steroid hormones, somatostatin, adipose cell membranes and other antigens that influence growth and metabolism also holds promise for regulating animal growth in a desired manner. Great potential is envisioned for use of gene manipulation to alter cellular aspects of tissue growth or to provide a unique property or ability to target cells that will enhance growth performance or composition of growth.

The objective of this chapter is not to provide an accurate detailed history or comparison of all strategies capable of enhancing animal growth. The goal will be to summarize the relative contributions of the most widely used strategies and to provide a detailed discussion of the more recently discovered methods by which animal growth is improved.

2. Steroid Hormone and Xenobiotic Regulation of Animal Growth

Sex differences in growth rate, feed efficiency, and composition of gain are readily apparent in both ruminant and nonruminant farm animals. These differences have been attributed to secretion of steroid hormones from the testis and ovary. Intact males exhibit faster rates of gain and a greater proportion of lean tissue gain (both greater protein mass and less fat content) than castrates or females. However, age and stage of sexual development at slaughter, management strategies, and traditional attitudes regarding meat quality attributes have indicated the continued practice of male castration in beef, swine, and lambs produced for meat in the United States. Intact males are used extensively for meat production in Europe and other parts of the world. Increased consumer preference and demand for leaner meat in recent years in the United States has sparked greater interest in providing leaner animals for slaughter and prompted greater trimming of fat from retail meat cuts. The conventional pork and beef carcass produced in the United States today contains approximately 28–32% or more lipid. A shift toward leaner animals is warranted on both an economic and a nutritional composition basis.

Manipulation of growth performance and composition of gain can be achieved with implants of natural or synthetic sex steroids. Dinusson et al. (1948, 1950) demonstrated that subcutaneous implantation of the synthetic estrogen diethylstilbestrol stimulated the growth of heifers. Oral administration was also found to be effective in cattle (Burroughs et al., 1954; Clegg and Caroll, 1954) and lambs (Andrews et al., 1949; Hale et al., 1953). Estradiol, estradiol benzoate plus progesterone, and estradiol benzoate plus testosterone proprionate, commercially available forms of natural estrogens and androgens, have been shown to increase daily gain by 8–15% and improve feed efficiency by 5–10% in finishing beef cattle or lambs (see reviews by Galbraith and Topps, 1981; Schanbacher, 1984; Roche and Quirke, 1986). Zeranol, derived from the fungal metabolite zeralenone, has estrogenic properties that provide similar growth promotion as that seen with the natural estrogens. The estrogenic compounds are most effective in males; growth promotion in females is more variable and less consistent (Galbraith and Topps, 1981).

The synthetic androgen trenbolone acetate is similar in structure to testosterone. Improved growth performance in finishing steers, heifers and lambs has been observed when trenbolone acetate is implanted alone or in combination with estradiol (Roche and Quirke, 1986). Rate of gain was increased by 10–20% and feed efficiency was improved by 10–20% in these studies.

Schanbacher (1984) noted that the magnitude of response for increased rate of gain and improved feed efficiency achieved with anabolic steroids decreases with time, probably reflecting the declining circulating concentration of the agent after the first few weeks of implantation. Anabolic steroids have not been shown to consistently improve growth performance in swine.

Carcass yeild and total carcass protein are generally increased with anabolic steroid implants in ruminants. Results vary with age of animals implanted, energy density of the diet, and length of time implanted before slaughter. Large increases in total carcass protein ranging from 20 to 40% have been achieved with combination implants in finishing lambs and cattle fed high-energy or concentrate diets (Galbraith and Topps, 1981; Sinnet-Smith et al., 1983). However, the body of data is not large and some studies report only 10–15% improvement in carcass protein content. Composition of gain has not been shown to be consistently improved by administration of anabolic steroids to swine (Galbraith and Topps, 1981; Roche and Quirke, 1986).

Research on the pharmacokinetics of these anabolic agents in beef cattle and other farm animals demonstrated that circulating concentrations of the steroids are maintained above preimplantation levels for 8 to 16 weeks following a single implant (Heitzman et al., 1984). Under the conditions recommended for the use of anabolic steroid implants, treatment of animals resulted in residues in meat that were magnitudes lower than those found occurring naturally in bulls and pregnant cows.

Use of anabolic steroid implants in ruminants has been widely adopted in the United States and Europe. Anabolic steroid implants are contraindicated in animals selected for breeding because testicular development may be retarded with estrogen implants in males and delay of onset of puberty and reduced ovulation rate may result in females (Roche and Quirke, 1986). Use of diethylstilbestrol either as an implant or as a feed additive has been banned, and it should be noted that regulatory status for individual anabolic steroids is different among the countries where their use is allowed. Anabolic androgen implants are not approved for use in the United States.

3. Manipulation of Animal Growth with Exogenous Somatotropin

Early work with farm animals demonstrated that exogenous somatotropin, or growth hormone (GH), given by injection, markedly alters nutrient use to improve growth rate and feed conversion of growing–finishing pigs (Turman and Andrews, 1955; Machlin, 1972), calves (Brumby, 1959) and lambs (Wagner and Veenhuizen, 1978). The studies with pigs and lambs demonstrate that the normal allometric pattern of tissue growth is also altered resulting in greater rates of skeletal muscle growth and reduced rates of adipose tissue accretion. Somatotropin influences on lipid metabolism in domestic livestock has recently been reviewed by Etherton and Walton (1986).

Turman and Andrews (1955) observed a 16% increase in rate of gain and a 24% improvement in feed conversion with GH administration in finishing swine. They also found that carcass fat was reduced by 21% and carcass protein was increased by 25%. Machlin's studies, also with finishing swine, produced similar results. Wagner and Veenhuizen (1978), using a relatively high dose of 15 mg oGH/day, found that GH increased rate of gain by 20% and improved feed conversion by 14% with 14 to 16 weeks of treatment. Carcasses from treated lambs contained 25% more protein and 37% less fat.

Because Machlin reported some toxicity effects with the doses he used, lower doses or shorter treatment periods than we now know to be effective in swine were used in early subsequent studies (Baile *et al.*, 1983; Chung *et al.*, 1985). However, the relative impurity of the early preparations of pituitary-derived GH may have also contributed to some of the toxicity effects. Improved purification techniques and greater availability of pituitary-derived GH, the advent of availability of recombinantly produced GH, and the demonstrated effects of GH to increase lactation in various species have all sparked a renewed interest in defining the potential of exogenous GH administration for improving efficiency and composition of growth in meat animals. Recent studies, particularly those with swine, have confirmed the earlier observations and advanced our understanding of the dose–response relationships for growth performance and composition variables.

It is now apparent that the optimum GH dose for finishing pigs is different for different growth variables. Boyd *et al.* (1986a) have shown that increased rate of gain is maximized at a lower dose of highly purified pituitary-derived porcine GH (ppGH) than is improvement in feed efficiency (Table I). Rate of gain was greatest (+ 16%) with 60 μg ppGH/kg, but feed efficiency was best

Table I
Dose–Response Relationships between Porcine Growth Hormone, Growth Performance, and Composition of Gain in Swine[a,b]

| | Dose, pGH (μg/kg body wt) | | | | | $S_{\bar{x}}$ | Regression response probability level | |
	0	30	60	120	200		Linear	Quadratic
Initial weight (kg)	47.3	46.8	46.8	46.4	46.8	0.60	NS	0.01
Average daily gain (kg)	0.95	1.0	1.10	1.07	1.02	0.05	NS	0.01
Feed/gain	3.08	2.72	2.41	2.21	2.17	0.05	0.01	0.01
Carcass protein (%)	14.7	15.9	17.0	18.1	18.8	0.17	0.01	0.01
Carcass lipid (%)	32.1	27.7	21.6	17.2	13.8	0.74	0.01	0.01
Carcass water (%)	49.6	52.6	56.4	60.3	62.8	0.56	0.01	0.01
Carcass ash (%)	2.6	2.9	3.0	3.3	3.6	0.25	0.01	0.06

[a] From Boyd et al. (1986a and unpublished data).
[b] Pituitary-derived pGH was administered daily by injection; dose was adjusted weekly. Animals were slaughtered at 100 kg live weight.

($+28\%$) at 120 μg/kg body weight. The reduced growth rate observed with the highest dose (200 μg/kg) was associated with a 22% reduction in feed intake and no further improvement in feed efficiency. Using barrows treated for 35 days with 0, 10, 30, and 70 μg/kg pGH, Etherton *et al.* (1987) found that neither growth performance nor carcass composition changes had begun to plateau at the highest dose. Studies in which ppGH and recombinantly produced porcine growth hormone (rpGH) were compared (Etherton *et al.*, 1986a) confirmed that feed efficiency is maximally improved at a higher dose (140 μg/kg) than is rate of gain (70 μg/kg). These studies and others (Ivy *et al.*, 1986) demonstrated that depression of feed intake is linearly related to pGH dose in swine, limiting the increased rate of gain at the higher doses.

GH improved carcass composition, with both linear and quadratic effects being significant for all but carcass ash content (Boyd *et al.*, 1986a; Table I). Because animals were both started and finished at a predetermined constant live weight in this study, analysis of the data using live weight or carcass weight as the covariate had no effect on the outcome. Interpretation of the observed differences is greatly facilitated with this experimental design. Carcass lipid content was reduced by 55% with the highest dose, and carcass protein and water content were reciprocally increased by 27 and 24%, respectively. Comparison of carcass composition of ppGH treatment groups with that of the baseline group revealed the truly dynamic nature of the effects of GH on nutrient use for protein and lipid gain (see Boyd and Bauman, this volume). Protein gain per day was a maximum of 55% greater with the 120 μg/kg dose, whereas the lipid gain per day was maximally reduced by 88% with 200 μg/kg ppGH.

Both mass and composition of skeletal muscle were altered in a dose-dependent manner with ppGH administration (Boyd *et al.*, 1986a). Muscle weights were 14–17% larger with 120 μg/kg ppGH; there was no additional increase with the 200 μg/kg dose (Boyd *et al.*, 1986b). Lipid content of individual hind leg muscles was reduced linearly with increasing ppGH dose, achieving a 50% reduction at the highest dose (data not shown). Carcass quality characteristics and Instron shear values of cooked longissimus muscle were not significantly affected by ppGH, nor was any dose–response relationship apparent for any carcass quality trait.

It should be noted that in studies that show the magnitude of response described above, diets have been fortified in excess of minimum National Research Council (NRC) requirements for protein and other nutrients. Data that describe specific nutrient requirements for pGH-treated swine have not been published. Recent studies have shown that restricting feed intake by as much as 20–30% did not impair the relative amount of improvement in rate of gain ($+25$–28%), feed efficiency ($+20\%$), or protein gain ($>35\%$) in young pigs given 100 μg/kg ppGH from 22 to 55 kg live weight (Campbell *et al.*, 1987).

Ruminants also show improved growth rate and feed conversion with daily administration of GH (Table II). There are essentially no published data for

Table II
Anabolic Effects of Growth Hormone in Ruminants

Treatment	No. per treatment	Initial weight (kg)	Treatment period (days)	ADG[a] (g/day)	Feed/ gain[b]	Carcass composition		Source
						Lean	Fat	
Sheep — 15 mg oGH/day (0.25 IU/day)	4–5	40	98–112	186 +20	8.8 −14	1.02 kg[c] +25	6.19 kg[c] −37	Wagner and Veenhuizen (1978)
— 7 mg oGH/day (0.16 IU/kg)	16	28	56	268 +4	5.64 −7.4*	4.02 kg[d] +6	6.49 kg[d] −9*	Muir et al. (1983)
— 0.1 mg bGH/kg (0.13 IU/kg) 0 0.16 mg oGH/day	8	17	84	284 +22**	4.49 −12**	11.4 kg[e] +24**	7.0 kg[e] −13	Johnson et al. (1985)
Cattle — 0.6 mg/kg$^{3/4}$	6	90	147	489 +8.6**	7.16 −2	19.5%[f] 0	5.27%[f] −17.5	Sandles and Peel (1987)
— 15 mg/day (0.11 IU/kg)	9	180	109	877 +8.1%	NA[g] NA[g]	90.98 kg[e] +5.1%**	17.25 kg[e] −9.5%**	Sejrsen and Bauman (unpublished data)

Heading for the composite columns: Control values and proportional response (%)

[a] Average daily live weight gain.
[b] Kilograms of feed per kg live weight gain.
[c] Chemical composition of gain (protein and lipid).
[d] Chemical composition of total carcass (protein and lipid).
[e] Muscle and adipose tissue weight based on carcass dissection.
[f] $N=4$ for composition data; chemical composition of soft tissues.
[g] NA, not applicable; pair fed identical twins, fixed intake.
* $p < 0.05$; ** $p < 0.01$.

studies with finishing cattle, but the average response for finishing lambs and young female cattle indicate a 10–20% increase in growth rate and a 10–20% improvement in feed conversion. Carcass protein content or muscle mass was consistently increased and carcass lipid content or dissected fat consistently reduced in lambs. Carcass quality characteristics were apparently not adversely affected, but as in swine, muscle lipid content was reduced in lambs and cattle. Dose–response data have not been published.

Although use of GH provides a remarkable opportunity to increase the rate of lean tissue growth while also reducing adipose tissue accretion in swine, lambs, and cattle, exogenous GH has not been approved for use in farm animal species. It is anticipated that commercial application could be achieved within 2 years if approval is granted.

Based on these remarkable improvements in growth performance and composition achieved with exogenous GH, it is tempting to speculate that similar responses might be achieved by increasing circulating somatomedin concentration or activity. The somatomedins or insulinlike growth factors (IGF) are produced by several tissues in the body. Circulating GH concentrations may, in some but not all cases, influence blood levels of somatomedin C or IGF-I. Because IGF-I has been shown to stimulate mitotic activity in many tissues, it has been postulated that some growth-promoting properties of GH may be mediated through IGF-I. Recombinantly produced IGF-I administered by subcutaneous implanted osmotic pumps at a dose of 120 μg/day for 7 days significantly increased body weight gain and tibia epiphyseal plate width in normal growing rats (Hizuka et al., 1986). The close association between insulin and IGF-I in the regulation of growth has also been demonstrated. Scheiwiller et al. (1986) showed that IGF-I administration to streptozotocin-induced diabetic rats restored normal growth without normalizing blood glucose levels, whereas GH treatment did not. The authors also demonstrated that insulin acts via an increase in synthesis of IGF-I in restoring growth in diabetic rats. A detailed discussion of this topic is not warranted here, because it has been addressed elsewhere in this book, but growth regulation by the somatomedins will continue to be investigated vigorously in the future. Sufficient quantities simply have not been available to allow many in vivo farm animal growth trials to be conducted, but based on recent studies it must be concluded that both insulin and GH influence growth by mediating IGF-I synthesis or activity.

4. Use of Growth Hormone-Releasing Factor (GRF) to Alter Animal Growth

The practical nature of improving growth performance and composition of gain with exogenous GH strongly suggests that it should be possible to emulate these responses with enhancement of endogenous secretion of GH. Many studies have shown that administration of exogenous human GRF (hGRF) by intravenous or subcutaneous injection or infusion transiently increases secretion and circulating concentration of endogenous GH in sheep, swine, and calves (Moseley *et al.*, 1984; Hart *et al.*, 1985; Kraft *et al.*, 1985; Trenkle and Plouzek, 1985; Al-Raheem *et al.*, 1986; Della-Ferra *et al.*, 1986a,b; Etherton *et al.*, 1986b). Mode of administration of bolus or short-term infusions does not appear to significantly affect the magnitude of the response, but variation between animals is large. Moseley et al. (1985) demonstrated that raised serum GH concentration could be sustained for 5 days with frequent microinjections of human pancreatic GRF, but the dose dependence of this response has not been reported.

Dose–response relationships have not been consistently obtained in all studies, but Della-Ferra *et al.* (1986b) and Kensinger *et al.* (1987) have observed consistent dose-dependent elevation of GH with intravenous and subcutaneous injections in sheep. Kensinger *et al.* also demonstrated that frequency of injection (40 μg/kg body weight two, four, or eight times daily) had little effect on the magnitude of the individual peak elevations. These authors showed that continuous subcutaneous infusion of hGRF was able to sustain a continuous 24-hr elevation of circulating GH, although it was smaller than that achieved with multiple injections.

Nitrogen retention data provide evidence that hGRF enhancement of endogenous GH secretion may enhance growth in lambs (Plouzek et al., 1984) and cattle (Moseley, et al., 1987). Moseley *et al.* (1987) administered by continuous intravenous infusion 0 or 3.6mg $GRF_{1-44}NH_2$ per day to Dutch–Friesian bull calves (148 kg) for 20 days. Urinary N decreased by 14% and N balance increased 18% during days 9–14 of treatment. Nitrogen intake, fecal N, and coefficient of N digestibility were similar for control and treated calves. GRF infusion increased weight gain 19% during days 13–21. The continuous infusion of GRF sustained elevated levels of circulating GH throughout the 20-day period, increasing the area under the curve 67%. The latter response was greater at day 20 than at day 1 or 10.

Etherton *et al.* (1986a) have compared the effects of daily injection of hGRF and pGH on growth performance and composition in finishing pigs. Equal doses (30 μg/kg body weight) administered for 30 days resulted in a nonsignificant increase in rate of gain (5.5%) and reduction in feed/gain (8.6%) with GRF, whereas both were significantly altered with pGH (Table III). Percent lipid in the soft tissues was reduced 13.3% with GRF and 18% with pGH (both $p < 0.05$). Percent protein and absolute muscle mass were both significantly increased with pGH; a trend toward increases in both was observed with GRF.

Similar results were observed by Beermann *et al.* (1988) when four times per day subcutaneous injection of 40 μg/kg oGH was compared with 5 μg/kg

Table III

Effects of pGH and GRF on Growth Performance[a] and Carcass Composition[b]
in Swine

	Control	GRF	pGH	$S_{\bar{x}}$
Initial weight (kg)	50.8	49.5	49.0	1.4
Average daily gain (kg)	0.90[c]	0.95[c,d]	1.00[d]	0.03
Feed/gain	3.01[c]	2.75[c,d]	2.44[d]	0.11
Protein (%)	14.7[c]	15.6[c,d]	15.9[d]	0.3
Lipid (%)	29.4[c]	25.5[d]	24.1[d]	0.9
Water (%)	54.4[c]	57.3[d]	58.4[d]	0.7
Ash (%)	0.9[c]	0.9[c]	0.9[c]	0.02
Adipose tissue (kg)	11.9[c]	10.3[c]	11.0[c]	0.4
Muscle mass (kg)	24.5[c]	28.5[c]	33.2[d]	0.6

[a]Pigs were treated daily by injection of 30 μg/kg body weight GRF or pGH for 30 days. There were 12 pigs per treatment for growth performance data and 8 pigs per treatment for composition data. (Data from Etherton *et al.*, 1986a.)
[b]Composition values are percentage of soft tissue mass of the carcass.
[c,d]Means in the same row that do not have a common superscript differ ($p < 0.05$).

and 10 μg/kg human GH releasing factor (hGRF) (1–44 NH$_2$) in lambs. The lower dose of hGRF increased average daily gain 17.7% (Table IV). The higher dose reduced feed intake resulting in no significant improvement in gain. Feed efficiency was improved 21% with oGH and 13–14% with both doses of hGRF. Growth hormone administration was more effective than hGRF treatment in improving carcass protein and water gain, however both significantly increased semitendinosus muscle weight 14–18%. Administration of hGRF was as effective as oGH in significantly reducing carcass lipid gain. Both oGH and hGRF significantly increased bone growth as indicated by ash gain differences.

These and other data clearly support the concept that use of exogenous GRF to sustain elevated endogenous GH concentrations may be an effective way to enhance growth performance and composition.

5. Use of β-Adrenergic Agonists to Manipulate Animal Growth

Oral or parenteral administration of substituted phenylethanolamines, which share some structural and pharmacological properties with the naturally occurring catecholamines epinephrine and norepinephrine, has been found to have profound effects on skeletal muscle growth and fat deposition in farm animals (Asato, 1984; Baker and Kiernan, 1983). The first reports of the repartitioning of tissue growth by this type of compound centered on the effects of clenbuterol in lambs (Baker *et al.*, 1984), poultry (Dalrymple *et al.*, 1984a), cattle (Ricks *et al.*, 1984) and swine (Dalrymple *et al.*, 1984b). Subsequently, similar results were observed with cimaterol in swine (Jones *et al.*, 1985; Moser *et al.*, 1986), lambs (Beermann *et al.*, 1986) and cattle (Hanrahan *et al.*, 1986). Other synthetic compounds exhibit similar effects on growth performance and composition in these species. The data available to date would seem to indicate that the β-adrenergic agonists are more effective in ruminants than nonruminants.

5.1. Effects of Adrenergic Agonists in Sheep

Several studies with clenbuterol and cimaterol and a single study with the Merck compound L-644,969 (Duquette *et al.*, 1987b) have demonstrated that rate of gain may be increased by about 15%, although the range is large (0–24%) with little indication of a dose–response relationship (Table V). Feed conversion was more consistently improved, by as much as 15–20% in these and other studies for which data are not shown (Kim *et al.*, 1986; Hanrahan *et al.*, 1986), with significant indications of a dose–response relationship.

Individual muscle or total muscle weight was increased by 20–30% in several studies. When two or more treatment intervals have been compared, the full magnitude of treatment differences was achieved within 3–6 weeks of initiating treatment and maintained for the duration of the experiment (Beermann *et al.*, 1986; Hanrahan *et al.*, 1986; O'Connor and Beermann, 1988).

Total carcass fat or lipid was reduced by 20–30% in several studies. Dose–response relationships were not apparent for either reduced fat accretion or increased muscle growth. Carcass yeild or dressing percent was significantly greater with β-agonist treatment.

Shear values of cooked meat from β-agonist-treated lambs do not appear to be adversely affected, although significant differences have been observed in some instances (Beermann *et al.*, 1985b; Hamby *et al.*, 1986). Rate of post-

Table IV
Effects of oGH and hGRF on Composition of Carcass Gain[a,b]

Treatment	N	Protein	Water	Lipid	Ash
			Component gain (g/day)		
Control	14	19.3*	62.4*	84.7*	4.3*
			Percent response		
oGH	18	+26.2†	+26†	−25†·‡	+27*·†
Lo hGRF	19	+18.7*·†	+16*·†	−18.3†	+43†
Hi hGRF	20	+13.1*·†	+7*·†	−33‡	+35†

[a]Ewe and wether lambs were treated four times per day by subcutaneous injection of 40 μg/kg oGH, 5 μg/kg hGRF, 10 μg/kg hGRF, or excipent. Initial weights averaged 25.4 kg. One half of the lambs were treated for 42 days, one half were treated for 56 days. Six lambs were slaughtered at 25 kg to provide baseline carcass composition. Composition values are percentage of the half-carcass.
[b]Means within a column with different superscripts differ ($p < 0.05$).

mortem pH decline is reduced and ultimate muscle pH is significantly higher in longissimus muscle of lambs fed cimaterol (Beermann et al., 1985a; Wang and Beermann, unpublished data). Shear values of cooked longissimus and biceps femoris muscle were not different in one study, but have been found to be higher in treated lambs in others. Lipid content of longissimus and two hind leg muscles is reduced by nearly 50% in lambs fed 10 ppm cimaterol (Beermann, unpublished data).

5.2. Effects of Adrenergic Agonists in Cattle

Very few data have been published for efficacy of the β-agonists in cattle, but the results indicate a response similar to that observed in lambs. Ricks et al. (1984) found 10 ppm clenbuterol in the diet to be effective in improving carcass composition in steers, but growth performance was not improved (Table VI). The very high does of 500 ppm also caused a marked improvement in composition, but a significant reduction in feed intake and rate of gain was also observed.

Friesian steers exhibited 18 and 30% increases in rate of gain and 23 and 30% improvement in feed conversion in response to a 91-day treatment with 3.5 and 5.1 ppm cimaterol in the diet (Hanrahan et al., 1986). Total muscle mass in the carcass was increased by 24–30%, while total dissected fat was reduced by 28–35%. Total bone and connective tissue were not different. Growth performance and carcass composition were improved more by the intermediate dose than by the low dose, but the highest dose (7.0 ppm) had a lesser effect on growth performance than the two lower doses.

Very young ruminants do respond to β-agonist administration. Williams et al. (1986, 1987) fed veal calves 0.18 and 1.8 ppm clenbuterol in milk replacer for 105 days and found significant improvement in skeletal muscle growth accompanied by a reduction of 20 and 42% in percent lipid in the carcass (Table VI). Suckling lambs administered cimaterol via an ear implant to deliver 1.8 mg/day for 6 weeks, exhibited a 45% reduction in fat gain/day and an 87% increase in protein gain per day determined by comparative slaughter (J. C. Wolff, personal communication).

5.3. Effects of Adrenergic Agonists in Swine

The effects of cimaterol on growth performance and carcass characteristics have been studied in finishing and young swine (Jones et al., 1985; Moser et al., 1986; Mersmann et al., 1987). Growth rate was not significantly increased at any of the three cimaterol doses used in either study with finishing pigs (Table VII). Feed conversion was significantly improved in one study but not the other. Dynamic changes in both growth rate and feed conversion were observed with the β-agonist L-644,969. Wallace *et al.* (1987) observed a large (23.5%) increase in growth rate after 1 week of feeding 1 ppm L-644,969, but a diminished response was observed thereafter. The 4 ppm dose exhibited an intermediate response when compared to 0.25 and 1 ppm. Both the 1 and 4 ppm doses significantly improved feed conversion through the sixth week, but the improvements were not significant by week seven.

Both cimaterol and L-644,969 significantly improved carcass composition (Table VII), and the magnitude of change was similar to that observed with clenbuterol in finishing swine diets (Dalrymple *et al.*, 1984b). Longissimus cross-sectional area was increased by 4–13% with cimaterol and by up to 29% with the 4 ppm dose of L-644,969. Individual muscle weights were 8–12% greater with cimaterol and carcass protein was increased in a dose-dependent manner with L-644,969.

Studies with another phenyl-ethanolamine, Ractopamine, indicate that the growth performance and the tissue repartitioning effects are dependent on protein content in the diet (Anderson *et al.*, 1987). Significant interactions were observed when 12, 15, and 18% protein diets containing 0, 5, or 20 ppm Ractopamine were fed to finishing pigs. Similar results were obtained when 12 and 16% protein diets were used and composition of gain was estimated by measuring N retention.

Contrary to the demonstrated repartitioning effects of the β-adrenergic agonists in very young ruminants, Mersmann *et al.* (1987) found that cimaterol did not improve growth performance or carcass composition in young pigs. Crossbred pigs weighing 8.7 kg were fed diets containing 14 or 18% protein and 0, 0.25, or 0.5 ppm cimaterol for 9 weeks. Measurements of adipose tissue lipogenesis and lipolysis were also unaffected by cimaterol in this study. Summarization of the data suggests that swine may be more sensitive to the β-agonists than ruminants. An alternative to continuous ad libitum administration may be required for optimal response.

6. Use of Immunization to Manipulate Animal Growth

Most attempts to manipulate animal growth through immunoneutralization have been directed toward altering the circulating concentration of specific hormones known to influence tissue growth processes. The objectives of the studies were either to demonstrate the important anabolic contribution of the hormone(s) or to improve growth performance or composition through decreasing the amount or effectiveness of a specific releasing factor or hormone (i.e., somatostatin). Somatostatin not only influences somatotropin secretion, and

Table V
The Anabolic Effects of Adrenergic Agonists in Sheep

| Treatment dose (ppm) | N | Treatment period (days) | Initial weight (kg) | Control values and proportional response (%) | | | | Source |
| | | | | ADG[a] (g/day) | Feed/ gain[b] | Carcass composition | | |
						Lean	Fat	
Clenbuterol								
0	20	56	32.3	196	8.13	17.5%[c]	21.1%[c]	Baker et al. (1984)
1				−8	−2.7	+9.7**	−20.4**	
10				0	−7.6	+12.0**	−27.0**	
100				+6.6	−17.2*	+9.7**	−22.7**	
0	30	56	37.4	212	8.31	143.8 g[d]	5.9 mm[e]	Baker et al. (1984)
2				+24.1**	−19.1**	+23**	−37.3**	
Cimaterol								
0	6	42, 84	17	328, 319	3.08, 4.61	191, 259[f]	4.4 mm, 6.5 mm[e]	Beermann et al. (1986)
10		42		−3	−5.8	+30.5**	−66**	
		84		0	−8.2	+20.6**	−32**	

0	8	35, 70	411, 396	3.39, 3.71	227, 291[f]	2.8 mm, 4.6 mm[e]	Beermann et al. (1986)
10		35	+13.4**	−17.4**	+23.2**	−67**	
		70	−3.0	−1	+25.4**	−24.2**	
0	10	45	352	4.94	66.9%[g]	16.6%[g]	Hanrahan et al. (1986)
0.57			+3.7	0	+6.4	−16.7	
2.29			+17.9*	−7.3	+5.2	−16.3	
11.42			+19.3*	−14.7*	+9.0	−33.1	
L-644,969							
0	20	42	265	8.64	15.5%[h]	26.6%[h]	Duquette et al. (1987b)
0.25			+15.5*	−9.2	+1.5	+1.1	
1			+14.3	−13.7*	+9.5**	−8.5**	
4			+13.9	−18.3**	+11.3**	−11.9**	

[a] Average daily live weight gain.
[b] Kilograms of feed per kg live weight gain.
[c] Chemical composition of the hindquarters (protein and lipid), $n = 10$.
[d] Weight of the semitendinosus muscle.
[e] Fat thickness at the 12th rib.
[f] Expressed as average weight increase (g) for three hind leg muscles.
[g] Percent physically dissected muscle and fat.
[h] Chemical composition of the half carcass.
$*p < 0.05$; $**p < 0.01$.

Table VI
The Anabolic Effects of Adrenergic Agonists in Cattle

| | Treatment dose (ppm) | No. per treatment | Treatment period (days) | Initial weight (kg) | Control values and proportional response (%) | | | | Source |
| | | | | | ADG[a] (g/day) | Feed/ gain[b] | Carcass composition | | |
							Lean	Fat	
Cattle	Clenbuterol								
	0	8	98	340	1007	11.83	15.4%[c]	35.2%[c]	Ricks et al. (1984)
	10				−8	−2.1	+13**	−20**	
	500				−20	0	+14**	−30**	
	Cimaterol								
	0	15	91	530	816	11.7	203 kg[d]	75.1 kg[d]	Hanrahan et al. (1986)
	3.5				+18*	−23.5*	+24.5*	−28.1*	
	5.1				+30.4*	−30.6*	+30.0*	−30.4*	
	7.0				+6.0	−8.2	+27.6*	−35.0*	
Veal	Clenbuterol								
	0	8	105	43.8	1100	1.67[e]	20.8%[f]	18.5%[f]	Williams et al. (1987)
	0.177	5			−1.8	−1.2	+3.8*	−19.8**	
	1.85	5			−3.6	+4.0	+5.4*	−41.9**	

[a] Average daily live weight gain.
[b] Kilograms of feed per kg live weight gain.
[c] Chemical composition of 9th–11th rib section (protein and lipid).
[d] Total carcass tissue weights based on half-carcass physical dissection.
[e] Kilograms of milk solids per kg gain.
[f] Chemical composition of the whole carcass (protein and lipid).
* $p < 0.05$; ** $p < 0.01$.

Table VII

Anabolic Effects of Adrenergic Agonists in Swine

Treatment dose (ppm)	No. per treatment	Treatment period (days)	Initial weight (kg)	ADG[a] (g/day)	Feed/ gain	Carcass composition		Source
						Lean	Fat	
Cimaterol								
0	40	51	64	760	3.93	1.726 kg[b]	2.58 cm[c]	Jones et al. (1985)
0.25				+5.3	−9.7*	+8.2*	−10.5*	
0.50				+1.3	−8.4*	+11.9*	−12.0*	
1.0				+3.9	−12.1*	+11.2*	−17.4*	
0	48	47–70 ($\bar{x}=57$)	61	770	3.47	36.7 cm²[d]	2.96 cm[c]	Moser et al. (1986)
0.25				+1.3	−5.7	+4.1	−9.1*	
0.50				−1.3	0	+3.8	−16.6*	
1.0				−2.6	−5.4	+8.7	−14.2*	
L-644,969								
0	18	49	65	898	3.37	14.07%[f]	38.51%[f]	Wallace et al. (1987)
0.25				0	−2.4	+4.1	−3.2	
1.0				0[e]	−4.6	+9.9*	−2.3	
4.0				−8[e]	−4.8	+12.0**	−2.7	

[a] Average daily live weight gain.
[b] Expressed on basis of semimembranosus and biceps femoris muscle weights.
[c] Expressed on basis of 10th rib fat thickness measures.
[d] Expressed on basis of longissimus cross-sectional area at the 10th rib.
[e] Both levels caused an increase in ADG at 1 week, but declined steadily each week thereafter.
[f] Expressed on basis of chemical composition (protein or lipid) of half carcass.
* $p < 0.05$; ** $p < 0.01$.

thereby may influence somatomedin production or secretion, but it is known to inhibit the release of other hormones including insulin (Koerker et al., 1974) and thyroid-stimulating hormone (Vale et al., 1974).

Dutch moorsheep were immunized against somatostatin linked to human serum α-globulin at 21 days of age and at 2-week intervals thereafter until the lambs were 20 weeks of age (Spencer and Garssen, 1983). Treated lambs exhibited significantly greater weight and height at the shoulders after 5 weeks of treatment and thereafter, but the divergence between groups occurred between 3 and 5 weeks for both traits. Average daily gain was increased by about 21% over the entire period and this weight was reflected in heavier carcasses in treated lambs. No significant difference between groups was found for body composition or basal levels of plasma GH or insulin. Feed intake was greater in lambs immunized against somatostatin, but feed conversion was also more efficient (Spencer et al., 1983). These results were confirmed by Laarveld et al. (1986) using similar immunization procedures and schedule in crossbred (Cheviot × Suffolk and Finn × Suffolk) lambs. Divergence in growth rate and body weight occurred after the first immunization (between 4 and 6 weeks) and growth was parallel thereafter. These authors also found that GH and insulin secretory responses to glucose and arginine, glucose tolerance, and plasma somatomedin levels were not different between groups. The lack of a measurable increase in blood GH, insulin, or somatomedins makes it difficult to speculate on the mechanisms by which immunization against somatostatin enhances growth rate. Immunization against somatostatin has also been shown to improve growth performance in calves (Lawrence et al., 1986; Vicini et al., 1986).

Immunoneutralization of endogenous steroid hormones allows one to selectively regulate an animal's exposure to anabolic steroids. Schanbacher (1984) has demonstrated the potency of this technique in immunizing ram lambs against testosterone or the hypothalamic releasing factor (LHRH) that regulates testosterone production in the ovine testis. Average daily gain of immunized rams was equivalent to that of wethers, demonstrating the important role of the testis in anabolic growth regulation.

A similar approach was taken to investigate the role of estradiol in mediation of lean tissue growth in bulls (Schanbacher, 1984). Bull calves were immunized against estradiol–serum albumin conjugate or estradiol–hemocyanin conjugate at birth and received occasional boosters thereafter. The bulls were finished on a silage-concentrate diet and slaughtered to compare carcass merit. Rate of gain and final body weight were slightly but not significantly increased with estradiol immunization, but hot carcass weight, backfat thickness, and longissimus cross-sectional area were significantly greater in treated bulls.

Wise and Ferrell (1984) found that immunization of heifers against KLH–estradiol or BSA–estradiol enhanced daily gain by 14–16% and improved feed conversion by 7 and 20%, respectively. Carcass yield and composition were unaffected by immunization against estradiol. Further work is required to establish the potential of immunoneutralization of steroids for enhancing meat production.

A novel approach to reducing fat accretion in animals has been demonstrated by Flint *et al.* (1986). Injecting sheep antibodies raised against rat adipocyte plasma membranes into the peritoneal cavity or subcutaneous fat depots was shown to reduce fat cell number by lysis or degradation of fat cells by large numbers of white blood cells (Flint and Futter, 1986). The process was also found to be complement dependent. Refinement of this technique may also lead to better regulation of adipose accretion in farm animals.

7. Summary and Perspectives

Potential for manipulation of growth of farm animals has never been greater than at present. The recent discovery of the repartitioning effects of select β-adrenergic agonists, the confirmation of the growth-promoting and repartitioning effects of somatotropin, the demonstrated growth-promoting effects of the somatomedins, effects of immunization of animals against target circulating hormones or releasing factors, cell constituents or receptors, and the advent of gene manipulation techniques offer a wider range of strategies than ever before available. The immediate application of these techniques must await required safety evaluation and full evaluation of the effects, if any, on the palatability characteristics of the meat derived from animals in which growth has been manipulated.

The additive effects of the rumen modifiers and anabolic steroids have been demonstrated, but the effects of combined administration of anabolic implants with somatotropin or the β-agonists, or of combining somatotropin and β-agonist administration have not been reported. Much work lies ahead to determine the mechanism of action of growth manipulation via GH or β-agonist administration, which will allow us to develop other logical strategies for beneficial growth manipulation. The role of the somatomedins in regulation of fetal and postnatal growth and metabolism is and will continue to be a fruitful area of investigation. We are still a long way from fully understanding the integrated physiology of adipogenesis, myogenesis, mammogenesis, and lactogenesis and the limitations on placental and fetal growth. We need to better understand the regulation of myofibrillar protein gene expression, protein turnover, and food intake. The other great challenge to animal scientists is to apply the powerful probes available in recombinant DNA technology to the rapidly expanding information we have on the important growth processes listed above.

References

Al-Raheem, S. N., J. E. Wheaton, Y. G. Massri, J. M. Marcek, R. D. Goodrich, W. Vale and J. Rivier. 1986. Effects of human pancreatic and rat hypothalamic growth hormone-releasing factors on growth hormone secretion in steers. Domest. Anim. Endocrinol. 3(2):87.

Anderson, D. B., E. L. Veenhuizen, W. P. Waitt, R. E. Paxton and S. S. Young. 1987. The effects of dietary protein on nitrogen metabolism, growth performance and carcass composition of finishing pigs fed Ractopamine. Fed. Proc. 46:1021 (Abstr.).

Andrews, F. N., W. M. Beeson and C. Harper. 1949. The effect of stilbestrol and testosterone on the growth and fattening of lambs. J. Anim. Sci. 8:578.

Asato, G., P. K. Baker, R. T. Bass, T. J. Bentley, S. Chari, R. H. Dalrymple, D. J. France, P. E. Gingher, B. L. Lences, J. J. Pascavage, J. M. Pensack and C. A. Ricks. 1984. Repartitioning agents; 5-[1-hydroxy-2-(isopropylamino)ethyl] anthranilonitrile and related phenethanolamines; agents for promoting growth, increasing muscle accretion and reducing fat deposition in meat producing animals. Agr. Biol. Chem. 48:2883.

Baile, C. A., M. A. Della-Ferra and C. L. MacLaughlin. 1983. Performance and carcass quality of swine injected daily with bacterially-synthesized human growth hormone. Growth 47:225.

Baker, P. K. and J. A. Kiernan. 1983. Phenylethanolamine derivatives and acid addition salts thereof for enhancing the growth rate of meat producing animals and improving the efficiency of feed utilization thereby. US Patent #4,404,222.

Baker, P. K., R. H. Dalrymple, D. L. Ingle and C. A. Ricks. 1984. Use of a β-adrenergic agonist to alter muscle and fat deposition in lambs. J. Anim. Sci. 59:1256.

Beermann, D. H., D. R. Campion and R. H. Dalrymple. 1985a. Mechanisms responsible for partitioning tissue growth in meat animals Proc. Recip. Meat Conf. 38:105.

Beermann, D. H., V. K. Fishell, D. E. Hogue and R. H. Dalrymple. 1985b. Effects of cimaterol (CL 263,780) and fishmeal on postmortem pH, tenderness and color in lamb skeletal muscle. J. Anim. Sci. 61 (Suppl. 1):217 (Abstr.).

Beermann, D. H., D. E. Hogue, V. K. Fishell, R. H. Dalrymple and C. A. Ricks. 1986. Effects of cimaterol and fishmeal on performance, carcass characteristics and muscle growth in lambs. J. Anim. Sci. 62:370.

Beermann, D. H., D. E. Hogue, V. K. Fishell, H. W. Dickson, S. Aronica, D. Dwyer and B. R. Schricker. 1988. Effects of exogenous ovine growth hormone (oGH) and human GH releasing factor (hGRF) on plasma oGH concentration and composition of gain in lambs. J. Anim. Sci. 66 (Suppl. 1):282.

Boyd, R. D., D. E. Bauman, D. H. Beermann, A. F. DeNeergard, L. Souza and W. R. Butler. 1986a. Titration of the porcine growth hormone dose which maximizes growth performance and lean deposition in swine. J. Anim. Sci. 63 (Suppl. 1):218 (Abstr.).

Boyd, R. D., D. E. Bauman, D. H. Beermann, A. F. DeNeergard, L. Souza and H. T. Kuntz. 1986b. Porcine somatotropin: dose titration for maximum growth rate, feed efficiency and lean deposition in swine. 1986 Proc. Cornell Nutr. Conf. pp 24–28.

Brumby, P. J. 1959. The influence of growth hormone on growth in young cattle. N.Z. J. Agr. Res. 2:683.

Burroughs, W., C. C. Culbertson, J. Kastelic, E. Cheng and W. H. Hale. 1954. The effects of trace amounts of diethylstilbestrol in rations of fattening steers. Science 120:66.

Campbell, R. G., T. J. Caperna, N. C. Steele and A. D. Mitchell. 1987. Effects of procine pituitary growth hormone (pGH) administration and energy intake on growth performance of pigs from 22–55 kg body weight. J. Anim. Sci. 65 (Suppl. 1):244 (Abstr.).

Chung, C. S., T. D. Etherton and J. P. Wiggins. 1985. Stimulation of swine growth by porcine growth hormone. J. Anim. Sci. 60:118.

Clegg, M. T. and F. D. Caroll. 1954. Further studies on the anabolic effects of stilbestrol in steers as indicated by carcass composition. J. Anim. Sci. 13:968.

Dalrymple, R. H., P. K. Baker, P. E. Gingher, D. L. Ingle, J. M. Pensack and C. A. Ricks. 1984a. A repartitioning agent to improve performance and carcass composition of broilers. Poult. Sci. 63:2376.

Dalrymple, R. H., P. K. Baker and C. A. Ricks. 1984b. Repartitioning agents to improve performance and body composition. Proc. Georgia Nutr. Conf. pp 111–118.

Della-Ferra, M. A., F. C. Buonomo and C. A. Baile. 1986a. Growth hormone releasing factors and secretion of growth hormone in sheep, calves and pigs. Domest. Anim. Endocrinol. 3(3):165.

Della-Ferra, M. A., F. C. Buonomo and C. A. Baile. 1986b. Growth hormone secretory responsiveness to multiple injections of growth hormone-releasing factor in sheep. Domest. Anim. Endocrinol. 3(3):153.

Dinusson, W. E., F. N. Andrews and W. M. Beeson. 1948. The effects of stilbestrol, testosterone and thyroid alteration on the growth and fattening of beef heifers. J. Anim. Sci. 7:523 (Abstr.).

Dinusson, W. E., F. N. Andrews and W. M. Beeson. 1950. The effects of stilbestrol, testosterone, thyroid alterations and spaying on growth and fattening of beef heifers. J. Anim. Sci. 9:321.

Duquette, P. F., E. L. Rickes, G. Olsen, H. Hedrick, T. P. Capizzi and E. M. Convey. 1987b. L-644-969 improves growth and carcass composition of lambs. In: Beta-agonists and their effects on animal growth and carcass quality (Ed. J. P. Hanrahan). Elsevier, Applied Science, London and New York, pp 119–126.

Etherton, T. D., C. M. Evock, C. S. Chung, P. E. Walton, M. N. Sillence, K. A. Magri and R. E. Ivy. 1986a. Stimulation of pig growth performance by long-term treatment with pituitary porcine growth hormone (pGH) and a recombinant pGH. J. Anim. Sci. 63 (Suppl. 1):219 (Abstr.).

Etherton, T. D. and P. E. Walton. 1986. Hormonal and metabolic regulation of lipid metabolism in domestic livestock. Proceedings of the 1984 ASAS Symposium "Current Concepts in Animal Growth". J. Anim. Sci. 63 (Suppl. 2):72.

Etherton, T. D., J. P. Wiggins, C. S. Chung, C. M. Evock, J. F. Rebhun and P. E. Walton. 1986b. Stimulation of pig growth performance by porcine growth hormone and growth hormone releasing factor. J. Anim. Sci. 63:1389.

Etherton, T. D., J. P. Wiggins, C. M. Evock, C. S. Chung, J. F. Rebhun, P. E. Walton and N. C. Steele. 1987. Stimulation of pig performance by porcine growth hormone: determination of the dose–response relationship. J. Anim. Sci. 64:443.

Flint, D. J., H. Coggrave, C. E. Futter, M. J. Gardner and T. J. Clarke. 1986. Stimulatory and cytotoxic effects of an antiserum to adipocyte plasma membranes on adipose tissue metabolism in vitro and in vivo. Int. J. Obesity 10:69.

Flint, D. J. and C. E. Futter. 1986. Immunological manipulation of body fat. Hannah Res. pp 123–127.

Galbraith, H. and J. H. Topps. 1981. Effects of hormones on the growth and body composition of animals. Nutr. Abstr. Rev. Ser. B51:521.

Hale, W. H., C. D. Story, C. C. Culbertson and W. Burroughs. 1953. The value of low levels of stilbestrol in the rations of fattening lambs. J. Anim. Sci. 12:918 (Abstr.).

Hamby, P. L., J. R. Stouffer and S. B. Smith. 1986. Muscle metabolism and real-time ultrasound measurement of muscle and subcutaneous adipose tissue growth in lambs fed diets containing a beta-agonist. J. Anim. Sci. 63:1410.

Hanrahan, J. P., J. F. Quirke, W. Bowman, P. Allen, J. McEwan, J. Fitzsimons, J. Kotzian and J. F. Roche. 1986. Beta-agonists and their effects on growth and carcass quality. In: W. Haresign and D. J. A. Cole (Ed.) Recent Advances in Animal Nutrition. pp 125–138. Butterworths, London.

Hart, I. C., P. M. E. Chadwick, A. Coert, S. James and A. D. Simmonds. 1985. Effects of different growth hormone releasing factors on the concentrations of growth hormone, insulin, and metabolites in the plasma of sheep maintained in positive and negative energy balance. J. Endocrinol. 105:113.

Heitzman, R. J., A. Carter, S. N. Dixon, D. J. Harwood and M. Phillips. 1984. Recent studies on pharmacokinetics and residues of anabolic agents in beef cattle and other farm animals. In: Roche and O'Callaghan (Ed.) Manipulation of Growth in Farm Animals. pp 1–14. Martinus Nijhoff Publ., The Hague.

Hizuka, N., K. Takano, K. Shizume, K. Asakawa, M. Miyakawa, I. Tanaka and R. Hirukawa. 1986. Insulin-like growth factor I stimulates growth in normal growing rats. Eur. J. Pharmacol. 125:143.

Ivy, R. E., C. D. Baldwin, G. W. Wolfrom and D. E. Mouzin. 1986. Effects of various levels of recombinant porcine growth hormone (pGH) injected intramuscularly in barrows. J. Anim. Sci. 63 (Suppl. 1):218 (Abstr.).

Johnsson, I. D., I. C. Hart and B. W. Butler-Hogg. 1985. The effects of exogenous bovine growth hormone and bromocriptine on growth, body development, fleece weight and plasma concentrations of growth hormone, insulin and prolactin in female lambs. Anim. Prod. 41:207.

Jones, R. W., R. A. Easter, F. K. Mckeith, R. H. Dalrymple, H. M. Maddock and P. J. Bechtel. 1985. Effects of the β-adrenergic agonist cimaterol (CL 263,780) on the growth and carcass characteristics of finishing swine. J. Anim. Sci. 61:905.

Kensinger, R. S., L. M. McMunn, R. K. Stover, B. R. Schricker, M. L. Maccecchini, H. W. Harpster and J. F. Kavanaugh. 1987. Plasma somatotropin response to exogenous growth hormone releasing factor in lambs. J. Anim. Sci. 64:1002.

Kim, Y. S., Y. B. Lee, C. R. Ashmore and R. H. Dalrymple. 1986. Effects of the repartitioning agent cimaterol (CL 263,780) on growth, carcass characteristics and skeletal muscle cellularity of lambs. J. Anim. Sci. 63 (Suppl. 1):221 (Abstr).

Koerker, D. J., W. Ruch, E. Chickedel, J. Palmer, C. Goodner, J. Ensinck and C. C. Gale. 1974. Somatostatin: hypothalamic inhibitor of endocrine pancreas. Science 184:482.

Kraft, L. A., P. K. Baker, C. A. Ricks, V. A. Lance, W. A. Murphy and D. H. Coy. 1985. Stimulation of growth hormone release in anesthetized and conscious pigs by synthetic human pancreatic growth hormone-releasing factor [hpGRF (1–29)-NH$_2$]. Domest. Anim. Endocrinol. 2(3):133.

Laarveld, B., R. K. Chaplin and D. E. Kerr. 1986. Somatostatin immunization and growth of lambs. Can. J. Anim. Sci. 66:77.

Lawrence, M. E., G. T. Schelling, F. M. Byers and L. W. Greene. 1986. Improvement of growth and feed efficiency in cattle by active immunization against somatostatin. J. Anim. Sci. 63 (Suppl. 1):215 (Abstr.).

Machlin, L. J. 1972. Effects of porcine growth hormone on growth and carcass composition of the pig. J. Anim. Sci. 35:794.

Mersmann, H. J., C. Y. Hu, W. G. Pond, D. C. Rule, J. E. Novakofski and S. B. Smith. 1987. Growth and adipose tissue metabolism in young pigs fed cimaterol with adequate or low dietary protein. J. Anim. Sci. 64:1384.

Moseley, W. M., J. Huisman and E. J. VanWeerden. 1987. Serum growth hormone and nitrogen metabolism responses in young bull calves infused with growth hormone-releasing factor for 20 days. Domest. Anim. Endocrinol. 4(1):51.

Moseley, W. M., L. F. Krabill, A. R. Friedman and R. F. Olsen. 1984. Growth hormone response of steers injected with synthetic human pancreatic growth hormone releasing factors. J. Anim. Sci. 58:430.

Moseley, W. M., L. F. Krabill, A. R. Friedman and R. F. Olsen. 1985. Administration of synthetic human pancreatic growth hormone-releasing factor for five days sustains raised serum concentration of growth hormone in steers. J. Endocrinol. 104:433.

Moser, R. L., R. H. Dalrymple, S. G. Cornelius, J. E. Pettigrew and C. E. Allen. 1986. Effects of cimaterol (CL 263,780) as a repartitioning agent in the diet for finishing pigs. J. Anim. Sci. 62:21.

Muir, L. A., S. Wien, P. F. Duquette, E. L. Rickes and E. H. Cordes. 1983. Effects of exogenous growth hormone and diethyl-stilbestrol on growth and carcass composition of growing lambs. J. Anim. Sci. 56:1315.

O'Connor, R. M. and D. H. Beermann. 1988. Week cimaterol treatment is adequate for maximum alteration of skeletal muscle growth and fat deposition in lambs. Fed. Proc. 2(4):A847.

Plouzek, C. A., A. Trenkle, W. Vale and J. Rivier. 1984. Effects of pulsatile intravenous injections of growth hormone-releasing factor on nitrogen retention in sheep. J. Anim. Sci. 59 (Suppl. 1):223 (Abstr.).

Ricks, C. A., R. H. Dalrymple, P. K. Baker and D. L. Ingle. 1984. Use of a β-agonist to alter fat and muscle deposition in steers. J. Anim. Sci. 59:1247.

Roche, J. F. and J. F. Quirke. 1986. The effects of steroid hormones and xenobiotics on growth of farm animals. In: P. J. Buttery, N. B. Haynes and D. B. Lindsay (Ed.) Control and Manipulation of Animal Growth. pp 39–51. Butterworths, London.

Sandles, L. D. and C. J. Peel. 1987. Growth and carcass composition of pre-pubertal dairy heifers treated with bovine growth hormone. Anim. Prod. 44:21.

Schanbacher, B. D. 1984. Manipulation of endogenous and exogenous hormones for red meat production. J. Anim. Sci. 59:1621.

Scheiwiller, E., H. P. Guler, J. Merryweather, C. Scandella, W. Maeriki, J. Zapf and E. R. Froesch. 1986. Growth restoration of insulin-deficient diabetic rats by recombinant human insulin-like growth factor I. Nature 323:169.

Sinnet-Smith, P. A., N. W. Dumelow and P. J. Buttery. 1983. Effects of trenbolone acetate and zeranol on protein metabolism in male castrate and female lambs. Brit. J. Nutr. 50:225.

Spencer, G. S. G. and G. J. Garssen. 1983. A novel approach to growth promotion using autoimmunisation against somatostatin. I. Effects on growth and hormone levels in lambs. Livestock Prod. Sci. 10:25.

Spencer, G. S. G., G. J. Garssen and P. L. Bergstrom. 1983. A novel approach to growth promotion using auto-immunisation against somatostatin. II. Effects on appetite, carcass composition and food utilisation in lambs. Livestock Prod. Sci. 10:469.

Trenkle, A. and C. Plouzek. 1985. Influence of feeding on growth hormone secretion and growth hormone-releasing factor response in sheep. Endocrinology 116 (Suppl. 1):194 (Abstr.).

Vale, W., C. Rivier, P. Brazeau and R. Guillemin. 1974. Effects of somatostatin on the secretion of thyrotropin and prolactin. Endocrinology 95:968.

Turman, E. J. and F. N. Andrews. 1955. Some effects of purified anterior pituitary growth hormone in swine. J. Anim. Sci. 14:7.

Vicini, J. L., J. H. Clark, W. L. Hurley and J. M. Bahr. 1986. Effect of immunization against somatostatin on growth of young dairy calves. J. Anim. Sci. 63(Suppl. 1):242 (Abstr.).

Wagner, J. F. and E. L. Veenhuizen. 1978. Growth performance, carcass deposition and plasma hormone levels in wether lambs when treated with growth hormone and thyroprotein. J. Anim. Sci. 47 (Suppl. 1):397 (Abstr.).

Wallace, D. H., H. B. Hedrick, R. L. Seward, C. P. Daurio and E. M. Convey. 1987. Growth and efficiency of feed utilization of swine fed a beta-adrenergic agonist (L-644,969). In: Beta agonists and their effects on animal growth and carcass quality (Ed. J. P. Hanrahan). Elsevier Applied Science, London and New York. pp 143–151.

Williams, P. E. V., L. Pagliani and G. M. Innes. 1986. Effects of a β-agonist (*clenbuterol*) on the heart rate, nitrogen balance and some carcass characteristics of veal calves. Livestock Prod. Sci. 15:289.

Williams, P. E. V., L. Pagliani, G. M. Innes, K. Pennie, C. I. Harris and P. Garthwaite. 1987. Effects of a β-agonist (clenbuterol) on growth, carcass composition, protein and energy metabolism of veal calves. Brit. J. Nutr. 57:417.

Wise, T. and C. Ferrell. 1984. Effects of immunization of heifers against estradiol on growth, reproductive traits and carcass characteristics. Proc. Soc. Exp. Biol. Med. 176:243.

Index